T0238287

Lecture Notes in Mathematics

Hirotaka Akiyoshi · Makoto Sakuma
Masaaki Wada · Yasushi Yamashita

Punctured Torus Groups and 2-Bridge Knot Groups (I)

Springer

Authors

Hirotaka Akiyoshi
Osaka City University
Advanced Mathematical Institute
3-3-138 Sugimoto, Sumiyoshi-ku,
Osaka 558-8585
Japan
e-mail: akiyoshi@sci.osaka-cu.ac.jp

Makoto Sakuma
Department of Mathematics
Hiroshima University
1-3-1, Kagamiyama
Higashi-Hiroshima 739-8526
Japan
e-mail: sakuma@math.sci.hiroshima-u.ac.jp

Masaaki Wada
Yasushi Yamashita
Department of Information
and Computer Sciences
Nara Women's University
Kita-uoya Nishimachi
Nara 630-8506
Japan
e-mail: wada@ics.nara-wu.ac.jp
yamasita@ics.nara-wu.ac.jp

Library of Congress Control Number: 2007925679

Mathematics Subject Classification (2000): 57M50, 30F40, 57M25, 20H10

ISSN print edition: 0075-8434
ISSN electronic edition: 1617-9692
ISBN-10 3-540-71806-0 Springer Berlin Heidelberg New York
ISBN-13 978-3-540-71806-2 Springer Berlin Heidelberg New York
DOI 10.1007/978-3-540-71807-9

Springer is a part of Springer Science+Business Media
springer.com

Typesetting by the authors and SPi using a Springer LaTeX macro package

Cover design: *design & production* GmbH, Heidelberg

Printed on acid-free paper SPIN: 12042029 VA41/3100/SPi 5 4 3 2 1 0

Preface

The main purpose of this monograph is to give a full description of Jorgensen's theory on the space $\mathcal{QF}$ of quasifuchsian (once) punctured torus groups with a complete proof. Our method is based on Poincare's theorem on fundamental polyhedra. This geometric approach enabled us to extend Jorgensen's theory beyond the quasifuchsian space and apply to knot theory.

1. History

By the late 70's Troels Jorgensen had made a series of detailed studies on the space $\mathcal{QF}$ of quasifuchsian (once) punctured torus groups from the view point of their Ford fundamental domains. These studies are summarized in his famous unfinished paper [40]. In it, he gave a complete description of the combinatorial structure of the Ford domain of every quasifuchsian punctured torus group, and showed that the space $\mathcal{QF}$ can be described in terms of the combinatorics of the faces of the Ford domain. This led to the description of $\mathcal{QF}$ in terms of the Farey triangulation, or the modular diagram. As a byproduct, the first examples of surface bundles over the circle with complete hyperbolic structures were obtained (cf. [41] and [43]).

To date, most of Jorgensen's work has not been published, yet it became widely known, motivated various research projects, and was successfully applied. His work, together with Riley's construction [67] of the complete hyperbolic structure on the figure-eight knot complement, has motivated Thurston's uniformization theorem of surface bundles over the circle [77] (cf. [63]). It had also motivated the experimental study by Mumford, McMullen and Wright [60] of the limit sets of quasifuchsian punctured torus groups. This work was sublimated into the beautiful book [61] by Mumford, Series and Wright, which displays deeply hidden fractal shapes of the space $\mathcal{QF}$ and the limit sets of punctured torus Kleinian groups.

2. Motivation

The authors' interest in Jorgensen's work grew from knot theory. We are interested in hyperbolic knots, and more generally hyperbolic links, i.e., mutually disjoint circles embedded in the 3-sphere S^3 whose complements admit complete hyperbolic structures of finite volume. Recall that the Ford domain of a complete cusped hyperbolic manifold of finite volume is the geometric dual to the canonical ideal polyhedral decomposition introduced by Epstein and Penner [27] (cf. [81]). Thus, by virtue of Mostow rigidity, the combinatorial structure of the Ford domain is a complete invariant of the topological type of such a manifold. In particular, by the knot complementary theorem due to Gordon and Luecke [32], this gives a complete invariant of a hyperbolic knot. In the joint work [71] with Weeks, the second author gave certain topological decompositions of 2-bridge link complements into topological ideal tetrahedra, by imitating Jorgensen's decomposition of punctured torus bundles over the circle (cf. [29]), and conjectured that they are combinatorially equivalent to the canonical decompositions. Here, a 2-bridge link is a link which can be drawn with only two local maxima and minima in the vertical direction (see Fig. 0.1). We had thought that if we could understand Jorgensen's work, then we would be able to prove the conjecture.

3. Extending of Jorgensen's theory beyond the quasifuchsian space and application to 2-bridge links

Fortunately, this turned out to be the case. Namely, we found a very natural way to understand the hyperbolic structures and the canonical decompositions of the 2-bridge link complements in the context of Jorgensen's work. To describe the idea, recall that the 2-bridge links are parametrized by pairs (p, q) of relatively prime integers (see [22, Chap. 12]) and that the complement of the 2-bridge link of type (p, q) is homeomorphic to (the interior of) the manifold obtained from $S \times [-1, 1]$, with S a 4-times punctured sphere, by attaching 2-handles along $\alpha \times (-1)$ and $\beta \times 1$, where α and β are simple loops on S of slopes $1/0$ and q/p, respectively (see Sect. 2.1, p. 16, for the definition of a slope); in particular, the link group (i.e., the fundamental group of the complement of the link) is isomorphic to the quotient group $\pi_1(S)/\langle\langle \alpha, \beta \rangle\rangle$, where $\langle\langle \cdot \rangle\rangle$ denotes the normal closure. The extended Jorgensen's theory realizes the operation of attaching 2-handles by a continuous family of hyperbolic cone-manifolds, whose cone axes are the union of the *upper* and *lower tunnels*, i.e., the co-cores of the 2-handles (see Fig. 0.1).

According to Keen-Series' theory of pleating varieties [44, 45, 46, 47, 49], $\mathcal{QF}$ is foliated by the pleating varieties, $\mathcal{P}(\lambda^-, \lambda^+)$, where (λ^-, λ^+) runs over (ordered) pairs of distinct projective measured laminations of the punctured torus T. By extending Jorgensen's theory beyond the quasifuchsian space (cf.

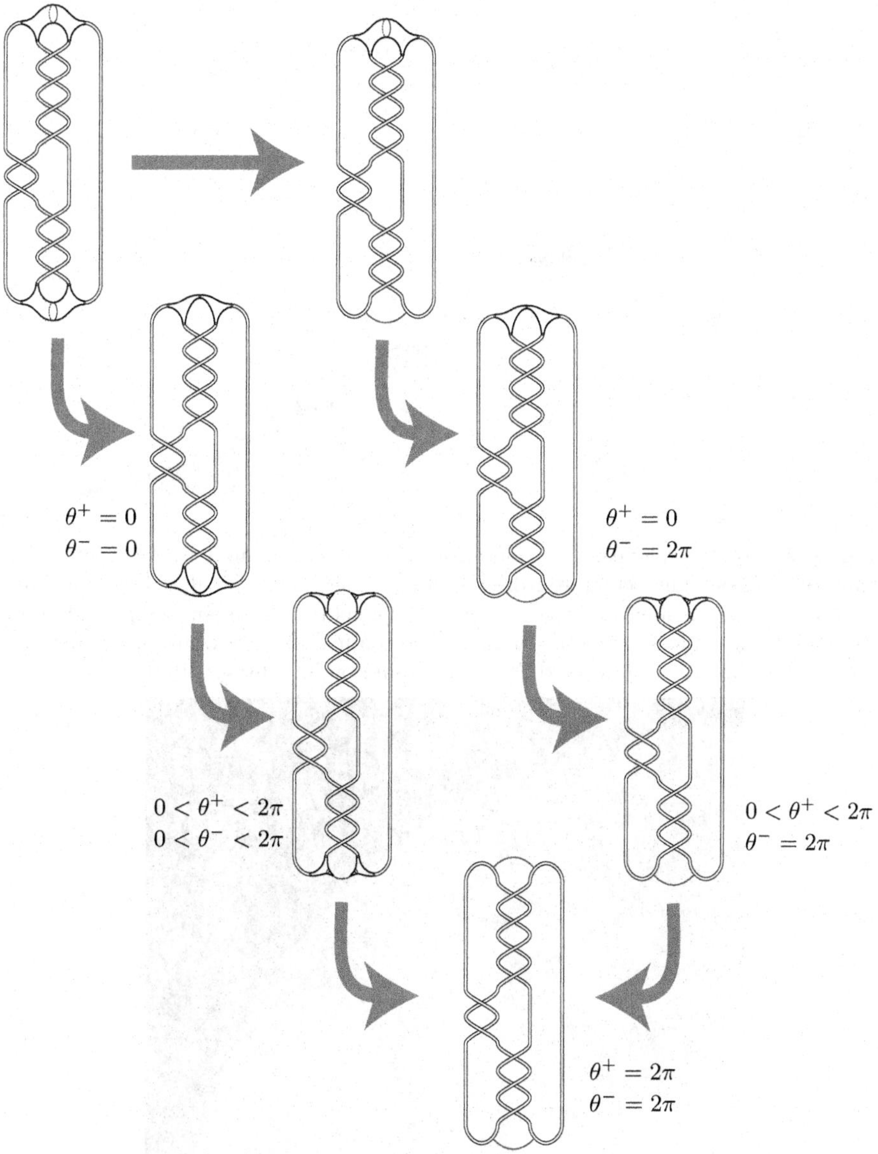

Fig. 0.1. Continuous family of hyperbolic cone-manifolds $M(\theta^-, \theta^+)$

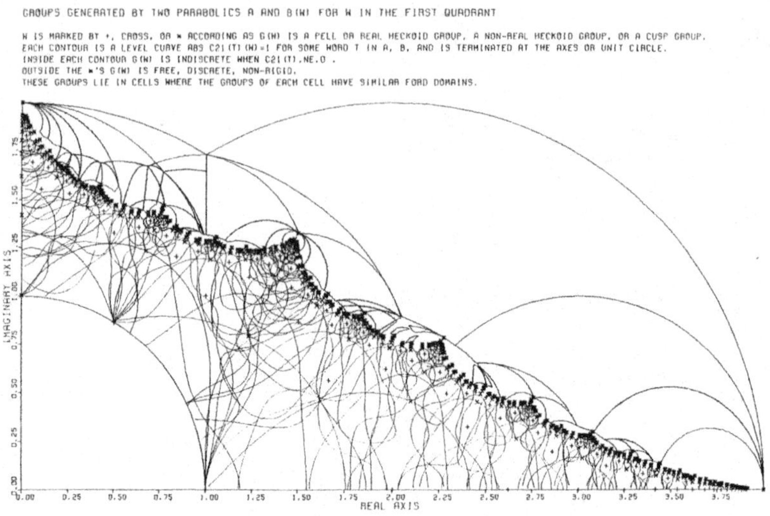

Fig. 0.2a. Riley's pioneering exploration of groups generated by two parabolic transformations. This computer-drawn picture has been circulated among the experts and has inspired many researchers in the fields of Kleinian groups and knot theory. This specific copy of the picture was obtained directly from Prof. Riley by the third author when he visited SUNY Binghamton in February 1991.

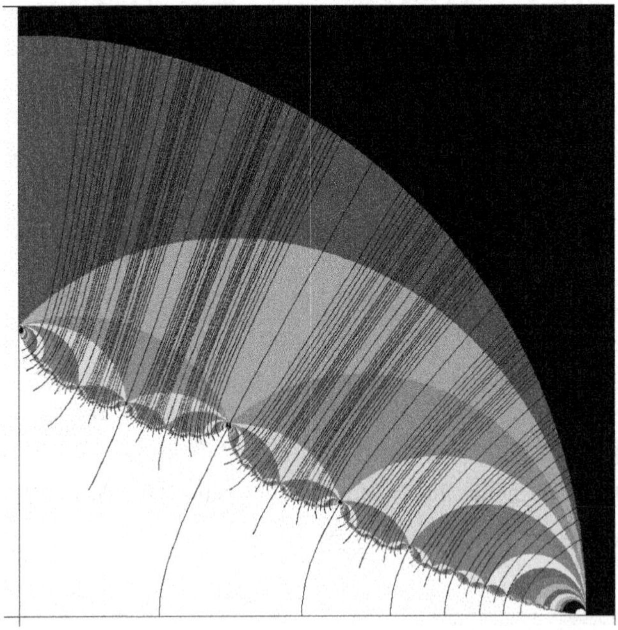

Fig. 0.2b. Riley slice of Schottky space together with pleating rays and their extensions

[66]), we found that if (λ^-, λ^+) is rational, i.e., each of $\lambda^\pm$ are rational, then the following hold:

1. The pleating variety $\mathcal{P}(\lambda^-, \lambda^+)$ has a natural extension to the outside of $\mathcal{QF}$ in the space of type-preserving representations of the fundamental group $\pi_1(T)$.
2. Each point in the extension is the holonomy representation of a certain hyperbolic cone manifold, which is commensurable with the hyperbolic cone manifold, $M(\theta^-, \theta^+)$, whose underlying space is the complement of a 2-bridge link and whose singular set is the union of the upper and lower tunnels, which have the cone angles θ^+ and θ^-, respectively. Moreover the 2-bridge link is of type (p, q), or of slope q/p, if (λ^-, λ^+) is equivalent to $(1/0, q/p)$ by a modular transformation.
3. If the (edge path) distance $d(1/0, q/p)$ in the Farey triangulation is ≥ 3, namely if $q \not\equiv \pm 1 \pmod p$, then the hyperbolic cone manifold $M(\theta^-, \theta^+)$ exists for every pair of cone angles in $[0, 2\pi]$. Thus we have a continuous family of hyperbolic cone manifolds connecting $M(0,0)$, the quotient hyperbolic manifold of a doubly cusped group, with $M(2\pi, 2\pi)$, the complete hyperbolic structure of the 2-bridge link complement.
4. If $1 \leq d(1/0, q/p) \leq 2$, namely if $q \equiv \pm 1 \pmod p$ and $p \neq 0$, then the hyperbolic cone manifold $M(\theta^-, \theta^+)$ exists for every pair of cone angles in $[0, 2\pi]$, except the pair $(2\pi, 2\pi)$. In addition, if $p \geq 3$, $M(\theta^-, \theta^+)$ collapses to the base orbifold of the Seifert fibered structure of the link complement as both cone angles approach 2π.
5. The holonomy group of $M(\theta^-, \theta^+)$ is discrete if and only if $\theta^\pm \in \{2\pi/n \,|\, n \in \mathbb{N}\} \cup \{0\}$. In particular, that of $M(2\pi, 2\pi/n)$ is generated by two parabolic transformations, which Riley called a *Heckoid group* in [68].

Actually, we have constructed these hyperbolic cone manifolds by explicitly constructing "Ford fundamental polyhedra". In other words, we have extended Jorgensen's description of the Ford fundamental polyhedra for quasifuchsian punctured torus groups to those of the hyperbolic cone manifolds arising from the 2-bridge links. In particular, we have shown that the canonical decompositions of hyperbolic 2-bridge link complements are isotopic to the topological ideal tetrahedral decompositions constructed in [71], proving the conjecture which motivated our project.

The above result also enables us to locate the 2-bridge link groups in the representation space (Fig. 0.2b). The shaded region of the figure illustrates (the first quadrant of) the *Riley slice* of the Schottky space, i.e., the subspace of $\mathbb{C}$ consisting of those complex numbers ω such that the group

$$G_\omega = \left\langle \begin{pmatrix} 1 & 1 \\ 0 & 1 \end{pmatrix}, \begin{pmatrix} 1 & 0 \\ \omega & 1 \end{pmatrix} \right\rangle$$

is discrete and free and such that the quotient $\Omega(G_\omega)/G_\omega$ of the domain of discontinuity is homeomorphic to the 4-times punctured sphere S (Definition

5.3.5). Each shaded region represents groups whose Ford domains have the same combinatorics. The lines in the shaded region are pleating rays of the Riley slice ([45]) and their extensions to the outside of the Riley slice correspond to the hyperbolic cone-manifolds $M(2\pi, \theta)$. In particular the end-points with positive imaginary parts represent hyperbolic 2-bridge link groups and those on the real line represent the orbifold fundamental groups of the base orbifolds of the Seifert fibered structures of non-hyperbolic 2-bridge link complements.

We think this realizes what Riley had in mind, for he devoted time and effort to identify 2-bridge link groups in the space of non-elementary two parabolic groups, yielding the mysterious output in Fig. 0.2a ([69]).

This describes a relation between the hyperbolic structure and the bridge structure of a 2-bridge link complement. Since a bridge structure is a kind of Heegaard structure, it is naturally expected that a similar relation holds between the hyperbolic structures and the Heegaard structures of hyperbolic manifolds. In particular, we conjecture that this is the case for tunnel number 1 hyperbolic knots and their unknotting tunnels. An *unknotting tunnel* for a knot K is an arc τ in S^3 with $\tau \cap K = \partial\tau$ such that the complement of an open regular neighborhood is homeomorphic to a genus 2 handlebody. A knot which admits an unknotting tunnel is said to have *tunnel number* 1. For example, a 2-bridge knot has tunnel number 1 and each of the upper and lower tunnels is an unknotting tunnel. Tunnel number 1 knots have been extensively studied, and in particular, non-hyperbolic tunnel number 1 knots were classified by [59]. An unknotting tunnel τ of a tunnel number 1 knot K gives a *Heegaard structure* of the knot complement $S^3 - K$, in the sense that $S^3 - K$ is homeomorphic to (the interior) of the manifold obtained from the genus 2 handlebody by adding a 2-handle, where τ corresponds to the co-core of the 2-handle. We would like to propose the following conjecture.

Conjecture. Let K be a tunnel number 1 hyperbolic knot and let τ be an unknotting tunnel for K. Then there is a continuous family of hyperbolic cone manifolds whose underlying space is the knot complement and whose cone axis is the unknotting tunnel τ, where the cone angle varies from 0 to 2π. In particular, τ is isotopic to a geodesic in the hyperbolic manifold $S^3 - K$.

4. Related results

Some of these results were announced in [8, 9, 10], and our original plan was to write a single paper or a book which contains the whole story. However, we found it very difficult to explain the whole theory at once, and thus decided to divide it into a few papers. This monograph is the first part of the series, and its main purpose is to give a full description of Jorgensen's theory on the space $\mathcal{QF}$ with a complete proof.

For Jorgensen's theory on the space $\mathcal{QF}$, supervised by Dunfield and partially influenced by [9] and [78], Schedler [72] gave a treatment based on the theory of holomorphic motions. Though the bijectivity of the *side parameter* map is not proved in his paper, his approach using holomorphic motions is natural and further development is expected in the future.

Our approach in turn is based on Poincare's theorem on fundamental polyhedra. This geometric approach enables us to extend Jorgensen's theory beyond the quasifuchsian space, where we need to treat indiscrete groups.

For (attempts of) expositions of Jorgensen's theory without proof, see [75, 8, 9, 65, 70].

The first author has extended Jorgensen's theory to the closure of $\mathcal{QF}$ in [2]. In particular, a rigorous proof was given to the well-known description of the Ford domain of the punctured torus bundles over the circle (cf. [12, 64]). We note that Lackenby [52] gave a topological proof to the fact that Jorgensen's ideal triangulations of punctured torus bundles are genuine geometric decompositions. Gueritaud [33] also gave an alternative proof to this fact by using the angle structure. In the appendix of the paper, Futer proves by modifying Gueritaud's argument that the topological ideal triangulations of the 2-bridge link complements in [71] are also geometric. Moreover, Gueritaud [34] also proved that these geometric decompositions are canonical.

In [3], the first author has found a nice relation between Jorgensen's parameter of $\mathcal{QF}$ and the conformal end invariant of elements of $\mathcal{QF}$. This together with Brock's results [21] leads to an estimate of the convex core volume in terms of Jorgensen's parameter. He has also found interesting applications of Jorgensen's theory to knot theory in [4].

The computer program, OPTi [78] (cf. [79]), has been developed by the third author for the project, and it has been a driving force for our work. It is our pleasure that it has now become a favorite tool for various colleagues in the world.

Collaborating with Komori and Sugawa, the third and last authors launched a project to draw Bers' slices of $\mathcal{QF}$, and various mysterious pictures have been produced ([50] and [82]).

5. A quick trip through Jorgensen's theory and its generalization

Jorgensen's theory enables us to intuitively understand how a simple fuchsian group evolves into complicated quasifuchsian punctured torus groups and boundary groups, by looking at their Ford domains (see Figs. 0.3–0.10, 0.17, 0.19–0.21 and 1.2). Jorgensen expresses this phenomenon as follows. *The Ford domain records the history of how the quasifuchsian group evolved from a simple fuchsian group.*

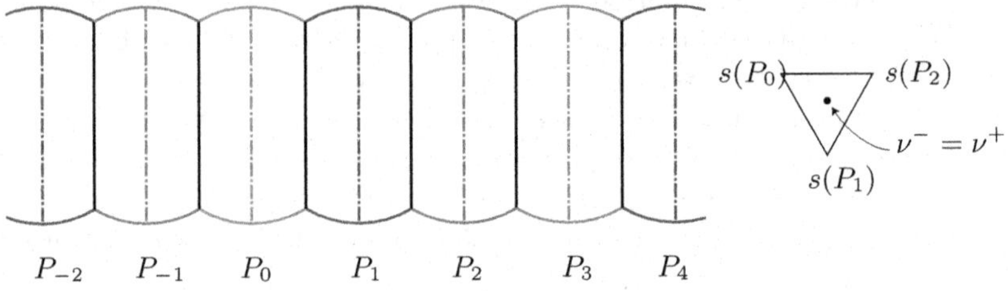

Fig. 0.3. $(1/3, 1/3, 1/3)$

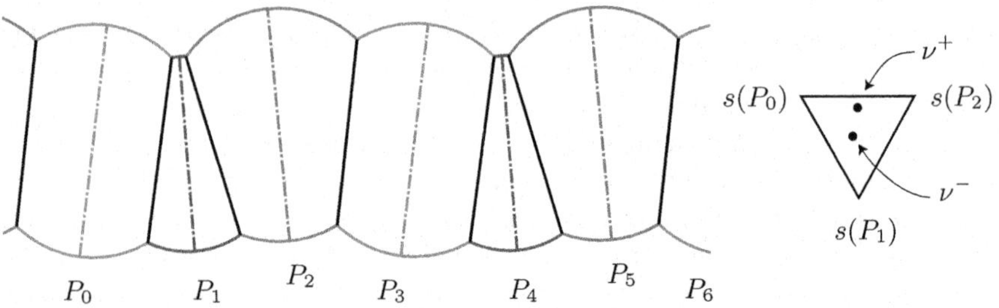

Fig. 0.4. $(0.421397 - 0.0483593\mathrm{i}, 0.295605 - 0.0422088\mathrm{i}, 0.282998 + 0.0905681\mathrm{i})$

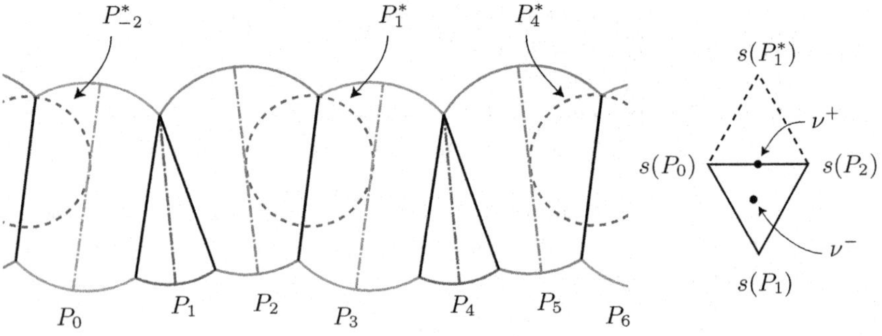

Fig. 0.5. $(0.433791 - 0.0551654\mathrm{i}, 0.290295 - 0.0481496\mathrm{i}, 0.275914 + 0.103315\mathrm{i})$

5.1. A fuchsian punctured torus group

The starting point of Jorgensen's theory is the fuchsian group illustrated in Fig. 0.11. For each integer j, let L_j be the geodesic in the upper half plane model $\mathbb{H}^2$ of the hyperbolic plane, represented by the Euclidean half circle with center $j/3$ and radius $1/3$. Let P_j be the order 2 elliptic transformation whose fixed point is equal to the highest point $(j + \mathrm{i})/3$ of L_j where $\mathrm{i} = \sqrt{-1}$. Then P_j interchanges the inside and outside of L_j and acts on L_j as a Euclidean isometry. The product $P_{j+2}P_{j+1}P_j$ is equal to the parabolic transformation $K(z) = z + 1$. Note that $P_{j+3n} = K^n P_j K^{-n}$ for every j, $n \in \mathbb{Z}$. Let Γ be

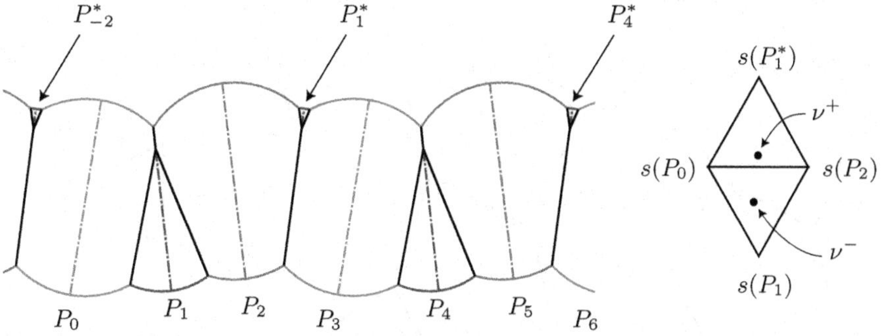

Fig. 0.6. (0.444228 − 0.0608968i, 0.285823 − 0.0531522i, 0.269949 + 0.114049i)

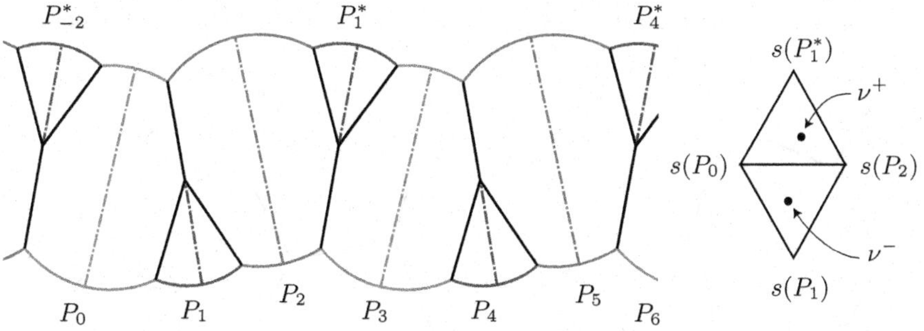

Fig. 0.7. (0.496414 − 0.0895542i, 0.263465 − 0.0781648i, 0.240121 + 0.167719i)

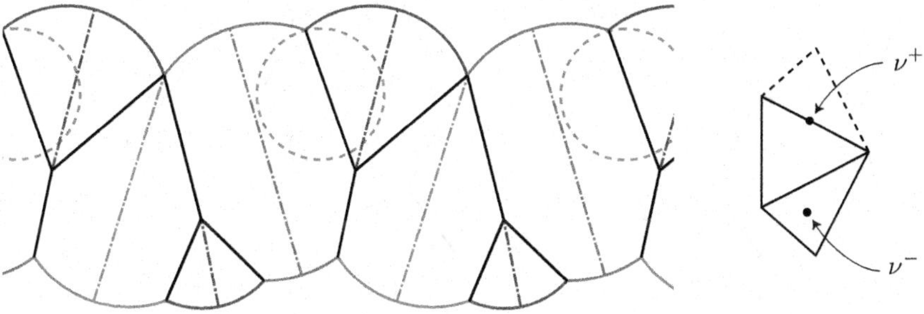

Fig. 0.8. (0.549741 − 0.118838i, 0.240619 − 0.103725i, 0.20964 + 0.222563i)

the group generated by $\{P_j \mid j \in \mathbb{Z}\}$. Then it is generated by three successive elements, say P_0, P_1 and P_2. Consider the shaded region R in Fig. 0.11. Then the edges of R are paired by P_0, P_1, P_2 and K. By applying Poincare's theorem on fundamental polyhedra to this setting, we see that R is a fundamental domain of the group Γ and

$$\Gamma \cong \langle P_0, P_1, P_2 \mid P_0^2 = P_1^2 = P_2^2 = 1 \rangle \cong (\mathbb{Z}/2\mathbb{Z}) * (\mathbb{Z}/2\mathbb{Z}) * (\mathbb{Z}/2\mathbb{Z}).$$

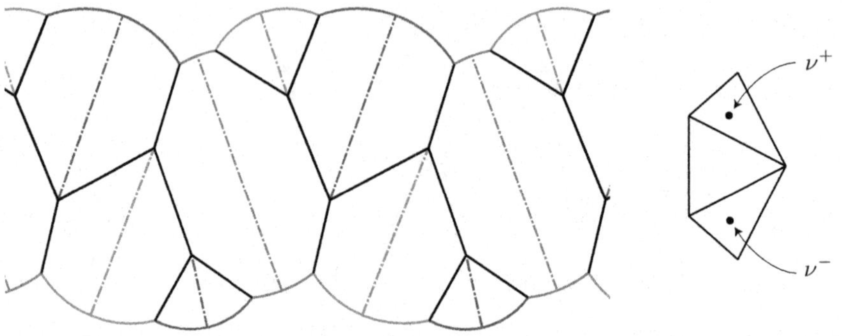

Fig. 0.9. $(0.594262 - 0.143287i, 0.221545 - 0.125063i, 0.184193 + 0.26835i)$

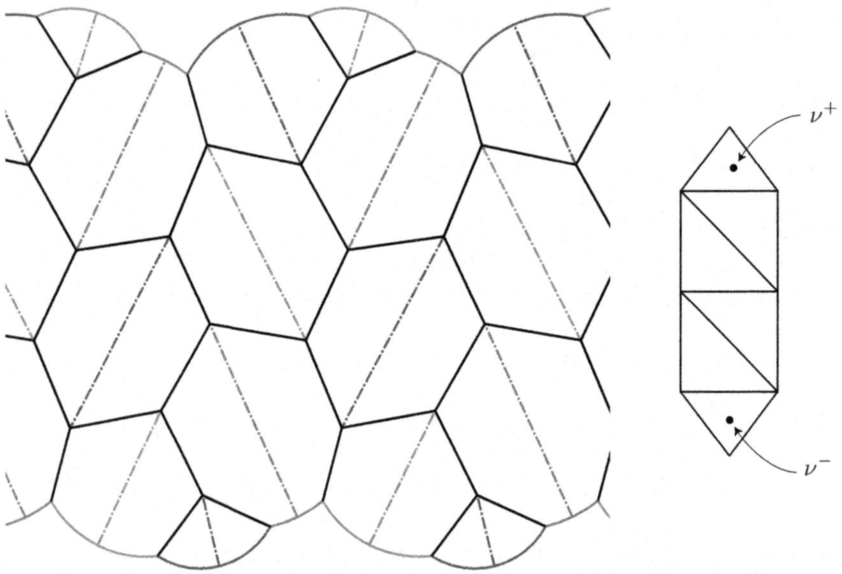

Fig. 0.10. $(0.652971 - 0.175526i, 0.196392 - 0.153203i, 0.150637 + 0.328729i)$

As shown in Fig. 0.12, the quotient $\mathbb{H}^2/\Gamma$ is a hyperbolic orbifold, $\mathcal{O}$, with underlying space once-punctured sphere and with three cone points of cone angle π. The subgroup Γ_0 of Γ of index 2, obtained as the kernel of the homomorphism $\Gamma \to \mathbb{Z}/2\mathbb{Z}$ sending each generator P_j to the generator of $\mathbb{Z}/2\mathbb{Z}$, is a rank 2 free group generated by $A := KP_0 = P_2P_1$ and $B := K^{-1}P_2 = P_0P_1$. The union $R \cup K(R)$ is a fundamental domain of Γ_0, and the quotient $\mathbb{H}^2/\Gamma_0$ is homeomorphic to the once-punctured torus, T, where the puncture corresponds to the commutator $[A, B] = K^2$. Thus Γ_0 is a *fuchsian*

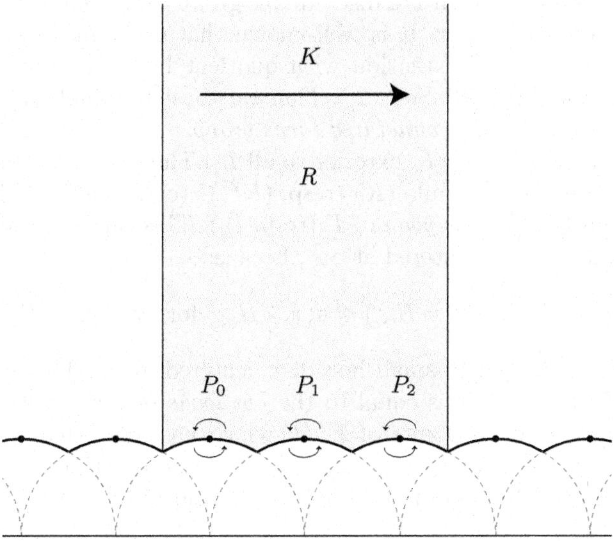

Fig. 0.11. Fuchsian group $\Gamma = \langle P_0, P_1, P_2 \rangle$ and its fundamental region R

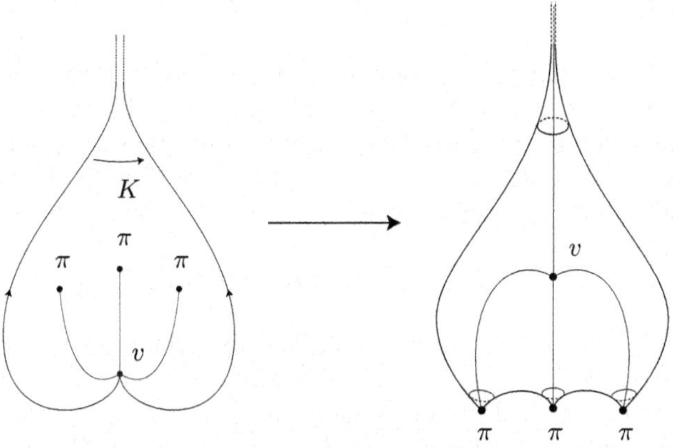

Fig. 0.12. By applying the edge pairings P_0, P_1 and P_2 to the fundamental region R, we obtained the surface on the left hand side. By further applying the edge pairing K to this surface, we obtain the orbifold $\mathcal{O}$ with underlying space once-punctured sphere and with three cone points of cone angle π.

punctured torus group, i.e., it is a discrete free group generated by two elements with parabolic commutator. It is well-known that every fuchsian punctured torus group has a $\mathbb{Z}/2\mathbb{Z}$-extension with quotient homeomorphic to $\mathcal{O}$ as a topological orbifold (cf. [40, Sect. 2]). Thus we abuse terminology and call the extended group a *fuchsian punctured torus group.*

Now look at the region, F, exterior to all L_j. Then this is a "fundamental domain of Γ (resp. Γ_0) modulo $\langle K\rangle$ (resp. $\langle K^2\rangle$)" (cf. Proposition 1.1.3). This region is called the *Ford polygon* of Γ (resp. Γ_0). This can be regarded as the "Dirichlet domain of Γ centered at ∞", because

$$F = \{x \in \mathbb{H}^2 \mid d(x, H_\infty) \leq d(x, ZH_\infty) \text{ for every } Z \in \Gamma\},$$

where H_∞ is a sufficiently small horodisk centered at ∞. This implies that the image of ∂F in $\mathbb{H}^2/\Gamma$ is equal to the *cut locus* of $\mathbb{H}^2/\Gamma$ with respect to the cusp, i.e., the set of points of $\mathbb{H}^2/\Gamma$ which has more than two shortest geodesics to the cusp. See Proposition 5.1.3, for a description of the Ford polygons of general fuchsian punctured torus groups.

5.2. 3-dimensional picture of the fuchsian punctured torus group

Figure 0.3 gives a 3-dimensional picture of the group Γ in Fig. 0.11. The elliptic transformation P_j acts on the upper half space model $\mathbb{H}^3$ of the hyperbolic 3-space as the π-rotation around the geodesic joining the two points $(j \pm \mathrm{i})/3$, where $\mathrm{i} = \sqrt{-1}$. (Here we identify the complex plane $\mathbb{C}$ with the boundary of the closure $\overline{\mathbb{H}}^3 = \mathbb{H}^3 \cup \mathbb{C}$.) The *isometric circle*

$$I(P_j) = \{z \in \mathbb{C} \mid |P_j'(z)| = 1\}$$

has center $c(P_j) = j/3$ and radius $1/3$. The hyperplane of $\mathbb{H}^3$ bounded by the isometric circle $I(P_j)$ is called the *isometric hemisphere* of P_j and is denoted by $Ih(P_j)$. Then P_j interchanges the exterior $Eh(P_j)$ and the interior $Dh(P_j)$ of the isometric hemisphere $Ih(P_j)$, and acts on $Ih(P_j)$ as a Euclidean isometry. By the argument in Subsection 5.1, we see that the common exterior $\cap_j Eh(P_j)$, where j runs over $\mathbb{Z}$, is a "fundamental domain of the action of Γ (resp. Γ_0) on $\mathbb{H}^3$ modulo $\langle K\rangle$ (resp. $\langle K^2\rangle$)". Thus it is equal to the *Ford domain* $Ph(\Gamma)$ of Γ, which is defined to be the common exteriors of the isometric hemispheres of all elements of Γ that do not fix ∞ (see Definition 1.1.2 and Proposition 1.1.3). As in the previous subsection, the Ford domain can be regarded as the "Dirichlet domain of Γ centered at ∞", namely

$$Ph(\Gamma) = \{x \in \mathbb{H}^3 \mid d(x, H_\infty) \leq d(x, ZH_\infty) \text{ for every } Z \in \Gamma\},$$

where H_∞ is a sufficiently small horoball centered at ∞. Thus the image of $\partial Ph(\Gamma)$ in $\mathbb{H}^3/\Gamma \cong \mathcal{O} \times (-1, 1)$ is equal to the *cut locus* of $\mathbb{H}^3/\Gamma$ with respect to the cusp, i.e., the set of points of $\mathbb{H}^3/\Gamma$ which has more than two shortest geodesics to the cusp: We call it the *Ford complex* of Γ.

Let $\overline{Ph}(\Gamma)$ be the closure of the Ford domain $Ph(\Gamma)$ in $\overline{\mathbb{H}}^3$. Then the intersection $P(\Gamma) := \overline{Ph}(\Gamma) \cap \mathbb{C}$ has precisely two connected components, the *upper polygon* $P^+(\Gamma)$ and the *lower polygon* $P^-(\Gamma)$. $P^\pm(\Gamma)$ is a fundamental domain of the action of Γ on $\Omega^\pm(\Gamma)$ modulo $\langle K \rangle$, where $\Omega^+(\Gamma)$ and $\Omega^-(\Gamma)$ are the upper and lower components of the domain of discontinuity $\Omega(\Gamma) = \mathbb{C} - \mathbb{R}$.

5.3. Parameters for punctured torus groups

Jorgensen's theory describes what happens to the Ford domain when we deform the group Γ keeping the condition that $K = P_2P_1P_0$ is the parabolic transformation $z \mapsto z+1$, or equivalently, deform the group Γ_0 keeping the condition that $[A, B]$ is the parabolic transformation $z \mapsto z+2$. Thus we first need to describe the space of all such groups. Let $\mathcal{X}$ be the space of the equivalence classes of marked subgroups Γ of $PSL(2,\mathbb{C})$ generated by an *elliptic generator triple* (P_0, P_1, P_2), i.e., a triple of order 2 elliptic transformations P_0, P_1 and P_2 such that the product $K := P_2P_1P_0$ is a parabolic transformation (cf. Definition 2.1.1). Two such marked groups Γ and Γ', endowed with elliptic generator triples (P_0, P_1, P_2) and (P_0', P_1', P_2'), respectively, are *equivalent* if they are conjugate in $PSL(2,\mathbb{C})$ and if the conjugation maps (P_0, P_1, P_2) to (P_0', P_1', P_2') (cf. Definition 2.2.6). We do not distinguish between an element of $\mathcal{X}$ and its representative. The space $\mathcal{X}$ is identified with a quotient of a subspace of the cartesian product $PSL(2,\mathbb{C})^3$ and thus is endowed with the quotient topology.

By a *marked punctured torus group*, we mean an element Γ of $\mathcal{X}$ such that Γ is discrete and isomorphic to $(\mathbb{Z}/2\mathbb{Z}) * (\mathbb{Z}/2\mathbb{Z}) * (\mathbb{Z}/2\mathbb{Z})$.

Then $\mathcal{X}$ is identified with the space of the equivalence classes of the non-trivial *Markoff triples*. Here a Markoff triple is a triple $(x, y, z) \in \mathbb{C}^3$ satisfying the Markoff equation

$$x^2 + y^2 + z^2 = xyz.$$

It is non-trivial if it is different from $(0, 0, 0)$. Two Markoff triples (x, y, z) and (x', y', z') are equivalent if the latter is equal to (x, y, z), $(x, -y, -z)$, $(-x, y, -z)$ or $(-x, -y, z)$. We associate to each marked group $\Gamma \in \mathcal{X}$, endowed with an elliptic generator triple (P_0, P_1, P_2), the equivalence class of a Markoff triple (x, y, z) by the following rule.

$$(x, y, z) = (\mathrm{tr}(KP_0), \mathrm{tr}(KP_1), \mathrm{tr}(KP_2)) = (\mathrm{tr}(A), \mathrm{tr}(AB), \mathrm{tr}(B)),$$

where $A = KP_0$ and $B = K^{-1}P_2$. Note that the right hand side is defined only up to sign, because it depends on the lifts to $SL(2,\mathbb{C})$. If we take the lifts so that the lift of AB is the product of the lift of A and that of B, then (x, y, z) satisfies the Markoff equation and its equivalence class is uniquely determined by the equivalence class of the marked group Γ, and vice versa (Propositions 2.3.6 and 2.4.2). For example, the group in Fig. 0.11 corresponds to the Markoff triple $(3,3,3)$.

More geometric parameter of (a subspace of) $\mathcal{X}$ is the *complex probability* defined as follows. Let (x,y,z) be a Markoff triple such that none of the entries are zero. Then its equivalence class is completely determined by the triple $(a_0,a_1,a_2) \in (\mathbb{C}-\{0\})^3$ defined by

$$(a_0,a_1,a_2) = (\frac{x}{yz}, \frac{y}{zx}, \frac{z}{xy}).$$

The only constraint on this triple is the identity

$$a_0 + a_1 + a_2 = 1,$$

and thus this triple is called a *complex probability.* This parameter has the geometric meaning that each entry gives the difference between the centers of the isometric circles of the elliptic generators. Namely,

$$a_0 = c(P_2)-c(P_1), \quad a_1 = c(P_3)-c(P_2), \quad a_2 = c(P_4)-c(P_3) = c(P_1)-c(P_0).$$

Here $P_3 = KP_0K^{-1}$ and $P_4 = KP_1K^{-1}$. Moreover, there is a nice geometric construction of a marked group from the corresponding complex probability (see Sect. 2.4, p. 29). The complex probability for the marked group in Fig. 0.3, for example, is $(\frac{1}{3}, \frac{1}{3}, \frac{1}{3})$. The triples in the captions of Figs. 0.3–0.10, 0.17, 0.19–0.25 are the complex probabilities of the corresponding marked punctured torus groups.

5.4. Small deformation of the fuchsian punctured torus group

Now let us study what happens to the Ford domain if we deform the group Γ in Fig. 0.3 a little in the space $\mathcal{X}$, namely we perturb the complex probability from $(\frac{1}{3}, \frac{1}{3}, \frac{1}{3})$ a little. The answer is that nothing happens to the combinatorial structure of the Ford domain (see Fig. 0.4). Namely, the polyhedron $\cap_j Eh(P_j)$ continues to be the Ford domain of the (deformed) group Γ. This fact can be proved by using Poincare's theorem on fundamental polyhedra and the following facts for the (deformed) group Γ, which are consequences of the "chain rule for isometric circles" (Lemmas 4.1.2 and 4.1.3).

1. Each face $Ih(P_j) \cap (\cap_j Eh(P_j))$ of the polyhedron $\cap_j Eh(P_j)$ is symmetric with respect to P_j.
2. The sum of the dihedral angles of the polyhedron $\cap_j Eh(P_j)$ along any three successive edges $Ih(P_j) \cap Ih(P_{j+1})$ is equal to 2π.

The above conclusion on the Ford domain corresponds to a special case of Proposition 6.2.1 (Openness), whose rigorous proof is given in Chap. 7. We note that Schedler [72] explains this phenomenon by developing a general theory of Ford domains based on the theory of holomorphic motions. A key fact behind both proofs is that $Ph(\Gamma)$ has no *hidden isometric hemispheres*, that is, for every element A of Γ which does not fix ∞, $\overline{Ph}(\Gamma) \cap \overline{Ih}(A) \neq \emptyset$ only if $Ih(A)$ support a 2-dimensional face of the polyhedron $\cap_j Eh(P_j)$, i.e., $A = P_j$ for some j for the group in Fig. 0.4 (see Lemma 7.1.6). Here $\overline{Ih}(A)$ denotes the closure of $Ih(A)$ in $\overline{\mathbb{H}}^3$.

The deformed group Γ is a marked *quasifuchsian punctured torus group*, i.e., it is a quasi-conformal deformation of the fuchsian punctured torus group in Fig. 0.3, or equivalently, the limit set of Γ continues to be a Jordan circle in the Riemann sphere $\partial\mathbb{H}^3$ and the quotient $(\mathbb{H}^3 \cup \Omega(\Gamma))/\Gamma$ continues to be homeomorphic to the product $\mathcal{O} \times [-1, 1]$.

The *quasifuchsian punctured torus space* $\mathcal{QF}$ is the subspace of $\mathcal{X}$ consisting of all (marked) quasifuchsian punctured torus groups. $\mathcal{QF}$ is an open subset of the 2-dimensional complex manifold $\mathcal{X}$. Bers' simultaneous uniformization theorem says that a quasifuchsian punctured torus group Γ is uniquely determined by the pair $(\Omega^-(\Gamma_0)/\Gamma_0, \Omega^+(\Gamma_0)/\Gamma_0)$ of punctured torus Riemann surfaces (see, for example, [38] or [55]). This correspondence implies a holomorphic isomorphism between the quasifuchsian space $\mathcal{QF}$ and the product $\mathrm{Teich}(T) \times \mathrm{Teich}(T) \cong \mathbb{H}^2 \times \mathbb{H}^2$ of the Teichmuller space of T. Jorgensen's theory also gives yet another parameterization of $\mathcal{QF}$ in terms of $\mathbb{H}^2 \times \mathbb{H}^2$ (see Main Theorem 1.3.5).

Though the space $\mathcal{QF}$ itself has a simple structure, its location in $\mathcal{X}$ is very complicated. Jorgensen's theory enables us to plot the shape of $\mathcal{QF}$. See Fig. 0.13 illustrating a slice of $\mathcal{QF}$ in $\mathcal{X}$. This is an output of OPTi [78], which in turn is based on Jorgensen's theory. See also the beautiful pictures in [61].

5.5. Birth of a new face in the Ford domain

We can continue the deformation in the previous subsection until some of the circular edges of the upper/lower polygons $P^{\pm}(\Gamma)$ shrinks to a point (see Fig. 0.5). Assume for simplicity that the circular edge $I(P_1) \cap P^+(\Gamma)$ of the upper polygon $P^+(\Gamma)$ shrinks to a point, v. (The existence of such a deformation is guaranteed by Jorgensen's theory.) What happens to the Ford domain under further deformation? The answer is given by Fig. 0.6.

To describe it, we note that the "chain rule for isometric circles" (Lemmas 4.1.2 and 4.1.3) implies that the isometric circle $I(P_1^*)$ of $P_1^* := P_2P_1P_2$ passes through the vertex $P_2(v)$ of the upper polygon $P^+(\Gamma)$ in Fig. 0.5. Thus $Ih(P_1^*)$ is a hidden isometric hemisphere of $Ph(\Gamma)$. Moreover this isometric hemisphere and its translates by powers of K are the only hidden isometric hemispheres of $Ph(\Gamma)$ (Lemma 7.1.6). By using this fact we see that if we

Fig. 0.13. $\mathcal{QF}$ in the slice of $\mathcal{X}$ at $a_0 = 1/3$

deform the group further then the Ford domain undergoes the following changes (see Fig. 0.6).

1. The hidden isometric hemisphere $Ih(P_1^*)$ breaks out through the vertex $P_2(v)$ and becomes to support a 2-dimensional face of $Ph(\Gamma)$.
2. The vertices v and $P_2(v)$ of the old upper polygon $P^+(\Gamma)$ lying in the complex place $\mathbb{C}$ are lifted to vertices of the new Ford domain $Ph(\Gamma)$ in the hyperbolic space $\mathbb{H}^3$.
3. The new upper polygon $P^+(\Gamma)$ is described by the sequence $\{P_j'\}$ defined by

$$P_0' := P_0, \quad P_1' := P_2, \quad P_2' := P_1^* = P_2P_1P_2,$$
$$P_{j+3n}' := K^nP_j'K^{-n} \quad (j \in \{0,1,2\},\ n \in \mathbb{Z}).$$

Namely, the edges of $\partial P^+(\Gamma)$ are $I(P_j') \cap P^+(\Gamma)$ $(j \in \mathbb{Z})$ in this order.

Moreover, the new Ford domain $Ph(\Gamma)$ is equal to the polyhedron $(\cap_j Eh(P_j)) \cap (\cap_j Eh(P_j'))$, and it is combinatorially dual to the *elliptic generator complex* in Fig. 0.14, which describes the relation between the two sequences $\{P_j\}$ and $\{P_j'\}$ (Definition 3.2.3).
The above description of the transition of the Ford domain is proved by using the following facts.

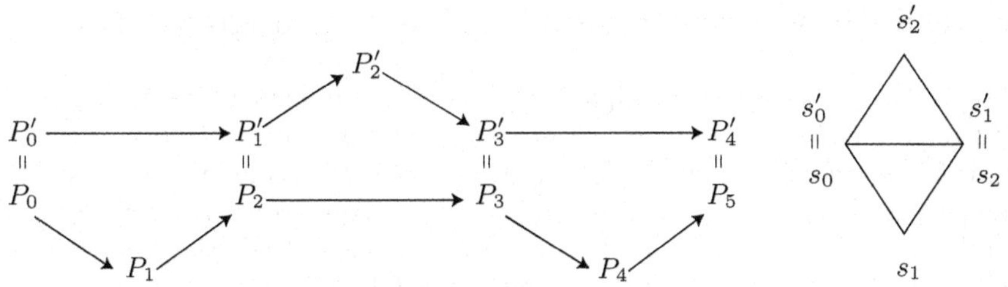

Fig. 0.14. Adjacent sequences of elliptic generators

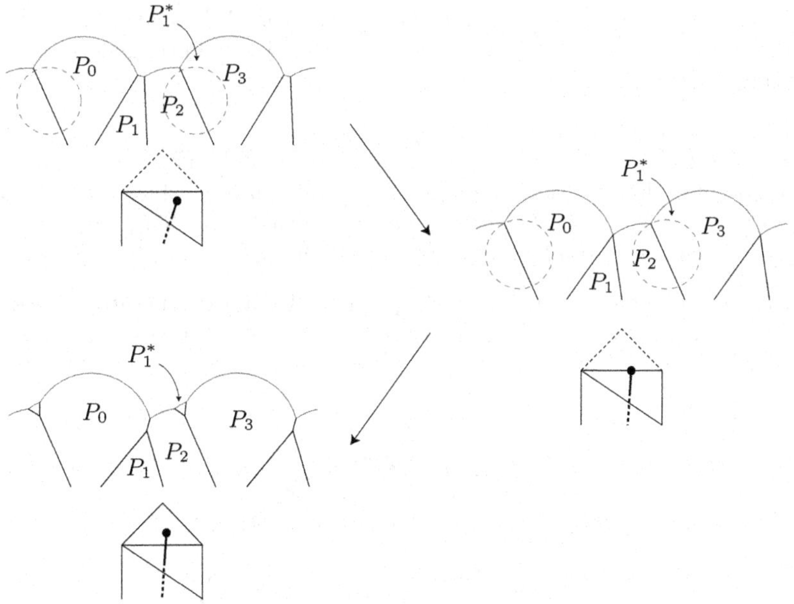

Fig. 0.15. Birth of a new face

1. The local behavior of the hidden isometric hemisphere $Ih(P_1^*)$ under small deformation is controlled by the "side parameter", which we explain in Subsection 5.7, and the only possible local behaviors are those illustrated in Fig. 0.15 (see Lemma 4.6.2).
2. The chain rule of the isometric circles (Lemma 4.1.3) guarantees that the above polyhedron $(\cap_j Eh(P_j)) \cap (\cap_j Eh(P'_j))$ satisfies the conditions of Poincare's theorem on fundamental polyhedra.

Jorgensen's theory shows that the transition described in the above is essentially the only possible transition of the combinatorial structure of the Ford domain when the group Γ is deformed in $\mathcal{QF}$ and that one can deform the group so that the Ford domain becomes arbitrary complicated (see Figs. 0.3–0.10).

5.6. Elliptic generators and the Farey triangulation

Observe that any successive triple $(P'_j, P'_{j+1}, P'_{j+2})$ in the bi-infinite sequence $\{P'_j\}$ introduced in the previous subsection is also an elliptic generator triple, i.e., it forms a generator system of Γ and $P'_{j+2}P'_{j+1}P'_j$ is equal to the parabolic transformation $K(z) = z + 1$. Such a bi-infinite sequence is called a *sequence of elliptic generators*, and an element of a sequence of elliptic generators is called an *elliptic generator* (cf. Definitions 2.1.1 and 2.1.13). By using the facts that the orbifold $\mathcal{O}$ is commensurable with the punctured torus T and that an elliptic generator corresponds to an essential simple loop in T, we can define the *slope* $s(P)$ for each elliptic generator P (Proposition 2.1.2 and Definition 2.1.3). Moreover, for any sequence $\{P_j\}$ of elliptic generators, the following hold.

1. $\{s(P_j)\}$ is a periodic sequence of period 3.
2. The set $\{s(P_0), s(P_1), s(P_2)\}$ spans a triangle of the Farey triangulation, or the modular diagram, $\mathcal{D}$ of the hyperbolic plane $\mathbb{H}^2$ (see Fig. 0.16).
3. $s(P'_2) = s(P_2P_1P_2)$ is the vertex opposite to the vertex $s(P_1)$ with respect to the edge $\langle s(P_0), s(P_2)\rangle = \langle s(P'_0), s(P'_1)\rangle$ in $\mathcal{D}$.

This gives a bijection between the sequences of elliptic generators (modulo shifts of indices by a multiple of 3) and the triangles of $\mathcal{D}$ (Proposition 2.1.10).

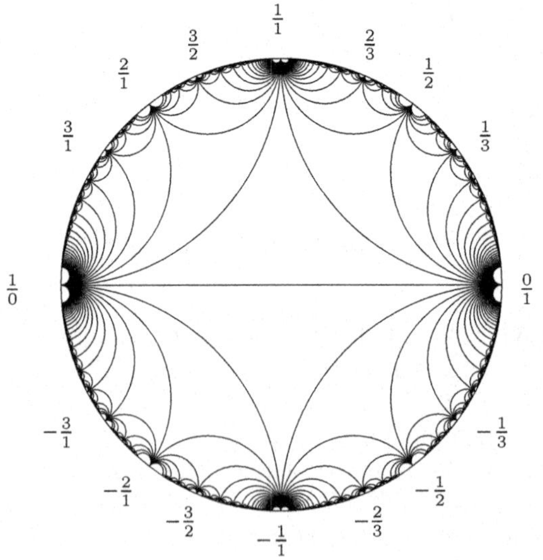

Fig. 0.16. Farey triangulation

Jorgensen's theory associates to each quasifuchsian punctured torus group $\Gamma \in \mathcal{QF}$ a pair (σ^-, σ^+) of triangles of $\mathcal{D}$, so that the combinatorial structure of the Ford domain $Ph(\Gamma)$ is completely determined by the pair (σ^-, σ^+)

(see Theorem 1.2.2). To be precise, (the boundaries of) the upper and lower polygons $P^+(\Gamma)$ and $P^-(\Gamma)$, respectively, are described by the sequences of elliptic generators associated with σ^+ and σ^-, and the Ford domain $Ph(\Gamma)$ is dual to the *elliptic generator complex* associated with the chain of triangles in $\mathcal{D}$ joining σ^- and σ^+ (Definition 3.2.3). If $\sigma^- = \sigma^+$, then the elliptic generator complex is identified with the real line $\mathbb{R}$ with vertex set $\mathbb{Z}$, where the vertex $j \in \mathbb{Z}$ corresponds to the j-th elliptic generator P_j. If σ^- and σ^+ are adjacent, then the elliptic generator complex is as illustrated in Fig. 0.14.

5.7. The side parameters of quasifuchsian punctured torus groups

The correspondence $\Gamma \mapsto (\sigma^-, \sigma^+)$ is refined to a (combinatorial) homeomorphism $\boldsymbol{\nu} : \mathcal{QF} \to \mathbb{H}^2 \times \mathbb{H}^2$, resembling the holomorphic isomorphism $\mathcal{QF} \cong \mathbb{H}^2 \times \mathbb{H}^2$ via the Bers' simultaneous uniformization theorem described in Subsection 5.4. To explain this homeomorphism, recall that the edges of the upper and lower polygons $P^+(\Gamma)$ and $P^-(\Gamma)$, respectively, are supported by the isometric circles of the sequences of elliptic generators $\{P_j^+\}$ and $\{P_j^-\}$ associated with σ^+ and σ^-. For $\epsilon \in \{-, +\}$ and $j \in \{0, 1, 2\}$, let θ_j^ϵ be the half of the angle of the circular edge of the polygon $P^\epsilon(\Gamma)$ contained in the isometric circle $I(P_j^\epsilon)$. Then the following identity holds (Proposition 4.2.16):

$$\theta_0^\epsilon + \theta_1^\epsilon + \theta_2^\epsilon = \frac{\pi}{2}.$$

Thus the triple $(\theta_0^\epsilon, \theta_1^\epsilon, \theta_2^\epsilon)$ can be regarded as $\pi/2$ times the barycentric coordinate of a point in the triangle $\sigma^\epsilon = \langle s(P_0^\epsilon), s(P_1^\epsilon), s(P_2^\epsilon)\rangle$ of (the abstract simplicial complex having the combinatorial structure of) the Farey triangulation $\mathcal{D}$. We denote the point by $\nu^\epsilon(\Gamma)$. Then $\nu^\epsilon(\Gamma)$ is not equal to a vertex of σ^ϵ and hence it is identified with a point in $\mathbb{H}^2 \cong |\mathcal{D}| - |\mathcal{D}^{(0)}|$, where $\mathcal{D}^{(0)}$ denotes the 0-skeleton of $\mathcal{D}$ and $|\cdot|$ denotes the underlying space of an abstract simplicial complex (cf. Sect. 1.3, p. 12). Set $\boldsymbol{\nu}(\Gamma) = (\nu^-(\Gamma), \nu^+(\Gamma))$ and call it the *side parameter* of Γ. Then Jorgensen's theory asserts that $\boldsymbol{\nu} : \mathcal{QF} \to \mathbb{H}^2 \times \mathbb{H}^2$ is a homeomorphism and that the combinatorial structure of the Ford domain $Ph(\Gamma)$ of $\Gamma \in \mathcal{QF}$ is completely described in terms of $\boldsymbol{\nu}(\Gamma)$ (see Main Theorem 1.3.5).

5.8. Jorgensen's theory for boundary groups.

A *marked punctured torus group* is (a representative of) an element of $\mathcal{X}$ which is discrete and isomorphic to the free product $(\mathbb{Z}/2\mathbb{Z}) * (\mathbb{Z}/2\mathbb{Z}) * (\mathbb{Z}/2\mathbb{Z})$. It is classically known that every marked group in the closure $\overline{\mathcal{QF}}$ of $\mathcal{QF}$ in $\mathcal{X}$ is a marked punctured torus group (see, for example, [55, Proposition 4.18]).

A group in $\overline{\mathcal{QF}} - \mathcal{QF}$ is called a *boundary group*. Moreover it is a consequence of Minsky's ending lamination theorem for punctured torus groups that every Kleinian punctured torus group is contained in the closure $\overline{\mathcal{QF}}$ (see [58]).

Jorgensen's theorem and its generalization enable us to get a visual understanding of these complicated groups (see Figs. 0.17, 0.19–0.21).

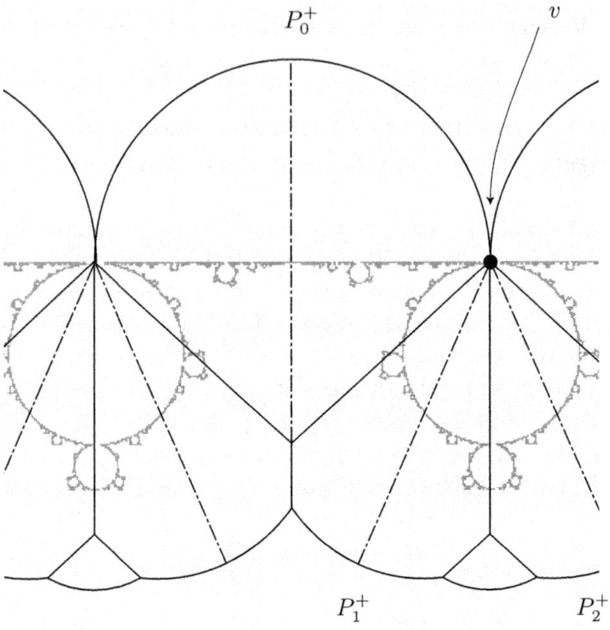

Fig. 0.17. $(\frac{1}{3}, \frac{1}{3} + \frac{\sqrt{5}}{6}\mathrm{i}, \frac{1}{3} - \frac{\sqrt{5}}{6}\mathrm{i})$ – Singly cusped group

A *singly cusped group*, Γ^*, is obtained from a quasifuchsian group Γ by a deformation which shrinks two successive edges of, say the upper polygon $P^+(\Gamma)$, into a single point, while fixing $\nu^-(\Gamma)$ (see Fig. 8 in Jorgensen [40] and Fig. 0.17). If the edges of $P^+(\Gamma)$ supported by the isometric circles $I(P_1^+)$ and $I(P_2^+)$ are shrunk into a point v, then the two fixed points of the loxodromic transformation KP_0^+ of Γ are united into the point v, and the corresponding element KP_0^+ of Γ^* becomes a parabolic transformation with parabolic fixed point v; this transformation is called an *accidental parabolic transformation*. The limit set of Γ^* is obtained from the Jordan curve $\Lambda(\Gamma)$ by pinching the two fixed points of each conjugate of (the old) KP_0^+ into a single point. Thus the "upper part" $\Omega^+(\Gamma^*)$ of the domain of discontinuity is not connected anymore; it consists of infinitely many (round) disks, whereas the lower component $\Omega^-(\Gamma^*)$ remains to be an open disk. Accordingly the quotient orbifold $\Omega^+(\Gamma^*)/\Gamma^*$ becomes an orbifold with underlying space a twice-punctured sphere with a cone point of cone angle π, or equivalently,

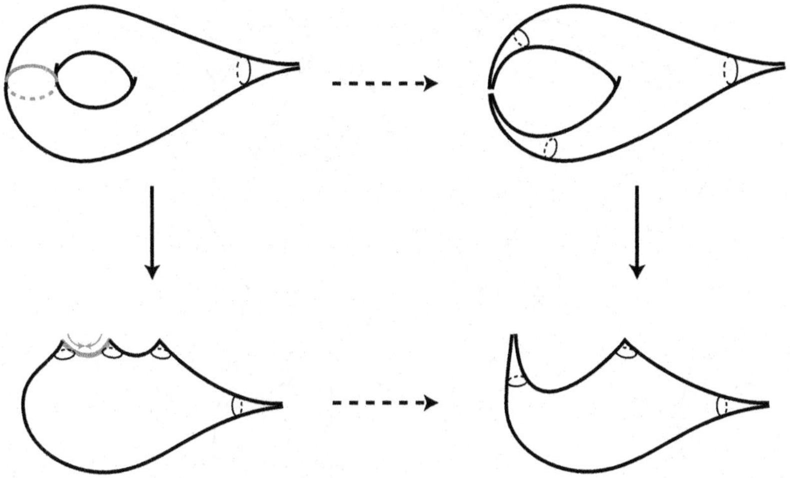

Fig. 0.18. Pinching the punctured torus $\Omega^+(\Gamma_0)/\Gamma_0$ and the quotient orbifold $\Omega^+(\Gamma)/\Gamma$

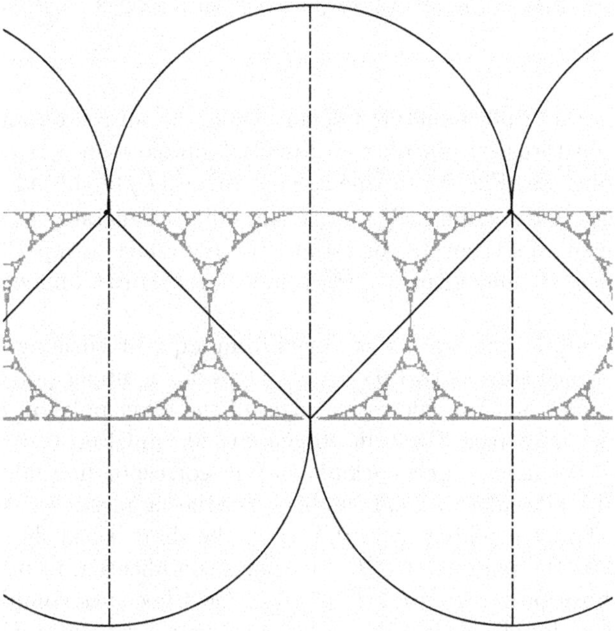

Fig. 0.19. $(\frac{1+i}{2}, \frac{1-i}{4}, \frac{1-i}{4})$ – Doubly cusped group

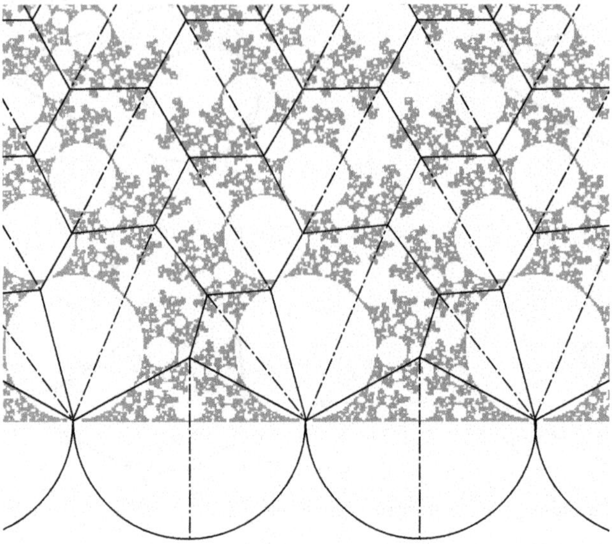

Fig. 0.20. (0.198264+0.376931i, 0.528473+0.142584i, 0.273263−0.519515i) – Singly degenerate group

the quotient of a triply punctured sphere which in turn is obtained from a once-punctured torus by pinching an essential simple loop corresponding to KP_0^+ to a point (see Fig. 0.18). The component $\nu^+(\Gamma)$ of the side parameter $\boldsymbol{\nu}(\Gamma)$ converges to the slope $s(P_0^+) \in \hat{\mathbb{Q}}$ of the elliptic generator P_0^+. The (generalized) Jorgensen's theory associates to the boundary group Γ^* the point $\boldsymbol{\nu}(\Gamma^*) := \lim \boldsymbol{\nu}(\Gamma)$ and gives a description of the Ford domain $Ph(\Gamma^*)$ in terms of $\boldsymbol{\nu}(\Gamma^*)$.

A *doubly cusped group*, Γ^*, is obtained from a quasifuchsian group Γ by a deformation which shrinks two successive edges of the upper polygon $P^+(\Gamma)$ into a single point and two successive edges of the lower polygon $P^-(\Gamma)$ into another single point (see Fig. 6 in Jorgensen [40] and Fig. 0.19). Then the limit set of Γ^* becomes a circle packing, and the corresponding side parameter $\boldsymbol{\nu}(\Gamma^*)$ is a pair of distinct rational numbers. For the simplest case where $\boldsymbol{\nu}(\Gamma^*)$ consists of a Farey neighbor, say $(0/1, 1/0)$, the limit set is the Apollonian packing, and as the parameter $\boldsymbol{\nu}(\Gamma^*)$ becomes complicated, to be precise, as the relative position between $\nu^-(\Gamma^*)$ and $\nu^+(\Gamma^*)$ becomes complicated, the limit set of the doubly cusped group becomes more and more intricate (see the beautiful Figs. 7.3, 9.15. 9.16 and 9.18 in [61]).

A singly or doubly cusped group remains to be *geometrically finite*, i.e., there is a finite volume submanifold of the quotient hyperbolic manifold which contains all closed geodesics, or equivalently, the quotient of the convex hull

Fig. 0.21. $(\frac{1}{2}+\frac{1}{2\sqrt{3}}\mathrm{i}, -\frac{1}{\sqrt{3}}\mathrm{i}, \frac{1}{2}+\frac{1}{2\sqrt{3}}\mathrm{i})$ – Doubly degenerate group

of the limit set has finite volume. Constructions of more complicated *geometrically infinite* groups are presented below.

A *singly degenerate group*, Γ^*, is obtained as the limit of a sequence of quasifuchsian (or geometrically finite) groups $\{\Gamma_n\}$ such that $\nu^+(\Gamma_n)$ tends to an irrational boundary point, whereas $\nu^-(\Gamma_n)$ is fixed (see Fig. 0.20). Then the upper component $\Omega^+(\Gamma_n)$ of the domain of discontinuity becomes "smaller and smaller" as $n \to \infty$, and finally disappears at the limit. So, $\Omega(\Gamma^*)/\Gamma^*$ consists of only one component. A singly degenerate group is not geometrically finite anymore and hence is geometrically infinite.

A *doubly degenerate group* Γ^* is obtained as the limit of a sequence of quasifuchsian (or geometrically finite) groups $\{\Gamma_n\}$ such that $\boldsymbol{\nu}(\Gamma_n)$ tends to a pair of mutually distinct irrational boundary points (see Fig. 0.21). The limit set Γ^* is the whole Riemann sphere, and it gives rise to a sphere filling Peano curve (cf. [24], [13], [56], [18]). The special case, where $\boldsymbol{\nu}(\Gamma^*) := \lim \boldsymbol{\nu}(\Gamma_n)$ is the pair of the fixed points of the linear fractional transformation determined by a matrix $M \in SL(2,\mathbb{Z})$ with $|\operatorname{tr}(M)| \geq 3$, is particularly interesting to topologists. Because it gives the infinite cyclic cover of a hyperbolic punctured torus bundle over the circle with monodromy M. Figure 0.21 illustrates the doubly degenerate group corresponding to the matrix $M = \begin{pmatrix} 2 & 1 \\ 1 & 1 \end{pmatrix}$. This gives the infinite cyclic cover of the figure-eight knot complement. Explicit constructions of this historically important group were given by Jorgensen and Marden [43] (cf. [40] and [41]) and Riley [67] independently and by completely different methods.

Recall that Minsky's ending lamination theorem [58] proves that the map assigning each $\Gamma \in \overline{\mathcal{QF}}$ with its end invariant gives a bijection from $\overline{\mathcal{QF}}$ to $\overline{\mathbb{H}^2} \times \overline{\mathbb{H}^2} - \mathrm{diagonal}(\partial\mathbb{H}^2)$, extending the holomorphic isomorphism $\mathcal{QF} \cong \mathbb{H}^2 \times \mathbb{H}^2$. The first author proved, by using and resembling Minsky's theorem, that the combinatorial homeomorphism $\boldsymbol{\nu} : \mathcal{QF} \to \mathbb{H}^2 \times \mathbb{H}^2$ extends to a surjective map $\boldsymbol{\nu} : \overline{\mathcal{QF}} \to \overline{\mathbb{H}^2} \times \overline{\mathbb{H}^2} - \mathrm{diagonal}(\partial\mathbb{H}^2)$ so that $\boldsymbol{\nu}(\Gamma)$ determines the combinatorial structure of the Ford domain for each $\Gamma \in \overline{\mathcal{QF}}$ (see [2]).

5.9. Extension of Jorgensen's theory beyond the quasifuchsian space.

Consider a doubly cusped group Γ with $\boldsymbol{\nu}(\Gamma) = (s^-, s^+)$, where s^- and s^+ are distinct vertices of $\mathcal{D}$. See Fig. 0.22 where $(s^-, s^+) = (\infty, 2/5)$. Assume, for simplicity, that s^- and s^+ are not Farey neighbors, i.e., the edge-path distance $d(s^-, s^+)$ is greater than 1. Let $\Sigma = (\sigma_1, \cdots, \sigma_m)$ be the chain of triangles of $\mathcal{D}$ intersecting the geodesic with endpoints s^- and s^+ in this order, and let $\Sigma^{(0)}$ be the set of the vertices of the triangles in Σ. Then, by (generalized) Jorgensen's theory, the Ford domain $Ph(\Gamma)$ is equal to the polyhedron

$$Eh(\Gamma, \Sigma) := \cap\{Eh(P) \,|\, P \text{ is an elliptic generator with } s(P) \in \Sigma^{(0)}\}.$$

For each $\epsilon \in \{-, +\}$, let P^ϵ be an elliptic generator with $s(P^\epsilon) = s^\epsilon$. Then the transformation $A^\epsilon := KP^\epsilon$ is an accidental parabolic transformation, whose parabolic fixed point is the point of tangency between $I(A^\epsilon) = I(P^\epsilon)$ and $I(KA^\epsilon K^{-1}) = K(I(P^\epsilon))$.

Now, perturb the group Γ in $\mathcal{X}$ so that each A^ϵ becomes an elliptic transformation (see Fig. 0.23). Note that its axis $\mathrm{Axis}\, A^\epsilon$ is equal to $Ih(A^\epsilon) \cap Ih(KA^\epsilon K^{-1})$ and its rotation angle θ^ϵ is equal to the exterior angle between $I(A^\epsilon)$ and $I(KA^\epsilon K^{-1})$. Generically, the resulting group Γ is not discrete anymore. However, we can see that the combinatorial structure of the corresponding polyhedron $Eh(\Gamma, \Sigma)$ is unchanged, except that a new edge contained in $\mathrm{Axis}\, A^\epsilon$ appears. Moreover the pairing transformations for the Ford domain of the original group continue to pair the faces of the new polyhedron $Eh(\Gamma, \Sigma)$. (To be precise, face pairings are defined for the quotient of $Eh(\Gamma, \Sigma)$ by $\langle K \rangle$.) The space obtained from $Eh(\Gamma, \Sigma)$ by pairing the faces turns out to be a hyperbolic cone manifold commensurable with the hyperbolic cone manifold $M(2\theta^-, 2\theta^+)$ in Sect. 3. We can show that this remains valid as long as $0 \le \theta^\epsilon < \pi$ (see Fig. 0.24).

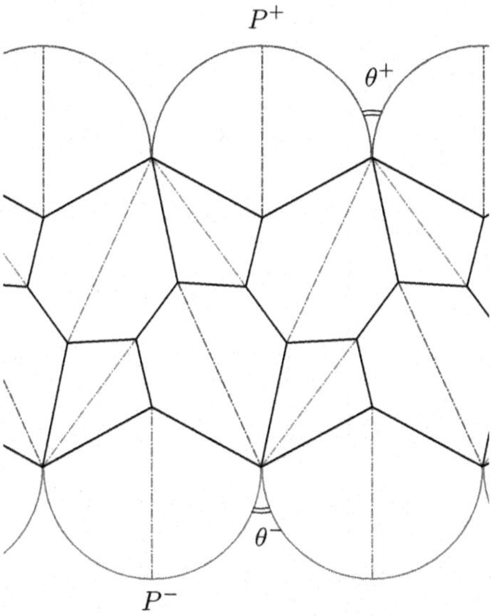

Fig. 0.22. $(0.525833 + 0.110676i, 0.207579 + 0.389324i, 0.266588 - 0.5i)$

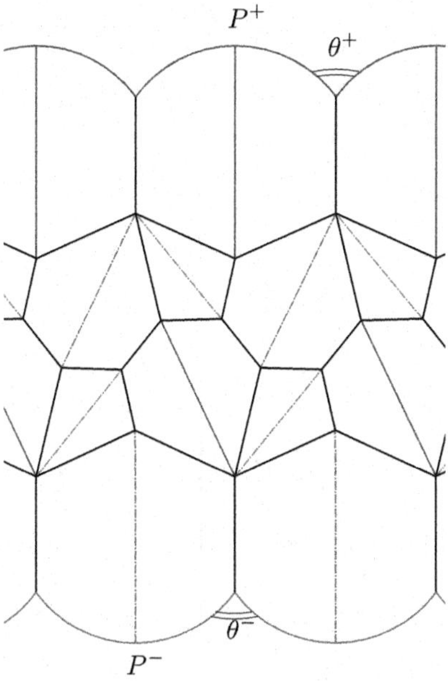

Fig. 0.23. $(0.483147 + 0.115832i, 0.232288 + 0.514691i, 0.284565 - 0.630523i)$

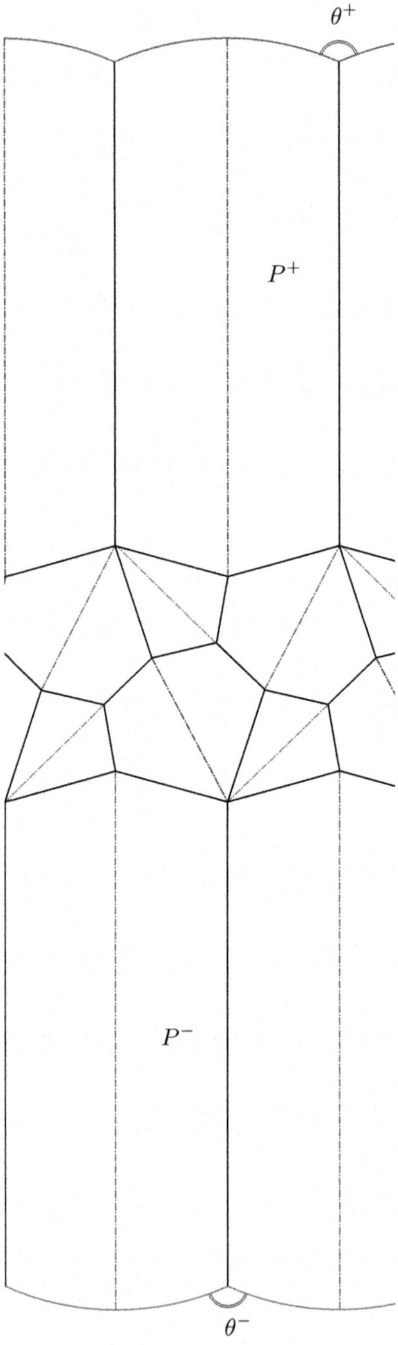

Fig. 0.24. $(0.348619 + 0.115197i, 0.310165 + 1.1507i, 0.341216 - 1.265900i)$

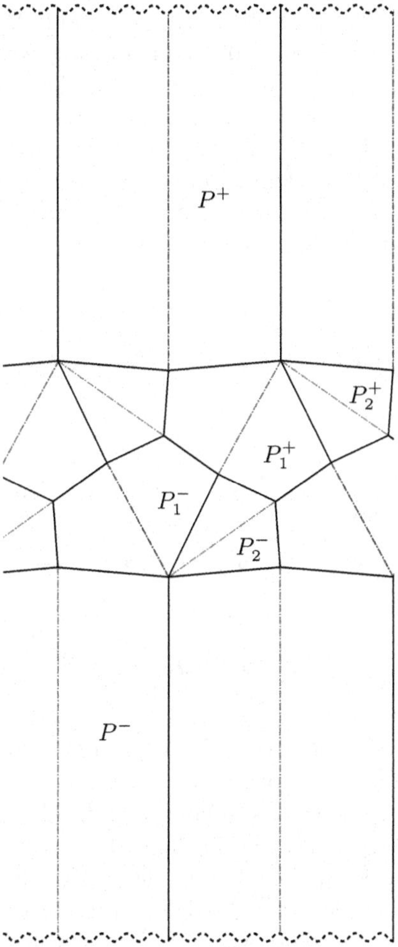

Fig. 0.25. $(0.136324 + 0.064967i, 0.429022 + 4.94894i, 0.434654 - 5.0139100i)$

As the angle θ^ϵ approaches π, the radius of $Ih(A^\epsilon)$ increases to ∞ and $Ih(A^\epsilon)$ converges to the vertical plane above a line in $\mathbb{C}$ parallel to the real line (see Fig. 0.25). So $Eh(P^\epsilon) = Eh(A^\epsilon)$ converges to one of the two half spaces bounded by the limit vertical plane. When θ^ϵ becomes π, the transformations A^ϵ and P^ϵ become π-rotations around vertical geodesics. Though the isometric hemisphere of P^ϵ is not defined, we continue to denote the above limit half space by $Eh(P^\epsilon)$.

We now explain what happens to the corresponding $Eh(\Gamma, \Sigma)$ when (θ^-, θ^+) becomes (π, π).

Suppose $d(s^-, s^+) \geq 3$, and assume $(s^-, s^+) = (1/0, q/p)$ for simplicity. Then $Eh(\Gamma, \Sigma)$ continues to be a 3-dimensional polyhedron and the pairing transformations for the Ford domain of the original group continue to pair the faces of $Eh(\Gamma, \Sigma)$. The space obtained from $Eh(\Gamma, \Sigma)$ by pairing the faces turns out to be a hyperbolic orbifold commensurable with the complement of the 2-bridge link of type (p, q). The only change in the combinatorial structure of $Eh(\Gamma, \Sigma)$, except the above change in $Eh(P^\epsilon)$, is that the faces P_1^+ and P_2^+ (resp. P_1^- and P_2^-) in Fig. 0.25 are united into the single face Q^+ (resp. Q^-) in Fig. 0.26. The actual Ford domain of the group Γ is equal to the union of the images of $Eh(\Gamma, \Sigma)$ by the infinite dihedral group $\langle P^-, P^+\rangle$, which acts on $\mathbb{H}^3$ as Euclidean isometries (see Fig. 0.26).

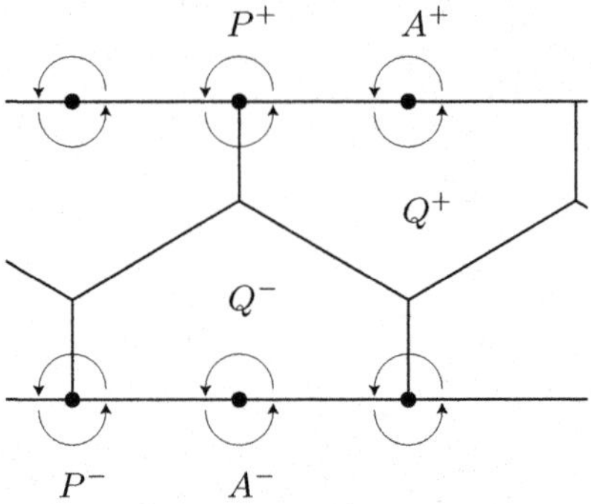

Fig. 0.26.

Suppose $d(s^-, s^+) = 2$, and assume $(s^-, s^+) = (1/0, 1/p)$ with $p \geq 3$ for simplicity. Then, as (θ^-, θ^+) approaches (π, π), $Eh(P^-) \cap Eh(P^+)$ degenerates into (a subset of) a vertical plane, and $Eh(\Gamma, \Sigma)$ degenerates into a 2-dimensional polyhedron contained in the vertical plane. See Figs. 0.27–0.29 where $p = 4$. The limit group Γ preserves the vertical plane, and the polygon $Eh(\Gamma, \Sigma)$ in the vertical plane is regarded as the Ford polygon of the action of Γ on the vertical plane. Moreover Γ is commensurable with the orbifold fundamental group of the base orbifold of the Seifert fibered structure of the complement of the 2-bridge link of type $(p, 1)$.

6. Reformulation of Jorgensen's theory for quasifuchsian punctured torus groups

For convenience, we regard a marked group Γ representing an element of $\mathcal{X}$ as the image of a *type-preserving representation* $\rho : \pi_1(\mathcal{O}) \to PSL(2, \mathbb{C})$, and identify the space $\mathcal{X}$ with the space of equivalence classes of type-preserving representations (Definitions 2.2.1 and 2.2.6). Here $\pi_1(\mathcal{O})$ is the orbifold fundamental group of the orbifold $\mathcal{O}$ and is isomorphic to the free product $(\mathbb{Z}/2\mathbb{Z}) * (\mathbb{Z}/2\mathbb{Z}) * (\mathbb{Z}/2\mathbb{Z})$. The fundamental group $\pi_1(T)$ of the once-punctured torus T is an index 2 subgroup of $\pi_1(\mathcal{O})$, and there is a one-to-one

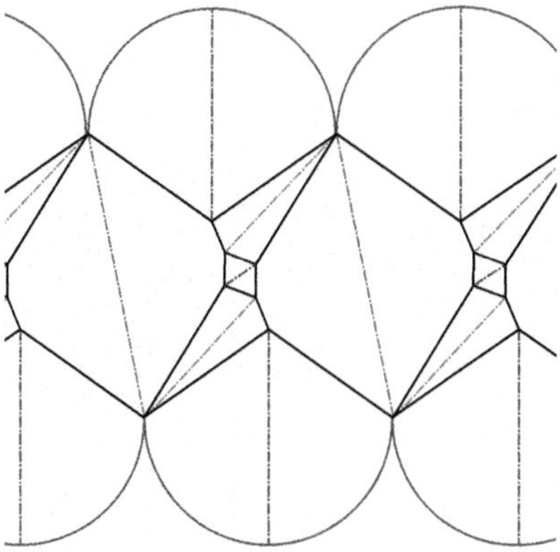

Fig. 0.27. (0.404256 − 0.254426i, 0.209822 − 0.303145i, 0.385922 + 0.557571i) – Doubly cusped group

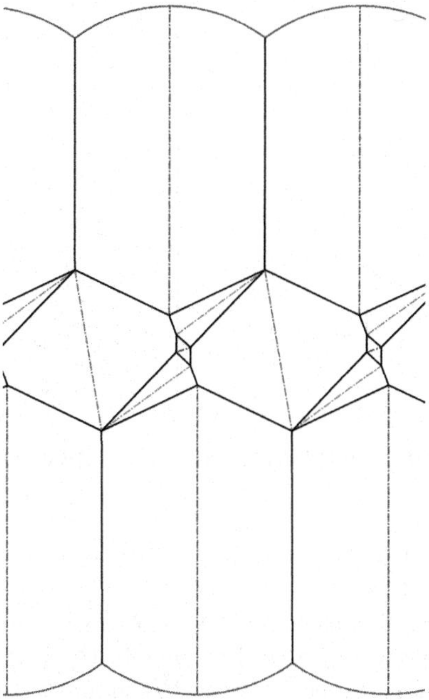

Fig. 0.28. (0.273239 − 0.269885i, 0.30051 − 0.644994i, 0.426251 + 0.914879i)

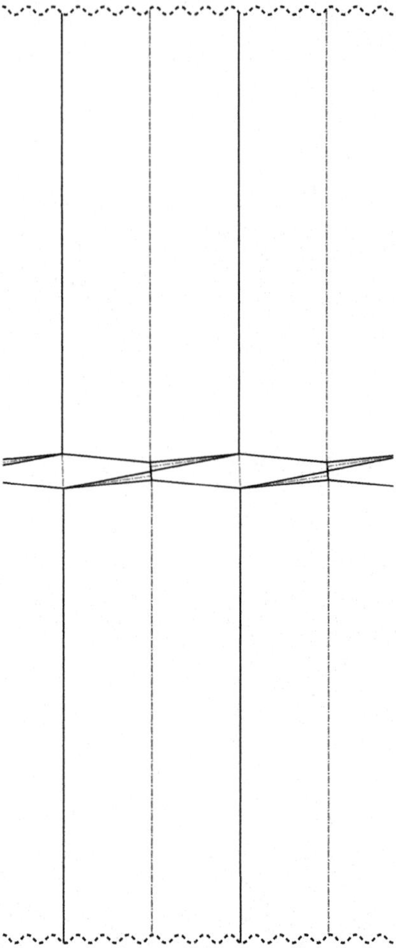

Fig. 0.29. (0.0186119 – 0.0948854i, 0.486063 – 4.97939i, 0.495325 + 5.07428i)

correspondence between type-preserving representations of $\pi_1(T)$ and those of $\pi_1(\mathcal{O})$ (Proposition 2.2.2).

Then a marked quasifuchsian punctured torus group Γ is identified with a quasifuchsian representation ρ, and we denote the Ford domain $Ph(\Gamma)$ and the side parameter $\boldsymbol{\nu}(\Gamma)$, respectively, by $Ph(\rho)$ and $\boldsymbol{\nu}(\rho)$. Jorgensen's theory says that $\boldsymbol{\nu} : \mathcal{QF} \to \mathbb{H}^2 \times \mathbb{H}^2$ is a homeomorphism and $\boldsymbol{\nu}(\rho)$ encodes the combinatorial structure of the Ford domain $Ph(\rho)$. By abuse of notation, we often denote the image $\boldsymbol{\nu}(\rho) = (\nu^-(\rho), \nu^+(\rho)) \in \mathbb{H}^2 \times \mathbb{H}^2$ by the symbol $\boldsymbol{\nu} = (\nu^-, \nu^+)$. Under this notation, let $\Sigma(\boldsymbol{\nu})$ be the chain of triangles of the Farey triangulation intersecting the geodesic segment $[\nu^-, \nu^+]$ joining ν^- with ν^+ (Definition 3.3.3). Then Jorgensen's theory says that the combinatorial structure of the Ford domain $Ph(\rho)$ is dual to a certain abstract simplicial complex $\mathcal{L}(\boldsymbol{\nu})$, called the *elliptic generator complex* associated with $\boldsymbol{\nu}$, which is constructed from $\Sigma(\boldsymbol{\nu})$ in Sect. 3.3 (p. 44). In particular, the faces of the Ford domain are supported by the isometric hemispheres of the images by ρ of the elliptic generators of $\pi_1(\mathcal{O})$ whose slopes are vertices of $\Sigma(\boldsymbol{\nu})$ (cf. Definitions 2.1.1 and 2.1.3).

By abuse of notation again, we often denote a point in $\mathbb{H}^2 \times \mathbb{H}^2$ by the symbol $\boldsymbol{\nu} = (\nu^-, \nu^+)$, and we call it a *label.* Then we are led to the concept of a *labeled representation*, which is defined to be a pair $\boldsymbol{\rho} = (\rho, \boldsymbol{\nu})$ of a type-preserving representation $\rho : \pi_1(\mathcal{O}) \to PSL(2, \mathbb{C})$ and a label $\boldsymbol{\nu} = (\nu^-, \nu^+) \in \mathbb{H}^2 \times \mathbb{H}^2$ (Definition 3.3.2). We say that $\boldsymbol{\rho}$ is *quasifuchsian* if (i) ρ is quasifuchsian, (ii) the Ford domain $Ph(\rho)$ is as described by Jorgensen, and (iii) $\boldsymbol{\nu}$ is equal to the side parameter of ρ (Definition 6.1.1). Our first task is to give a characterization of quasifuchsian labeled representations, which is suitable for the proof of Main Theorem 1.3.5. This is done by Theorem 6.1.8 (Good implies quasifuchsian), which shows that if a labeled representation is "good", then it is quasifuchsian. Here a labeled representation is defined to be *good* if it satisfies the three conditions, Nonzero, Frontier and Duality (Definition 6.1.7). Though these conditions are rather complicated, it is not difficult to check if a given labeled representation satisfies them. Thus Theorem 6.1.8 gives us a practical method for checking if a given labeled representation is quasifuchsian. (Actually, Modified Main Theorem 6.1.11 implies that the converse to Theorem 6.1.8 is also valid, and hence a labeled representation is quasifuchsian if and only if it is good.) The proof of Theorem 6.1.8 is done by using Poincare's theorem on fundamental polyhedra in Sects. 6.3 (p. 138), 6.4 (p. 142) and 6.5 (p. 144).

Let $\mathcal{J}[\mathcal{QF}] \subset \mathcal{QF} \times (\mathbb{H}^2 \times \mathbb{H}^2)$ be the space of all good labeled representations. Then Jorgensen's result is equivalent to Modified Main Theorem 6.1.11 which asserts that the projections $\boldsymbol{\mu}_1 : \mathcal{J}[\mathcal{QF}] \to \mathcal{QF}$ and $\boldsymbol{\mu}_2 : \mathcal{J}[\mathcal{QF}] \to \mathbb{H}^2 \times \mathbb{H}^2$ are homeomorphisms.

7. Idea of the proof that $\boldsymbol{\mu}_1$ is a homeomorphism

The injectivity of $\boldsymbol{\mu}_1$ is guaranteed by Proposition 6.2.5 (Uniqueness of good label), which is an easy consequence of the definition and is proved in Sect. 6.4 (p. 142). So the main task is the proof of surjectivity. Namely, we need to show that the combinatorial structure of the Ford domain of a given quasifuchsian representation is as described by Jorgensen. Roughly speaking, the proof amounts to showing that under a continuous deformation of the quasifuchsian representations, combinatorial transitions of the Ford domain happen only at their frontier in the ideal boundary, i.e., in the complex plane $\mathbb{C}$, as illustrated in Fig. 4.15, and the combinatorial structure in the hyperbolic space is stable. The actual proof consists of the following two steps.

1. Proposition 6.2.1 (Openness), which guarantees the openness of the image of the projection $\boldsymbol{\mu}_1 : \mathcal{J}[\mathcal{QF}] \to \mathcal{QF}$ in $\mathcal{QF}$

2. Propositions 6.2.2 (SameStratum) and 6.2.4 (Closedness), which guarantee the closedness of the image of the projection $\boldsymbol{\mu}_1 : \mathcal{J}[\mathcal{QF}] \to \mathcal{QF}$ in $\mathcal{QF}$

Since we can easily see that the image of $\boldsymbol{\mu}_1$ is non-empty, i.e., there exists a good labeled representation (Proposition 5.1.3), and since $\mathcal{QF}$ is connected, the above results imply that $\boldsymbol{\mu}_1 : \mathcal{J}[\mathcal{QF}] \to \mathcal{QF}$ is surjective.

Step 1 is completed in Chap. 7, and a rough idea for this step is as follows. To show the openness of the image $\boldsymbol{\mu}_1(\mathcal{J}[\mathcal{QF}])$ in $\mathcal{QF}$ around $\rho = \boldsymbol{\mu}_1(\rho, \boldsymbol{\nu})$, we pick up a set of elliptic generators of $\pi_1(\mathcal{O})$, which is finite modulo conjugation by the peripheral element, and study how the pattern of the corresponding isometric hemispheres change after a small perturbation of the representation ρ. By virtue of the key Lemmas 4.6.1, 4.6.2 and 4.6.7 (see Fig. 4.15) and Lemma 7.2.3 (disjointness), we see that each nearby representation ρ' has a label $\boldsymbol{\nu}'$ such that $(\rho', \boldsymbol{\nu}')$ is good. We thus obtain the openness of $\boldsymbol{\mu}_1(\mathcal{J}[\mathcal{QF}])$. We note that Schedler [72] obtains the corresponding result, by establishing a general stability theorem of the Ford domains using the theory of holomorphic motions. However, we do not know if the idea of holomorphic motions can be applied in the proof of the generalization of Jorgensen's theory to hyperbolic cone-manifolds.

Step 2 is completed in Chap. 8. To describe a rough idea, let $\{(\rho_n, \boldsymbol{\nu}_n)\}$ be a sequence of good labeled representations such that $\lim \rho_n \in \mathcal{QF}$. Then Proposition 6.2.2 (SameStratum) guarantees that some subsequence satisfies the condition SameStratum (Definition 6.2.2). Roughly speaking, this conditions says that the Ford domains $Ph(\rho_n)$ have the same combinatorial structure and therefore we can talk about the "behavior of a face (or an edge or a vertex) of $Ph(\rho_n)$ as $n \to \infty$". The proof of Proposition 6.2.2 is done in Sect. 8.1 (p. 172), and it is based on the fact that the convergence $\rho_n \to \rho_\infty$ is strong and a certain lemma (Lemma 8.1.1) due to Jorgensen, which is a prototype of Minsky's pivot theorem in [58].

Proposition 6.2.4 (Closedness) is proved in Sects. 8.3 (p. 180) - 8.12 (p. 209) in Chap. 8, and its main ingredient is to show that no unexpected degeneration of a face of $Ph(\rho_n)$ happens as $n \to \infty$. This is the most involved part of this monograph. A reason why it is so involved is that we have to list all possible degenerations, before showing degenerations do not happen. However, as is found in Jorgensen's original argument [40], the idea to prohibit degenerations consists of only a few simple observations (see the introduction to Chap. 8). Another reason for the complication in this step (and the previous step) lies in the treatment of the 'thin' case, i.e., the case when both components of $\boldsymbol{\nu}_\infty = \lim \boldsymbol{\nu}_n$ belong to the interior of a single edge τ of the Farey triangulation. However, we can treat this special case by using the results established in Sect. 5.2 (p. 106).

8. Idea of the proof that μ_2 is a homeomorphism

The proof consists of the following two steps.

1. Proposition 6.2.7 (Convergence), which shows that, for a sequence of good labeled representations $\{(\rho_n, \nu_n)\}$, if $\lim \nu_n \in \mathbb{H}^2 \times \mathbb{H}^2$ exists (and if it satisfies the condition SameStratum), then it has a subsequence such that the corresponding subsequence of $\{\rho_n\}$ converges in $\overline{\mathcal{QF}}$.
2. Chapter 9, where the proof of bijectivity of the map μ_2 is established by using Propositions 6.2.2 (SameStratum), 6.2.4 (Closedness) and 6.2.7 (Convergence) and an elementary intersection theory in algebraic geometry.

We describe a rough idea of Step 2. In Lemma 4.2.18, we see that if (ρ, ν) is a good labeled representation which belongs to the inverse image of a label $\nu \in \mathbb{H}^2 \times \mathbb{H}^2$ by μ_2, then ρ (to be precise, a *Markoff map* inducing ρ) satisfies a certain algebraic equation. What we need to do is to single out a unique 'geometric' root among numerous roots of an algebraic equation associated with ν. To this end, we introduce the concept of the *geometric multiplicity* $d_G(\nu)$ for each $\nu = (\nu^-, \nu^+)$ by using the elementary intersection theory in algebraic geometry (Definition 9.2.1). Then the bijectivity of the projection $\mu_2 : \mathcal{J}[\mathcal{QF}] \to \mathbb{H}^2 \times \mathbb{H}^2$ is equivalent to the assertion that $d_G(\nu) = 1$ for every label ν. After making a detailed study of the algebraic varieties associated with the labels (Sect. 9.1, p. 215), we show in Corollary 9.2.7 that $d_G(\nu)$ does not depend on ν by using the "continuity of roots" (Lemmas 9.3.1 and 9.3.3) and Propositions 6.2.2 (SameStratum), 6.2.4 (Closedness) and 6.2.7 (Convergence). On the other hand, it is easy to see that $d_G(\nu) = 1$ if ν belongs to the diagonal set, i.e., $\nu^- = \nu^+$ (Proposition 5.1.5). We thus obtain the bijectivity of $\mu_2 : \mathcal{J}[\mathcal{QF}] \to \mathbb{H}^2 \times \mathbb{H}^2$, and hence that of $\nu : \mathcal{QF} \to \mathbb{H}^2 \times \mathbb{H}^2$. We note that, though the bijectivity is claimed in [40], no indication of the proof is presented. The arguments outlined here is new in this sense. Actually, we were able to complete this step only fairly recently.

9. Organization of the monograph

The monograph consists of nine chapters and an appendix.

In Chap. 1, we reformulate Jorgensen's theory from the 3-dimensional viewpoint, and give a conceptual description of Jorgensen's theorem (Theorems 1.2.2, 1.3.2 and Main Theorem 1.3.5). We also give a description of the ideal tetrahedral complex which is a geometric dual to the Ford domain (Theorem 1.4.2). This chapter is essentially equal to the announcement in [10].

In Chap. 2, we describe the intimate relations among the Fricke surfaces, namely the punctured torus T, the four-times punctured sphere S, and the $(2, 2, 2, \infty)$-orbifold $\mathcal{O}$. We first give a complete description of the 'geometric' generator systems of the fundamental groups of the Fricke surfaces (Propositions 2.1.6 and 2.1.9) and describe their relation with the Farey triangulation (Proposition 2.1.10). Then we show the equivalence among the spaces of the type-preserving representations (Definition 2.2.1) of the fundamental groups of Fricke surfaces (Proposition 2.2.2). The concepts of the Markoff maps and complex probabilities are introduced in Sects. 2.3 and 2.4, respectively, and explicit matrices for the type-preserving representations and its intuitive description are given (Lemma 2.3.7 and Proposition 2.4.4). Though almost all contents in this chapter seem to be known to the experts, we could not find explicit reference for some of the contents. Thus we included full proofs for all propositions in this chapter.

In Chap. 3, we introduce the definitions of a labeled representation $\boldsymbol{\rho} = (\rho, \boldsymbol{\nu})$, the complex $\mathcal{L}(\boldsymbol{\nu})$, which is a combinatorial dual to the Ford domain, and the virtual Ford domain $Eh(\boldsymbol{\rho})$. These are used in Chap. 6 to reformulate Main Theorem 1.3.5.

In Chap. 4, we first give a detailed proof to the chain rule for isometric circles (Lemma 4.1.2), on which the whole argument is based. We then introduce Jorgensen's side parameter (Definition 4.2.9), and prove various properties of the side parameter.

In Chap. 5, we give a detailed study of some special examples. In particular, complete descriptions of real representations and *isosceles* representations are given in Sects. 5.1 and 5.2. Though isosceles representations themselves are simple objects, their neighborhood in $\mathcal{QF}$ is rich in variety. However, they are essentially controlled by side parameters (Proposition 5.2.13). These representations also form the starting point toward the proof of Jorgensen's theory. In Sect. 5.3, we describe how two-parabolic groups arise as the images of type-preserving representations of certain kind, and explain the reason why the generalized Jorgensen's theory is useful to the study of the 2-bridge links.

In Chap. 6, we give a 2-dimensional reformulation of Main Theorem 1.3.5 (Modified Main Theorem 6.1.11). This is more similar to Jorgensen's original description than the three dimensional picture described in Chap. 1, and is rather complicated. However, it fits with the proof given in this monograph. The equivalence between the two formulations is guaranteed by Theorem 6.1.8, which is proved by using (a variation of) Poincare's theorem on fundamental polyhedron. In Sect. 6.2, we give a route map for the proof of Modified Main Theorem 6.1.11. At the end of this chapter, we prove certain properties of the elements in $\mathcal{QF}$ which are useful in the actual computation of the Ford domains (see Propositions 6.7.1 and 6.7.2). These properties are also used in the Bers' slice project in [50] (cf. [82]).

The remaining Chaps. 7, 8 and 9 are devoted to the proof of Modified Main Theorem 6.1.11. To be precise, Steps 1 and 2 , respectively, presented in Sect. 7 are completed in Chaps. 7 and 8, and Steps 1 and 2 , respectively, presented in Sect. 8 are completed in Chaps. 8 and 9. (See also Sect. 6.2 for the route map.)

In the Appendix, we give a proof to some of the basic facts concerning the Ford domain.

Acknowledgment

We would like to express our deepest thanks to Troels Jorgensen for his explanations of his results, invaluable suggestions, and for the beautiful mathematics he had produced. We were very much encouraged when he told us he had got an inspiration that there should be some relationship between his work and 2-bridge links. We also thank his collaborator, Al Marden, for his interest on our work and encouragement. It was great pleasure to explain our approach to him and Caroline Series at University of Warwick. We sincerely thank Caroline Series for her kind hospitality at University of Warwick. Comparison of our work and her joint works with Linda Keen on punctured torus groups lead us to the related works [7, 10]. We were stimulated very much by attending the wonderful workshops organized by her. We also thank members and visitors of University of Warwick, including David Epstein, Brian Bowditch, Young Eun Choi, Raquel Diaz and Mark Lackenby for their interest and encouragement. We thank Brian Bowditch for sending us his mysterious paper [17] and for fruitful discussion, which brought the first and second authors to the related works with Hideki Miyachi [5, 6]. We thank Travis Schedler for sending us his beautiful paper [72] on Jorgensen's work, which was partially influenced by our preliminary announcement [9]. Interplay with his paper has improved the monograph very much. In particular, Chap. 7 and Appendix A.1 were much influenced by his paper. We thank Norbert A'Campo, Michael Heusener, Eriko Hironaka and Kazuhiro Konno for helpful discussion concerning Chap. 9. We thank Stephan Hamperies, Heinz Helling, Richard Weidmann and Heiner Zieschang for their helpful suggestions concerning Sect. 2.1. Stephan Hamperies taught us a neat proof of Proposition 2.1.6, which greatly simplifies our original proof. We thank Alexander Mednykh for discussion concerning Proposition 2.2.2. We thank Norbert A'Campo, Gerhard Burde, Shungbok Hong, Yoichi Imayoshi and Masaaki Yoshida for inviting the second author to series of lectures on our work at University Basel, Goethe University Frankfurt am Main, Pohang University, Osaka City University and Kyushu University. Yohei Komori encouraged us to organize a workshop to explain the first version of the monograph. The workshop was held at Osaka University during 12-15, September, 2005. We thank the participants of the workshop, in particular to Michihiko Fujii, Yohei Komori, Hideki Miyachi, Toshihiro Nakanishi, Yoshihide Okumura, Ken'ichi Ohshika, Hiroki Sato, Kenneth Shackleton, Masahiko Taniguchi, Akira Ushijima and Xiantao Wang for their devotion, patience and suggestions. We thank Kenneth Shackleton and Yoshihide Okumura for their careful reading of the first version. We also thank Colin Adams, Iain Aitchison, Michel Boileau, Warren Dicks, Craig Hodgson, Sadayoshi Kojima, John Parker, Joan Porti, Alan Reid, Toshiyuki

Sugawa, Hyam Rubinstein and Claude Weber for stimulating conversations and encouragements. Finally, we thank the referees for their very appropriate comments on the first version, which enabled us to improve this monograph drastically.

Contents

1

Jorgensen's picture of quasifuchsian punctured torus groups

In [40, Theorems 3.1, 3.2 and 3.3], Jorgensen describes the combinatorial structure of the Ford domain of a quasifuchsian punctured torus group. It was very difficult for the authors to get a conceptual understanding of the statement, because it consists of nine assertions, each of which describes some property of the Ford domain, and it does not explicitly present a topological or combinatorial model of the Ford domain. In this chapter, we construct an explicit model of the Ford domain, and reformulate Jorgensen's theorem in terms of the model. In short, we present a 3-dimensional picture to Jorgensen's theorem. We note that this chapter is essentially equal to the announcement [10].

In Sect. 1.1, we recall the definition of quasifuchsian punctured torus groups and their Ford domains, and then explain the geometric meaning of the Ford domain as the canonical spine of the quotient hyperbolic manifold. We also give a quick review of the duality between the Ford domains and the canonical ideal polyhedral decompositions of cusped hyperbolic manifolds introduced by Epstein and Penner [27], and recall the EPH-decomposition of a possibly infinite volume cusped hyperbolic manifold, which was introduced by the first and second authors motivated by this project. Those readers who are interested only in the Ford domains can skip this part of the section.

In Sect. 1.2, we construct a family of spines of $T \times [-1, 1]$, where T is the topological once-punctured torus, and reformulate Jorgensen's Theorems 3.1, 3.2 and 3.3 in [40] into the single Theorem 1.2.2, which essentially asserts that the Ford complex (= the canonical spine determined by the Ford domain) of the quotient quasifuchsian punctured torus hyperbolic manifold is isotopic to a spine in the family.

In Sect. 1.3, we recall Jorgensen's side parameter of a quasifuchsian punctured torus group and reformulate Jorgensen's Theorem 4.6 in [40], which assert that the side-parameter is actually a parameter of $\mathcal{QF}$, as Theorem 1.3.2.

In Sect. 1.4, we present a topological model of the canonical ideal polyhedral decomposition dual to the Ford domain of a quasifuchsian punctured torus manifold (Theorem 1.4.2).

1.1 Punctured torus groups, Ford domains and EPH-decompositions

Let T be the topological (once) punctured torus. A *marked punctured torus group* is the image of a discrete faithful representation $\rho : \pi_1(T) \to PSL(2, \mathbb{C})$ satisfying the following condition:

- If $\omega \in \pi_1(T)$ is represented by a simple loop around the puncture, then $\rho(\omega)$ is parabolic.

Two marked punctured torus groups $\Gamma = \rho(\pi_1(T))$ and $\Gamma' = \rho'(\pi_1(T))$ are *equivalent* if ρ is conjugate to ρ' by an element of $PSL(2, \mathbb{C})$. A *marked fuchsian punctured torus group* is a marked punctured torus group which is determined by a *fuchsian* representation, i.e., a faithful and discrete $PSL(2, \mathbb{R})$-representation. A *marked quasifuchsian punctured torus group* is a marked punctured torus group Γ which is a quasiconformal deformation of a fuchsian punctured torus group. (For standard terminologies and facts in Teichmüller theory and Kleinian groups, see [38] and [55].) This is equivalent to the condition that the domain of discontinuity $\Omega(\Gamma)$ consists of exactly two simply connected components $\Omega^{\pm}(\Gamma)$, whose quotient $\Omega^{\pm}(\Gamma)/\Gamma$ are each homeomorphic to T. We employ a sign convention so that there is an orientation-preserving homeomorphism f from $T \times [-1, 1]$ to the quotient manifold $\bar{M}(\Gamma) = (\mathbb{H}^3 \cup \Omega(\Gamma))/\Gamma$ such that $f(T \times \{\pm 1\}) = \Omega^{\pm}(\Gamma)/\Gamma$ and that the isomorphism $f_* : \pi_1(T \times [-1, 1]) = \pi_1(T) \to \pi_1(M(\Gamma)) = \Gamma < PSL(2, \mathbb{C})$ is identified with ρ. Since such a homeomorphism is unique up to isotopy, we can identify the topological triple $(\bar{M}(\Gamma), \Omega^-(\Gamma)/\Gamma, \Omega^+(\Gamma)/\Gamma)$ with $(T \times [-1, 1], T \times \{-1\}, T \times \{1\})$. The *quasifuchsian punctured torus space* $\mathcal{QF}$ is the space of the equivalence classes of marked quasifuchsian punctured torus groups. By Bers' simultaneous uniformization theorem, $\mathcal{QF}$ is identified with the product $\mathrm{Teich}(T) \times \mathrm{Teich}(T) \cong \mathbb{H}^2 \times \mathbb{H}^2$ of two copies of the Teichmüller space of T via the correspondence $\Gamma \mapsto (\Omega^-(\Gamma)/\Gamma, \Omega^+(\Gamma)/\Gamma))$. In particular, it is an open connected subset in the space of type-preserving $PSL(2, \mathbb{C})$-representations of $\pi_1(T)$ modulo conjugation (Definitions 2.2.1 and 2.2.6). As a consequence of Minsky's celebrated theorem [58], the space of marked punctured torus groups is equal to the closure $\overline{\mathcal{QF}}$ of $\mathcal{QF}$ in the representation space.

We now recall the definition of the isometric circles and the Ford domains.

Definition 1.1.1. *Let* $A = \begin{pmatrix} a & b \\ c & d \end{pmatrix}$ *be an element of* $PSL(2, \mathbb{C}) = \mathrm{Isom}^+(\mathbb{H}^3)$ *such that* $A(\infty) \neq \infty$, *namely* $c \neq 0$.

1. *The* isometric circle $I(A)$ *of* A *is defined by*

$$I(A) = \{z \in \mathbb{C} \mid |A'(z)| = 1\} = \{z \in \mathbb{C} \mid |cz + d| = 1\}.$$

Thus $I(A)$ *is the circle in the complex plane with center* $c(A) = -d/c = A^{-1}(\infty)$ *and radius* $r(A) = 1/|c|$. $D(A)$ *denotes the disk in* $\mathbb{C}$ *bounded by* $I(A)$, *and* $E(A)$ *denotes the closed exterior* $\mathbb{C} - \operatorname{int} D(A)$ *of* $D(A)$.
2. *The* isometric hemisphere $Ih(A)$ *is the hyperplane of the upper half space* $\mathbb{H}^3$ *bounded by* $I(A)$. $Dh(A)$ *denotes the half space of* $\mathbb{H}^3$ *bounded by* $Ih(A)$ *whose closure contains* $D(A)$, *and* $Eh(A)$ *denotes the closed exterior* $\mathbb{H}^3 - \operatorname{int} Dh(A)$ *of* $Dh(A)$.
3. $\overline{Ih}(A)$, $\overline{Dh}(A)$ *and* $\overline{Eh}(A)$, *respectively, denote the closure of* $Ih(A)$, $Dh(A)$ *and* $Eh(A)$ *in the closure* $\overline{\mathbb{H}}^3 = \mathbb{H}^3 \cup \mathbb{C}$ *of the upper half space model of hyperbolic 3-space.*

Let Γ be a non-elementary Kleinian group such that the stabilizer Γ_∞ of ∞ contains parabolic transformations.

Definition 1.1.2. *The* Ford domain $Ph(\Gamma)$ *of* Γ *is the subset of the upper half space* $\mathbb{H}^3$ *which consists of all points lying exterior to each of isometric hemispheres defined by* Γ. *The* Ford polygon $P(\Gamma)$ *is the subset of the complex plane which consists of all points lying exterior to each of the isometric circles defined by* Γ. *We also define* $\overline{Ph}(\Gamma)$ *to be the subset of* $\overline{\mathbb{H}}^3$ *similarly. Namely*

$$Ph(\Gamma) = \bigcap\{Eh(A) \mid A \in \Gamma - \Gamma_\infty\},$$
$$P(\Gamma) = \bigcap\{E(A) \mid A \in \Gamma - \Gamma_\infty\},$$
$$\overline{Ph}(\Gamma) = \bigcap\{\overline{Eh}(A) \mid A \in \Gamma - \Gamma_\infty\}$$

The above notations follow those of [39]: P stands for *p*olygon and *p*olyhedron and h stands for (3-dimensional) *h*yperbolic space. But they are slightly different from those in [40], where the same sets are denoted by $\widetilde{P}(\Gamma)$ and $\widetilde{Ph}(\Gamma)$ respectively.

We describe a geometric meaning of the Ford domain. To this end, pick a small horoball, H_∞, centered at ∞ which is *precisely invariant* by (Γ, Γ_∞), that is, for any element $A \in \Gamma$, $A(H_\infty) \cap H_\infty \neq \emptyset$ if and only if $A \in \Gamma_\infty$. The existence of such a horoball is guaranteed by the Shimizu-Leutbecher inequality (see Lemma 2.5.2(1) or [55, Lemma 0.6]). Then for each element $A \in \Gamma - \Gamma_\infty$, the isometric hemisphere $Ih(A)$ is equal to the set of points in $\mathbb{H}^3$ which are equidistant from H_∞ and $A^{-1}(H_\infty)$. This implies that $Ph(\Gamma)$ can be regarded as the "Dirichlet domain of Γ centered at ∞", because

$$\begin{aligned} Ph(\Gamma) &= \{x \in \mathbb{H}^3 \mid d(x, H_\infty) \le d(x, AH_\infty) \text{ for every } A \in \Gamma\} \\ &= \{x \in \mathbb{H}^3 \mid d(x, H_\infty) = d(x, \Gamma H_\infty)\}. \end{aligned}$$

Just as the (usual) Dirichlet domain is a fundamental polyhedron, we have the following proposition. See Appendix (Sect. A.1) for the proof of the proposition and related facts.

Proposition 1.1.3. *The Ford domain $Ph(\Gamma)$ is a "fundamental polyhedron of Γ modulo Γ_∞", in the following sense.*

1. $\mathbb{H}^3 = \cup\{A(Ph(\Gamma)) \mid A \in \Gamma\}$.
2. *If $A \in \Gamma_\infty$ then $A(Ph(\Gamma)) = Ph(\Gamma)$, whereas if $A \in \Gamma - \Gamma_\infty$ then $A(\operatorname{int} Ph(\Gamma)) \cap \operatorname{int} Ph(\Gamma) = \emptyset$.*
3. *$Ph(\Gamma)$ is a convex polyhedron (Definition 3.4.1(2)).*
4. *For any compact set K of $\mathbb{H}^3$, the set $\{A\Gamma_\infty \in \Gamma/\Gamma_\infty \mid A(Ph(\Gamma)) \cap K \neq \emptyset\}$ is a finite set.*

By the above proposition, the intersection of $Ph(\Gamma)$ with a fundamental domain of Γ_∞ is a fundamental domain of Γ. As is noted in [27, Sect. 4], it is more natural to work with the quotient $Ph(\Gamma)/\Gamma_\infty$ in $\mathbb{H}^3/\Gamma_\infty$. In fact the hyperbolic manifold $M(\Gamma)$ is obtained from $Ph(\Gamma)/\Gamma_\infty$ by identifying pairs of faces by isometries.

Note that H_∞ projects to a cuspidal region, C, in the quotient hyperbolic 3-manifold $M(\Gamma) = \mathbb{H}^3/\Gamma$. Then by the preceding description of the Ford domain, the image of $\partial Ph(\Gamma)$ in $M(\Gamma)$ is equal to the cut locus of $M(\Gamma)$ with respect to C, which is defined as follows.

Definition 1.1.4. *(1) The* cut locus, *$Cut(M(\Gamma), C)$, of $M(\Gamma)$ with respect to C is the subspace of $M(\Gamma)$ consisting of those points which have more than one shortest geodesics to a fixed horospherical neighborhood.*

(2) The Ford complex, Ford(Γ), *of Γ is the closure of the cut locus $Cut(M(\Gamma), C)$ in the Kleinian manifold $\bar{M}(\Gamma) = (M \cup \Omega(\Gamma))/\Gamma$.*

In the remainder of this section, we describe the ideal polyhedral complex, denoted by $\Delta_{\mathbb{E}}(\Gamma)$, which is the geometric dual to Ford(Γ). Let $\widetilde{\mathrm{Ford}}(\Gamma)$ be the 2-dimensional complex in the hyperbolic space obtained as the inverse image of Ford$(\Gamma) \cap M(\Gamma)$. It should be noted that $\widetilde{\mathrm{Ford}}(\Gamma)$ is the cut locus of the disjoint union, ΓH_∞, of the images of the horoball H_∞ by Γ, namely, $\widetilde{\mathrm{Ford}}(\Gamma)$ consists of the points in $\mathbb{H}^3$ which have more than two shortest geodesics to ΓH_∞. Let p be a vertex of $\partial Ph(\Gamma) \subset \widetilde{\mathrm{Ford}}(\Gamma)$. Then, generically, p is the intersection of three isometric hemispheres, $Ih(A_j)$ $(j \in \{1,2,3\})$, and hence it is equidistant from the four horoballs, H_∞ and $A_j^{-1}(H_\infty)$ $(j \in \{1,2,3\})$. We regard the ideal tetrahedron spanned by the centers of these horoballs as the geometric *dual* to the vertex p. Similarly, for each edge (resp. face) of $\widetilde{\mathrm{Ford}}(\Gamma)$, we can associate an ideal triangle (an ideal edge) as its geometric dual. The family of these ideal tetrahedra, ideal triangles and ideal edges descends to an ideal polyhedral complex embedded in $M(\Gamma)$; this is the desired ideal polyhedral complex $\Delta_{\mathbb{E}}(\Gamma)$.

A more precise description of $\Delta_{\mathbb{E}}(\Gamma)$ is given as follows. For a point p in the cut locus $\widetilde{\mathrm{Ford}}(\Gamma)$ of ΓH_∞, let $H_{x_0}, H_{x_1}, \cdots, H_{x_k}$ be the horoballs of ΓH_∞ such that:

1. H_{x_i} is centered at x_i.
2. $d(p, H_{x_0}) = d(p, H_{x_1}) = \cdots = d(p, H_{x_k}) = d(p, \Gamma H_\infty)$.
3. $d(p, H') > d(p, \Gamma H_\infty)$ for any horoball H' in $\Gamma H_\infty - \{H_{x_0}, H_{x_1}, \cdots, H_{x_k}\}$.

Let $\widetilde{\Delta}_p = \langle x_0, x_1, \cdots, x_k \rangle$ be the ideal polyhedron spanned by the centers $x_0, x_1, \cdots, x_k$ of the horoballs $H_{x_0}, H_{x_1}, \cdots, H_{x_k}$. If two points p and p' belong to the same (open) cell of $\widetilde{\mathrm{Ford}}(\Gamma)$, then $\widetilde{\Delta}_p = \widetilde{\Delta}_{p'}$. Thus $\widetilde{\Delta}_p$ is regarded as the geometric dual to the cell, e, of $\widetilde{\mathrm{Ford}}(\Gamma)$ in which p belongs to, and we denote it by $\widetilde{\Delta}_e$. Then the family

$$\widetilde{\Delta}_{\mathbb{E}}(\Gamma) := \{\widetilde{\Delta}_e \mid e \text{ is a cell of } \widetilde{\mathrm{Ford}}(\Gamma)\}$$

determines a Γ-invariant ideal polyhedral decomposition of $\mathbb{H}^3$. This descends to an ideal polyhedral complex $\Delta_{\mathbb{E}}(\Gamma)$ embedded in $M(\Gamma)$. We note that some member of $\Delta_{\mathbb{E}}(\Gamma)$ is not a true ideal polyhedron but a quotient of an ideal polyhedron if Γ has a nontrivial torsion.

Following the argument of Epstein and Penner [27, Sect. 10] (cf. [7, Sect. 10]), we show that $\Delta_{\mathbb{E}}(\Gamma)$ arises from the Epstein-Penner convex hull construction in the Minkowski space. Let $\mathbb{E}^{1,3}$ be the 4-dimensional Minkowski space with the Minkowski product

$$\langle x, y \rangle = -x_0 y_0 + x_1 y_1 + x_2 y_2 + x_3 y_3.$$

Then

$$\mathbb{H}^3 = \{x \in \mathbb{E}^{1,3} \mid \langle x, x \rangle = -1, \ x_0 > 0\},$$

together with the restriction of the Minkowski product to the tangent space, gives a hyperboloid model of the 3-dimensional hyperbolic space. Any horoball H in this model is represented by a vector, v, in the positive light cone (i.e., $\langle v, v \rangle = 0$ and $v_0 > 0$) as

$$H = \{x \in \mathbb{H}^3 \mid \langle v, x \rangle \geq -1\}.$$

The center of the horoball H corresponds to the ray thorough v, and as v moves away from the origin along the ray, the horoball contracts towards its center. Let v_∞ denote the light-like vector representing the horoball H_∞. Then its orbit Γv_∞ is the set of light-like vectors corresponding to the horoballs in ΓH_∞. Let $\mathcal{C}$ be the closed convex hull of Γv_∞ in $\mathbb{E}^{1,3}$. Now consider the ideal polyhedron $\widetilde{\Delta}_p = \langle x_0, x_1, \cdots, x_k \rangle$ in $\mathbb{H}^3$ which is dual to a vertex p of $\widetilde{\mathrm{Ford}}(\Gamma)$. Then the horoballs $H_{x_0}, H_{x_1}, \cdots, H_{x_k}$ in the orbit ΓH_∞ centered at $x_0, x_1, \cdots, x_k$ attain the minimal distance from p among the members in ΓH_∞. Let v_{x_i} be the light-like vector representing the horoball H_{x_i}. After coordinate change, we may assume the vertex p corresponds to the vector

$(1, 0, 0, 0)$ in the hyperboloid model. Then the points $v_{x_0}, v_{x_1}, \cdots, v_{x_k}$ lie in a horizontal hyperplane $W : x_0 =$ constant, because of the second condition. Moreover, by the third condition, all points in $\Gamma v_\infty - \{v_{x_0}, v_{x_1}, \cdots, v_{x_k}\}$ lie above the hyperplane W. Hence we see that the hyperplane W is a support plane of the convex hull $\mathcal{C}$ (i.e., $\mathcal{C}$ is contained in one of the two closed half space bounded by W), and $W \cap \partial\mathcal{C}$ in the polyhedron $\langle v_{x_0}, v_{x_1}, \cdots, v_{x_k}\rangle$. In other words, $\langle v_{x_0}, v_{x_1}, \cdots, v_{x_k}\rangle$ is a (top-dimensional) face of $\partial\mathcal{C}$. Moreover, it is *Euclidean* in the sense that the restriction of the Minkowski product to the hyperplane W is positive-definite. The ideal polyhedron $\langle x_0, x_1, \cdots, x_k\rangle$ dual to p is equal to the image of the Euclidean face $\langle v_{x_0}, v_{x_1}, \cdots, v_{x_k}\rangle$ of $\partial\mathcal{C}$ by the radial projection form the origin to $\mathbb{H}^3$. In conclusion, the ideal polyhedral complex $\Delta_{\mathbb{E}}(\Gamma)$ is obtained as follows. Consider the collection of faces of $\partial\mathcal{C}$ which has a Euclidean support plane, that is, the collection of the subset of $\mathbb{E}^{1,3}$ which is of the form $W \cap \mathcal{C}$ for some Euclidean support plane W of $\mathcal{C}$. Then their images by the radial projection compose a Γ-invariant ideal polyhedral complex embedded in $\mathbb{H}^3$, and $\Delta_{\mathbb{E}}(\Gamma)$ is equal to its image in $M(\Gamma)$ (see [7, Sect. 10]).

Though $\Delta_{\mathbb{E}}(\Gamma)$ has the nice geometric meaning that it is dual to the *Ford complex*, its underlying space $|\Delta_{\mathbb{E}}(\Gamma)|$ looks far from nice. In general, it is not convex and is strictly smaller than the convex core $M_0(\Gamma)$. This is because we take only Euclidean faces of $\partial\mathcal{C}$ into account in the construction of $\Delta_{\mathbb{E}}(\Gamma)$. If the group Γ were a finitely generated Kleinian group of *cofinite volume* with parabolic transformations, then as is proved by Epstein and Penner [27], the above construction gives a finite ideal polyhedral decomposition of the whole quotient hyperbolic manifold. Moreover, every face of $\partial\mathcal{C}$ is Euclidean and hence each piece of the decomposition admits a natural Euclidean structure. Thus it is called the *Euclidean decomposition*. However, in general, $\partial\mathcal{C}$ can have non-Euclidean faces. So, it is natural to try to construct an ideal polyhedral complex by taking all faces of $\partial\mathcal{C}$ into account. This was made explicit in [7], and we call it the *EPH-decomposition* and denote it by $\Delta(\Gamma)$. Here the letters E, P and H, respectively, stand for Euclidean (or elliptic), parabolic and hyperbolic.

At the end of this section, we note that $\mathrm{Ford}(\Gamma) \cap M(\Gamma)$ is a *spine* of $M(\Gamma)$, that is, it is a strong deformation retract of $M(\Gamma)$. Moreover, this spine is *canonical* in the sense that it is uniquely determined from the cusped hyperbolic manifold $M(\Gamma)$. We can apply the same construction to every cusped hyperbolic manifold, and in the special case when the manifold is of finite volume and has only one cusp (e.g. the complement of a hyperbolic knot), the combinatorial structure of the Ford complex is a complete invariant of the underlying topological 3-manifold by virtue of the Mostow rigidity theorem. This fact motivated us to study Jorgensen's work [40], which determines the combinatorial structure of the Ford domain of punctured torus groups.

1.2 Jorgensen's theorem for quasifuchsian punctured torus groups (I)

In this section, we give a description of Jorgensen's theorem in [40] from the view point of 3-manifolds, which describes the combinatorial structures of the Ford domains of quasifuchsian punctured torus groups. Before presenting the precise statement of Jorgensen's theorem, we give a brief intuitive description of the idea. Let Γ be a quasifuchsian punctured torus group. Then the quotient manifold $\bar{M}(\Gamma)$ is identified with $T \times [-1,1]$. Since $P(\Gamma) \subset \mathbb{C}$ is a fundamental domain of the action of Γ on $\Omega(\Gamma) = \Omega^-(\Gamma) \cup \Omega^+(\Gamma)$, modulo Γ_∞, $P(\Gamma)$ is a disjoint union of $P^-(\Gamma)$ and $P^+(\Gamma)$ where $P^\pm(\Gamma) = P(\Gamma) \cap \Omega^\pm(\Gamma)$. The first assertion of Jorgensen's theorem is that each of $P^\pm(\Gamma)$ is simply connected. This implies that the image of $\partial P^\pm(\Gamma)$ in $T \times \{\pm 1\}$ is a spine of the punctured torus T. If Γ is fuchsian, then these two spines are identical, and the Ford complex $\mathrm{Ford}(\Gamma)$ is equal to the product of the spine with the interval $[-1,1]$. In general, these two spines are not isotopic to each other. However, they are related by a canonical sequence of Whitehead moves as described later. (This is largely why the punctured torus is so special.) The "trace" of the canonical sequence of Whitehead moves form a spine of $T \times [-1,1]$. The main assertion of Jorgensen's theorem is that this spine is isotopic to the Ford complex $\mathrm{Ford}(\Gamma)$. Thus we can say that $\mathrm{Ford}(\Gamma)$ records the history of how the two boundary spines evolved.

Now let's give the precise statement. We begin by recalling basic topological facts on the punctured torus T. To this end, we identify T with the quotient space $(\mathbb{R}^2 - \mathbb{Z}^2)/\mathbb{Z}^2$. A simple loop in T is said to be *essential*, if it bounds neither a disk nor a once-punctured disk. Similarly, a simple arc in T having the puncture as endpoints is said to be *essential*, if it does not cut off a disk with a point on the boundary removed. Then the isotopy classes of essential simple loops (resp. essential simple arcs) in T are in one-to-one correspondence with $\hat{\mathbb{Q}} := \mathbb{Q} \cup \{1/0\}$: A representative of the isotopy class corresponding to $r \in \hat{\mathbb{Q}}$ is the projection of a line in $\mathbb{R}^2$ (the line being disjoint from $\mathbb{Z}^2$ for the loop case, and intersecting $\mathbb{Z}^2$ for the arc case). The element $r \in \hat{\mathbb{Q}}$ associated to a circle or an arc is called its *slope*. The representative of the isotopy class of an essential arc of slope r is denoted by β_r.

Consider the ideal triangle in the hyperbolic plane $\mathbb{H}^2 = \{z \in \mathbb{C} \mid \Im(z) > 0\}$ spanned by the ideal vertices $\{0/1, 1/1, 1/0\}$. Then the translates of this ideal triangle by the action of $SL(2,\mathbb{Z})$ form a tessellation of $\mathbb{H}^2$. This is called the *modular diagram* or the *Farey triangulation* and is denoted by $\mathcal{D}$. The abstract simplicial complex having the combinatorial structure of $\mathcal{D}$ is also denoted by the same symbol. The set of (ideal) vertices of $\mathcal{D}$ is equal to $\hat{\mathbb{Q}}$, and a typical (ideal) triangle σ of $\mathcal{D}$ is spanned by $\{\frac{p_1}{q_1}, \frac{p_1+p_2}{q_1+q_2}, \frac{p_2}{q_2}\}$ where $\begin{pmatrix} p_1 & p_2 \\ q_1 & q_2 \end{pmatrix} \in SL(2,\mathbb{Z})$.

Let $\sigma = \langle r_0, r_1, r_2 \rangle$ be a triangle of $\mathcal{D}$. Then the essential arcs β_{r_0}, β_{r_1}, β_{r_2} are mutually disjoint, and their union determines a *topological ideal triangulation* $\mathrm{trg}(\sigma)$ of T, in the sense that T cut open along $\beta_{r_0} \cup \beta_{r_1} \cup \beta_{r_2}$ is the disjoint union of two 2-simplices with all vertices deleted. Let $\mathrm{spine}(\sigma)$ be a 1-dimensional cell complex embedded in T which is dual to the 1-skeleton of $\mathrm{trg}(\sigma)$. Thus $\mathrm{spine}(\sigma)$ consists of two vertices and three edges γ_i $(i = 0, 1, 2)$, such that γ_i intersects the 1-skeleton of $\mathrm{trg}(\sigma)$ transversely precisely at a point of β_{r_i}. Note that $\mathrm{spine}(\sigma)$ is a deformation retract of T and hence is a *spine* of T. We define the *slope* of an edge γ_i of $\mathrm{spine}(\sigma)$ to be $r_i \in \hat{\mathbb{Q}}$, the slope of the ideal edge β_{r_i} of $\mathrm{trg}(\sigma)$ dual to γ_i.

Let $\tau = \langle r_0, r_1 \rangle$ be an edge of $\mathcal{D}$. Then the union $\beta_{r_0} \cup \beta_{r_1}$ determines a *topological ideal polygonal decomposition* of T, in the sense that T cut open along it is homeomorphic to a quadrilateral with all vertices deleted. Let $\mathrm{spine}(\tau)$ be a 1-dimensional cell complex embedded in T which is dual to the 1-skeleton of $\mathrm{trg}(\tau)$. Then $\mathrm{spine}(\tau)$ consists of a single vertex and two edges, and it is also a spine of T. The slope of an edge of $\mathrm{spine}(\tau)$ is also defined as explained in the preceding paragraph.

Let $\mathcal{D}^{(i)}$ denote the set of i-simplices of $\mathcal{D}$. Then we have the following well-known fact.

Lemma 1.2.1. *For any spine C of T, there is a unique element δ of $\mathcal{D}^{(1)} \cup \mathcal{D}^{(2)}$ such that C is isotopic to* $\mathrm{trg}(\delta)$.

If $\tau = \langle r_0, r_1 \rangle$ is an edge of a triangle $\sigma = \langle r_0, r_1, r_2 \rangle$ of $\mathcal{D}$, then $\mathrm{spine}(\tau)$ is obtained from $\mathrm{spine}(\sigma)$ by collapsing the edge γ_2 of $\mathrm{spine}(\sigma)$ of slope r_2 to a point (see Fig. 1.1). By an *elementary transformation*, we mean this transformation or its converse.

Let (δ^-, δ^+) be a pair of elements of $\mathcal{D}^{(1)} \cup \mathcal{D}^{(2)}$. Then, since the 1-skeleton of the dual to $\mathcal{D}$ is a tree, there is a unique sequence $\delta^- = \delta_0, \delta_1, \delta_2, \cdots, \delta_m = \delta^+$ in $\mathcal{D}^{(1)} \cup \mathcal{D}^{(2)}$ satisfying the following conditions.

1. For each $i \in \{0, 1, \cdots, m-1\}$, either δ_i is an edge of δ_{i-1} or δ_{i+1} is an edge of δ_i.
2. $\delta_i \neq \delta_j$ whenever $i \neq j$.

Thus we obtain a canonical sequence of elementary transformations

$$\mathrm{spine}(\delta^-) = \mathrm{spine}(\delta_0) \to \mathrm{spine}(\delta_1) \to \cdots \to \mathrm{spine}(\delta_m) = \mathrm{spine}(\delta^+).$$

Regard the sequence as a continuous family $\{C_t\}_{t \in [-1,1]}$ of spines of T, and set

$$\mathrm{Spine}(\delta^-, \delta^+) = \cup_{t \in [-1,1]} C_t \subset T \times [-1, 1].$$

In the special case when $\delta^- = \delta^+$, we set

$$\mathrm{Spine}(\delta^-, \delta^+) = \delta^- \times [-1, 1] = \delta^+ \times [-1, 1] \subset T \times [-1, 1].$$

Then $\mathrm{Spine}(\delta^-, \delta^+)$ is a 2-dimensional subcomplex of $T \times [-1, 1]$ satisfying the following conditions.

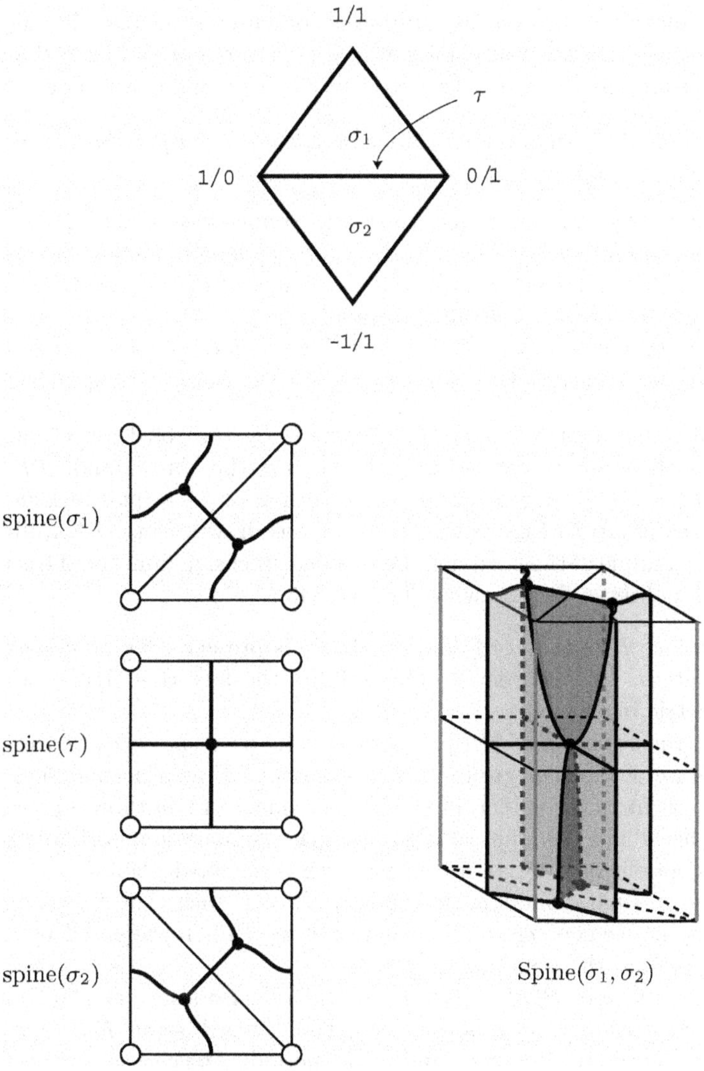

Fig. 1.1. Elementary transformation

1. $\mathrm{Spine}(\delta^-, \delta^+) \cap (T \times \{\epsilon 1\}) = \mathrm{spine}(\delta^\epsilon) \times \{\epsilon 1\}$ for each $\epsilon = \pm$.
2. There is a level-preserving deformation retraction from $T \times [-1, 1]$ to $\mathrm{Spine}(\delta^-, \delta^+)$.

Figure 1.1 illustrates $\mathrm{Spine}(\delta^-, \delta^+)$, where δ^- and δ^+ are elements of $\mathcal{D}^{(2)}$ sharing a common edge. We note that it has a natural cellular structure, consisting of a unique inner-vertex, four inner-edges and six 2-dimensional faces.

The following theorem paraphrases Jorgensen's results [42, Theorems 3.1-3.3], and describes the combinatorial structures of the Ford domains of quasifuchsian punctured torus groups (see [65, Sect. 3] for another exposition).

Theorem 1.2.2 (Jorgensen). *For any quasifuchsian punctured torus group Γ, the following hold:*

1. *$P(\Gamma)$ consists of two simply connected components $P^{\pm}(\Gamma) \subset \Omega^{\pm}(\Gamma)$. In particular, for each $\epsilon = \pm$, $P^{\epsilon}(\Gamma)$ is a fundamental domain for the action of Γ on $\Omega^{\epsilon}(\Gamma)$ modulo Γ_{∞}, and the image of $\partial P^{\epsilon}(\Gamma)$ in $\Omega^{\epsilon}(\Gamma)/\Gamma$ is a spine of T, which we denote by* $\mathrm{spine}^{\epsilon}(\Gamma)$.
2. *Let δ^{ϵ} be the element of $\mathcal{D}^{(1)} \cup \mathcal{D}^{(2)}$ such that* $\mathrm{spine}^{\epsilon}(\Gamma)$ *is isotopic to* $\mathrm{spine}(\delta^{\epsilon})$. *Then the Ford complex* $\mathrm{Ford}(\Gamma)$ *is isotopic to* $\mathrm{Spine}(\delta^{-}, \delta^{+})$.

The computer program *OPTi* [78] made by the third author visualizes the above theorem: we can see in real time how the Ford domain $Ph(\Gamma)$ and the limit set $\Lambda(\Gamma)$ vary according to deformation of a quasifuchsian punctured torus group Γ. Figure 1.2(a), which was drawn using OPTi, illustrates a typical example of the Ford domain of a quasifuchsian punctured torus group Γ. We can observe the following (cf. [40], [43]).

1. Each face F of the Ford domain $Ph(\Gamma)$ is preserved by an elliptic transformation, P_F, of order 2. This reflects the fact that $M(\Gamma)$ admits an isometric involution.
2. The transformations $\{P_F\}$, where F runs over the faces of the $Ph(\Gamma)$, generate a Kleinian group $\tilde{\Gamma}$ which contains Γ as a normal subgroup of index 2. In fact Γ is identified with (the image of a faithful representation of) the orbifold fundamental group of the 2-dimensional orbifold which is the quotient of T by an involution with three fixed points.
3. There is a parabolic transformation, K, of $\tilde{\Gamma}$ such that $K(\infty) = \infty$ and K^2 is the element $\rho(\omega) \in \Gamma$, where $\omega \in \pi_1(T)$ is represented by a simple loop around the puncture.
4. If F is a face of $Ph(\Gamma)$, then $F' = K(F)$ is also a face of $Ph(\Gamma)$ and the transformation $K \circ P_F$ is the element of Γ which sends F to F'.
5. There is a continuous family of periodic piecewise geodesic lines $\{L_t\}_{t\in(-1,1)}$ contained in $\partial Ph(\Gamma)$ such that $F \cap L_t$ is a (possibly degenerate) geodesic segment orthogonal to $F \cap \mathrm{Axis}(P_F)$ for each face F. Let S_t be the vertical piecewise totally geodesic plane in $Ph(\Gamma)$ lying above L_t. Then S_t projects to a punctured torus, T_t, in $M(\Gamma) = T \times (-1, 1)$ isotopic to a level surface, and the image, C_t, of ∂S_t forms a spine of T_t. So $\{C_t\}$ gives a continuous family of spines of T. Moreover, C_t is generic or non-generic according as L_t is disjoint from the projections of the vertices of $Ph(\Gamma)$. This family $\{C_t\}$ (with certain modification near $t = \pm 1$) realizes the canonical sequence of elementary moves transforming $\mathrm{spine}^{-}(\Gamma)$ to $\mathrm{spine}^{+}(\Gamma)$. (See Sect. 6.5 for detailed explanation.)

(a) $\mathrm{Ford}(\Gamma) = \mathrm{Spine}(\nu^-(\Gamma), \nu^+(\Gamma))$

(b) Horospherical cross section of $\Delta_{\mathbb{E}}(\Gamma)$

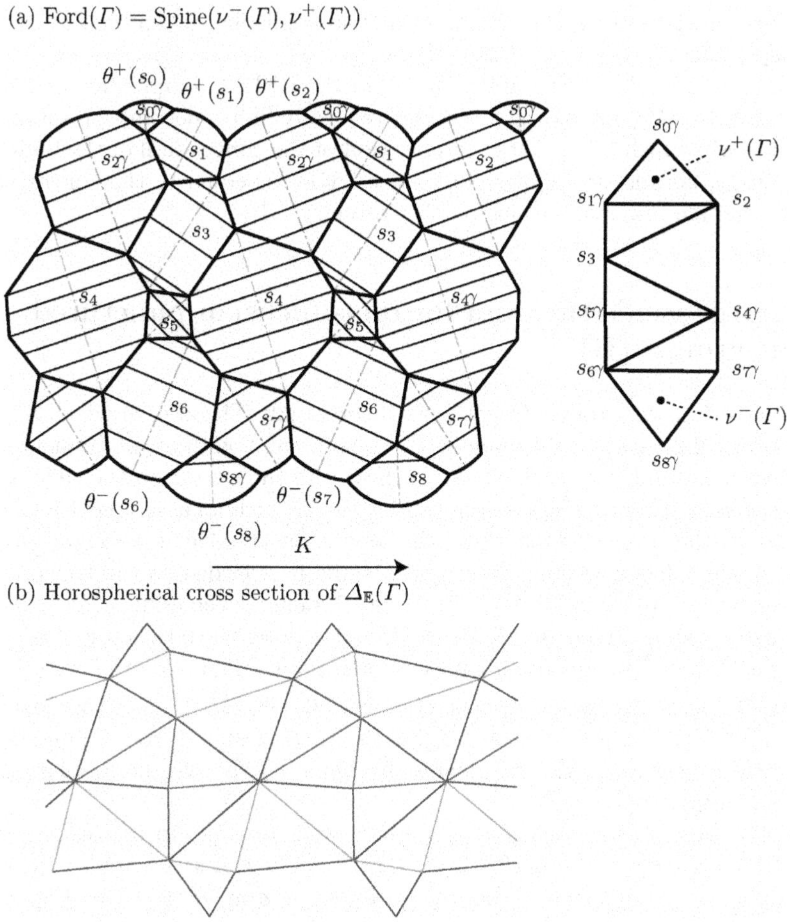

Fig. 1.2.

Figure 1.2(b) illustrates the cross section of $\tilde{\Delta}_{\mathbb{E}}(\Gamma)$ along a small horosphere H_∞ centered at ∞, where $\tilde{\Delta}_{\mathbb{E}}(\Gamma)$ is the Γ-invariant ideal polyhedral complex in $\mathbb{H}^3$ obtained as the inverse image of the dual ideal polyhedral complex $\Delta_{\mathbb{E}}(\Gamma)$. It is also regarded as the projection of $\tilde{\Delta}_{\mathbb{E}}(\Gamma) \cap \partial H_\infty$ to the complex plane $\mathbb{C}$. Then the vertices are identified with the centers of the isometric hemispheres supporting faces of the Ford domain $Ph(\Gamma)$. To be more explicit, if v is a projection of a vertex of $\tilde{\Delta}_{\mathbb{E}}(\Gamma) \cap \partial H_\infty$, then the vertical geodesic $[v, \infty)$ joining v to ∞ is an edge of $\tilde{\Delta}_{\mathbb{E}}(\Gamma)$, and v is the center of the isometric hemisphere supporting a face of $Ph(\Gamma)$ dual to the edge $[v, \infty)$ of $\tilde{\Delta}_{\mathbb{E}}(\Gamma)$. Similarly, each triangle $\langle x_0, x_1, x_2 \rangle$ of $\tilde{\Delta}_{\mathbb{E}}(\Gamma) \cap \partial H_\infty$ is a horospherical cross section of an ideal tetrahedron in $\tilde{\Delta}_{\mathbb{E}}(\Gamma)$ dual to the vertex of $Ph(\Gamma)$

obtained as the intersection of the isometric hemispheres centered at x_0, x_1 and x_2.

Let $\tilde{\Delta}(\Gamma)$ be the inverse image in $\mathbb{H}^3$ of the EPH-decomposition of $\Delta(\Gamma)$. Then $\tilde{\Delta}(\Gamma)\cap\partial H_\infty$ gives a (not necessarily locally finite) decomposition of the infinite strip which arises as the intersection of the convex hull of the limit set with ∂H_∞, because the underlying space $|\Delta(\Gamma)|$ is equal to the convex core minus the bending laminations by [7, Corollary 1.1].

1.3 Jorgensen's theorem for quasifuchsian punctured torus groups (II)

In this section, we explain Jorgensen's theorem [39, Theorem 4] which refines Theorem 1.2.2. A *weighted spine* of T is a spine of T with an assignment of a positive real number to each edge, which we call the *weight* on the edge, such that the sum of the weights is equal to 1. We call such an assignment a *weight system* on the spine. By regarding the weight on an edge as a weight on the slope of the edge, a weight system on a spine is regarded as the barycentric coordinate of a point in $|\mathcal{D}|-|\mathcal{D}^{(0)}|$. Here $|\mathcal{D}|$ denotes the underlying space of the abstract simplicial complex $\mathcal{D}$, and $|\mathcal{D}^{(0)}|$ denotes the 0-skeleton of $|\mathcal{D}|$. By fixing a $PSL(2,\mathbb{Z})$-equivariant bijective continuous map from the underlying space $|\mathcal{D}|$ (of the abstract simplicial complex $\mathcal{D}$) onto $\mathbb{H}^2\cup\hat{\mathbb{Q}}\subset\overline{\mathbb{H}}^2$, we identify $|\mathcal{D}|-|\mathcal{D}^{(0)}|$ with $\mathbb{H}^2$. Thus each weighted spine corresponds to a unique point of $\mathbb{H}^2$. For each $\nu\in\mathbb{H}^2$, we denote by spine(ν) the weighted spine of T corresponding to ν.

Let Γ be a marked quasifuchsian punctured torus group. For each $\epsilon=\pm$ and for each edge e of spine$^\epsilon(\Gamma)$, let $t^\epsilon(e)$ be $1/\pi$ times the angle, $\theta^\epsilon(e)$, of a circular arc component of the inverse image of e in $\partial P^\epsilon(\Gamma)$ (see Fig. 1.2). Then we have the following (see Proposition 4.2.16):

Lemma 1.3.1. *The sum of* $t^\epsilon(e)$ *where* e *runs over the edges of* spine$^\epsilon(\Gamma)$ *is equal to* 1.

Thus spine$^\epsilon(\Gamma)$ has the structure of a weighted spine of T where the weight of an edge e is $t^\epsilon(e)$. Let $\nu^\epsilon(\Gamma)$ be the point of $\mathcal{D}$ corresponding to the weighted spine spine$^\epsilon(\Gamma)$, and put $\boldsymbol{\nu}(\Gamma)=(\nu^-(\Gamma),\nu^+(\Gamma))$. We call it the *side parameter* of Γ following [39]. The following refinement of Theorem 1.2.2(1) justifies the terminology (see [39, Theorem 4]):

Theorem 1.3.2 (Jorgensen). *The map* $\boldsymbol{\nu}:\mathcal{QF}\to\mathbb{H}^2\times\mathbb{H}^2$ *is a homeomorphism.*

Remark 1.3.3. See [3] for a relation between the side-parameter $\boldsymbol{\nu}(\Gamma)$ and the usual end invariant of Γ (see [58]), which records the conformal structure $(\Omega^-(\Gamma)/\Gamma,\Omega^+(\Gamma)/\Gamma)\in Teich(T)\times Teich(T)=\mathbb{H}^2\times\mathbb{H}^2$.

To combine Theorems 1.2.2 and 1.3.2, we introduce the following concept:

Definition 1.3.4. *(1) A* weighted relative spine *of $T \times [-1,1]$ is a 2-dimensional subcomplex C of $T \times [-1,1]$ satisfying the following conditions.*

1. *There is a level-preserving deformation retraction from $T \times [-1,1]$ to* Spine(δ^-, δ^+) *for some $\delta^\pm \in \mathcal{D}^{(2)} \cup \mathcal{D}^{(1)}$.*
2. *A weight system is specified on each of $\partial^\pm C := C \cap T \times \{\pm 1\}$.*

Two weighted relative spines are equivalent, *if the underlying relative spines are isotopic and the weight systems coincide (after the isotopy).*

(2) For $\boldsymbol{\nu} = (\nu^-, \nu^+) \in \mathbb{H}^2 \times \mathbb{H}^2$, Spine$(\boldsymbol{\nu})$ *denotes the weighted relative spine, such that the underlying relative spine is* Spine$(\delta^-(\boldsymbol{\nu}), \delta^+(\boldsymbol{\nu}))$, *where $\delta^\epsilon(\boldsymbol{\nu})$ denotes the element of $\mathcal{D}^{(2)} \cup \mathcal{D}^{(1)}$ whose interior contains ν^ϵ, and the weight system on $\partial^\pm$*Spine$(\boldsymbol{\nu})$ *is given by $\nu^\pm$.*

Then we can summarize Theorems 1.2.2 and 1.3.2 as follows:

Main Theorem 1.3.5 (Jorgensen). *For each $\Gamma \in \mathcal{QF}$, the Ford complex* Ford(Γ) *has a structure of weighted relative spine of $T \times [-1,1]$. Moreover, there is a homeomorphism $\boldsymbol{\nu} : \mathcal{QF} \to \mathbb{H}^2 \times \mathbb{H}^2$, such that the weighted spine* Ford(Γ) *is equivalent to* Spine$(\boldsymbol{\nu}(\Gamma))$ *for any $\Gamma \in \mathcal{QF}$.*

1.4 The topological ideal polyhedral complex Trg($\boldsymbol{\nu}$) dual to Spine($\boldsymbol{\nu}$)

As explained in Sect. 1.1, the Ford complex Ford(Γ) is a dual to the subcomplex $\Delta_{\mathbb{E}}(\Gamma)$ of $\Delta(\Gamma)$ consisting of the Euclidean faces. In this section, we describe the structure of $\Delta_{\mathbb{E}}(\Gamma)$ following the exposition by Floyd-Hatcher [29] of Jorgensen's ideal triangulation of punctured torus bundles over S^1.

For each element $\boldsymbol{\nu} = (\nu^-, \nu^+)$ of $\mathbb{H}^2 \times \mathbb{H}^2 = (|\mathcal{D}| - |\mathcal{D}^{(0)}|) \times (|\mathcal{D}| - |\mathcal{D}^{(0)}|)$, we construct a topological ideal simplicial complex Trg$(\boldsymbol{\nu})$. To describe the construction, we introduce the following definition.

Definition 1.4.1 (Chain). *By the* chain $\Sigma(\boldsymbol{\nu})$ *spanned by $\boldsymbol{\nu} = (\nu^-, \nu^+) \in \mathbb{H}^2 \times \mathbb{H}^2$, we mean the sequence $(\sigma_1, \sigma_2, \cdots, \sigma_n)$ of triangles of $\mathcal{D}$ whose interiors intersect the oriented geodesic segment joining ν^- with ν^+ in this order. Note that $n = 0$ if and only if $\nu^\pm$ is contained in a single edge τ of $\mathcal{D}$. In this case we redefine $\Sigma(\boldsymbol{\nu}) = \{\tau\}$. $\Sigma(\boldsymbol{\nu})$ and $\boldsymbol{\nu}$ is said to be* thick *or* thin *according as $n \geq 1$ or $n = 0$.*

Case 1. $n \geq 2$. Let $\widetilde{\mathrm{trg}}(\sigma_i)$ be the ideal triangulation of $\mathbb{R}^2 - \mathbb{Z}^2$ obtained as the lift of $\mathrm{trg}(\sigma_i)$. By superimposing $\widetilde{\mathrm{trg}}(\sigma_{i+1})$ upon $\widetilde{\mathrm{trg}}(\sigma_i)$, we obtain an array of ideal tetrahedra whose bottom faces compose $\widetilde{\mathrm{trg}}(\sigma_i)$ and whose top faces compose $\widetilde{\mathrm{trg}}(\sigma_{i+1})$. We denote this array of ideal tetrahedra by $\widetilde{\mathrm{Trg}}(\sigma_i, \sigma_{i+1})$. By stacking $\widetilde{\mathrm{Trg}}(\sigma_1, \sigma_2), \cdots, \widetilde{\mathrm{Trg}}(\sigma_{n-1}, \sigma_n)$ up in order, we obtain a set of layers whose bottom faces compose $\widetilde{\mathrm{trg}}(\sigma_1)$ and whose top faces compose $\widetilde{\mathrm{trg}}(\sigma_n)$. The covering transformation group $\mathbb{Z}^2$ of the covering $\mathbb{R}^2 - \mathbb{Z}^2 \to T$ freely acts on the above topological ideal simplicial complex, and we define $\mathrm{Trg}(\boldsymbol{\nu})$ to be the quotient topological ideal simplicial complex.

Case 2. $n = 1$. Then $\mathrm{Trg}(\boldsymbol{\nu})$ is defined to be the 2-dimensional topological ideal triangulation $\mathrm{trg}(\sigma_1)$.

Case 3. $n = 0$. Then $\delta^-(\boldsymbol{\nu}) = \delta^+(\boldsymbol{\nu})$ is an edge τ of $\mathcal{D}$. We define $\mathrm{Trg}(\boldsymbol{\nu})$ to be the 2-dimensional topological ideal cellular complex $\mathrm{Trg}(\delta^{\pm}(\boldsymbol{\nu}))$. It should be noted that $\mathrm{Trg}(\boldsymbol{\nu})$ is not a topological ideal triangulation, but a topological ideal cellular complex.

Note that the underlying space $|\mathrm{Trg}(\boldsymbol{\nu})|$ is homeomorphic to the the quotient space of $T \times [-1, 1]$ by an equivalence relation $\sim$ such that $(x, s) \sim (y, t)$ only if $x = y$ (cf. [71, Sect. II.2]). In particular, $|\mathrm{Trg}(\boldsymbol{\nu})|$ is homotopy equivalent to T and has a natural embedding into $T \times (-1, 1)$. We then have the following theorem by Theorem 1.2.2.

Theorem 1.4.2. *For any $\Gamma \in \mathcal{QF}$, $\Delta_{\mathbb{E}}(\Gamma)$ is isotopic to $\mathrm{Trg}(\boldsymbol{\nu}(\Gamma))$ in the convex core $M_0(\Gamma)$ of $M(\Gamma)$.*

Figure 1.2(b) illustrates the cross section of $\Delta_{\mathbb{E}}(\Gamma)$ along a horosphere centered at ∞ for the quasifuchsian punctured torus group Γ in Fig. 1.2(a).

2

Fricke surfaces and $PSL(2, \mathbb{C})$-representations

The topological once-punctured torus T, the 4-times punctured sphere S and the $(2, 2, 2, \infty)$-orbifold $\mathcal{O}$ are commensurable, and are called *Fricke surfaces* (see [74]). In this chapter, we give a detailed study of the fundamental groups of Fricke surfaces and their representations to $PSL(2, \mathbb{C})$.

In Sect. 2.1, we classify the "geometric" generator systems of the fundamental groups of Fricke surfaces (Propositions 2.1.6 and 2.1.9), and describe the space of all geometric generators in terms of the Farey triangulation (Proposition 2.1.10). Though these results seem to be well-known, we could not find an explicit proof in the literature. For this reason, we include a full proof, following an idea suggested to the authors by Stephan Hamperies that greatly simplified our original proof.

In Sect. 2.2, we study *type-preserving* representations of the fundamental groups of Fricke surfaces to $PSL(2, \mathbb{C})$, i.e., those irreducible representations which send peripheral elements to parabolic transformations. We show that the spaces of type-preserving $PSL(2, \mathbb{C})$-representations for Fricke surfaces are essentially identical (Proposition 2.2.2). This fact is well-known and easily proved for T and $\mathcal{O}$. However, we could not find the result for S in the literature. Thus we include the proof, because it relates the study of 2-bridge link groups to that of punctured torus groups (cf. Sect. 5.3). We also point out Proposition 2.2.8, which reduces the study of Ford domains of punctured torus groups to that of Ford domains of Kleinian groups obtained as the image of $\pi_1(\mathcal{O})$ by type-preserving representations.

In Sect. 2.3, we describe the space of type-preserving representations in terms of the affine algebraic variety determined by the Markoff equation $x^2 + y^2 + z^2 = xyz$, following [17] (Proposition 2.3.4 and 2.3.6).

In Sect. 2.4, we recall the *complex probability* parameter of type-preserving representations, and describe a conceptual geometric construction of a type-preserving representation from a given complex probability, i.e., a triple of complex numbers (a_0, a_1, a_2) such that $a_0 + a_1 + a_2 = 1$ (Proposition 2.4.4).

In the last section, Sect. 2.5, we collect several well-known properties of discrete groups, which we use in this paper.

2.1 Fricke surfaces and their fundamental groups

Let T, S and $\mathcal{O}$, respectively, be the once-punctured torus, the 4-times punctured sphere, and the $(2,2,2,\infty)$-orbifold (i.e., the orbifold with underlying space a punctured sphere and with three cone points of cone angle π). They have $\mathbb{R}^2 - \mathbb{Z}^2$ as the common covering space. To be precise, let G and $\tilde{G}$, respectively, be the groups of transformations on $\mathbb{R}^2 - \mathbb{Z}^2$ generated by π-rotations about points in $\mathbb{Z}^2$ and $(\frac{1}{2}\mathbb{Z})^2$. Then $T = (\mathbb{R}^2 - \mathbb{Z}^2)/\mathbb{Z}^2$, $S = (\mathbb{R}^2 - \mathbb{Z}^2)/G$ and $\mathcal{O} = (\mathbb{R}^2 - \mathbb{Z}^2)/\tilde{G}$. In particular, there is a $\mathbb{Z}_2$-covering $T \to \mathcal{O}$ and a $\mathbb{Z}_2 \oplus \mathbb{Z}_2$-covering $S \to \mathcal{O}$: the pair of these coverings is called the *Fricke diagram* and each of T, S, and $\mathcal{O}$ is called a *Fricke surface* (see [74]).

A simple loop in a Fricke surface is said to be *essential* if it does not bound a disk, a disk with one puncture, or a disk with one cone point. Similarly, a simple arc in a Fricke surface joining punctures is said to be *essential* if it does not cut off a *"monogon"*, i.e., a disk minus a point on the boundary. Then the isotopy classes of essential simple loops (resp. essential simple arcs with one end in a given puncture) in a Fricke surface are in one-to-one correspondence with $\hat{\mathbb{Q}} := \mathbb{Q} \cup \{1/0\}$: A representative of the isotopy class corresponding to $r \in \hat{\mathbb{Q}}$ is the projection of a line in $\mathbb{R}^2$ (the line being disjoint from $\mathbb{Z}^2$ for the loop case, and intersecting $\mathbb{Z}^2$ for the arc case). The element $r \in \hat{\mathbb{Q}}$ associated to a circle or an arc is called its *slope*. An essential loop of slope r in T or $\mathcal{O}$ (resp. S) is denoted by α_r (resp. $\tilde{\alpha}_r$). The notation reflects the following fact: After an isotopy, the restriction of the projection $T \to \mathcal{O}$ to α_r $(\subset T)$ gives a homeomorphism from α_r $(\subset T)$ to α_r $(\subset \mathcal{O})$, while the restriction of the projection $S \to \mathcal{O}$ to $\tilde{\alpha}_r$ gives a two-fold covering from $\tilde{\alpha}_r$ $(\subset S)$ to α $(\subset \mathcal{O})$.

Since T and S are finite regular coverings of the orbifold $\mathcal{O}$, the fundamental groups of T and S are regarded as normal subgroups of the orbifold fundamental group of $\mathcal{O}$ of finite index. These groups have the following group presentations:

$$\pi_1(T) = \langle A_0, B_0 \rangle, \tag{2.1}$$
$$\pi_1(S) = \langle K_0, K_1, K_2, K_3 \mid K_0K_1K_2K_3 = 1 \rangle, \tag{2.2}$$
$$\pi_1(\mathcal{O}) = \langle P_0, Q_0, R_0 \mid P_0^2 = Q_0^2 = R_0^2 = 1 \rangle, \tag{2.3}$$

Here the generators satisfy the following conditions: Set $K = (P_0Q_0R_0)^{-1}$, then K is represented by the puncture of $\mathcal{O}$ and satisfies the relations

$$K^2 = [A_0, B_0], \quad A_0 = KP_0 = R_0Q_0, \quad B_0 = K^{-1}R_0 = P_0Q_0,$$
$$K_0 = K, \quad K_1 = K^{P_0}, \quad K_2 = K^{Q_0}, \quad K_3 = K^{R_0},$$

where X^Y denotes YXY^{-1}. *Throughout this paper, we reserve the symbol K to denote the element of $\pi_1(\mathcal{O})$ defined in the above.*

Definition 2.1.1 ((Elliptic) generator triple). *(1) An ordered pair (A, B) of elements in $\pi_1(T)$ is a* generator pair *of $\pi_1(T)$ if they generate $\pi_1(T)$ and* *satisfy* $[A, B] = K^2$. *In this case,* A *and* B *are, respectively, called the* left *and* right generators, *and* (A, AB, B) *is called a* generator triple.

(2) An ordered triple (P, Q, R) *of elements of* $\pi_1(\mathcal{O})$ *is called an* elliptic generator triple *if they generate* $\pi_1(\mathcal{O})$ *and satisfy* $P^2 = Q^2 = R^2 = 1$ *and* $(PQR)^{-1} = K$. *A member of an elliptic generator triple is called an* elliptic generator. $\mathcal{EG}$ *denotes the set of all elliptic generators.*

We can easily see the following proposition.

Proposition 2.1.2. *If* (P, Q, R) *is an elliptic generator triple of* $\pi_1(\mathcal{O})$, *then* $(KP, KQ, K^{-1}R)$ *is a generator triple of* $\pi_1(T)$. *Conversely, if* (A, AB, B) *is a generator triple of* $\pi_1(T)$, *then* $(K^{-1}A, K^{-1}AB, KB)$ *is an elliptic generator triple.*

Definition 2.1.3. *(1) If an elliptic generator triple* (P, Q, R) *and a generator triple* (A, AB, B) *are related as in Proposition 2.1.2, then we say that they are* associated with *each other.*

(2) The slope, $s(A)$, *of a generator* A *of* $\pi_1(T)$ *is defined to be the slope of an essential loop in* T *representing the conjugacy class of the generator.*

(3) The slope, $s(P)$, *of an elliptic generator* P *of* $\pi_1(\mathcal{O})$ *is defined as the slope of the generator* KP *of* $\pi_1(T)$.

Remark 2.1.4. If (P, Q, R) is an elliptic generator triple, then it follows from Proposition 2.1.6(1.1) below that (R, P^K, Q^K) is also an elliptic generator triple. Thus KR is a (left) generator of $\pi_1(T)$, and hence the slope of R is also well-defined.

Convention 2.1.5. Throughout this paper, we assume

$$(s(A_0), s(A_0B_0), s(B_0)) = (s(P_0), s(Q_0), s(R_0)) = (0/1, 1/1, 1/0) = (0, 1, \infty).$$

for the generators in the group presentations (2.1) and (2.3).

We have the following classification of elliptic generator triples.

Proposition 2.1.6. *(1) For any elliptic generator triple* (P, Q, R), *the following holds:*

(1.1) The triple of any three consecutive elements in the following bi-infinite sequence is also an elliptic generator triple.

$$\cdots, P^{K^{-1}}, Q^{K^{-1}}, R^{K^{-1}}, P, Q, R, P^K, Q^K, R^K, \cdots$$

(1.2) (P, R, Q^R) *and* (Q^P, P, R) *are also elliptic generator triples.*

(2) Conversely, any elliptic generator triple is obtained from a given elliptic generator triple by successively applying the operations in (1).

Proof. Since (1) can be proved by direct calculation, we give proof of (2). We present a proof indicated to us by Stephan Hamperies (cf. [23, Notation 2.3]), and greatly simplifying our original proof. Let (P_0, P_1, P_2) be an elliptic generator triple, and let σ_0 and σ_1 be the "braid" automorphism of $\pi_1(\mathcal{O})$ defined by

$$(\sigma_0(P_0), \sigma_0(P_1), \sigma_0(P_2)) = (P_0P_1P_0, P_0, P_2),$$
$$(\sigma_1(P_0), \sigma_1(P_1), \sigma_1(P_2)) = (P_0, P_1P_2P_1, P_1).$$

Then σ_0 and σ_1 preserve K and hence they map elliptic generator triples to elliptic generator triples. Moreover, we have the following lemma.

Lemma 2.1.7. *The group of automorphisms of $\pi_1(\mathcal{O})$ preserving K is generated by σ_0 and σ_1.*

Proof. Though the proof of this lemma is parallel to [22, Proof of Proposition 10.7], we include it the proof for completeness. Let f be an automorphism of $\pi_1(\mathcal{O})$ which preserves K. Since $f(P_j)$ has order 2 and since $\pi_1(\mathcal{O})$ is isomorphic to the free product of three cyclic groups $\langle P_j \rangle$ of order 2, $f(P_j)$ is conjugate to $P_{\tau(j)}$ for some $\tau(j) \in \{0, 1, 2\}$. Since $f(P_2P_1P_0) = P_2P_1P_0$, τ must be a permutation on the set $\{0, 1, 2\}$. Hence we have $f(P_j) = W_jP_{\tau(j)}W_j^{-1}$, for some element $W_j \in \pi_1(\mathcal{O})$. Express W_j as a reduced word in $\{P_0, P_1.P_2\}$, and let $l(W_j)$ be the minimal word length of W_j. Set $\lambda(f) = l(W_1)+l(W_2)+l(W_3)$. We prove the lemma by induction on $\lambda(f)$. If $\lambda(f) = 0$, then we have $f = id$. Suppose $\lambda(f) > 0$. Then there will be reduction in the product

$$W_2P_{\tau(2)}W_2^{-1}W_1P_{\tau(1)}W_1^{-1}W_0P_{\tau(0)}W_0^{-1} = P_2P_1P_0$$

such that some $P_{\tau(j)}$ is canceled out by $P_{\tau(j)}$ contained in W_{j+1}^{-1} or W_{j-1}. (Otherwise all $W_{j+1}^{-1}W_j$ must be reduced to 1, and hence all W_j must be trivial words.) Suppose W_{j+1}^{-1} cancels $P_{\tau(j)}$. Then we see $l(W_{j+1}^{-1}W_jP_{\tau(j)}W_j^{-1}) < l(W_{j+1})$. Apply $f\sigma_jf^{-1}$ to the elliptic generator triple $(f(P_0), f(P_1), f(P_2))$. Then

$$\begin{aligned}(f\sigma_jf^{-1})(f(P_j)) &= f(P_j)f(P_{j+1})f(P_j) \\ &= (W_jP_{\tau(j)}W_j^{-1}W_{j+1})P_{\tau(j+1)}(W_{j+1}^{-1}W_jP_{\tau(j)}W_j^{-1}).\end{aligned}$$

This implies $\lambda(f) > \lambda((f\sigma_jf^{-1})f) = \lambda(f\sigma_j)$. By using the inductive hypothesis, we see that f is contained in the group generated by σ_0 and σ_1. Similarly, if W_{j-1} cancels $P_{\tau(j)}$, then we obtain the same conclusion by applying $f\sigma_{j-1}^{-1}f^{-1}$.

By the above lemma, we have only to show that σ_0 and σ_1 are contained in the group generated by the following automorphisms:

$$f_1 : (P_0, P_1, P_2) \mapsto (P_1, P_2, P_0^K), \qquad f_2 : (P_0, P_1, P_2) \mapsto (P_0, P_2, P_1^{P_2}).$$

But we can check that $\sigma_0 = f_1f_2^{-1}f_1^{-1}$ and $\sigma_1 = f_2^{-1}$. This completes the proof of Proposition 2.1.6.

The following lemma can be verified by simple calculation.

Lemma 2.1.8. *Let* $(A, AB, B) = (KP, KQ, K^{-1}R)$ *be the generator triple associated with an elliptic generator triple* (P, Q, R). *Set* $(s_0, s_1, s_2) = (s(A), s(AB), s(B)) = (s(P), s(Q), s(R))$. *Then the following hold:*

1. (AB, ABA^{-1}, A^{-1}) *and* (B^{-1}, A, BA), *respectively, are the generator triples associated with* (Q, R, P^K) *(resp.* $(P^{K^{-1}}, Q, R)$*), and we have:*

$$(s(AB), s(ABA^{-1}), s(A^{-1})) = (s(Q), s(R), s(P^K)) = (s_1, s_2, s_0),$$
$$(s(B^{-1}), s(A), s(BA)) = (s(P^{K^{-1}}), s(Q), s(R)) = (s_2, s_0, s_1).$$

2. (A, ABA^{-1}, BA^{-1}) *and* (AB^{-1}, A, B), *respectively, are the generator triples associated with* (P, R, Q^R) *and* (Q^P, P, R), *and we have:*

$$(s(A), s(ABA^{-1}), s(BA^{-1})) = (s(P), s(R), s(Q^R)) = (s_0, s_2, s_1'),$$
$$(s(AB^{-1}), s(A), s(B)) = (s(Q^P), s(P), s(R)) = (s_1', s_0, s_2),$$

where s_1' *is the vertex of* $\mathcal{D}$ *which is opposite to* s_1 *with respect to the edge* $\langle s_0, s_2 \rangle$.

By Proposition 2.1.6 and Lemma 2.1.8, we obtain the following proposition.

Proposition 2.1.9. *(1) For any generator triple* (A, AB, B), *the following hold.*

(1.1) Both (AB, ABA^{-1}, A^{-1}) *and* (B^{-1}, A, BA) *are generator triples.*

(1.2) Both (A, ABA^{-1}, BA^{-1}) *and* (AB^{-1}, A, B) *are generator triples.*

(2) Conversely, any generator triple is obtained from the standard generator triple by successively applying the operations in (1).

For the slopes of elliptic generator triples, we have the following proposition.

Proposition 2.1.10. *(1) For any elliptic generator triple* (P, Q, R), *the oriented triangle* $\langle s(P), s(Q), s(R) \rangle$ *of* $\mathcal{D}$ *is coherent with the oriented triangle* $\langle 0/1, 1/1, 1/0 \rangle = \langle s(P_0), s(Q_0), s(R_0) \rangle$.

(2) For any oriented triangle $\langle s_0, s_1, s_2 \rangle$ *of* $\mathcal{D}$ *coherent with the oriented triangle* $\langle 0/1, 1/1, 1/0 \rangle$, *there is a bi-infinite sequence* $\{P_j\}_{j \in \mathbb{Z}}$ *of elliptic generators satisfying the following conditions:*

1. *For each* $j \in \mathbb{Z}$, *we have* $s(P_j) = s_{[j]}$, *where* $[j]$ *denotes the integer in* $\{0, 1, 2\}$ *such that* $[j] \equiv j \pmod 3$.
2. *The triple of any three consecutive elements* P_{j-1}, P_j, P_{j+1} *is an elliptic generator triple.*
3. $P_j^{K^m} = P_{j+3m}$.
4. *For any elliptic generator* P *with* $s(P) = s_k \in \{s_0, s_1, s_2\}$, *there is a unique integer* j *with* $[j] = k$ *such that* $P = P_j$.

Furthermore, such a sequence is unique modulo shift of indices by multiples of 3.

Proof. (1) This follows from Proposition 2.1.6 and Lemma 2.1.8.

(2) By Proposition 2.1.6(1.1) and Lemma 2.1.8, the bi-infinite sequence

$$\cdots, P_0^{K^{-1}}, Q_0^{K^{-1}}, R_0^{K^{-1}}, P_0, Q_0, R_0, P_0^K, Q_0^K, R_0^K, \cdots$$

obtained by *expanding* the standard generator triple (P_0, Q_0, R_0) satisfies the conditions 1, 2 and 3 for the oriented triangle $\langle 0, 1, \infty \rangle$. By Proposition 2.1.6(1) and Lemma 2.1.8, we can also construct such a bi-infinite sequences for each of the three oriented triangles adjacent to $\langle 0, 1, \infty \rangle$. By repeating this procedure, we can construct a bi-infinite sequence $\{P_j\}$ of elliptic generators satisfying the conditions 1, 2 and 3 for every oriented triangle $\langle s_0, s_1, s_2 \rangle$ of $\mathcal{D}$. To show that this sequence also satisfies the condition 4, pick an elliptic generator P such that $s(P) = s_k \in \{s_0, s_1, s_2\}$. Then by using Proposition 2.1.6(2) and Lemma 2.1.8, we can find a bi-infinite sequence $\{P'_j\}$ of elliptic generators which contains P and satisfies the conditions 1, 2 and 3. We have only to show that $\{P'_j\}$ coincides with $\{P_j\}$ modulo shift of indices by multiples of 3. This is a consequence of the following Lemma 2.1.11, which shows that there is no ambiguity in the construction of such a sequence for a triangle from that for an adjacent triangle. This also proves the uniqueness of such a sequence.

Lemma 2.1.11. *For an elliptic generator triple (P, Q, R), we have*

$$(Q^{K^n})^{(R^{K^n})} = (Q^R)^{K^n}, \quad Q^P = (Q^R)^{K^{-1}}.$$

Hence the bi-infinite sequences obtained by expanding the elliptic generator triples (P, R, Q^K), $(P^{K^n}, R^{K^n}, (Q^{K^n})^{(R^{K^n})})$ and (Q^P, P, R) all coincide modulo shift of indices.

Proof. This can be proved by simple calculation.

Convention 2.1.12 (Ordered triangle). When we mention to a triangle $\sigma = \langle s_0, s_1, s_2 \rangle$ of $\mathcal{D}$, we always assume that the order of the vertices is fixed so that the orientation is coherent with that of $\langle s(P_0), s(Q_0), s(R_0) \rangle = \langle 0, 1, \infty \rangle$ (cf. Convention 2.1.5).

Definition 2.1.13. *(1) For a triangle $\sigma = \langle s_0, s_1, s_2 \rangle$ of $\mathcal{D}$, the bi-infinite sequence $\{P_j\}$ in Proposition 2.1.10(3) is called the* sequence of elliptic generators associated with σ , *and it is denoted by $\mathcal{EG}(\sigma)$.*

(2) An ordered pair (P, Q) of elliptic generators are called an elliptic generator pair *if P and Q appear successively in this order in a sequence of elliptic generators.*

(3) More generally, for a subcomplex Σ of $\mathcal{D}$, $\mathcal{EG}(\Sigma)$ denotes the set of elliptic generators defined by $\mathcal{EG}(\Sigma) = \{P \in \mathcal{EG} \mid s(P) \in \Sigma^{(0)}\}$.

By Proposition 2.1.10(1), we may assume the following when considering the sequence of elliptic generators associated with adjacent triangles of $\mathcal{D}$.

Assumption 2.1.14. (Adjacent triangles) The symbols $\sigma = \langle s_0, s_1, s_2 \rangle$ and $\sigma' = \langle s'_0, s'_1, s'_2 \rangle$ denote triangles of $\mathcal{D}$ sharing the edge $\tau := \langle s_0, s_2 \rangle = \langle s'_0, s'_1 \rangle$. The symbols $\{P_j\}$ and $\{P'_j\}$, respectively, denote the sequences of elliptic generators associated with σ and σ', such that $P'_0 = P_0$, $P'_1 = P_2$ and $P'_2 = P_1^{P_2}$ (see Fig. 2.1)

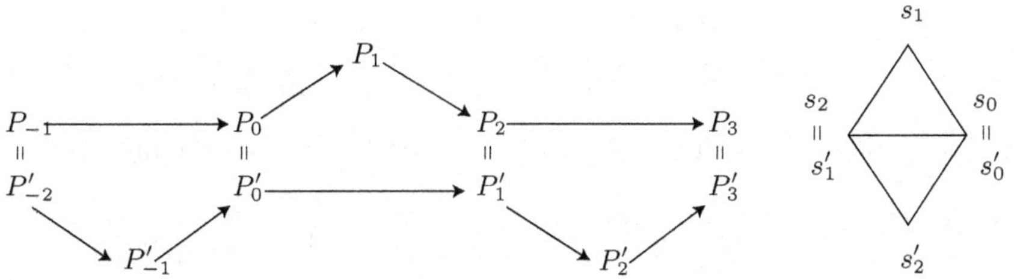

Fig. 2.1. Adjacent sequences of elliptic generators

2.2 Type-preserving representations

In this section, we introduce the family of $PSL(2,\mathbb{C})$-representations of the fundamental groups of the Fricke surfaces which are studied in this paper.

Definition 2.2.1 (Type-preserving $PSL(2,\mathbb{C})$-representation).
(1) A $PSL(2,\mathbb{C})$-representation of $\pi_1(T)$ (resp. $\pi_1(\mathcal{O})$) is type-preserving *if it is irreducible (equivalently, it does not have a common fixed point in $\partial\mathbb{H}^3$) and sends peripheral elements to parabolic transformations.*

(2) A $PSL(2,\mathbb{C})$-representation ρ of $\pi_1(S)$ is type-preserving *if it is irreducible, sends peripheral elements to parabolic transformations, and satisfies the following condition: If $\mathrm{Fix}(\rho(K_i)) = \mathrm{Fix}(\rho(K_j))$ for some $0 \le i < j \le 3$, then $\rho(K_i) = \rho(K_j^{-1})$.*

(3) For $X = T$, $\mathcal{O}$ and S, $\mathrm{Hom_{tp}}(\pi_1(X), PSL(2,\mathbb{C}))$ denotes the space of all type-preserving $PSL(2,\mathbb{C})$-representations of $\pi_1(X)$.

Proposition 2.2.2. *The spaces $\mathrm{Hom_{tp}}(\pi_1(X), PSL(2,\mathbb{C}))$ with $X = T$, $\mathcal{O}$ and S are all identical in the following sense.*

1. *The restriction of any type-preserving $PSL(2,\mathbb{C})$-representation of $\pi_1(\mathcal{O})$ to $\pi_1(T)$ (resp. $\pi_1(S)$) is type-preserving.*
2. *Conversely, every type-preserving $PSL(2,\mathbb{C})$-representation of $\pi_1(T)$ (resp. $\pi_1(S)$) extends to a unique type-preserving $PSL(2,\mathbb{C})$-representation of $\pi_1(\mathcal{O})$.*

Remark 2.2.3. In this paper, we need the assertion only for T, which is well-known. We give a proof for S for completeness and for application to 2-bridge links in the forthcoming paper [11].

Proof. (1) Since the assertion for $\pi_1(T)$ is well-known (see e.g. [76, Sect. 5.4]), we prove the assertion for $\pi_1(S)$. Let $\rho : \pi_1(\mathcal{O}) \to PSL(2,\mathbb{C})$ be a type-preserving representation. Suppose $\mathrm{Fix}(\rho(K_0)) = \mathrm{Fix}(\rho(K_1))$. Then

$$\begin{aligned}\mathrm{Fix}(\rho(K_2)) &= \rho(Q)(\mathrm{Fix}(\rho(K_0))) = \rho(Q)(\mathrm{Fix}(\rho(K_1)))\\ &= \mathrm{Fix}(\rho(K_1^Q)) = \mathrm{Fix}(\rho(K_3)).\end{aligned}$$

Since ρ is irreducible, we must have

$$\mathrm{Fix}(\rho(K_0)) = \mathrm{Fix}(\rho(K_1)) \neq \mathrm{Fix}(\rho(K_2)) = \mathrm{Fix}(\rho(K_3)).$$

Suppose $\rho(K_0K_1) \neq 1$. Then $\rho(K_0K_1) = \rho((K_2K_3)^{-1})$ is parabolic, and we have

$$\mathrm{Fix}(\rho(K_0)) = \mathrm{Fix}(\rho(K_0K_1)) = \mathrm{Fix}(\rho((K_2K_3)^{-1})) = \mathrm{Fix}(\rho(K_3)),$$

This is a contradiction. Hence $\rho(K_0) = \rho(K_1^{-1})$.

Suppose, in general, that $\mathrm{Fix}(\rho(K_i)) = \mathrm{Fix}(\rho(K_j))$ for some $0 \le i < j \le 3$. Then, after cyclic permutation, we may assume $(i,j) = (0,1)$ or $(0,2)$. The former case is settled by the above argument. The second case is settled by applying the same argument to the quadruple $(K_0, K_2, K_3, K_0K_1K_0^{-1})$.

(2) Since the assertion for $\pi_1(T)$ is well-known (cf. [40, Sect. 2]), we prove the assertion for $\pi_1(S)$ by using the arguments in [35, Proof of Proposition 1.1]. Throughout the proof, we employ the following convention: For a geodesic L in $\mathbb{H}^3$, we denote the π-rotation about L by the same symbol L.

Let ρ be an element of $\mathrm{Hom}_{\mathrm{tp}}(\pi_1(S), PSL(2,\mathbb{C}))$, and set $g_i = \rho(K_i)$. We show that ρ extends to a type-preserving representation of $\pi_1(\mathcal{O})$. As in the proof of Lemma 2.2.4, after cyclic permutation of indices, we may assume $\mathrm{Fix}(g_0) \neq \mathrm{Fix}(g_1)$. Let L_{01}, L_0 and L_1 be the geodesics in $\mathbb{H}^3$ satisfying the following conditions.

1. L_{01} joins $\mathrm{Fix}(g_0)$ and $\mathrm{Fix}(g_1)$.
2. $\mathrm{Fix}(g_0)$ (resp. $\mathrm{Fix}(g_1)$) is an endpoint of L_0 (resp. L_1).
3. $g_0 = L_0L_{01}$ and $g_1 = L_{01}L_1$.

Then we can find a geodesic H_{01} intersecting L_{01} orthogonally and the π-rotation H_{01} maps the geodesic L_0 to the geodesic $L_{01}(L_1)$, the image of L_1 by the π-rotation L_{01}. Since the π-rotation around the geodesic $L_{01}(L_1)$ is equal to $L_{01}L_1L_{01}$, we have $L_{01}L_1L_{01} = H_{01}L_0H_{01}$. Hence

$$H_{01}g_0H_{01} = H_{01}L_0L_{01}H_{01} = H_{01}L_0H_{01}L_{01} = L_{01}L_1L_{01}L_{01} = L_{01}L_1 = g_1.$$

The assumption $\mathrm{Fix}(g_0) \neq \mathrm{Fix}(g_1)$ implies $\mathrm{Fix}(g_2) \neq \mathrm{Fix}(g_3)$. So there are geodesics L_{23}, L_2, L_3 and H_{23} such that $g_2 = L_2L_{23}$, $g_3 = L_{23}L_3$ and $H_{23}g_2H_{23} = g_3$. Set $f_{01} = g_0H_{01}$ and $f_{23} = g_2H_{23}$. Then

$$f_{01}^2 = g_0 H_{01} g_0 H_{01} = g_0 g_1 = (g_2 g_3)^{-1} = (g_2 H_{23} g_2 H_{23})^{-1} = f_{23}^{-2}.$$

If ρ is fuchsian, then both f_{01} and f_{23} are purely hyperbolic transformations and hence the above equality implies $f_{01} = f_{23}^{-1}$ (see [35, Proof of Proposition 1.1]). To show that the same identity holds for the general case, we need the following lemma.

Lemma 2.2.4. *There is a continuous path* $\{\rho_t\}_{t\in[0,1]}$ *in* $\mathrm{Hom}_{\mathrm{tp}}(\pi_1(S), PSL(2, \mathbb{C}))$ *satisfying the following conditions.*

1. $\rho_1 = \rho$ *and* ρ_0 *is fuchsian.*
2. $\mathrm{Fix}(\rho_t(K_0)) \neq \mathrm{Fix}(\rho_t(K_1))$ *and hence the transformations* $f_{01}^{(t)}$ *and* $f_{23}^{(t)}$ *are defined as in the above for every for every* $t \in [0,1]$.
3. $(f_{01}^{(t)})^2 = (f_{23}^{(t)})^{-2}$ *is loxodromic except possibly at* $t = 1$.

Proof. Pick a fuchsian representation $\rho' \in \mathrm{Hom}_{\mathrm{tp}}(\pi_1(S), PSL(2,\mathbb{C}))$. We show that there is a continuous path $\{\rho_t\}_{t\in[0,1]}$ in $\mathrm{Hom}_{\mathrm{tp}}(\pi_1(S), PSL(2,\mathbb{C}))$ satisfying the required conditions with $\rho_0 = \rho'$. To this end, set $g_i' = \rho'(K_i)$, and let L_i', L_{01}' and L_{23}' be the geodesics in $\mathbb{H}^3$ determined by g_i' $(0 \le i \le 3)$ as in the preceding paragraph. Then it is obvious that there is a continuous family of triples of mutually distinct geodesics $\{(L_{01}^{(t)}, L_0^{(t)}, L_1^{(t)})\}_{t\in[0,1]}$ connecting (L_{01}', L_0', L_1') with (L_{01}, L_0, L_1) which satisfy the following conditions.

1. $L_{01}^{(t)}$ shares an endpoint with each of $L_0^{(t)}$ and $L_1^{(t)}$.
2. $\bar{L}_0^{(t)}$ and $\bar{L}_1^{(t)}$ are disjoint, except possibly at $t = 1$.
3. Let $H_{01}^{(t)}$ be the geodesic as in the preceding paragraph, and let $G_{01}^{(t)}$ be the axis of the π-rotation $L_{01}^{(t)} H_{01}^{(t)}$. Then $\bar{L}_0^{(t)}$ and $\bar{G}_{01}^{(t)}$ are disjoint, except possibly at $t = 1$.

Set $g_0^{(t)} := L_0^{(t)} L_{01}^{(t)}$ and $g_1^{(t)} := L_{01}^{(t)} L_1^{(t)}$. Then $g_0^{(t)}$ and $g_1^{(t)}$ are parabolic, and $g_0^{(t)} g_1^{(t)} = L_0^{(t)} L_1^{(t)}$ is loxodromic except possibly at $t = 1$. Moreover, $f_{01}^{(t)} := g_0^{(t)} H_{01}^{(t)} = L_0^{(t)} G_{01}^{(t)}$ is loxodromic except possibly at $t = 1$.

Pick a continuous family of geodesics $\{L_2^{(t)}\}_{t\in[0,1]}$ connecting L_2' with L_2, such that $L_2^{(t)}$ intersects the axis of the loxodromic transformation $g_0^{(t)} g_1^{(t)}$ orthogonally for each $t \in [0,1)$. Then there is a continuous family of geodesics $\{L_3^{(t)}\}_{t\in[0,1]}$ connecting L_3' with L_3, such that $L_3^{(t)} L_2^{(t)} = g_0^{(t)} g_1^{(t)}$ for every $t \in [0,1]$. Let $\{L_{23}^{(t)}\}_{t\in[0,1]}$ be a continuous family of geodesics connecting L_{23}' with L_{23}, such that $L_{23}^{(t)}$ shares an endpoint with each of $L_2^{(t)}$ and $L_3^{(t)}$. Set $g_2^{(t)} = L_2^{(t)} L_{23}^{(t)}$ and $g_3^{(t)} = L_{23}^{(t)} L_3^{(t)}$. Then there is a representation $\rho_t : \pi_1(S) \to PSL(2,\mathbb{C})$ such that $\rho_t(K_i) = g_i^{(t)}$ $(0 \le i \le 3)$, and $\{\rho_t\}_{t\in[0,1]}$ gives the desired path in $\mathrm{Hom}_{\mathrm{tp}}(\pi_1(S), PSL(2,\mathbb{C}))$.

Let $\{\rho_t\}_{t\in[0,1]}$ be a continuous path in Lemma 2.2.4. By the last condition in the lemma, the transformation $(f_{01}^{(t)})^2$, with $t \in [0,1)$, is loxodromic and hence it has precisely two square roots which differ by the composition of π-rotation about its axis. Since $f_{01}^{(t)} = f_{23}^{(t)}$ at $t = 0$, we have $f_{01}^{(t)} = f_{23}^{(t)}$ for every $t \in [0,1]$ by continuity. Hence we have $f_{01} = f_{23}$.

Since H_{01} is contained in the normalizer of the group $\langle g_0, g_1 \rangle$, so is f_{01}. Similarly f_{23} is contained in the normalizer of $\langle g_2, g_3 \rangle$. Hence $f_{01} = f_{23}^{-1}$ is contained in the normalizer of $\langle g_0, g_1, g_2, g_3 \rangle$, and so is H_{01}. In fact the action of H_{01} on $\langle g_0, g_1, g_2, g_3 \rangle$ by conjugation is given by

$$(g_0, g_1, g_2, g_3) \mapsto (g_1, g_0, g_0^{-1} g_3 g_0, g_0^{-1} g_3^{-1} g_2 g_3 g_0).$$

On the other hand, the action of P on $\pi_1(S)$ by conjugation is given by

$$(K_0, K_1, K_2, K_3) \mapsto (K_1, K_0, K_0^{-1} K_3 K_0, K_0^{-1} K_3^{-1} K_2 K_3 K_0).$$

Hence the representation ρ of $\pi_1(S)$ extends to a representation of the subgroup $\langle P, K_0, K_1, K_2, K_3 \rangle$ which maps P to H_{01}.

Suppose that $\mathrm{Fix}(g_i)$ $(0 \le i \le 3)$ are mutually distinct. Then by the above argument, we can find a π-rotation H_{02} (resp. H_{03}) such that the representation ρ of $\pi_1(S)$ extends to a representation of the subgroup $\langle Q, K_0, K_1, K_2, K_3 \rangle$ (resp. $\langle R, K_0, K_1, K_2, K_3 \rangle$) which maps Q to H_{02} (resp. R to $H_{0,3}$). This can be done by applying the argument to the generator system $(K_0, K_2, K_3, K_3^{-1} K_2^{-1} K_1, K_2, K_3)$ (resp. (K_3, K_0, K_1, K_2)). Since $\pi_1(\mathcal{O})$ is the free product of three cyclic groups $\langle P \rangle$, $\langle Q \rangle$ and $\langle R \rangle$, we have a representation $\rho^* : \pi_1(\mathcal{O}) \to PSL(2,\mathbb{C})$ such that ρ^* maps the triple (P, Q, R) to (H_{01}, H_{02}, H_{03}). To show that the restriction of ρ^* to $\pi_1(S)$ is equal to the original representation ρ, pick an element W of $\pi_1(S)$. Then by the choice of H_{01}, H_{02} and H_{03}, $\rho^*(W)$ normalizes $\rho(\pi_1(S))$ and the action of $\rho^*(W)$ on $\rho(\pi_1(S))$ by conjugation is equal to that of $\rho(W)$. On the other hand, since ρ is irreducible there are infinitely many points in $\partial\mathbb{H}^3$ which are fixed by elements of $\rho(\pi_1(S))$. Hence we have $\rho^*(W) = \rho(W)$. Thus ρ^* is an extension of ρ, and this completes the proof of Proposition 2.2.2(2) for the generic case.

Suppose $\mathrm{Fix}(g_i) = \mathrm{Fix}(g_j)$ for some $0 \le i < j \le 3$. By changing the generator system of $\rho(\pi_1(S))$ if necessary, we may assume $\mathrm{Fix}(g_0) = \mathrm{Fix}(g_1)$. Since ρ is type-preserving, we have $g_0 = g_1^{-1}$ and hence $g_2 = g_3^{-1}$. Since ρ is irreducible, $\mathrm{Fix}(g_0) \ne \mathrm{Fix}(g_2)$. Let H be the geodesic joining $\mathrm{Fix}(g_0)$ and $\mathrm{Fix}(g_2)$. Then $H g_0 H = g_0^{-1} = g_1$ and $H g_2 H = g_2^{-1} = g_3$. Set $H_{01} = g_0^{-1} H$. Then H_{01} is an elliptic transformation of order 2 and we can easily check

$$H_{01} g_0 H_{01} = g_1, \quad H_{01} g_2 H_{01} = g_0^{-1} g_3 g_0.$$

Hence the action of H_{01} on $\langle g_0, g_1, g_2, g_3 \rangle$ by conjugation is compatible with that of P on $\pi_1(S)$. Moreover, since $\mathrm{Fix}(g_0) \ne \mathrm{Fix}(g_2) = \mathrm{Fix}(g_3)$, the elements H_{02} and H_{03} are defined. By the argument for the generic case, we see that the representation $\rho^* : \pi_1(\mathcal{O}) \to PSL(2,\mathbb{C})$ which maps the triple (P, Q, R) to (H_{01}, H_{02}, H_{03}) is the desired extension of ρ. This completes the proof of Proposition 2.2.2.

Remark 2.2.5. In Definition 2.2.1(2) of a type-preserving $PSL(2,\mathbb{C})$-representation of $\pi_1(S)$, the last condition is essential. To see this, let $\rho : \pi_1(S) \to PSL(2,\mathbb{C})$ be a representation defined by

$$\rho(K_0) = \begin{pmatrix} 1 & \omega_0 \\ 0 & 1 \end{pmatrix}, \qquad \rho(K_1) = \begin{pmatrix} 1 & \omega_1 \\ 0 & 1 \end{pmatrix},$$
$$\rho(K_2) = \begin{pmatrix} 1 & 0 \\ -4/(\omega_0+\omega_1) & 1 \end{pmatrix}, \quad \rho(K_3) = \begin{pmatrix} 1 & -\omega_0-\omega_1 \\ 4/(\omega_0+\omega_1) & -3 \end{pmatrix},$$

where ω_0 and ω_1 are non-zero complex numbers such that $\omega_0 + \omega_1 \neq 0$. Then ρ does not satisfy the last condition, though it is irreducible and sends peripheral elements to parabolic transformations. Thus ρ does not extend to a representation of $\pi_1(\mathcal{O})$ by Proposition 2.2.2.

By Proposition 2.2.2, the following is well-defined.

Definition 2.2.6. *The space of the equivalence classes of type-preserving $PSL(2,\mathbb{C})$-representations of $\pi_1(X)$ where $X = T$, $\mathcal{O}$ or S is denoted by $\mathcal{X}$. Namely,*

$$\mathcal{X} = \mathrm{Hom}_{\mathrm{tp}}(\pi_1(X), PSL(2,\mathbb{C}))/PSL(2,\mathbb{C}).$$

Throughout this paper, we employ the following convention.

Convention 2.2.7. (1) We do not distinguish between an element of $\mathcal{X}$ and its representative: they are denoted by the same symbol so long as there is no fear of confusion.

(2) When we choose a representative $\rho : \pi_1(\mathcal{O}) \to PSL(2,\mathbb{C})$ of an element of $\mathcal{X}$, we always assume that ρ is *normalized* so that the following identity is satisfied.

$$\rho(K) = \begin{pmatrix} 1 & 1 \\ 0 & 1 \end{pmatrix}$$

(3) Let $\{\rho_n\}$ be a sequence in $\mathcal{X}$ which converges to ρ_∞ in $\mathcal{X}$. Then it is well-known that there are normalized representatives $\rho_n : \pi_1(\mathcal{O}) \to PSL(2,\mathbb{C})$ which converge to a representative $\rho_\infty : \pi_1(\mathcal{O}) \to PSL(2,\mathbb{C})$ in $\mathrm{Hom}_{\mathrm{tp}}(\pi_1(X), PSL(2,\mathbb{C}))$ (cf. Corollary 2.3.8). We always assume that representatives of the elements ρ_n and ρ_∞ of $\mathcal{X}$ satisfy this condition.

At the end of this section, we note the following basic fact, which reduces the study of Ford domains of punctured torus groups to that of Ford domains of the Kleinian groups obtained as the images of $\pi_1(\mathcal{O})$ by type-preserving representations.

Proposition 2.2.8. *Let $\rho : \pi_1(\mathcal{O}) \to PSL(2,\mathbb{C})$ be a type-preserving representation which is discrete. Then, under Convention 2.2.7, the following hold.*

1. *The Ford domain $Ph(\rho(\pi_1(\mathcal{O})))$ of the Kleinian group $\rho(\pi_1(\mathcal{O}))$ is equal to the Ford domain $Ph(\rho(\pi_1(T)))$ of the Kleinian subgroup $\rho(\pi_1(T))$.*

2. The Ford complex $\mathrm{Ford}(\rho(\pi_1(\mathcal{O}))) \subset \bar{M}(\rho(\pi_1(\mathcal{O})))$ *is equal to the image of* $\mathrm{Ford}(\rho(\pi_1(T))) \subset \bar{M}(\rho(\pi_1(T)))$ *in* $\bar{M}(\rho(\pi_1(\mathcal{O})))$.
3. $\widetilde{\Delta}_{\mathbb{E}}(\rho(\pi_1(T))) = \widetilde{\Delta}_{\mathbb{E}}(\rho(\pi_1(\mathcal{O})))$. *Thus* $\Delta_{\mathbb{E}}(\rho(\pi_1(\mathcal{O})))$ *is equal to the image of* $\Delta_{\mathbb{E}}(\rho(\pi_1(T)))$ *in in* $M(\rho(\pi_1(\mathcal{O})))$.

Proof. To show the first assertion, we have only to show that, for any element $\gamma \in \pi_1(\mathcal{O})$ such that $\rho(\gamma)$ does not fix ∞, $Ih(\rho(\gamma)) = Ih(\rho(\gamma_0))$ for some $\gamma_0 \in \pi_1(T)$. If $\gamma \in \pi_1(T)$, this is obvious. So assume that $\gamma \in \pi_1(\mathcal{O}) - \pi_1(T)$. Then $\gamma = K\gamma_0$ for some $\gamma_0 \in \pi_1(T)$. Thus we have $Ih(\rho(\gamma)) = Ih(\rho(K)\rho(\gamma_0)) = Ih(\rho(\gamma_0))$ (cf. Lemma 4.1.1(2)). Hence we obtain the first assertion. The second assertion is a consequence of the first one.

Definition 2.2.9. *Under Convention 2.2.7, suppose* ρ *is discrete. Then* $Ph(\rho)$ *denotes the Ford domain* $Ph(\rho(\pi_1(T))) = Ph(\rho(\pi_1(\mathcal{O})))$. *If there is no fear of confusion, then we occasionally denote* $\mathrm{Ford}(\rho(\pi_1(\mathcal{O})))$ *or* $\mathrm{Ford}(\rho(\pi_1(\mathcal{O})))$ *by the symbol* $\mathrm{Ford}(\rho)$. *We also occasionally denote* $\Delta_{\mathbb{E}}(\rho(\pi_1(\mathcal{O})))$ *or* $\Delta_{\mathbb{E}}(\rho(\pi_1(T)))$ *by the symbol* $\Delta_{\mathbb{E}}(\rho)$.

Remark 2.2.10. The Ford domain of the Kleinian group $\rho(\pi_1(\mathcal{O}))$ is not necessarily equal to the Ford domain of $\rho(\pi_1(S))$.

2.3 Markoff maps and type-preserving representations

In this section, we give a description of the type-preserving $PSL(2,\mathbb{C})$-representations of $\pi_1(\mathcal{O})$ following [17] and [40]. We begin with the following lemma.

Lemma 2.3.1. *(1) Every type-preserving representation* $\rho : \pi_1(T) \to PSL(2,\mathbb{C})$ *lifts to a representation* $\tilde{\rho} : \pi_1(T) \to SL(2,\mathbb{C})$ *such that* $\mathrm{tr}(\tilde{\rho}(K^2)) = -2$. *Moreover, there are precisely four such lifts for each* ρ.

(2) Conversely, every representation $\tilde{\rho} : \pi_1(T) \to SL(2,\mathbb{C})$ *such that* $\mathrm{tr}(\tilde{\rho}(K^2)) = -2$ *and* $\tilde{\rho}(K^2) \neq -I$ *descends to a type-preserving* $PSL(2,\mathbb{C})$*-representation of* $\pi_1(T)$.

Proof. Since $\pi_1(T)$ is a rank 2 free group, every $PSL(2,\mathbb{C})$-representation of $\pi_1(T)$ has precisely four lifts. Moreover we can easily see that a representation $\tilde{\rho} : \pi_1(T) \to SL(2,\mathbb{C})$ is irreducible if and only if $\mathrm{tr}(\tilde{\rho}(K^2)) = -2$ and $\tilde{\rho}(K^2) \neq -I$. Hence we obtain the desired results.

Definition 2.3.2 (Type-preserving $SL(2,\mathbb{C})$-representation).

(1) A representation $\tilde{\rho} : \pi_1(T) \to SL(2,\mathbb{C})$ *is* type-preserving *if it is irreducible and sends peripheral elements to parabolic transformations.*

(2) The space of the equivalence classes of type-preserving $SL(2,\mathbb{C})$*-representations of* $\pi_1(T)$ *is denoted by* $\tilde{\mathcal{X}}$. *Namely,*

$$\tilde{\mathcal{X}} = \mathrm{Hom}_{\mathrm{tp}}(\pi_1(T), SL(2,\mathbb{C}))/SL(2,\mathbb{C}),$$

where $\mathrm{Hom}_{\mathrm{tp}}(\pi_1(X), SL(2,\mathbb{C}))$ *denotes the space of all type-preserving* $SL(2,\mathbb{C})$*-representations of* $\pi_1(T)$.

Following Bowditch [17], we introduce the following notion.

Definition 2.3.3 (Markoff map associated with a type-preserving representation). *For a type-preserving* $SL(2,\mathbb{C})$*-representation* $\tilde{\rho} : \pi_1(T) \to SL(2,\mathbb{C})$*, let* $\phi = \phi_{\tilde{\rho}}$ *be the map from* $\mathcal{D}^{(0)} = \hat{\mathbb{Q}}$ *to* $\mathbb{C}$ *define by* $\phi(r) = \mathrm{tr}(\tilde{\rho}(\alpha_r))$*, where* α_r *is an element of* $\pi_1(T)$ *represented by a simple loop of slope* r*. We call it the* Markoff map *associated with* $\tilde{\rho}$.

The following proposition describes the genesis of the above terminology.

Proposition 2.3.4. *For any type-preserving* $SL(2,\mathbb{C})$*-representation* $\tilde{\rho}$ *of* $\pi_1(T)$*, the corresponding Markoff map* $\phi = \phi_{\tilde{\rho}} : \mathcal{D}^{(0)} \to \mathbb{C}$ *satisfies the following conditions:*

(1) For any triangle $\langle s_0, s_1, s_2\rangle$ *of* $\mathcal{D}$*, the triple* $(\phi(s_0), \phi(s_1), \phi(s_2))$ *is a Markoff triple, that is, it is a solution of the Markoff equation,*

$$x^2 + y^2 + z^2 = xyz.$$

(2) For any pair of triangles $\langle s_0, s_1, s_2\rangle$ *and* $\langle s_1, s_2, s_3\rangle$ *of* $\mathcal{D}$ *sharing a common edge* $\langle s_1, s_2\rangle$*, we have:*

$$\phi(s_0) + \phi(s_3) = \phi(s_1)\phi(s_2).$$

Proof. This follows from the following well-known trace identities for matrices X and Y in $SL(2,\mathbb{C})$:

$$\mathrm{tr}(X)^2 + \mathrm{tr}(Y)^2 + \mathrm{tr}(XY)^2 - \mathrm{tr}(X)\,\mathrm{tr}(Y)\,\mathrm{tr}(XY) = 2 + \mathrm{tr}([X,Y]),$$

$$\mathrm{tr}(XY) + \mathrm{tr}(XY^{-1}) = \mathrm{tr}(X)\,\mathrm{tr}(Y).$$

Definition 2.3.5. *(1) By a* Markoff map, *we mean a map* $\phi : \mathcal{D}^{(0)} \to \mathbb{C}$ *satisfying the conclusions of the above lemma.*

(2) The trivial Markoff map *is the Markoff map which sends everything to* 0.

(3) Φ *denotes the space of all non-trivial Markoff maps.*

Note that there is a bijective correspondence between Markoff maps and Markoff triples, by fixing a triangle $\langle s_0, s_1, s_2\rangle$ of $\mathcal{D}$, and by associating to a Markoff map ϕ the triple $(\phi(s_0), \phi(s_1), \phi(s_2))$. This gives an identification of Φ with the variety in $\mathbb{C}^3$ (with 0 deleted) determined by the Markoff equation $x^2 + y^2 + z^2 = xyz$. In particular, Φ gets a topology as a subset of $\mathbb{C}^3$. The following proposition is proved by [17, Sect. 4] and [40, Sect. 2].

Proposition 2.3.6. *The correspondence* $\tilde{\rho} \mapsto \phi_{\tilde{\rho}}$ *induces a homeomorphism from* $\tilde{\mathcal{X}}$ *to* Φ.

The following lemma, obtained by Jorgensen [40], gives a (local) cross-section of the above correspondence:

Lemma 2.3.7. *Let ϕ be a Markoff map, and put $(x,y,z) = (\phi(0/1), \phi(1/1), \phi(1/0))$. Assume that $y \neq 0$.*

(1) Let $\tilde{\rho} : \pi_1(T) \to SL(2,\mathbb{C})$ be a representation determined by the following formula:

$$\tilde{\rho}(A_0) = \begin{pmatrix} x - z/y & x/y^2 \\ x & z/y \end{pmatrix}, \quad \tilde{\rho}(A_0B_0) = \begin{pmatrix} y & -1/y \\ y & 0 \end{pmatrix},$$
$$\tilde{\rho}(B_0) = \begin{pmatrix} z - x/y & -z/y^2 \\ -z & x/y \end{pmatrix}.$$

Then $\tilde{\rho}$ is type-preserving and the Markoff map $\phi_{\tilde{\rho}}$ associated with $\tilde{\rho}$ is equal to ϕ.

(2) Let ρ be the $PSL(2,\mathbb{C})$-representation of $\pi_1(T)$ induced by the above $\tilde{\rho}$. Then it extends to a type-preserving $PSL(2,\mathbb{C})$-representation of $\pi_1(\mathcal{O})$ satisfying the following identities:

$$\rho(P_0) = \begin{pmatrix} z/y & (yz - x)/y^2 \\ -x & -z/y \end{pmatrix}, \quad \rho(Q_0) = \begin{pmatrix} 0 & -1/y \\ y & 0 \end{pmatrix},$$
$$\rho(R_0) = \begin{pmatrix} -x/y & (xy - z)/y^2 \\ -z & x/y \end{pmatrix}, \quad \rho(K) = \begin{pmatrix} 1 & 1 \\ 0 & 1 \end{pmatrix}.$$

(3) The restriction of the above $PSL(2,\mathbb{C})$-representation ρ to $\pi_1(S)$ satisfies the following identities:

$$\rho(K_0) = \begin{pmatrix} 1 & 1 \\ 0 & 1 \end{pmatrix}, \quad \rho(K_1) = \begin{pmatrix} 1 + xz/y & z^2/y^2 \\ -x^2 & 1 - xz/y \end{pmatrix},$$
$$\rho(K_2) = \begin{pmatrix} 1 & 0 \\ -y^2 & 1 \end{pmatrix}, \quad \rho(K_3) = \begin{pmatrix} 1 - xz/y & x^2/y^2 \\ -z^2 & 1 + xz/y \end{pmatrix}.$$

Proof. This lemma can be verified by direct calculation. For recipes for finding the cross section, see [70, Sect. 3] and [80, Sect. 1]

By using the above lemma, we can prove the following well-known fact.

Corollary 2.3.8. *Let $\{\rho_n\}$ be a sequence in $\mathcal{X}$ which converges to ρ_∞ in $\mathcal{X}$. Then we can find normalized representatives $\rho_n : \pi_1(\mathcal{O}) \to PSL(2,\mathbb{C})$ which converges to a representative $\rho_\infty : \pi_1(\mathcal{O}) \to PSL(2,\mathbb{C})$ in $\mathrm{Hom}_{\mathrm{tp}}(\pi_1(X), PSL(2,\mathbb{C}))$.*

Proof. Let ϕ_∞ (resp. ϕ_n) be the Markoff map inducing ρ_∞ (resp. ρ_n). Suppose $\phi_\infty(1/1) \neq 0$. Then $\phi_n(1/1) \neq 0$ for all sufficiently large n, and the representations constructed by Lemma 2.3.7 from ϕ_n and ϕ_∞ satisfy the desired property. In the general case, pick an elliptic generator triple (P, Q, R)

such that $\phi_\infty(s(Q)) \neq 0$. Then we can construct representatives as in Lemma 2.3.7, by using (P, Q, R) instead of (P_0, Q_0, R_0), for each $\rho \in \mathcal{X}$ induced by a Markoff map ϕ such that $\phi(s(Q)) \neq 0$. By using this fact we obtain the desired result through the above argument.

2.4 Markoff maps and complex probability maps

Let ϕ be a Markoff map with $(x, y, z) = (\phi(0/1), \phi(1/1), \phi(1/0))$. We give a geometric description of the representation $\rho = \rho_\phi : \pi_1(\mathcal{O}) \to PSL(2, \mathbb{C})$ constructed from ϕ in Lemma 2.3.7.

Case 1. $xyz \neq 0$. Note that the Markoff identity $x^2 + y^2 + z^2 = xyz$ implies the identity,

$$a_0 + a_1 + a_2 = 1, \quad \text{where} \quad a_0 = \frac{x}{yz}, \quad a_1 = \frac{y}{zx}, \quad a_2 = \frac{z}{xy}. \tag{2.4}$$

Following Jorgensen, we call the triple (a_0, a_1, a_2) the *complex probability* associated with the Markoff triple (x, y, z). We note that the Markoff triple up to sign is recovered from the complex probability, because:

$$x^2 = \frac{1}{a_1 a_2}, \quad y^2 = \frac{1}{a_2 a_0}, \quad z^2 = \frac{1}{a_0 a_1}. \tag{2.5}$$

By the formula in Lemma 2.3.7, we have the following (see Fig. 2.2):

$$\begin{aligned}
\rho(P_0) &= \text{the } \pi\text{-rotation about the geodesic with endpoints } -a_0 \pm i/x,\\
\rho(Q_0) &= \text{the } \pi\text{-rotation about the geodesic with endpoints } 0 \pm i/y,\\
\rho(R_0) &= \text{the } \pi\text{-rotation about the geodesic with endpoints } a_1 \pm i/z.
\end{aligned}$$

The images by ρ of the members in the sequence of elliptic generators $\{P_j\}$ associated with $\langle 0/1, 1/1, 1/0 \rangle$, such that $(P_0, Q_0, R_0) = (P_0, P_1, P_2)$, are given as follows. By repeatedly drawing the vectors in the complex probability (a_0, a_1, a_2) in this cyclic order, we obtain a bi-infinite, possibly singular, broken line $\mathcal{L}$ on the complex plane $\mathbb{C}$ with vertices $\{c_j\}$ such that $c_{j+1} - c_j = a_{[j-1]}$ and $(c_0, c_1, c_2) = (-a_0, 0, a_1)$. Draw the vector i/x, i/y, or i/z from c_j according as j is congruent to 0, 1, or 2 modulo 3. Note that the vector emanating from c_j is equal to $\sqrt{(-a_{j+1})(a_{j-1})}$, and hence it bisects the angle between the edges of $\mathcal{L}$ incident on c_j and its length is equal to the multiplicative mean of the lengths of the two edges. Then $\rho(P_j)$ is the π-rotation about the geodesic with endpoints $c_j \pm i/x$, $c_j \pm i/y$, or $c_j \pm i/z$ according as $j \equiv 0$, 1, or 2 $\pmod 3$. In particular, the isometric circle $I(\rho(P_j))$ has center c_j and radius $1/|x|$, $1/|y|$ or $1/|z|$ accordingly. We can easily check, by using only elementary Euclidean geometry, that the product $\rho(K) = \rho(P_{j+1})\rho(P_j)\rho(P_{j-1})$ is the translation $(z, t) \mapsto (z + 1, t)$. This gives a geometric description of Jorgensen's cross section in Lemma 2.3.7. We note that $\rho(P_{j+3k})$ is the conjugate of $\rho(P_j)$ by the Euclidean isometry $\rho(K)^k = \begin{pmatrix} 1 & k \\ 0 & 1 \end{pmatrix}$. Thus the isometric circle $I(\rho(P_{j+3k}))$ is equal to the image of the isometric circle $I(\rho(P_j))$ by the translation $z \mapsto z+k$.

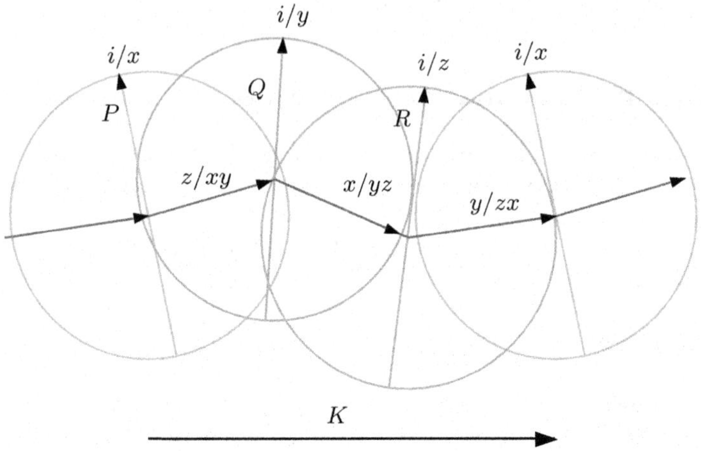

Fig. 2.2. Isometric circles of elliptic generators for the general case $xyz \neq 0$

Case 2. $xyz = 0$. We assume without losing generality that $x = 0$ and $z = iy$. Then we have the following (see Fig. 2.3):

$$\begin{aligned}\rho(P_0) &= \text{the } \pi\text{-rotation about the vertical geodesic above } -1/2,\\ \rho(Q_0) &= \text{the } \pi\text{-rotation about the geodesic with endpoints } \pm i/y,\\ \rho(R_0) &= \text{the } \pi\text{-rotation about the geodesic with endpoints } \pm i/z.\end{aligned}$$

Here a *vertical geodesic* above a point $z \in \mathbb{C}$ means the geodesic $\{z\} \times \mathbb{R}_+$ in $\mathbb{H}^3 = \mathbb{C} \times \mathbb{R}_+$. The axes of $\rho(Q_0)$ and $\rho(R_0)$ intersect orthogonally and $\rho(R_0Q_0)$ is the π-rotation about the vertical geodesic above 0. We see that $\rho(K) = \rho(R_0Q_0P_0)$ is the parabolic transformation $(z,t) \to (z+1,t)$.

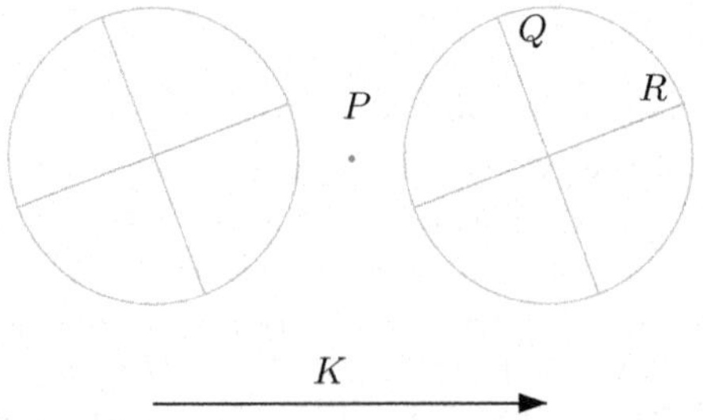

Fig. 2.3. Isometric circles of elliptic generators for the special case $x = 0$

In order to refine the above geometric description, we introduce the concept of a complex probability map following [17] and [40]. Let $\mathcal{T}$ be a binary tree (a countably infinite simplicial tree all of whose vertices have degree 3) properly embedded in $\mathbb{H}^2$ dual to $\mathcal{D}$. A *directed edge*, $\vec{e}$, of $\mathcal{T}$ can be thought of an ordered pair of adjacent vertices of $\mathcal{T}$, referred to as the *head* and *tail* of $\vec{e}$. We introduce the notation $\vec{e} \leftrightarrow (s_1, s_2; s_0, s_3)$ to mean that s_0, s_1, s_2 and s_3 are the ideal vertices of $\mathcal{D}$ such that (1) $\langle s_1, s_2 \rangle$ is the dual to $\vec{e}$ and that (2) $\langle s_0, s_1, s_2 \rangle$ (resp. $\langle s_1, s_2, s_3 \rangle$) is dual to the head (resp. tail) of $\vec{e}$. Let $\vec{E}(\mathcal{T})$ be the set of directed edges of $\mathcal{T}$. For a Markoff map ϕ, let $\vec{E}_\phi(\mathcal{T})$ be the subset of $\vec{E}(\mathcal{T})$ consisting of those directed edges $\vec{e}$ such that if $\vec{e} \leftrightarrow (s_1, s_2; s_0, s_3)$ then $\phi(s_1)\phi(s_2) \neq 0$. Then we define a map $\psi = \psi_\phi : \vec{E}_\phi(\mathcal{T}) \to \mathbb{C}$ by

$$\psi(\vec{e}) = \frac{\phi(s_0)}{\phi(s_1)\phi(s_2)}. \tag{2.6}$$

We call $\psi = \psi_\phi$ the *complex probability map* associated with the Markoff map ϕ. We extend ψ to a map $\psi : \vec{E}(\mathcal{T}) \to \widehat{\mathbb{C}}$ by setting $\psi(\vec{e}) = \infty$ if $\phi(s_1)\phi(s_2) = 0$. (Note that (i) if $\phi(s_1)\phi(s_2) = 0$ then $\phi(s_0) \neq 0$ and that (ii) if $\phi(s_0) = 0$ then $\phi(s_1)\phi(s_2) \neq 0$.) An ordered triple $(\vec{e}_0, \vec{e}_1, \vec{e}_2)$ of elements of $\vec{E}(\mathcal{T})$ is said to be *dual* to an oriented simplex $\sigma = \langle s_0, s_1, s_2 \rangle$ if $\vec{e}_j$ $(0 \le j \le 2)$ is dual to $\langle s_{j+1}, s_{j+2} \rangle$ (the indices are considered modulo 3) and has the vertex of $\mathcal{T}$ dual to σ as the head. Then we have the following:

Lemma 2.4.1. *(1) Let σ be an oriented triangle of $\mathcal{D}$ and $(\vec{e}_0, \vec{e}_1, \vec{e}_2)$ a triple of elements of $\vec{E}_\phi(\mathcal{T})$ dual to σ. Then:*

$$\psi(\vec{e}_0) + \psi(\vec{e}_1) + \psi(\vec{e}_2) = 1.$$

We call the ordered triple $(\psi(\vec{e}_0), \psi(\vec{e}_1), \psi(\vec{e}_2))$ the complex probability of ϕ (or the value of ψ) at σ.

(2) For each $\vec{e} \in \vec{E}_\phi(\mathcal{T})$, let $-\vec{e}$ be the element of $\vec{E}_\phi(\mathcal{T})$ obtained from $\vec{e}$ by reversing the direction. Then

$$\psi(\vec{e}) + \psi(-\vec{e}) = 1.$$

(3) Under Notation 2.1.14 (Adjacent triangles), let $(\vec{e}_0, \vec{e}_1, \vec{e}_2)$ and $(\vec{e}'_1, \vec{e}'_2, \vec{e}'_3)$, respectively, be cyclically ordered triples of elements of $\vec{E}_\phi(\mathcal{T})$ dual to $\sigma = \langle s_0, s_1, s_2 \rangle$ and $\sigma' = \langle s'_0, s'_1, s'_2 \rangle$. Then we have the following:

$$\psi(\vec{e}'_0) = \frac{\psi(\vec{e}_0)\psi(\vec{e}_1)}{1 - \psi(\vec{e}_1)}, \quad \psi(\vec{e}'_1) = \frac{\psi(\vec{e}_1)\psi(\vec{e}_2)}{1 - \psi(\vec{e}_1)},$$
$$\psi(\vec{e}'_2) = 1 - \psi(\vec{e}_1).$$

Proof. This can be proved by direct calculation (cf. [17]).

Let $\varPsi$ be the set of complex probability maps. Then we can put a topology on $\varPsi$ so that the map $\phi \mapsto \psi$ is a four-fold covering map $\varPhi \to \varPsi$. Indeed, the transformations $(x,y,z) \to (-x,-y,z)$ and $(x,y,z) \to (x,-y,-z)$ generate a free $\mathbb{Z}_2 \oplus \mathbb{Z}_2$-action on the space of Markoff triples and that on the space $\varPhi$: the quotient of $\varPhi$ by this action is identified with $\varPsi$. Thus we have the following proposition:

Proposition 2.4.2. *We have the following commutative diagram of $\mathbb{Z}_2 \oplus \mathbb{Z}_2$-coverings:*

$$\begin{array}{ccc} \varPhi & \xrightarrow{\cong} & \tilde{\mathcal{X}} \\ \downarrow & & \downarrow \\ \varPsi & \xrightarrow{\cong} & \mathcal{X} \end{array}$$

Convention 2.4.3. (1) When we give mention to one of the elements $\phi \in \varPhi$, $\psi \in \varPsi$, $\tilde{\rho} \in \tilde{\mathcal{X}}$ and $\rho \in \mathcal{X}$, the remaining symbols represent elements related to the mentioned one by the above commutative diagram so long as this does not create confusion. When we need to be precise, we use symbols ρ_ϕ, ψ_ρ and ϕ_ρ (even though ϕ is not uniquely determined by ρ).

(2) We do not distinguish between an element of $\tilde{\mathcal{X}}$ (resp. $\mathcal{X}$) and its representative: they are denoted by the same symbol so long as there is no fear of confusion. Moreover, When we choose a representative $\tilde{\rho} : \pi_1(T) \to SL(2,\mathbb{C})$ or $\rho : \pi_1(\mathcal{O}) \to PSL(2,\mathbb{C})$, we always assume that it is *normalized* so that the following identity is satisfied.

$$\tilde{\rho}(K^2) = \begin{pmatrix} -1 & -2 \\ 0 & -1 \end{pmatrix}, \qquad \rho(K) = \begin{pmatrix} 1 & 1 \\ 0 & 1 \end{pmatrix}.$$

The geometric description of a (normalized) type-preserving representation in the above is summarized as follows.

Proposition 2.4.4. *Let ϕ be a nontrivial Markoff map, and $\rho : \pi_1(\mathcal{O}) \to SL(2,\mathbb{C})$ the (normalized) type-preserving representation induced by $\tilde{\rho}$.*

(1) Let P be an elliptic generator of slope s.

(1.1) Suppose $\phi(s) \neq 0$. Then $\rho(P)$ is the π-rotation about the geodesic with endpoints $c(\rho(P)) \pm i/\phi(s)$, where $c(\rho(P)) := \rho(P)^{-1}(\infty)$ is the center of the isometric circle $I(\rho(P))$. In particular, the radius, $r(\rho(P))$, of $I(\rho(P))$ is equal to $|1/\phi(s)|$.

(1.2) Suppose $\phi(s) = 0$. Then $\rho(P)$ is the π-rotation about a vertical geodesic. In particular, $\rho(P)$ stabilizes ∞ and acts on $\mathbb{H}^3$ as a Euclidean isometry.

(2) Let $\sigma = \langle s_0, s_1, s_2\rangle$ be a triangle of $\mathcal{D}$ and let $\{P_j\}$ be the sequence of elliptic generators associated with σ. Then the following hold:

(2.1) Suppose that $\phi(s_0)\phi(s_1)\phi(s_2) \neq 0$. Then

$$c(\rho(P_{j+1})) - c(\rho(P_j)) = \frac{\phi(s_{[j-1]})}{\phi(s_{[j]})\phi(s_{[j+1]})} = \psi(\overrightarrow{e}_{[j-1]}),$$

where $(\vec{e}_0, \vec{e}_1, \vec{e}_2)$ *is the triple of elements of* $\vec{E}(\mathcal{T})$ *dual to* σ. *Moreover, the open line segment* $\mathrm{proj}(\mathrm{Axis}(\rho(P_j)))$ *bisects the angle* $\angle(c(\rho(P_{j-1})), c(\rho(P_j)), c(\rho(P_{j+1})))$, *where* $\mathrm{proj} : \mathbb{H}^3 \to \mathbb{C}$ *denotes the projection. In other words,*

$$\arg \frac{i/\phi(s_{[j]})}{c(\rho(P_{j+1})) - c(\rho(P_j))} = \arg \frac{c(\rho(P_{j-1})) - c(\rho(P_j))}{i/\phi(s_{[j]})} \pmod{2\pi}.$$

Furthermore $I(\rho(P_{j+3k}))$ *is equal to the image of* $I(\rho(P_j))$ *by the translation* $z \mapsto z + k$.

(2.2) Suppose that $\phi(s_0)\phi(s_1)\phi(s_2) = 0$, *say* $\phi(s_0) = 0$. *Then* $\rho(P_{3j+1})$ *and* $\rho(P_{3j+2})$ *share the same isometric hemisphere and their axes intersect orthogonally. The axis of* $\rho(P_{3j})$ *is the vertical geodesic above the midpoint of* $c(\rho(P_{3j-2})) = c(\rho(P_{3j-1}))$ *and* $c(\rho(P_{3j+1})) = c(\rho(P_{3j+2}))$. *Furthermore* $I(\rho(P_{j+3k}))$ *is equal to the image of* $I(\rho(P_j))$ *by the translation* $z \mapsto z + k$, *for every* j *with* $j \not\equiv 0 \pmod 3$.

Remark 2.4.5. Under the assumption of the above lemma, we have

$$d(\mathrm{Axis}(\rho(P_j)), \mathrm{Axis}(\rho(P_{j+1}))) = 2\cosh^{-1}(\phi(s_{[j-1]})/2),$$

where $\mathrm{Axis}(\cdot)$ denotes the axis and d denotes the complex distance.

In this paper, we often consider the situation where the assumption of (2.1) of Proposition 2.4.4 is satisfied, so we summarize it as the following assumption.

Assumption 2.4.6 (σ-NonZero). We assume the following:

1. $\sigma = \langle s_0, s_1, s_2 \rangle$ is an ordered triangle of $\mathcal{D}$ (cf. Convention 2.1.5 (Ordered triangle)), $\{P_j\}$ is the sequence of elliptic generators associated with σ, $(\vec{e}_0, \vec{e}_1, \vec{e}_2)$ is the ordered triple of elements of $\vec{E}(\mathcal{T})$ dual to σ.
2. $\rho : \pi_1(\mathcal{O}) \to PSL(2,\mathbb{C})$ is a type-preserving representation, and $\phi \in \Phi$, $\psi \in \Psi$, $\tilde{\rho} \in \tilde{\mathcal{X}}$ are as in Convention 2.4.3.
3. We assume that $\phi^{-1}(0) \cap \sigma^{(0)} = \emptyset$, and hence $(a_0, a_1, a_2) = (\psi(\vec{e}_0), \psi(\vec{e}_1), \psi(\vec{e}_2))$ is defined. We call it the *complex probability* of ρ at σ.

2.5 Miscellaneous properties of discrete groups

In this section we collect several well-known properties of discrete groups, which we use in this paper.

Lemma 2.5.1. *Let* $X, Y \in PSL(2,\mathbb{C})$ *with* $c(X) \neq c(Y)$. *Then* $r(XY^{-1}) = r(X)r(Y)/|c(X) - c(Y)|$.

Proof. Let $X = \begin{bmatrix} a & b \\ c & d \end{bmatrix}$ and $Y = \begin{bmatrix} a' & b' \\ c' & d' \end{bmatrix}$. Then $r(X) = 1/|c|$, $r(Y) = 1/|c'|$, $c(X) = -d/c$ and $c(Y) = -d'/c'$. Since $XY^{-1} = \begin{bmatrix} ad' - bc' & -ab' + ba' \\ cd' - dc' & -cb' + da' \end{bmatrix}$, we have

$$r(XY^{-1}) = \frac{1}{|cd' - dc'|} = \frac{1}{|c|}\frac{1}{|c'|}\frac{1}{|-d/c + d'/c'|} = r(X)r(Y)/|c(X) - c(Y)|.$$

Lemma 2.5.2. *Let Γ be a Kleinian group containing $\begin{bmatrix} 1 & 1 \\ 0 & 1 \end{bmatrix}$. Then the following hold:*

1. *For any $X \in \Gamma - \Gamma_\infty$, $r(X) \le 1$.*
2. *For any $X, Y \in \Gamma - \Gamma_\infty$ with $c(X) \neq c(Y)$, we have $|c(X) - c(Y)| \ge r(X)r(Y)$.*
3. *For any $X, Y \in \Gamma - \Gamma_\infty$ with $c(X) = c(Y)$, we have $r(X) = r(Y)$.*

Proof. The assertion (1) follows from the Shimizu-Leutbecher lemma or the Jorgensen's inequality (see, for example, [14, Theorem 5.4.1] or [54, Sect. II.C]). To prove (2) and (3), let $X, Y \in \Gamma - \Gamma_\infty$. Suppose $c(X) \neq c(Y)$. Then, by Lemma 2.5.1, $r(X)r(Y) = r(XY^{-1})|c(X) - c(Y)|$. Since $c(X) \neq c(Y)$, we have $XY^{-1} \in \Gamma - \Gamma_\infty$. Thus, by the conclusion (1), $r(XY^{-1}) \le 1$, and hence we obtain the assertion (2). Suppose $c(X) = c(Y)$. Then since $X^{-1}KX$ is a parabolic transformation fixing $c(X)$ and since $Y^{-1}X$ also fixes $c(X)$, $Y^{-1}X$ cannot be a loxodromic transformation (cf. [14, Theorem 5.1.2]). Hence we obtain the assertion (3).

Corollary 2.5.3. *Let Γ and Γ_∞ be as in Lemma 2.5.2, and let X and Y be elements of $\Gamma - \Gamma_\infty$. If $|c(X) - c(Y)| < r(X)r(Y)$, then $Ih(X) = Ih(Y)$.*

Lemma 2.5.4. *Let ρ be an element of $\mathcal{X}$ and ϕ a Markoff map inducing ρ.*

(1) If $\rho \in \mathcal{QF}$, then, for any elliptic generator P, $\rho(KP)$ is a loxodromic transformation. Thus $\phi^{-1}[-2,2]$ is empty.

(2) If $\rho \in \overline{\mathcal{QF}}$, then the following hold.

1. *$\phi^{-1}(-2,2) = \emptyset$.*
2. *For any elliptic generator P, the radius of $I(\rho(P))$ is bounded above by 1.*
3. *For any two distinct elliptic generators P and Q, the distance between $c(\rho(P))$ and $c(\rho(Q))$ is bounded below by $r(\rho(P))r(\rho(Q))$. In particular, $c(\rho(P)) \neq c(\rho(Q))$.*
4. *The complex probability of ρ at any triangle σ in $\mathcal{D}$ is well-defined and is contained in $(\mathbb{C} - \{0\})^3$.*

Proof. (2.1) Let ρ be an element of $\overline{\mathcal{QF}}$. Then $\tilde{\rho} : \pi_1(T) \to SL(2,\mathbb{C})$ is faithful. Let s be an element of $\hat{\mathbb{Q}}$ and A a generator of slope s. Then $\tilde{\rho}(A)$ is of infinite order, and hence $\phi(s) = \operatorname{tr}\tilde{\rho}(A) \notin (-2,2)$.

(2.2) Let ρ be an element of $\mathcal{QF}$ and P an elliptic generator. Then, by (2.1), $\phi(s(P)) \neq 0$ and hence $r(\rho(P)) = 1/|\phi(s(P))| < \infty$ by Proposition 2.4.4(1.1). In particular, $\rho(P)$ does not stabilize ∞. Thus, by Lemma 2.5.2(1), $r(\rho(P)) \leq 1$.

(2.3) By (2.2) both $\rho(P)$ and $\rho(Q)$ do not stabilize ∞. If $c(\rho(P)) = c(\rho(Q))$, then $r(\rho(P)) = r(\rho(Q))$ by Lemma 2.5.2(3), and hence $\rho(QP) \in \rho(\pi_1(T))$ stabilizes every point in the vertical geodesic connecting $c(\rho(P))$ and ∞. Thus $\rho(QP)$ is either elliptic or the identity. This contradicts the fact that $\rho|_{\pi_1(T)}$ is faithful. Therefore $c(\rho(P)) \neq c(\rho(Q))$. By Lemma 2.5.2(2), the distance between $c(\rho(P))$ and $c(\rho(Q))$ is bounded below by $r(\rho(P))r(\rho(Q))$.

(2.4) is a direct consequence of (2-1).

3

Labeled representations and associated complexes

In this chapter, we introduce the notation which we use to reformulate the main theorems in Chap. 6.

As explained in Sect. 2.4, the infinite broken line in $\mathbb{C}$ obtained by successively joining the centers $c(\rho(P_j))$ of the isometric circles $I(\rho(P_j))$, where $\{P_j\}$ is a sequence of elliptic generators, recovers the type-preserving representation ρ. Moreover, this broken line plays a key role in the description of the combinatorial structure of the Ford domain in the case ρ is quasifuchsian. Thus we introduce, in Sect. 3.1, the notation $\mathcal{L}(\rho, \sigma)$ to represent the broken line, where σ is the triangle of the Farey triangulation spanned by the slopes of $\{P_j\}$. Then we introduce the concept for a Markoff map to be *upward* at σ (Definition 3.1.3), and show that precisely one Markoff map among the four Markoff maps inducing a given representation is upward (Lemma 3.1.4). This concept is used in Sect. 4.2 to define the side parameter.

In Sect. 3.2, we generalize the definition of $\mathcal{L}(\rho, \sigma)$ to $\mathcal{L}(\rho, \Sigma)$, where $\Sigma = (\sigma_1, \cdots, \sigma_m)$ is a chain of triangles in the Farey triangulation. It may well be a singular 2-dimensional simplicial complex in $\mathbb{C}$, whose 1-skeleton is the union of $\mathcal{L}(\rho, \sigma_i)$'s. The main theorems imply that if ρ is quasifuchsian, then there is a unique chain Σ such that $\mathcal{L}(\rho, \Sigma)$ is dual to the Ford domain. If $\mathcal{L}(\rho, \Sigma)$ is non-singular, its combinatorial structure depends only on Σ. Actually, it can be described by using the description of the space of elliptic generators in terms of the Farey triangulation (Sect. 2.1). This leads to the definition of the *elliptic generator complex* $\mathcal{L}(\Sigma)$ (Definition 3.2.3).

In Sect. 3.3, we define a *labeled representation* as a pair $\boldsymbol{\rho} = (\rho, \boldsymbol{\nu})$ of a type-preserving representation ρ and a label $\boldsymbol{\nu} = (\nu^-, \nu^+) \in \mathbb{H}^2 \times \mathbb{H}^2 = (|\mathcal{D}| - |\mathcal{D}^{(0)}|) \times (|\mathcal{D}| - |\mathcal{D}^{(0)}|)$. What we actually have in mind are pairs $(\rho, \boldsymbol{\nu})$ of a quasifuchsian representation ρ and its side parameter $\boldsymbol{\nu} = \boldsymbol{\nu}(\rho)$. This concept is used in Chap. 6 to reformulate the main theorems.

In Sect. 3.4, we define the *virtual Ford domain* $Eh(\boldsymbol{\rho})$ for a labeled representation $\boldsymbol{\rho}$ and related notation. The main theorems imply that for every element $\rho \in \mathcal{QF}$ there is a unique label $\boldsymbol{\nu}$ such that the Ford domain is equal to the virtual Ford domain $Eh(\boldsymbol{\rho})$, where $\boldsymbol{\rho} = (\rho, \boldsymbol{\nu})$, as weighted complexes.

3.1 The complex $\mathcal{L}(\rho, \sigma)$ and upward Markoff maps

The geometric description of (normalized) type-preserving representations in Proposition 2.4.4 motivates the following definition.

Definition 3.1.1. *Under Assumption 2.4.6 (σ-NonZero):*

(1) $\mathcal{L}(\rho, \sigma)$ denotes the (possibly singular) bi-infinite broken line in $\mathbb{C}$ which is obtained by successively joining the centers $\{c(\rho(P_j))\}$ of the isometric circles, where $\{P_j\}$ is the sequence of elliptic generators associated with σ. We note that $\mathcal{L}(\rho, \sigma)$ is invariant by the transformation $z \mapsto z + 1$.

(2) $\overrightarrow{c}(\rho; P_j, P_{j+1})$ denotes the oriented edge with foot $c(\rho(P_j))$ and head $c(\rho(P_{j+1}))$ of $\mathcal{L}(\rho, \sigma)$. It is also regarded as the complex number defined by

$$\overrightarrow{c}(\rho; P_j, P_{j+1}) = c(\rho(P_{j+1})) - c(\rho(P_j)).$$

(3) $\mathcal{L}(\rho, \sigma)$ is said to be simple *if the underlying space $|\mathcal{L}(\rho, \sigma)|$ is homeomorphic to the real line $\mathbb{R}$ and $\{c(\rho(P_j))\}$ sit on it in the order of the suffix $j \in \mathbb{Z}$.*

(4) $\mathcal{L}(\rho, \sigma)$ is said to be weakly simple *if there is a sequence $\{\rho_n\}$ of type-preserving representations converging to ρ such that each $\mathcal{L}(\rho_n, \sigma)$ is simple.*

The following lemma describes the relationship between $\mathcal{L}(\rho, \sigma)$ and $\mathcal{L}(\rho, \sigma')$ for adjacent triangles σ and σ'.

Lemma 3.1.2. *Under Notation 2.1.14 (Adjacent triangles) and Assumption 2.4.6 (σ-NonZero), suppose that $\phi^{-1}(0) \cap (\sigma^{(0)} \cup \sigma'^{(0)}) = \emptyset$. Then the Euclidean triangle $\Delta(c(\rho(P_0)), c(\rho(P_1)), c(\rho(P_2)))$ is similar to $\Delta(c(\rho(P'_1)), c(\rho(P'_2)), c(\rho(P'_3)))$. In particular, the union of the two infinite broken lines $\mathcal{L}(\rho, \sigma)$ and $\mathcal{L}(\rho, \sigma')$ form a bi-infinite sequence of mutually similar triangles (see Fig. 3.1). Here the triangles are possibly degenerate and the interiors of the triangles possibly intersect.*

Proof. This is a direct consequence of Lemma 2.4.1(3).

Definition 3.1.3 (Upward). *Under Assumption 2.4.6 (σ-NonZero), suppose that $\mathcal{L}(\rho, \sigma)$ is simple. Then we say that the Markoff map ϕ is* upward *at σ, if for every $j \in \mathbb{Z}$, the vector $i/\phi(s_{[j]})$ at $c(\rho(P_j))$ is "upward" with respect to $\mathcal{L}(\rho, \sigma)$, that is, the following inequality holds (cf. Fig. 2.2):*

$$0 < \arg \frac{i/\phi(s_{[j]})}{\overrightarrow{c}(\rho; P_j, P_{j+1})} = \arg \frac{-\overrightarrow{c}(\rho; P_{j-1}, P_j)}{i/\phi(s_{[j]})} < \pi.$$

Here $\arg z \in (-\pi, \pi]$ denotes the argument of a non-zero complex number z, and the identity among the two angles follows from Proposition 2.4.4(2.1). This is also equivalent to the condition:

$$0 < \arg \frac{i/\phi(s_{[j]})}{\psi(\overrightarrow{e}_{[j-1]})} = \arg \frac{-\psi(\overrightarrow{e}_{[j+1]})}{i/\phi(s_{[j]})} < \pi.$$

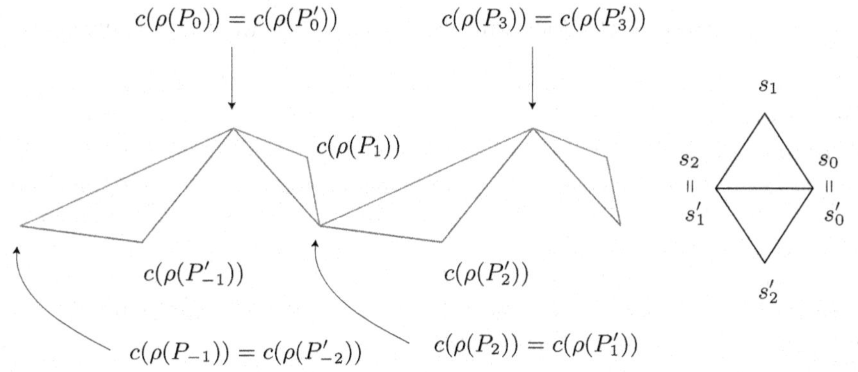

Fig. 3.1. $\mathcal{L}(\rho, \sigma)$ and $\mathcal{L}(\rho, \sigma')$

Lemma 3.1.4. *Under Assumption 2.4.6 (σ-NonZero), suppose that $\mathcal{L}(\rho, \sigma)$ is simple. Then precisely one Markoff map ϕ among the four Markoff maps inducing ρ is upward. Moreover, for any other Markoff map ϕ' inducing ρ, there is only one element $j \in \{0, 1, 2\}$ such that the vector $i/\phi'(s_j)$ is upward with respect to $\mathcal{L}(\rho, \sigma)$.*

Proof. Let ϕ_0 be the integral Markoff map determined by $\phi_0(s_0) = \phi_0(s_1) = \phi_0(s_2) = 3$, and let ρ_0 be the type-preserving representation determined by ϕ_0. Then $\mathcal{L}(\rho_0, \sigma)$ is simple and ϕ_0 is upward with respect to $\mathcal{L}(\rho_0, \sigma)$. Let ρ be a type-preserving representation such that $\mathcal{L}(\rho, \sigma)$ is simple.

Claim 3.1.5. There is a continuous family $\{\mathcal{L}_t\}_{0 \le t \le 1}$ of periodic broken lines in $\mathbb{C}$ satisfying the following conditions:

1. $\mathcal{L}_0 = \mathcal{L}(\rho_0, \sigma)$ and $\mathcal{L}_1 = \mathcal{L}(\rho, \sigma)$.
2. $\mathcal{L}_t$ is invariant by the map $z \mapsto z + 1$ and each period consists of three line segments.
3. $\mathcal{L}_t$ is simple for each $t \in [0, 1]$.

Proof. Set $c = \min\{\Im z \mid z \in \mathcal{L}_1\}$, and consider the horizontal line $L = \{z \in \mathbb{C} \mid \Im z = c\}$. Then $\mathcal{L}_1$ lies in the closed upper half space bounded by L. By using the periodic region bounded by L and $\mathcal{L}_1$, we can deform $\mathcal{L}_1$ to a periodic broken line whose underlying space is equal to L. Such a periodic broken line can be deformed to $\mathcal{L}_0$. Thus we obtain the desired result.

By the second condition of the claim, $\mathcal{L}_t$ corresponds to a complex probability, and hence $\mathcal{L}_t = \mathcal{L}(\rho_t, \sigma)$ for some type preserving representation ρ_t. By the continuity of $\{\mathcal{L}_t\}$, $\{\rho_t\}$ is continuous, and therefore it lifts to a continuous family of Markoff maps $\{\phi_t\}$. Since ϕ_0 is upward at σ, the third condition of the claim implies that ϕ_t is upward at σ for every $t \in [0, 1]$. Since $\rho_1 = \rho$ by the third condition of the claim, we obtain the first assertion of the lemma. The second assertion follows from the observation stated just before Proposition 2.4.2.

We now introduce the following definition, which we occasionally use in this paper.

Definition 3.1.6. *(1) Let ℓ be an oriented (topological) line in $\mathbb{C}$ which completes to a simple loop of $\widehat{\mathbb{C}}$. Then $H_L(\ell)$ and $H_R(\ell)$, respectively, denote the closure of the components of $\mathbb{C} - \ell$ which lies on the left and right hand side of ℓ. We say that a subset A of $\mathbb{C}$ lies on the left (resp. right) hand side of ℓ if $A \subset H_L(\ell)$ (resp. $A \subset H_R(\ell)$).*

(2) For an oriented line segment $\overrightarrow{z_0 z_1}$ in $\mathbb{C}$, $H_L(\overrightarrow{z_0 z_1})$ and $H_R(\overrightarrow{z_0 z_1})$, respectively, denotes $H_L(\ell)$ and $H_R(\ell)$, where ℓ is the oriented (straight) line in $\mathbb{C}$ obtained by extending $\overrightarrow{z_0 z_1}$ on both sides.

We note that the following is obtained as a corollary to Claim 3.1.5.

Lemma 3.1.7. *Under Assumption 2.4.6 (σ-NonZero), suppose that $\mathcal{L}(\rho, \sigma)$ is simple. Then the region of $\mathbb{C} - |\mathcal{L}(\rho, \sigma)|$ which lies in the above is equal to the region which lies on the left hand side of $|\mathcal{L}(\rho, \sigma)|$, where $|\mathcal{L}(\rho, \sigma)|$ is oriented so that it is coherent with the oriented line segment with initial and terminal points $c(\rho(P_0))$ and $c(\rho(P_1))$, respectively.*

Proof. The assertion is obvious in the case when $\phi = \phi_0$ in Claim 3.1.5. Thus we obtain the conclusion for every ϕ satisfying the assumption by Claim 3.1.5.

Definition 3.1.8. *Under Assumption 2.4.6 (σ-NonZero), suppose that $\mathcal{L}(\rho, \sigma)$ is simple. Then we say that $\mathcal{L}(\rho, \sigma)$ is* convex to the above *or* convex to the (+)-side *at $c(\rho(P_j))$ if*

$$\Im\left(\frac{\overrightarrow{c}(\rho; P_j, P_{j+1})}{\overrightarrow{c}(\rho; P_{j-1}, P_j)}\right) = \Im\left(\frac{\overrightarrow{e}_{[j-1]}}{\overrightarrow{e}_{[j+1]}}\right) < 0.$$

We say that $\mathcal{L}(\rho, \sigma)$ is convex to the below *or* convex to the (−)-side *at $c(\rho(P_j))$ if*

$$\Im\left(\frac{\overrightarrow{c}(\rho; P_j, P_{j+1})}{\overrightarrow{c}(\rho; P_{j-1}, P_j)}\right) = \Im\left(\frac{\overrightarrow{e}_{[j-1]}}{\overrightarrow{e}_{[j+1]}}\right) > 0.$$

Lemma 3.1.9. *Under Assumption 2.4.6 (σ-NonZero), suppose that $\mathcal{L}(\rho, \sigma)$ is simple. If $\Im c(\rho(P_1))$ is greater (resp. less) than both $\Im c(\rho(P_0))$ and $\Im c(\rho(P_2))$, then $\mathcal{L}(\rho, \sigma)$ is convex to the above (resp. below) at $\Im c(\rho(P_1))$.*

Proof. We prove the lemma only in the case when the inequality

$$\Im c(\rho(P_2)) \le \Im c(\rho(P_0)) < \Im c(\rho(P_1))$$

holds. (A similar argument works for the remaining cases.) Let (a_0, a_1, a_2) be the complex probability of ρ at σ. Then, from the assumption, it follows that $\Im a_2 > 0$ and that $\Im a_0 < 0$. Let $\overrightarrow{l}$ be the oriented line segment $\overrightarrow{c}(\rho; P_1, P_2)$. Suppose to the contrary that $\mathcal{L}(\rho, \sigma)$ is not convex to the above at $\Im c(\rho(P_1))$

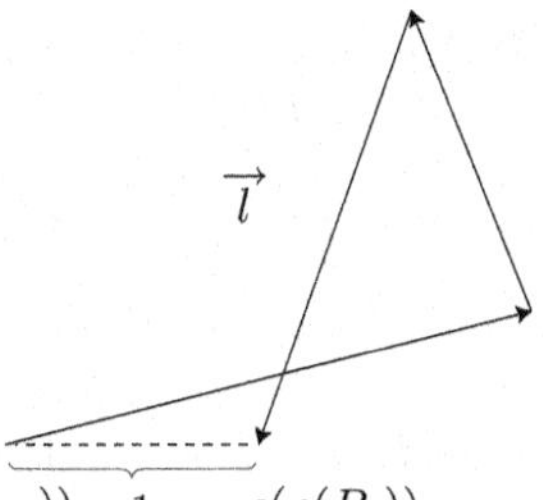

Fig. 3.2. $\Im c(\rho(P_2)) \leq \Im c(\rho(P_0)) < \Im c(\rho(P_1))$ and $\mathcal{L}(\rho, \sigma)$ is convex to the below at $\Im c(\rho(P_1))$

(see Fig. 3.2). Since $\Im(a_0/a_2) \neq 0$ from the assumption, $\mathcal{L}(\rho, \sigma)$ is convex to the below at $\Im c(\rho(P_1))$, and hence $\Im(a_0/a_2) > 0$. Thus $c(\rho(P_0))$ is contained in $\operatorname{int} H_L(\vec{l})$. On the other hand, since $c(\rho(P_{-1})) = c(\rho(P_2)) - 1$, $c(\rho(P_{-1}))$ is contained in $\operatorname{int} H_R(\vec{l})$. Let S be the horizontal strip $\{z \in \mathbb{C} \mid \Im c(\rho(P_2)) \leq \Im z \leq \Im c(\rho(P_1))\}$ in $\mathbb{C}$. Then $c(\rho(P_{-1}))$, $c(\rho(P_0))$ and $\vec{l}$ are contained in S by the assumption. Moreover $S - \vec{l}$ has the two components $S \cap H_L$ and $S \cap H_R$. Thus the edge $\vec{c}(\rho; P_{-1}, P_0)$ of $\mathcal{L}(\rho, \sigma)$ intersects the edge $\vec{l}$ of $\mathcal{L}(\rho, \sigma)$. This contradicts the assumption that $\mathcal{L}(\rho, \sigma)$ is simple.

3.2 The complexes $\mathcal{L}(\rho, \Sigma)$ and $\mathcal{L}(\Sigma)$

Definition 3.2.1 (Chain). *(1) By a* thick chain *in the Farey triangulation $\mathcal{D}$, we mean a (non-empty) finite sequence $\Sigma = (\sigma_1, \cdots, \sigma_m)$ of mutually distinct triangles in $\mathcal{D}$ such that σ_{i+1} is adjacent to σ_i for each i $(1 \leq i \leq m-1)$. We call σ_{i-1} (resp. σ_{i+1}) the* predecessor *(resp.* successor*) of σ_i in the chain. We also say that σ_{i-1} (resp. σ_{i+1}) lies on the $(-)$-side (resp. $(+)$-side) of σ_i.*

(2) By a thin chain *in $\mathcal{D}$, we mean a sequence $\Sigma = (\tau)$ of length 1 consisting of a single edge τ of $\mathcal{D}$.*

(3) Σ is called a chain *if it is either a thick chain or a thin chain. A chain Σ is also regarded as a subcomplex of $\mathcal{D}$, and $\Sigma^{(d)}$ denotes the set of d-simplices of the simplicial complex Σ.*

(4) A vertex $s \in \Sigma^{(0)}$ is called a pivot *of Σ if it is a vertex of more than two triangles of Σ.*

Definition 3.2.2 (Simple). *Let $\rho : \pi_1(\mathcal{O}) \to PSL(2, \mathbb{C})$ be a type-preserving representation, and let $\Sigma = (\sigma_1, \cdots, \sigma_m)$ be a thick chain in $\mathcal{D}$. Assume that $\phi^{-1}(0) \cap \Sigma^{(0)} = \emptyset$. Then $\mathcal{L}(\rho, \Sigma)$ denotes the union of the bi-infinite broken lines $\{\mathcal{L}(\rho, \sigma_i)\}$ in $\mathbb{C}$. We say that $\mathcal{L}(\rho, \Sigma)$ is* simple *if the following conditions are satisfied (see Fig. 1.2(b)):*

1. *$\mathcal{L}(\rho, \sigma_i)$ is simple for every i $(1 \leq i \leq m)$.*

2. *$\mathcal{L}(\rho, \sigma_i)$ and $\mathcal{L}(\rho, \sigma_j)$ ($i \neq j$) intersect only at the vertices $\{c(\rho(P)) \mid P \in \mathcal{EG}(\sigma_i \cap \sigma_j)\}$ (cf. Definition 2.1.13).*
3. *$\mathcal{L}(\rho, \sigma_{i+1})$ lies above $\mathcal{L}(\rho, \sigma_i)$ for each i. Namely, $|\mathcal{L}(\rho, \sigma_{i+1})| - |\mathcal{L}(\rho, \sigma_i)|$ is contained in the component of $\mathbb{C} - |\mathcal{L}(\rho, \sigma_i)|$ which contains the region $\{z \in \mathbb{C} \mid \Im(z) \geq c\}$ for sufficiently large number c.*

If $\mathcal{L}(\rho, \Sigma)$ is simple, then by adding the triangles in $\mathbb{C}$ spanned by edges of $\mathcal{L}(\rho, \Sigma)$, we obtain a 1- or 2-dimensional simplicial complex embedded in $\mathbb{C}$, which is invariant by the translation $z \mapsto z + 1$. We continue to denote it by the same symbol $\mathcal{L}(\rho, \Sigma)$.

If $\mathcal{L}(\rho, \Sigma)$ is simple, then the 2-dimensional complex $\mathcal{L}(\rho, \Sigma)$ is combinatorially equivalent to the abstract simplicial complex $\mathcal{L}(\Sigma)$ defined as follows.

Definition 3.2.3 (Elliptic generator complex). *Let $\Sigma = (\sigma_1, \cdots, \sigma_m)$ be a thick chain.*

(1) $\mathcal{L}(\Sigma)$ (or $\mathcal{L}(\sigma_1, \cdots, \sigma_m)$) denotes the simplicial complex constructed as follows, and we call it the elliptic generator complex *associated with the chain Σ:*

1. *The vertex set $\mathcal{L}(\Sigma)^{(0)}$ is identified with the set of elliptic generators whose slope is contained in $\Sigma^{(0)}$.*
2. *The edge set $\mathcal{L}(\Sigma)^{(1)}$ is identified with the set of the elliptic generator pairs (P, Q) (cf. Definition 2.1.13(2)), which appears (successively in this order) in a sequence of elliptic generators associated with a triangle of Σ.*
3. *The set $\mathcal{L}(\Sigma)^{(2)}$ of the 2-simplices is identified with the set of the elliptic generator triples (P, Q, R), such that (P, Q), (Q, R) and (P, R) are edges of $\mathcal{L}(\Sigma)$.*

(2) The self-map $P \mapsto P^K$ on $\mathcal{L}(\Sigma)^{(0)}$ induces a simplicial automorphism on $\mathcal{L}(\Sigma)$, and we denote it by the symbol K.

(3) $\mathcal{L}(\Sigma)/\langle K \rangle$ and $\mathcal{L}(\Sigma)/\langle K^2 \rangle$, respectively, denote the abstract cell complex obtained as the quotient of $\mathcal{L}(\Sigma)$ by the group $\langle K \rangle$ and $\langle K^2 \rangle$.

The 1-skeleton of $\mathcal{L}(\Sigma)$ is obtained as the union of the 1-dimensional simplicial complex $\mathcal{L}(\sigma_i)$ ($\sigma_i \in \Sigma^{(2)}$), which is obtained by joining the vertices $\{P_j\}$, the sequence of elliptic generators associated with σ_i, successively by edges.

The following lemma is obvious from the definition.

Lemma 3.2.4. *(1) For an elliptic generator pair (P, Q), the following conditions are equivalent.*

1. *(P, Q) determines a 1-simplex of $\mathcal{L}(\Sigma)$.*
2. *There is a triangle σ in Σ whose (oriented) boundary contains the oriented edge $\langle s(P), s(Q) \rangle$, where σ is oriented following Convention 2.1.12.*
3. *There is an elliptic generator R such that (P, Q, R) is an elliptic generator triple and the triangle $\langle s(P), s(Q), s(R) \rangle$ belongs to Σ.*

(2) For an elliptic generator triple (P, Q, R), the following conditions are equivalent.

1. *(P, Q, R) determines a 2-simplex of $\mathcal{L}(\Sigma)$.*
2. *Both $\sigma := \langle s(P), s(Q), s(R)\rangle$ and $\sigma' := \langle s(P), s(R), s(Q^R)\rangle$ belong to Σ.*
3. *The two triangles of $\mathcal{D}$ containing the edge $\langle s(P), s(R)\rangle$ belong to Σ.*

Let ρ and Σ be as in Definition 3.2.2, and assume that $\mathcal{L}(\rho, \Sigma)$ is simple. Then the map $c_\rho : \mathcal{L}(\Sigma)^{(0)} \to \mathcal{L}(\rho, \Sigma)^{(0)}$ sending P to $c(\rho(P))$ extends to a combinatorial isomorphism from $\mathcal{L}(\Sigma)$ to $\mathcal{L}(\rho, \Sigma)$, which is $(K, \rho(K))$-equivariant. Moreover, if (P, Q, R), σ and σ' are as in Lemma 3.2.4(2), and if σ lies on the ϵ-side of σ' (cf. Definition 3.2.1(1)), then the Euclidean 2-simplex $\langle c(\rho(P)), c(\rho(Q)), c(\rho(R))\rangle$ lies on the $(-\epsilon)$-side of $\mathcal{L}(\rho, \sigma)$, which contains the Euclidean edges, $\langle c(\rho(P)), c(\rho(Q))\rangle$ and $\langle c(\rho(Q)), c(\rho(R))\rangle$, whereas it lies on the (ϵ)-side of $\mathcal{L}(\rho, \sigma')$, which contains the Euclidean edge $\langle c(\rho(P)), c(\rho(R))\rangle$. Motivated by this observation, we introduce the following definition.

Definition 3.2.5. *In Lemma 3.2.4(2), assume that σ lies on the ϵ-side of σ'. Then we say that the 2-simplex (P, Q, R)* lies on the $(-\epsilon)$-side *of $\mathcal{L}(\sigma)$ and that (P, Q, R)* lies on the ϵ-side *of $\mathcal{L}(\sigma')$. We also say that the 2-simplex (P, Q, R)* lies on the $(-\epsilon)$-side *of the edges (P, Q) and (Q, R) and that (P, Q, R)* lies on the ϵ-side *of the edge (P, R).*

Definition 3.2.6. *When we mention a 2-simplex of $\mathcal{L}(\Sigma)$ whose vertex set is $\{X, Y, Z\}$, and when we do not need to specify the ordering of the vertices, we denote the 2-simplex by $((X, Y, Z))$. It should be noted that precisely one ordered triple among the six ordered triples consisting of the elements of $\{X, Y, Z\}$ is an elliptic generator triple. Similarly, the symbol $((X, Y))$ denotes the 1-simplex of $\mathcal{L}(\Sigma)$ whose vertex set is $\{X, Y\}$. We also note that precisely one of the two ordered pairs consisting of X and Y is an elliptic generator pair.*

For a thin chain Σ, $\mathcal{L}(\Sigma)$ is defined as follows.

Definition 3.2.7 (Elliptic generator complex associated with a thin chain). *Let $\Sigma = (\tau)$ be a thin chain.*

(1) $\mathcal{L}(\Sigma) = \mathcal{L}(\tau)$ denotes the abstract simplicial complex defined as follows. Set $\Sigma^* = (\sigma', \sigma)$ under Notation 2.1.14 (Adjacent triangles). Then $\mathcal{L}(\tau)$ is the 1-dimensional subcomplex of $\mathcal{L}(\Sigma^*)$ spanned by the vertices

$$\{P \in \mathcal{L}(\Sigma^*)^{(0)} \mid s(P) \in \tau^{(0)}\} = \{P_j \mid j \not\equiv 1 \pmod 3\} = \{P'_j \mid j \not\equiv 2 \pmod 3\}.$$

This complex is called the elliptic generator complex *associated with the thin chain $\Sigma = (\tau)$. The cellular complexes $\mathcal{L}(\tau)/\langle K\rangle$ and $\mathcal{L}(\tau)/\langle K^2\rangle$ are defined similarly.*

(2) Let $\rho : \pi_1(\mathcal{O}) \to PSL(2, \mathbb{C})$ be a type-preserving representation such that $\phi_\rho^{-1}(0) \cap \tau^{(0)} = \emptyset$. Then $\mathcal{L}(\rho, \Sigma) = \mathcal{L}(\rho, \tau)$ denotes the image of $\mathcal{L}(\tau)$ by the simplicial map $c_\rho : \mathcal{L}(\tau) \to \mathbb{C}$ such that $c_\rho(P) = c(\rho(P))$ for $P \in \mathcal{L}(\tau)^{(0)}$. It is said to be simple *if the simplicial map $c_\rho : \mathcal{L}(\tau) \to \mathbb{C}$ is an embedding.*

3.3 Labeled representation $\boldsymbol{\rho} = (\rho, \boldsymbol{\nu})$ and the complexes $\mathcal{L}(\boldsymbol{\rho})$ and $\mathcal{L}(\boldsymbol{\nu})$

Definition 3.3.1 (Thick/thin label). *(1) By a* label, *we mean a point* $\boldsymbol{\nu} = (\nu^-, \nu^+)$ *of* $(|\mathcal{D}| - |\mathcal{D}^{(0)}|) \times (|\mathcal{D}| - |\mathcal{D}^{(0)}|) = \mathbb{H}^2 \times \mathbb{H}^2$ *(see Sect. 1.3).*

(2) A label $\boldsymbol{\nu} = (\nu^-, \nu^+)$ *is said to be* thin *if there is an edge of* $\mathcal{D}$ *which contains both* ν^- *and* ν^+. *It is said to be* thick *if it is not thin.*

Definition 3.3.2 (Labeled representation). *(1)* A labeled representation *is a pair* $\boldsymbol{\rho} = (\rho, \boldsymbol{\nu})$ *of an element* $\rho \in \mathcal{X}$ *and a label* $\boldsymbol{\nu} = (\nu^-, \nu^+)$. *We call* $\boldsymbol{\nu}$ *the* label *of* $\boldsymbol{\rho}$.

(2) A labeled representation $\boldsymbol{\rho} = (\rho, \boldsymbol{\nu})$ *is said to be* thick *or* thin *according as* $\boldsymbol{\nu}$ *is thick or thin.*

Definition 3.3.3 (Chain and elliptic generator complex associated with a label).

(1) Let $\boldsymbol{\nu} = (\nu^-, \nu^+)$ *be a thick label. Then* $\Sigma(\boldsymbol{\nu})$ *denotes the thick chain* $(\sigma_1, \cdots, \sigma_m)$ *in* $\mathcal{D}$ *such that the* interior *of the oriented geodesic line segment in* $\mathbb{H}^2$ *joining* ν^- *to* ν^+ *intersects* $\sigma_1, \cdots, \sigma_m$ *in this order. The triangles* σ_1 *and* σ_m, *respectively, are denoted by* $\sigma^-(\boldsymbol{\nu})$ *and* $\sigma^+(\boldsymbol{\nu})$. *They are called the* $(-)$-terminal triangle *(resp.* $(+)$-terminal triangle*) of* $\Sigma(\boldsymbol{\nu})$*, respectively. They are often abbreviated to* σ^- *and* σ^+*, respectively.*

(2) Let $\boldsymbol{\nu} = (\nu^-, \nu^+)$ *be a thin label. Then* $\tau(\boldsymbol{\nu})$ *denotes the edge of* $\mathcal{D}$ *containing both* ν^- *and* ν^+*, and* $\Sigma(\boldsymbol{\nu})$ *denotes the thin chain* $(\tau(\boldsymbol{\nu}))$ *in* $\mathcal{D}$. *The triangles of* $\mathcal{D}$ *having* $\tau(\boldsymbol{\nu})$ *as an edge are denoted by* $\sigma^-(\boldsymbol{\nu})$ *and* $\sigma^+(\boldsymbol{\nu})$. *Here the sign* $\pm$ *are chosen arbitrarily.*

(3) For a label $\boldsymbol{\nu} = (\nu^-, \nu^+)$, $\mathcal{L}(\boldsymbol{\nu})$ *denotes* $\mathcal{L}(\Sigma(\boldsymbol{\nu}))$*, and we call it the* elliptic generator complex *associated with the label* $\boldsymbol{\nu}$.

Definition 3.3.4 (NonZero). *A labeled representation* $\boldsymbol{\rho} = (\rho, \boldsymbol{\nu})$ *is said to satisfy the condition* NonZero *if* $\phi_\rho^{-1}(0) \cap \Sigma(\boldsymbol{\nu})^{(0)} = \emptyset$.

Definition 3.3.5. *Let* $\boldsymbol{\rho} = (\rho, \boldsymbol{\nu})$ *be a thick labeled representation which satisfies the condition NonZero. Then* $\mathcal{L}(\boldsymbol{\rho}) = \mathcal{L}(\rho, \boldsymbol{\nu})$ *denotes* $\mathcal{L}(\rho, \Sigma(\boldsymbol{\nu}))$ *in* $\mathbb{C}$ *(see Definition 3.2.2).*

We will show that if $\rho \in \mathcal{QF}$ and $\boldsymbol{\nu} = \boldsymbol{\nu}(\rho)$, where $\boldsymbol{\nu} : \mathcal{QF} \to \mathbb{H}^2 \times \mathbb{H}^2$ is the map in Theorem 1.3.2 (see Theorem 6.1.12), then $\mathcal{L}(\boldsymbol{\rho})$ is simple. Thus Fig. 1.2 gives a typical example of $\mathcal{L}(\boldsymbol{\rho})$ which is simple.

3.4 Virtual Ford domain

We first recall the definition of convex polyhedra following [28].

Definition 3.4.1 (Convex polyhedron). *(1) A* totally geodesic subspace *of $\mathbb{H}^n$ (resp. $\overline{\mathbb{H}^n} = \mathbb{H}^n \cup \partial\mathbb{H}^n$) is a copy of $\mathbb{H}^d$ (resp. $\overline{\mathbb{H}^d}$) with $0 \le d \le n$ (resp. $1 \le d \le n$), embedded geodesically in $\mathbb{H}^n$ (resp. $\overline{\mathbb{H}^n}$). We also regard a singleton in $\partial\mathbb{H}^3$ as a totally geodesic subspace of dimension 0. A* hyper-subspace *of $\mathbb{H}^n$ (resp. $\overline{\mathbb{H}^n}$) is a totally geodesic subspace of of $\mathbb{H}^n$ (resp. $\overline{\mathbb{H}^n}$) of dimension $n-1$. A* closed half space *of $\mathbb{H}^n$ (resp. $\overline{\mathbb{H}^n}$) is the closure of a component of the complement of a hyper-subspace in $\mathbb{H}^n$ (resp. $\overline{\mathbb{H}^n}$).*

(2) A subspace F of $\mathbb{H}^n$ (resp. $\overline{\mathbb{H}^n}$) is a convex polyhedron *if it is the intersection of a family $\mathcal{H}$ of closed half spaces with the property that each point of F has a neighborhood meeting at most a finite number of boundaries of elements of $\mathcal{H}$. A singleton in $\partial\mathbb{H}^3$ is also regarded as a convex polyhedron of dimension 0.*

(3) A convex polyhedron F in $\mathbb{H}^n$ (resp. $\overline{\mathbb{H}^n}$) is said to be spanned *by a subset V of $\mathbb{H}^n$ (resp. $\overline{\mathbb{H}^n}$), if F is the intersection of the closed half spaces of $\mathbb{H}^n$ (resp. $\overline{\mathbb{H}^n}$) containing V.*

(4) Let F be a convex polyhedron in $\mathbb{H}^n$ (resp. $\overline{\mathbb{H}^n}$). The dimension $\dim F$ *is defined as the the dimension of the smallest totally geodesic subspace, H, containing P. The* boundary ∂F *of F is the topological boundary (or the frontier) of P in H. The* interior $\operatorname{int} F$ *of F is defined to be $F - \partial F$. A subset F' of ∂F is said to be a* codimension-one face *of F if $F' = F \cap H'$ for some hyperplane H' of H, and $\dim F' = dimF - 1$. If $i \ge 2$, the* codimension-i faces *of F are defined inductively as codimension-one faces of codimension $i-1$ faces of F.*

(5) A geodesic segment *in $\overline{\mathbb{H}^3}$ is a 1-dimensional convex polyhedron. For two (possibly identical) points a and b in $\overline{\mathbb{H}^3}$, we denote by $[a,b]$ the (possibly degenerate) geodesic segment in $\overline{\mathbb{H}^3}$ with endpoints a and b. A* half geodesic *(resp.* complete geodesic*) is a geodesic segment in $\overline{\mathbb{H}^3}$ spanned by a point in $\mathbb{H}^3$ and a point in $\partial\overline{\mathbb{H}^3}$ (resp. spanned by two points in $\partial\overline{\mathbb{H}^3}$.)*

Remark 3.4.2. It should be noted that for a given subset V of $\mathbb{H}^n$ (resp. $\overline{\mathbb{H}^n}$), the intersection of the closed half spaces of $\mathbb{H}^n$ (resp. $\overline{\mathbb{H}^n}$) containing V is not necessarily a convex polyhedron. Thus a 'convex polyhedron spanned by V' does not necessarily exist.

We now introduce the definition of a virtual Ford domain and related definitions.

Definition 3.4.3 (Virtual Ford domain). *Let $\boldsymbol{\rho} = (\rho, \boldsymbol{\nu})$ be a labeled representation which satisfies the condition NonZero.*

(1) $Eh(\boldsymbol{\rho})$ (resp. $E(\boldsymbol{\rho})$) denotes the subspace of $\mathbb{H}^3$ (resp. $\mathbb{C}$) defined as follows:

$$Eh(\boldsymbol{\rho}) := \bigcap \{Eh(\rho(P)) \mid P \in \mathcal{L}(\boldsymbol{\nu})^{(0)}\}$$

$$E(\boldsymbol{\rho}) := \bigcap \{E(\rho(P)) \mid P \in \mathcal{L}(\boldsymbol{\nu})^{(0)}\}$$

We call $Eh(\boldsymbol{\rho})$ the virtual Ford domain *of $\boldsymbol{\rho}$.*

(2) For each $\epsilon \in \{-,+\}$, $E^{\epsilon}(\boldsymbol{\rho})$ *denotes the component of* $E(\boldsymbol{\rho})$ *containing the region* $\{z \in \mathbb{C} \,|\, \epsilon\Im(z) \geq L\}$ *for sufficiently large positive number* L.

To describe the cellular structure of the (virtual) Ford domains, we need the following definitions.

Definition 3.4.4. *Let* ξ *be a simplex of* $\mathcal{L}(\boldsymbol{\nu})$.

1. *The* link $\mathrm{lk}(\xi, \mathcal{L}(\boldsymbol{\nu}))$ *of* ξ *in* $\mathcal{L}(\boldsymbol{\nu})$ *is the subcomplex of* $\mathcal{L}(\boldsymbol{\nu})$ *spanned by the vertices* $X \in \mathcal{L}(\boldsymbol{\nu})^{(0)} - \xi^{(0)}$ *such that there is a simplex of* $\mathcal{L}(\boldsymbol{\nu})$ *containing both* X *and* ξ.
2. *The* star $\mathrm{st}_0(\xi, \mathcal{L}(\boldsymbol{\nu}))$ *of* ξ *in* $\mathcal{L}(\boldsymbol{\nu})$ *is the subcomplex of* $\mathcal{L}(\boldsymbol{\nu})$ *spanned by* ξ *and* $\mathrm{lk}(\xi, \mathcal{L}(\boldsymbol{\nu}))$.

Remark 3.4.5. If $\dim \xi \geq 1$, then $\mathrm{st}_0(\xi, \mathcal{L}(\boldsymbol{\nu}))$ is smaller than the star neighborhood of ξ in $\mathcal{L}(\boldsymbol{\nu})$, in general.

Definition 3.4.6. *Let* $\boldsymbol{\rho} = (\rho, \boldsymbol{\nu})$ *be a labeled representation which satisfies the condition NonZero. For each simplex* $\xi \in \mathcal{L}(\boldsymbol{\nu})^{(\leq 2)}$, $F_{\boldsymbol{\rho}}(\xi)$ *denotes the convex polyhedron in* $\mathbb{H}^3$ *defined by*

$$F_{\boldsymbol{\rho}}(\xi) = \left(\bigcap\{Ih(\rho(P)) \,|\, P \in \xi^{(0)}\}\right) \cap Eh(\rho, lk(\xi, \mathcal{L}(\boldsymbol{\nu}))),$$

where

$$Eh(\rho, \mathrm{lk}(\xi, \mathcal{L}(\boldsymbol{\nu}))) = \bigcap\{Eh(\rho(X)) \,|\, X \in \mathrm{lk}(\xi, \mathcal{L}(\boldsymbol{\nu}))^{(0)}\}.$$

We abbreviate the image by $F_{\boldsymbol{\rho}}$ *of a 2-simplex* $\langle P, Q, R\rangle$ *(resp. a* 1*-cell* $\langle P, Q\rangle$*, a* 0*-cell* $\langle P\rangle$*) as* $F_{\boldsymbol{\rho}}(P, Q, R)$ *(resp.* $F_{\boldsymbol{\rho}}(P, Q)$*,* $F_{\boldsymbol{\rho}}(P)$*). Thus:*

$$\begin{aligned}
F_{\boldsymbol{\rho}}(P, Q, R) &= Ih(\rho(P)) \cap Ih(\rho(Q)) \cap Ih(\rho(R)),\\
F_{\boldsymbol{\rho}}(P, Q) &= Ih(\rho(P)) \cap Ih(\rho(Q)) \cap Eh(\rho, \mathrm{lk}(\langle P, Q\rangle, \mathcal{L}(\boldsymbol{\nu}))),\\
F_{\boldsymbol{\rho}}(P) &= Ih(\rho(P)) \cap Eh(\rho, \mathrm{lk}(\langle P\rangle, \mathcal{L}(\boldsymbol{\nu}))).
\end{aligned}$$

We note that among the vertices of $\mathcal{L}(\boldsymbol{\nu})$, only the vertices of $\mathrm{st}_0(\xi, \mathcal{L}(\boldsymbol{\nu}))$ are involved in the definition of $F_{\boldsymbol{\rho}}(\xi)$.

Definition 3.4.7. *Let* $\boldsymbol{\rho} = (\rho, \boldsymbol{\nu})$ *be a labeled representation which satisfies the condition NonZero. For each simplex* $\xi \in \mathcal{L}(\boldsymbol{\nu})^{(\leq 2)}$, $\overline{F}_{\boldsymbol{\rho}}(\xi)$ *denotes the closed convex polyhedron in* $\overline{\mathbb{H}}^3$ *defined by*

$$\overline{F}_{\boldsymbol{\rho}}(\xi) = \left(\bigcap\{\overline{Ih}(\rho(P)) \,|\, P \in \xi^{(0)}\}\right) \cap \overline{Eh}(\rho, lk(\xi, \mathcal{L}(\boldsymbol{\nu}))),$$

where

$$\overline{Eh}(\rho, \mathrm{lk}(\xi, \mathcal{L}(\boldsymbol{\nu}))) = \bigcap\{\overline{Eh}(\rho(X)) \,|\, X \in \mathrm{lk}(\xi, \mathcal{L}(\boldsymbol{\nu}))^{(0)}\}.$$

.

Lemma 3.4.8. *Let $\boldsymbol{\rho} = (\rho, \boldsymbol{\nu})$ be a labeled representation which satisfies the condition NonZero. Then for any simplex ξ of $\mathcal{L}(\boldsymbol{\nu})$, we have the following.*
(1) $\overline{F}_{\boldsymbol{\rho}}(\xi) \cap \mathbb{H}^3 = F_{\boldsymbol{\rho}}(\xi)$.
(2) $\overline{F}_{\boldsymbol{\rho}}(\xi)$ is equal to the closure $\overline{F_{\boldsymbol{\rho}}(\xi)}$ of $F_{\boldsymbol{\rho}}(\xi)$ in $\overline{\mathbb{H}}^3$ if $F_{\boldsymbol{\rho}}(\xi) \neq \emptyset$.

Proof. (1) For any element $A \in PSL(2,\mathbb{C})$ such that $A(\infty) \neq \infty$, we have $\overline{Ih}(A) \cap \mathbb{H}^3 = Ih(A)$ and $\overline{Eh}(A) \cap \mathbb{H}^3 = Eh(A)$. Hence we have $\overline{F}_{\boldsymbol{\rho}}(\xi) \cap \mathbb{H}^3 = F_{\boldsymbol{\rho}}(\xi)$.
(2)

$$\begin{aligned}\overline{F_{\boldsymbol{\rho}}(\xi)} &= \overline{(\cap\{Ih(\rho(P)) \mid P \in \xi^{(0)}\}) \cap (\cap\{Eh(\rho(X)) \mid X \in \mathrm{lk}(\xi, \mathcal{L}(\boldsymbol{\nu}))^{(0)}\})} \\ &\subset \left(\cap\{\overline{Ih}(\rho(P)) \mid P \in \xi^{(0)}\}\right) \cap \left(\cap\{\overline{Eh}(\rho(X)) \mid X \in \mathrm{lk}(\xi, \mathcal{L}(\boldsymbol{\nu}))^{(0)}\}\right) \\ &= \overline{F}_{\boldsymbol{\rho}}(\xi).\end{aligned}$$

To see the converse, let x be a point of $\overline{F}_{\boldsymbol{\rho}}(\xi)$. If $x \in \mathbb{H}^3$, then $x \in \overline{F}_{\boldsymbol{\rho}}(\xi) \cap \mathbb{H}^3 = F_{\boldsymbol{\rho}}(\xi) \subset \overline{F_{\boldsymbol{\rho}}(\xi)}$, and so we may assume $x \in \mathbb{C}$. Since $\boldsymbol{\rho}$ satisfies the condition Duality, $F_{\boldsymbol{\rho}}(\xi) \neq \emptyset$. Pick a point y of $F_{\boldsymbol{\rho}}(\xi)$, and let ℓ be the geodesic in $\mathbb{H}^3$ joining y to x. Then $\ell \subset \overline{F}_{\boldsymbol{\rho}}(\xi) \cap \mathbb{H}^3 = F_{\boldsymbol{\rho}}(\xi)$. Hence $x \in \overline{\ell} \subset \overline{F_{\boldsymbol{\rho}}(\xi)}$. This completes the proof of Lemma 3.4.8.

By an argument parallel to the above, we obtain the following lemma.

Lemma 3.4.9. *Let $\boldsymbol{\rho} = (\rho, \boldsymbol{\nu})$ be a labeled representation which satisfies the condition NonZero. Then the following hold.*
(1) $\overline{Eh}(\boldsymbol{\rho}) \cap \mathbb{H}^3 = Eh(\boldsymbol{\rho})$ and $\overline{Eh}(\boldsymbol{\rho}) \cap \mathbb{C} = E(\boldsymbol{\rho})$.
(2) $\overline{Eh}(\boldsymbol{\rho})$ is equal to the closure $\overline{Eh(\boldsymbol{\rho})}$ of $Eh(\boldsymbol{\rho})$ in $\overline{\mathbb{H}}^3$.

We need to compare the virtual Ford domains of two labeled representations sharing the same representation in Chap. 8. So, we introduce the following definition.

Definition 3.4.10. *Let $\mathcal{L}$ be an elliptic generator complex $\mathcal{L}(\boldsymbol{\nu})$ for some label $\boldsymbol{\nu}$, and let ρ be an element of $\mathcal{X}$ such that $\phi_\rho^{-1}(0) \cap \Sigma(\boldsymbol{\nu})^{(0)} = \emptyset$.*

1. *We define the* virtual Ford domain *$Eh(\rho, \mathcal{L})$ by*

$$Eh(\rho, \mathcal{L}) = \bigcap\{Eh(\rho(P)) \mid P \in \mathcal{L}^{(0)}\}.$$

2. *For a cell $\xi \in \mathcal{L}^{(i)}$, we define the* virtual face *$F_{(\rho,\mathcal{L})}(\xi)$ by*

$$F_{(\rho,\mathcal{L})}(\xi) = \left(\bigcap\{Ih(\rho(P)) \mid P \in \xi^{(0)}\}\right) \cap \left(\bigcap\{Eh(\rho(P)) \mid P \in \mathrm{lk}(\xi, \mathcal{L})^{(0)}\}\right).$$

4

Chain rule and side parameter

The essential ingredient of Jorgensen's work in [40] is a detailed analysis of how the pattern of isometric hemispheres bounding the Ford domain change as one varies the group. This idea can be found in his preceding work [39] on the infinite cyclic Kleinian groups. (See the work [25] due to Drumm and Poritz for its detailed exposition and generalization.) In this chapter we first describe the "chain rule for isometric circles" (Lemma 4.1.2), which affords a foundation on the analysis, and then we introduce the key notion of Jorgensen's *side parameter* (Definition 4.2.9) and prove various of its properties.

In Sect. 4.1, we give a detailed proof and an intuitive explanation of the chain rule for isometric circles (Lemma 4.1.2).

In Sect. 4.2, we give the definition of Jorgensen's *side parameter* $\theta^{\epsilon}(\rho,\sigma)$ for a pair (ρ,σ) and a sign $\epsilon \in \{-,+\}$ (Definition 4.2.9). If $\sigma = \langle s_0, s_1, s_2\rangle$, it is a triple $(\theta^{\epsilon}(\rho,\sigma;s_0), \theta^{\epsilon}(\rho,\sigma;s_1), \theta^{\epsilon}(\rho,\sigma;s_2))$ of real numbers whose sum is equal to $\pi/2$ (Proposition 4.2.16). Thus if all components of $\theta^{\epsilon}(\rho,\sigma)$ are non-negative, then $\theta^{\epsilon}(\rho,\sigma)$ is identified with a point in $\sigma \subset \mathcal{D}$ (Definition 4.2.17). Moreover, if $\rho \in \mathcal{QF}$ and if the Ford complex $\mathrm{Ford}(\Gamma)$ of $\Gamma = \rho(\pi_1(T))$ is isotopic to $\mathrm{Spine}(\sigma^-,\sigma^+)$ for a pair of triangles σ^- and σ^+ (see Theorem 1.2.2), then the point $\nu^{\epsilon}(\Gamma) \in \sigma^{\epsilon}$ defined in Sect. 1.3 is equal to $\theta^{\epsilon}(\rho,\sigma)$. In Lemma 4.2.18, we give an algebraic equation for a Markoff map ϕ to "realize" a given point $\boldsymbol{\nu} = (\nu^-,\nu^+) \in \mathbb{H}^2 \times \mathbb{H}^2$. This motivates the notion of an *algebraic root* for $\boldsymbol{\nu}$ (Definition 4.2.19). The last step of the proof of the Main Theorem 1.3.5 requires detailed study of the algebraic equation (Chap. 9).

In Sect. 4.3, we describe how the side parameters control relative positions of isometric circles and hemispheres (see Lemmas 4.3.1–4.3.6). Though these lemmas look obvious at a glance, their rigorous proofs require careful arguments. We then introduce the key notion of an *ϵ-terminal triangle* (Definition 4.3.8). If ρ is an element of $\mathcal{QF}$ and $\Sigma(\boldsymbol{\nu}(\rho)) = (\sigma_1, \cdots, \sigma_n)$ is the chain spanned by $\boldsymbol{\nu}(\rho)$ (cf. Main Theorem 1.3.5 and Definition 1.4.1), then σ_1 and σ_n, respectively, are a $(-)$-terminal triangle and a $(+)$-terminal triangle of ρ.

In Sect. 4.4, we prove a few basic properties concerning ϵ-terminal triangles. Though they also look obvious at a glance, their rigorous proofs again require careful arguments, and in particular, the proof of Proposition 4.4.4 is lengthy. However, the readers may skip the proofs, because of the following reason. Proposition 4.4.4 and its sister, Proposition 4.5.7, are used only in the proof of Lemmas 4.6.1, 4.6.2 and 4.6.7 (Proofs of Claims 4.6.3 and 4.6.5), which in turn are used in Sects. 7.2 and 7.3, in the proof of Proposition 6.2.1 (Openness). However, Propositions 4.4.4 and 4.5.7 for the quasifuchsian case are consequences of Lemma 7.1.6 (Hidden isometric hemisphere). Nevertheless, we decided to include the proofs, because the results in the section simplify the structure of the proof of the Main Theorem 1.3.5 and because they are necessary in the forthcoming paper [11], where we treat hyperbolic cone-manifolds.

In Sect. 4.5, we study relation between side parameters at adjacent triangles. The results in this section are used in the next Sect. 4.6.

In Sect. 4.6, we prove the key Lemmas 4.6.1, 4.6.2 and 4.6.7, which describe how the terminal triangles change according to small deformation of type-preserving representations. These lemmas hold the key to the proof of Proposition 6.2.1 (Openness in $\mathcal{X}$).

Section 4.7 is devoted to the proof of the technical Lemma 4.5.5, which is used in the proof of the key lemmas in Sect. 4.5. We note that the techniques in this section plays an important role in the first author's study [3] which compares Jorgensen's side parameter and the conformal invariants of quasifuchsian punctured torus groups.

In Sect. 4.8, we study those representations ρ such that $\mathcal{L}(\rho, \sigma)$ is weakly simple but is not simple. We show that this happens if and only if either $\mathcal{L}(\rho, \sigma)$ is *singly folded* or *doubly folded* (Lemma 4.8.1). We also prove Lemma 4.8.7, which implies Lemma 8.9.3, which in turn plays an important role in Chap. 8.

4.1 Chain rule for isometric circles

In this section, we recall fundamental facts concerning the isometric circles.

Lemma 4.1.1. *(1) Let A be an element of $PSL(2,\mathbb{C})$ which does not fix ∞. Then*

$$A(E(A) \cup \{\infty\}) = D(A^{-1}), \qquad A(D(A)) = E(A^{-1}) \cup \{\infty\},$$
$$A(Eh(A) \cup \{\infty\}) = Dh(A^{-1}), \qquad A(Dh(A)) = Eh(A^{-1}) \cup \{\infty\},$$

Moreover, if we orient circles in $\mathbb{C}$ counter-clockwise, then the restriction of A to $I(A)$ is an orientation-reversing Euclidean isometry onto $I(A^{-1})$.

(2) Let W be an element of $PSL(2,\mathbb{C})$ which preserves ∞ and acts on $\mathbb{C}$ as an Euclidean isometry. Then

$$I(AW) = W^{-1}(I(A)), \qquad I(WA) = I(A),$$
$$Ih(AW) = W^{-1}(Ih(A)), \qquad Ih(WA) = Ih(A).$$

In particular, $I(WAW^{-1}) = W(I(A))$.

(3) Let A *and* B *elements of* $PSL(2,\mathbb{C})$ *which do not fix* ∞*. Then* $I(A) = I(B)$ *if and only if* BA^{-1} *is a Euclidean isometry.*

Proof. Though these are well-known, we give proofs for completeness.

(1) For a point $z \in \mathbb{C}$, set $w = A(z)$. Then $z \in E(A)$ if and only if $|A'(z)| \geq 1$. The latter condition is equivalent to the condition that $|(A^{-1})'(w)| = 1/|A'(z)| \leq 1$, which in turn is equivalent to the condition $w \in D(A^{-1})$. Hence $A(E(A)) \subset D(A^{-1})$ and therefore we have $A(E(A) \cup \{\infty\}) = D(A^{-1})$. The remaining assertions are consequences of this identity.

(2) By the assumption, $|W'(z)| = 1$ for every $z \in \mathbb{C}$. Hence $|(AW)'(z)| = |A'(W(z))W'(z)| = |A'(W(z))|$. Thus $z \in I(AW)$ if and only if $W(z) \in I(A)$. So we have $I(AW) = W^{-1}(I(A))$. The second identity follows from the identity $|(WA)'(z)| = |W'(A(z))A'(z)| = |A'(z)|$.

(3) The "if" part follows from (2). To see the converse, assume that $I(A) = I(B)$. Let z be a point in $I(A) = I(B)$. Then

$$\begin{aligned}
|(BA^{-1})'(A(z))| &= |B'(A^{-1}(A(z))) \cdot (A^{-1})'(A(z))| \\
&= |B'(z) \cdot (A^{-1})'(A(z))| \\
&= |(A^{-1})'(A(z))| && \text{because } z \in I(B) \\
&= 1 && \text{because } A(z) \in I(A^{-1}).
\end{aligned}$$

Hence the restriction of BA^{-1} to $A(I(A)) = I(A^{-1})$ is a Euclidean isometry. Moreover $BA^{-1}(D(A^{-1})) = B(E(A)) = B(E(B)) = D(B)$. Thus we can conclude that BA^{-1} is a Euclidean isometry.

The following lemma is called the *chain rule for isometric circles* because it follows from the chain rule for differentials. Though it is well-known, we cannot find a proof in the literature. We give a proof for completeness, because it is essential to our whole argument.

Lemma 4.1.2 (Chain rule for isometric circles). *Let* A *and* B *be elements of* $PSL(2,\mathbb{C})$ *which do not fix* ∞*, such that* $c(A) \neq c(B)$*. Then the following hold.*

1. $A(Ih(A) \cap Ih(B)) = Ih(A^{-1}) \cap Ih(BA^{-1})$ *and* $B(Ih(A) \cap Ih(B)) = Ih(B^{-1}) \cap Ih(AB^{-1})$.
2. $A(I(A) \cap E(B)) = I(A^{-1}) \cap E(BA^{-1})$ *and* $B(I(B) \cap E(A)) = I(B^{-1}) \cap E(AB^{-1})$.
3. *Suppose* $I(A) \cap I(B) \neq \emptyset$ *and* $I(A) \neq I(B)$*, namely the dihedral angle,* $\theta(X,Y)$*, of* $E(X) \cap E(Y)$ *(at* $I(X) \cap I(Y)$*) is well-defined. Then the dihedral angles of* $E(A^{-1}) \cap E(BA^{-1})$ *and* $E(AB^{-1}) \cap E(B^{-1})$ *are well-defined, and the following identity holds:*

$$\theta(A,B) + \theta(B^{-1}, AB^{-1}) + \theta(BA^{-1}, A^{-1}) = 2\pi.$$

Proof. (1) We first note that BA^{-1} and AB^{-1} do not fix ∞ and hence $I(BA^{-1})$ and $I(AB^{-1})$ are defined. Let z be a point in $I(A) \cap I(B)$. Then by the proof of Lemma 4.1.1(3), we see $|(BA^{-1})'(A(z))| = 1$ and hence $A(z) \in I(BA^{-1})$. On the other hand, $A(z) \in I(A^{-1})$ because $z \in I(A)$. Hence $A(z) \in I(A^{-1}) \cap I(BA^{-1})$. Thus we have proved $A(I(A) \cap I(B)) \subset I(A^{-1}) \cap I(BA^{-1})$. This implies $A(I(A) \cap I(B)) = I(A^{-1}) \cap I(BA^{-1})$ and $A(Ih(A) \cap Ih(B)) = Ih(A^{-1}) \cap Ih(BA^{-1})$. The second identity follows from the first one by interchanging the roles of A and B.

(2) Let z be a point in $I(A) \cap E(B)$. The above calculation also implies $|(BA^{-1})'(A(z))| \leq 1$, and hence $A(z) \in E(BA^{-1})$. Thus we have $A(I(A) \cap E(B)) \subset E(BA^{-1})$. Since $A(I(A)) = I(A^{-1})$, this implies $A(I(A) \cap E(B)) \subset I(A^{-1}) \cap E(BA^{-1})$ and hence we obtain the first identity. The second identity follows from the first one by interchanging the roles of A and B.

(3) To prove this, we need to introduce some notation and notice some elementary fact, which we present below. For a point z of the isometric circle $I(X)$ of a Möbius transformation $X \in PSL(2, \mathbb{C})$, let $v_z^+(X)$ and $v_z^-(X)$, respectively, be the unit tangent vector to $I(X)$ at z, which is coherent (resp. anti-coherent) with the counter-clockwise orientation of $I(X)$. Let (X, Y) be an ordered pair of Möbius transformations $X, Y \in PSL(2, \mathbb{C})$ such that $I(X)$ and $I(Y)$ intersect transversely. Then we can see that there is a unique point, z, in $I(X) \cap I(Y)$ such that $v_z^+(X)$ points into $E(Y)$, $v_z^-(Y)$ points into $E(X)$, and the dihedral angle $\theta(X, Y)$ of $E(X) \cap E(Y)$ is equal to $\arg(v_z^+(X)/v_z^-(Y))$, where we identify tangent vectors with complex numbers. If z' is the other point of $I(X) \cap I(Y)$, then $v_{z'}^-(X)$ points into $E(Y)$, $v_{z'}^+(Y)$ points into $E(X)$, and $\theta(X, Y)$ is equal to $\arg(v_{z'}^+(Y)/v_{z'}^-(X))$, where we identify tangent vectors with complex numbers.

We may assume that $I(A)$ intersects $I(B)$ transversely, because the tangential case is obtained as limit of the generic case. Let p be the point of $I(A) \cap I(B)$ such that $v_p^+(A)$ points into $E(B)$, $v_p^-(B)$ points into $E(A)$, and $\theta(A, B) = \arg(v_p^+(A)/v_p^-(B))$. By Lemma 4.1.1(1), we see $dA(v_p^+(A)) = v_q^-(A^{-1})$, where $q = A(p)$. Moreover, Lemma 4.1.2(2) implies that $v_q^-(A^{-1})$ points into $E(BA^{-1})$. Thus we have

$$\theta(BA^{-1}, A^{-1}) = \arg \frac{v_q^+(BA^{-1})}{v_q^-(A^{-1})} = \arg \frac{dA(w)}{dA(v_p^+(A))} = \arg \frac{w}{v_p^+(A)},$$

where $w = dA^{-1}(v_q^+(BA^{-1}))$. Similarly, we see that the vector $dB(v_p^-(B)) = v_r^+(B^{-1})$ points into $E(AB^{-1})$, where $r = B(p)$. This implies

$$\theta(B^{-1}, AB^{-1}) = \arg \frac{v_r^+(B^{-1})}{v_r^-(AB^{-1})} = \arg \frac{dB(v_p^-(B))}{dB(w')} = \arg \frac{v_p^-(B)}{w'},$$

where $w' = dB^{-1}(v_r^-(AB^{-1}))$. On the other hand, we have $v_r^-(AB^{-1}) = d(BA^{-1})(v_q^+(BA^{-1}))$ by Lemma 4.1.1(1), and hence

$$w' = dB^{-1}(v_r^-(AB^{-1})) = dB^{-1}(d(BA^{-1})(v_q^+(BA^{-1}))) = dA^{-1}(dA(w)) = w.$$

Hence

$$\begin{aligned}\theta(A,B)+&\theta(B^{-1},AB^{-1})+\theta(BA^{-1},A^{-1})\\&=\arg\frac{v_p^+(A)}{v_p^-(B)}+\arg\frac{v_p^-(B)}{w}+\arg\frac{w}{v_p^+(A)}\\&\equiv\arg 1\pmod{2\pi}\\&\equiv 0\pmod{2\pi}\end{aligned}$$

Since all dihedral angles are positive and less than π, the above congruence implies the desired identity.

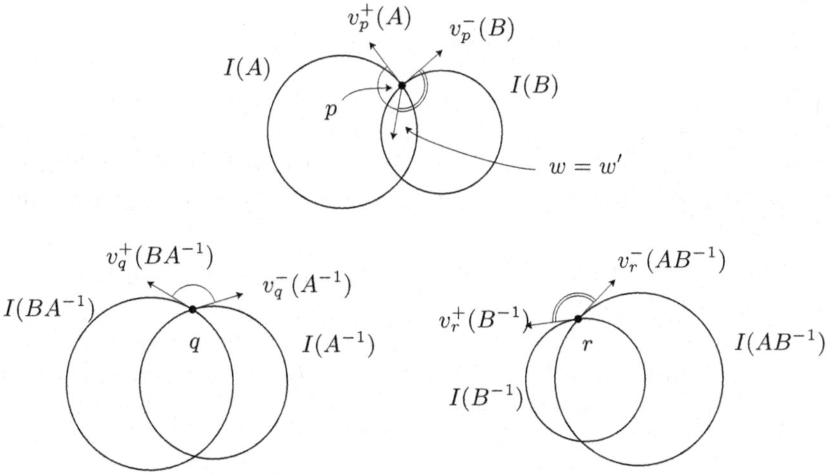

Fig. 4.1. Three pairs of circles

Since Lemma 4.1.2 (Chain rule) is so important, we also give an intuitive explanation of the lemma. To this end, consider a Kleinian group Γ such that Γ_∞ consists of Euclidean isometries. Let e be an edge of its Ford complex $\mathrm{Ford}(\Gamma) \subset M(\Gamma)$. Then the valency of e is generically equal to 3, and we assume this condition is satisfied. Let F_j $(j = 1, 2, 3)$ be the faces of $\mathrm{Ford}(\Gamma)$ sharing the edge e. Take a base point b of $M(\Gamma)$ in the horospherical neighborhood of the main cusp, i.e., the image of a precisely (Γ, Γ_∞)-invariant horoball H_∞ centered at ∞. Let α_j and β_j $(j \in \{1, 2, 3\})$ be oriented arcs in $M(\Gamma)$ satisfying the following conditions (see Fig. 4.2):

1. The endpoint of α_j is equal to the initial point of α_{j+1}, where the indices are considered modulo 3, and the union $\alpha_1 \cup \alpha_2 \cup \alpha_3$ forms a meridian around the edge e.
2. Each α_j intersects $\mathrm{Ford}(\Gamma)$ transversely at a single point in $\mathrm{int}\, F_j$.

3. Each β_j is disjoint from Ford(Γ) and joins the base point b to the initial point of α_j.

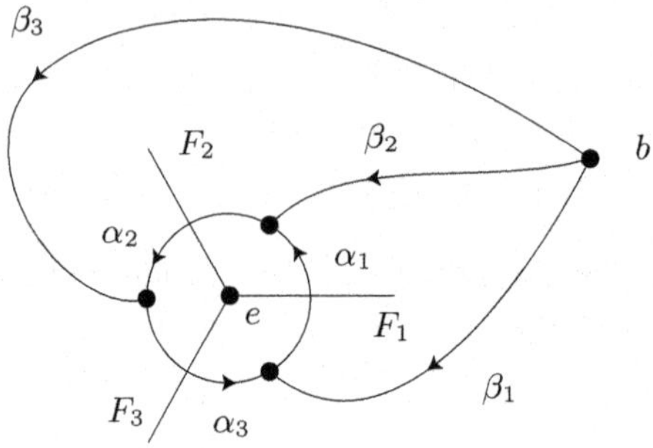

Fig. 4.2. Paths α_j and β_j

Let A_j be the element of $\pi_1(M(\Gamma), b) \cong \Gamma$ represented by the oriented loop $\beta_j \cdot \alpha_j \cdot \beta_{j+1}^{-1}$. Then $A_1A_2A_3 = 1$. Since A_j is dual to F_j in the sense that it is represented by a based loop intersecting Ford(Γ) at a single point in int F_j, the following hold.

1. The isometric hemisphere $Ih(A_j)$ supports a face of the Ford domain $Ph(\Gamma)$ which projects to F_j.
2. $Ih(A_j) \cap Ih(A_{j+1}^{-1})$ contains an edge of $Ph(\Gamma)$ which projects to e.
3. The dihedral angle θ_j between the faces F_j and F_{j+1} along e is equal to $\theta(A_j, A_{j+1}^{-1})$, the dihedral angle of $Eh(A_j) \cap Eh(A_{j+1}^{-1})$ along $Ih(A_j) \cap Ih(A_{j+1}^{-1})$.

In the above, indices are considered modulo 3. Since $\theta_1 + \theta_2 + \theta_3 = 2\pi$, we have

$$\theta(A_1, A_2^{-1}) + \theta(A_2, A_3^{-1}) + \theta(A_3, A_1^{-1}) = 2\pi.$$

If we set $A = A_1$, $B = A_2^{-1}$ then $A_3 = BA^{-1}$, because $A_1A_2A_3 = I$. The above identity is then nothing other than the identity in Lemma 4.1.2 (Chain rule). This gives a conceptual proof to the lemma in the special situation. As a matter of fact, this is the typical situation where we use the lemma. To be more precise, we use the chain rule in order to guarantee that the sum of the dihedral angles in a "cycle" is equal to 2π when we use Poincare's theorem on fundamental polyhedra (Sect. 6.4).

The following lemma is a consequence of Lemmas 4.1.1 and 4.1.2.

Lemma 4.1.3 (Chain rule for elliptic generators). *Let ρ be a type-preserving representation of $\pi_1(\mathcal{O})$, and let $\{P_j\}$ be the sequence of elliptic generators associated with a triangle $\sigma = \langle s_0, s_1, s_2 \rangle$ of $\mathcal{D}$. Assume that $\phi^{-1}(0) \cap \sigma^{(0)} = \emptyset$ and hence the isometric hemisphere $Ih(\rho(P_j))$ is well-defined for every j. Then the following hold.*

(1) $Ih(\rho(P_{j+3n})) = \rho(K)^n(Ih(\rho(P_j)))$, $Ih(\rho(P_{j-1}P_j)) = Ih(\rho(P_{j+1}))$ *and* $Ih(\rho(P_jP_{j-1})) = Ih(\rho(P_{j-2}))$.

(2) $\rho(P_j)$ *interchanges* $Ih(\rho(P_{j-1})) \cap Ih(\rho(P_j))$ *and* $Ih(\rho(P_j)) \cap Ih(\rho(P_{j+1}))$.

(3) Suppose $Ih(\rho(P_j)) \cap Ih(\rho(P_{j+1})) \neq \emptyset$ *for every integer* j. *Let* θ_j *be the dihedral angle of* $Eh(\rho(P_j)) \cap Eh(\rho(P_{j+1}))$. *Then* $\theta_{j-1} + \theta_j + \theta_{j+1} = 2\pi$.

(4) Suppose that $\phi(s'_{[j]}) \neq 0$, *where* $s'_{[j]}$ *is the vertex of* $\mathcal{D}$ *opposite to* $s_{[j]}$ *with respect to the edge* $\langle s_{[j-1]}, s_{[j+1]}\rangle$. *Then the isometric hemispheres* $Ih(P_j^{P_{j-1}})$ *and* $Ih(P_j^{P_{j+1}})$ *exist, and the following hold (see Fig. 4.3):*

$$\begin{aligned}
\rho(P_{j-1})(v(\rho; P_{j-1}, P_j, P_{j+1})) &= v(\rho; P_{j-2}, P_j^{P_{j-1}}, P_{j-1}) \\
\rho(P_j)(v(\rho; P_{j-1}, P_j, P_{j+1})) &= v(\rho; P_{j-1}, P_j, P_{j+1}) \\
\rho(P_{j+1})(v(\rho; P_{j-1}, P_j, P_{j+1})) &= v(\rho; P_{j+1}, P_j^{P_{j+1}}, P_{j+2}).
\end{aligned}$$

In the above, $v(\rho; X, Y, Z)$, *where* (X, Y, Z) *is a triple of elements of* $\pi_1(\mathcal{O})$, *denotes the set* $Ih(\rho(X)) \cap Ih(\rho(Y)) \cap Ih(\rho(Z))$.

Proof. (1) The first identity follows from Lemma 4.1.1(2) and the facts that $P_{j+3n} = K^n P_j K^{-n}$ and $\rho(K)$ is a Euclidean translation. The remaining identities also follow from the same lemma and the following identities.

$$\begin{aligned}
P_{j-1}P_j &= P_{j-1}P_jP_{j+1}P_{j+1} = K^{-1}P_{j+1}, \\
P_jP_{j-1} &= P_jP_{j-1}P_{j-2}P_{j-2} = KP_{j-2}.
\end{aligned}$$

(2) By Lemma 4.1.2(2), $\rho(P_j)(Ih(\rho(P_{j-1})) \cap Ih(\rho(P_j))) = Ih(\rho(P_{j-1}P_j)) \cap Ih(\rho(P_j))$. Since $Ih(\rho(P_{j-1}P_j)) = Ih(\rho(P_{j+1}))$ by (1), we obtain the desired result.

(3) We apply Lemma 4.1.2(3) for the pair $(A, B) := (\rho(P_j), \rho(P_{j+1}))$. Then $\theta(A, B) = \theta_j$. Since $I(A^{-1}) = I(\rho(P_j))$ and $I(BA^{-1}) = I(\rho(P_{j+1}P_j)) = I(\rho(P_{j-1}))$ by (1), we have $\theta(BA^{-1}, A^{-1}) = \theta_{j-1}$. Similarly, since $I(B^{-1}) = I(\rho(P_{j+1})))$ and $I(A^{-1}B) = I(\rho(P_jP_{j+1})) = I(\rho(P_{j+2}))$ by (1), we have $\theta(B^{-1}, AB^{-1}) = \theta_{j+1}$. Hence we obtain the desired result by Lemma 4.1.2(3).

(4) By Lemma 2.1.8(2), the bi-infinite sequence

$$\cdots, P_{j-2}, P_j^{P_{j-1}}, P_{j-1}, P_{j+1}, P_j^{P_{j+1}}, P_{j+2}, \cdots$$

form the sequence of elliptic generators associated with $\sigma' := \langle s_{[j+1]}, s'_{[j]}, s_{[j-1]}\rangle$. Hence the assumption $\phi(s'_{[j]}) \neq 0$ implies that the isometric hemispheres $Ih(P_j^{P_{j-1}})$ and $Ih(P_j^{P_{j+1}})$ exist. Moreover (2) implies

$$\begin{aligned}
\rho(P_{j-1})(Ih(\rho(P_{j-1})) \cap Ih(\rho(P_{j+1}))) &= Ih(\rho(P_j^{P_{j-1}})) \cap Ih(\rho(P_{j-1})), \\
\rho(P_{j+1})(Ih(\rho(P_{j-1})) \cap Ih(\rho(P_{j+1}))) &= Ih(\rho(P_{j+1})) \cap Ih(\rho(P_j^{P_{j+1}})).
\end{aligned}$$

Hence

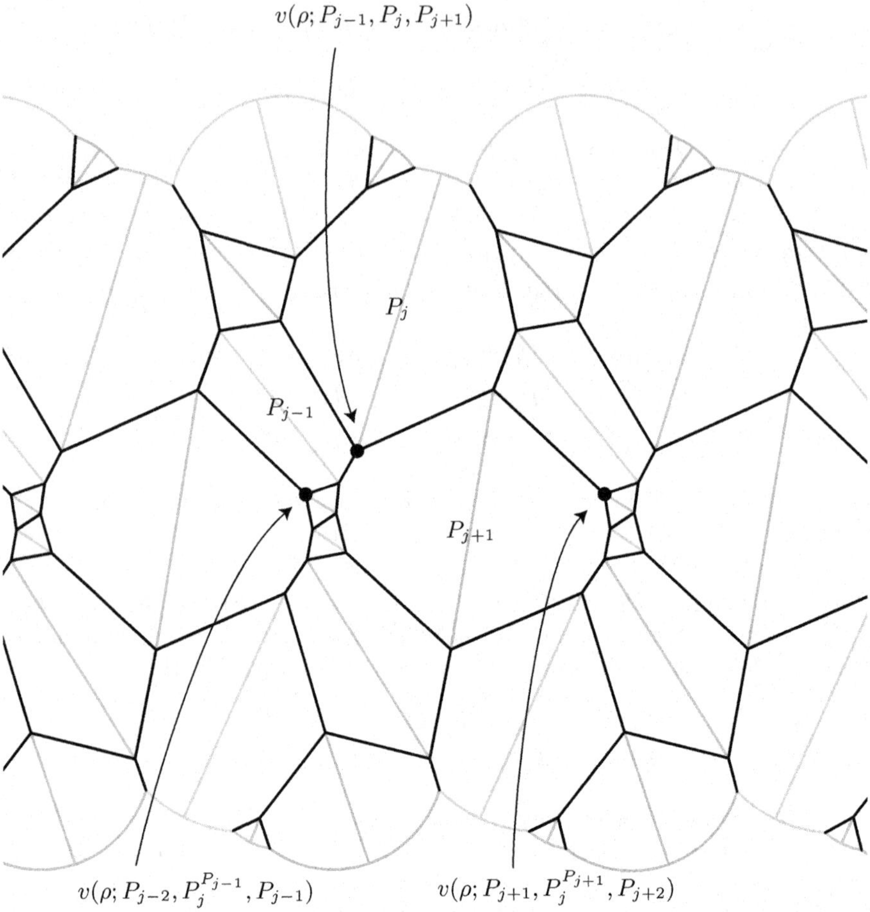

Fig. 4.3. Chain rule

$$\begin{aligned}
&\rho(P_{j-1})(v(\rho; P_{j-1}, P_j, P_{j+1})) \\
&= \rho(P_{j-1})(Ih(\rho(P_{j-1})) \cap Ih(\rho(P_j))) \cap \rho(P_{j-1})(Ih(\rho(P_j)) \cap Ih(\rho(P_{j+1}))) \\
&= (Ih(\rho(P_{j-2})) \cap Ih(\rho(P_{j-1}))) \cap (Ih(\rho(P_{j-1})) \cap Ih(\rho(P_j^{P_{j-1}}))) \\
&= v(\rho; P_{j-2}, P_j^{P_{j-1}}, P_{j-1}).
\end{aligned}$$

Thus we have the first identity. The remaining identities are proved similarly.

4.2 Side parameter

In this section, we introduce the side parameter defined by Jorgensen.

Lemma 4.2.1. *Under Assumption 2.4.6 (σ-NonZero), the following conditions are equivalent:*

1. *The positive real numbers* $|\phi(s_0)|$, $|\phi(s_1)|$, *and* $|\phi(s_2)|$ *satisfy the triangle inequality.*
2. *The positive real numbers* $\sqrt{|a_0|}$, $\sqrt{|a_1|}$, *and* $\sqrt{|a_2|}$ *satisfy the triangle inequality.*
3. $I(\rho(P_j))$ *and* $I(\rho(P_{j+1}))$ *intersect in two points for some* $j \in \mathbb{Z}$.
4. $I(\rho(P_j))$ *and* $I(\rho(P_{j+1}))$ *intersect in two points for every* $j \in \mathbb{Z}$.

Proof. It is obvious from the definition of the complex probability (see Sect. 2.4) that the first condition is equivalent to the second condition.

The equivalence between the third and the fourth conditions follows from Lemma 4.1.3(2) (Chain rule).

We prove the equivalence between the first and the last conditions. Recall that the radius of $I(\rho(P_j))$ is $|1/\phi(s_{[j]})|$ and that $d(c(\rho(P_j)), c(\rho(P_{j+1})) = |\phi(s_{[j-1]})/(\phi(s_{[j]})\phi(s_{[j+1]}))|$ (see Lemma 2.4.4). Hence $I(\rho(P_j))$ and $I(\rho(P_{j+1}))$ intersect in two points if and only if

$$\left|\frac{1}{\phi(s_{[j]})}\right| + \left|\frac{1}{\phi(s_{[j+1]})}\right| > \left|\frac{\phi(s_{[j-1]})}{\phi(s_{[j]})\phi(s_{[j+1]})}\right|.$$

This condition is equivalent to

$$|\phi(s_{[j]})| + |\phi(s_{[j+1]})| > |\phi(s_{[j-1]})|.$$

Hence we see that the third condition is equivalent to the first condition.

Definition 4.2.2 (Triangle inequality). *Under Assumption 2.4.6 (σ-NonZero), we say that ϕ (or ρ) satisfies the* triangle inequality *at σ if the (mutually equivalent) conditions of Lemma 4.2.1 are satisfied.*

Remark 4.2.3. See Lemma 4.2.11 for a geometric meaning of the similarity class of a triangle with edge lengths $|\phi(s_0)|$, $|\phi(s_1)|$, and $|\phi(s_2)|$.

In the following we presume the following two assumption:

Assumption 4.2.4 (σ-Simple). Under Assumption 2.4.6 (σ-NonZero), we assume that the following conditions are satisfied:

1. ϕ satisfies the triangle inequality at σ (Definition 4.2.2).
2. $\mathcal{L}(\rho, \sigma)$ is simple (Definition 3.1.1(2)).
3. ϕ is upward at σ (Definition 3.1.3 and Lemma 3.1.4).

Definition 4.2.5. *Under Assumption 4.2.4 (σ-Simple), we use the following notation (cf. Fig. 4.5).*

1. $\mathrm{Fix}^{\epsilon}_{\sigma}(\rho(P_j))$ *denotes the point* $c(\rho(P_j)) + \epsilon i/\phi(s_{[j]})$ *in* $\mathbb{C}$.
2. $v^{\epsilon}(\rho; P_j, P_{j+1})$ *($\epsilon = \pm$) denotes the points of* $I(\rho(P_j)) \cap I(\rho(P_{j+1}))$, *such that the vector* $\overrightarrow{v^-(\rho; P_j, P_{j+1})v^+(\rho; P_j, P_{j+1})}$ *is upward with respect to* $\mathcal{L}(\rho, \sigma)$, *i.e.,*

$$\arg \frac{v^+(\rho; P_j, P_{j+1}) - v^-(\rho; P_j, P_{j+1})}{c(\rho(P_{j+1})) - c(\rho(P_j))} = \pi/2.$$

This condition is equivalent to the following inequality.

$$\epsilon \Im \left(\frac{v^\epsilon(\rho; P_j, P_{j+1}) - c(\rho(P_j))}{c(\rho(P_{j+1})) - c(\rho(P_j))} \right) > 0.$$

It should be noted that the fixed point set of the Möbius transformation $\rho(P_j)$ on $\hat{\mathbb{C}}$ is equal to $\{\mathrm{Fix}_\sigma^-(\rho(P_j)), \mathrm{Fix}_\sigma^+(\rho(P_j))\}$.

Notation 4.2.6. If ρ, σ and $\{P_j\}$ are fixed under the setting in Definition 4.2.5, we employ the following abbreviations: $c(j) = c(\rho(P_j))$, $I(j) = I(\rho(P_j))$, $Ih(j) = Ih(\rho(P_j))$, $D(j) = D(\rho(P_j))$, $Dh(j) = Dh(\rho(P_j))$, $E(j) = E(\rho(P_j))$, $Eh(j) = Eh(\rho(P_j))$, $v^\epsilon(j, j+1) = v^\epsilon(\rho; P_j, P_{j+1})$, $\overrightarrow{c}(j, j+1) = \overrightarrow{c}(\rho; P_j, P_{j+1})$, $\mathrm{Fix}^\epsilon(j) = \mathrm{Fix}_\sigma^\epsilon(\rho(P_j))$, $\mathrm{Axis}(j) = \mathrm{Axis}(\rho(P_j))$ and $\overrightarrow{f}(j) = \overrightarrow{\mathrm{Fix}_\sigma^-(\rho(P_j))\,\mathrm{Fix}_\sigma^+(\rho(P_j))}$.

Lemma 4.2.7. *Under Assumption 4.2.4 (σ-Simple), we have:*

$$\rho(P_j)(v^\epsilon(\rho; P_{j-1}, P_j)) = v^\epsilon(\rho; P_j, P_{j+1}).$$

Proof. We may assume $j = 1$ without loss of generality. By Lemma 4.1.3(2) (Chain rule), $\rho(P_1)$ sends $v^+(0,1)$ to either $v^+(1,2)$ or $v^-(1,2)$. To prove the desired identity, let ℓ_- be the oriented line in $\mathbb{C}$ obtained by extending the oriented edge $\overrightarrow{c}(0,1)$. Then $v^+(0,1)$ and $v^-(0,1)$, respectively, lie on the left and the right of ℓ_-. Similarly, $v^+(1,2)$ and $v^-(1,2)$, respectively, lie on the left and the right of the oriented line ℓ_+ which is obtained by extending the oriented edge $\overrightarrow{c}(1,2)$. On the other hand, the transformation $\rho(P_1)$ maps the left hand side of the oriented line ℓ_- to that of the oriented line ℓ_+, because $\rho(P_1)$ is orientation-preserving and maps the oriented circle $\ell_- \cup \{\infty\}$ to the oriented circle $\ell_+ \cup \{\infty\}$. Hence it must send $v^+(0,1)$ to $v^+(1,2)$.

By using the above lemma, we obtain the following lemma.

Lemma 4.2.8. *Under Assumption 4.2.4 (σ-Simple), we have the following (see Fig. 4.4).*

$$\begin{aligned} &\epsilon \arg \frac{v^\epsilon(\rho; P_{j-1}, P_j) - c(\rho(P_j))}{\mathrm{Fix}_\sigma^\epsilon(\rho(P_j)) - c(\rho(P_j))} \\ &= \epsilon \arg \frac{\mathrm{Fix}_\sigma^\epsilon(\rho(P_j)) - c(\rho(P_j))}{v^\epsilon(\rho; P_j, P_{j+1}) - c(\rho(P_j))} \not\equiv \pi \pmod{2\pi}. \end{aligned}$$

Proof. We may assume $j = 1$ without loss of generality. Since $\rho(P_1)$ acts on $I(1)$ as an orientation-reversing Euclidean isometry, Lemma 4.2.7 implies that the above two angles are equal. In the remainder we prove that these angles are not equal to π modulo 2π. For simplicity we prove this only for the case $\epsilon = +$. By definition, $\mathrm{Fix}^-(1)$ lies on the right hand side of the oriented line ℓ_- in the proof of Lemma 4.2.7, and the point $v^+(0,1)$ lies on the left hand side of ℓ_-; in fact $\mathrm{Fix}^-(1) \in \operatorname{int} H_R(\ell_-)$ and $v^+(0,1) \in \operatorname{int} H_L(\ell_-)$ (cf. Definition 3.1.6). Hence $v^+(0,1) \neq \mathrm{Fix}^-(1)$. This implies that the complex number $v^+(0,1) - c(1)$ is not a negative real multiple of $\mathrm{Fix}^+(1) - c(1)$. Hence we obtain the conclusion.

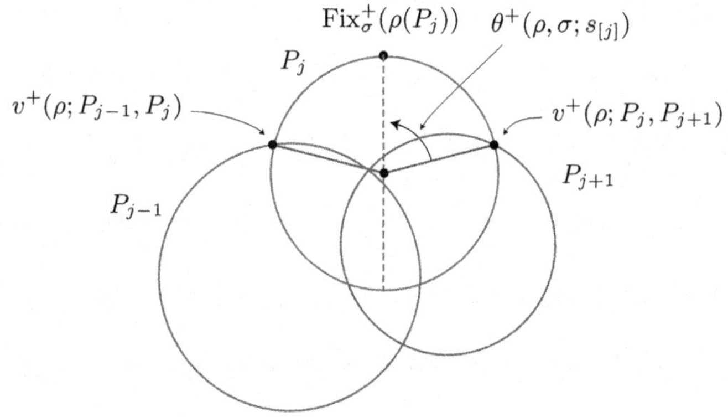

Fig. 4.4. Side parameter

We now introduce the following key definition.

Definition 4.2.9 (Side parameter). *Under Assumption 4.2.4 (σ-Simple):*

(1) The angle in Lemma 4.2.8 is denoted by $\theta^\epsilon(\rho, \sigma; s_{[j]}) \in (-\pi, \pi)$. (See Fig. 4.4.)

(2) The ϵ-side parameter *of ρ at σ, denoted by $\theta^\epsilon(\rho, \sigma)$, is the triple defined by*

$$\theta^\epsilon(\rho, \sigma) := (\theta^\epsilon(\rho, \sigma; s_0), \theta^\epsilon(\rho, \sigma; s_1), \theta^\epsilon(\rho, \sigma; s_2)).$$

We note that the terminology, "parameter", in the above is eventually justified by Theorem 1.3.2.

Notation 4.2.10. Under Assumption 4.2.4 (σ-Simple):

(1) $\Delta^\epsilon_j(\rho, \sigma)$, Δ^ϵ_j for short, denotes the Euclidean triangle

$$\Delta^\epsilon_j(\rho, \sigma) := \Delta(c(\rho(P_j)), c(\rho(P_{j+1})), v^\epsilon(\rho; P_j, P_{j+1}))$$

in $\mathbb{C}$ spanned by the three points $c(\rho(P_j))$, $c(\rho(P_{j+1}))$, and $v^\epsilon(\rho; P_j, P_{j+1})$, for each $j \in \mathbb{Z}$.

(2) $\Delta(\rho, \sigma)$ denotes the (similarity class) of a Euclidean triangle $\Delta(w_0, w_1, w_2)$ with vertices w_0, w_1 and w_2 such that

$$|w_2 - w_1| = |\phi(s_0)|, \quad |w_0 - w_2| = |\phi(s_1)|, \quad |w_1 - w_0| = |\phi(s_2)|,$$

and $\alpha(\rho, \sigma; s_j) \in (0, \pi)$ denotes the (inner) angle of the triangle at the vertex w_j.

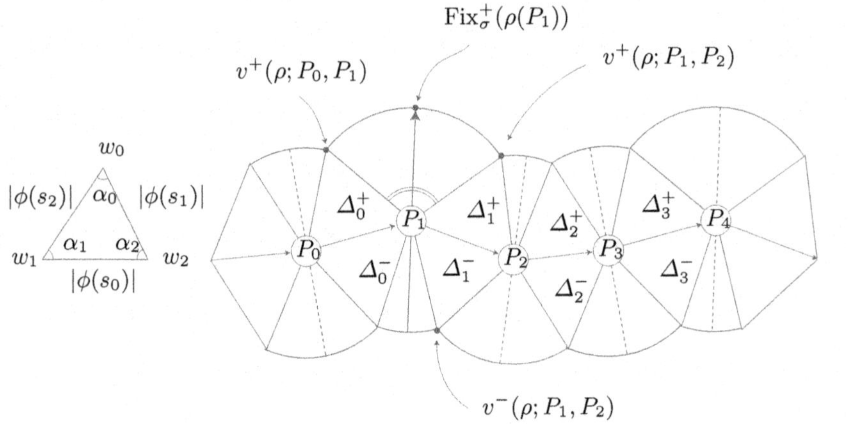

Fig. 4.5. $\Delta_j^\epsilon(\rho, \sigma)$ and $\Delta(\rho, \sigma)$

The following lemma is easily proved (see Fig. 4.5).

Lemma 4.2.11. *Under Assumption 4.2.4 (σ-Simple), the following hold.*

(1) $2\theta^\epsilon(\rho, \sigma; s_j)$ is equal to the "signed angle" at $c(\rho(P_j))$ between the two triangles $\Delta_{j-1}^\epsilon(\rho, \sigma)$ and $\Delta_j^\epsilon(\rho, \sigma)$, where the sign is defined so that $\theta^\epsilon(\rho, \sigma; s_j) \geq 0$ if and only if $\operatorname{int} \Delta_{j-1}^\epsilon(\rho, \sigma) \cap \operatorname{int} \Delta_j^\epsilon(\rho, \sigma) = \emptyset$.

(2) $\Delta_j^\epsilon(\rho, \sigma)$ is similar to $\Delta(\rho, \sigma)$ by a similarity which maps the ordered triple $(c(\rho(P_j)), c(\rho(P_{j+1})), v^\epsilon(\rho; P_j, P_{j+1}))$ to $(w_{[j]}, w_{[j+1]}, w_{[j+2]})$.

Here is an intuitive geometric description of the sequence of the triangles $\{\Delta_j^\epsilon(\rho, \sigma)\}_{j \in \mathbb{Z}}$. $\Delta_j^\epsilon(\rho, \sigma)$ is obtained from $\Delta_{j-1}^\epsilon(\rho, \sigma)$ by:

1. rotating about the vertex $c(j)$ so that the edge $\overline{c(j)v^\epsilon(j-1, j)}$ overlaps the edge $\overline{c(j)c(j+1)}$, and then
2. shrinking or expanding so that the image of $\overline{c(j)v^\epsilon(j-1, j)}$ coincides with $\overline{c(j)c(j+1)}$.

Lemma 4.2.12. *Under Assumption 2.4.6 (σ-NonZero), suppose that $\mathcal{L}(\rho, \sigma)$ is simple. Then we have*

$$\arg \frac{\psi(\vec{e}_{[j-1]})}{\psi(\vec{e}_{[j+1]})} = \arg \frac{\vec{c}(\rho; P_j, P_{j+1})}{\vec{c}(\rho; P_{j-1}, P_j)} \in (-\pi, \pi).$$

Moreover, the sum of any three successive arguments is equal to 0, namely

$$\sum_{j=0}^{2} \arg \frac{\psi(\vec{e}_{[j-1]})}{\psi(\vec{e}_{[j+1]})} = \arg \frac{\psi(\vec{e}_1)}{\psi(\vec{e}_0)} + \arg \frac{\psi(\vec{e}_2)}{\psi(\vec{e}_1)} + \arg \frac{\psi(\vec{e}_0)}{\psi(\vec{e}_2)} = 0.$$

Proof. By Proposition 2.4.4(2.1),

$$\frac{\psi(\overrightarrow{e}_{[j-1]})}{\psi(\overrightarrow{e}_{[j+1]})} = \frac{\overrightarrow{c}(j,j+1)}{\overrightarrow{c}(j-1,j)}$$

Since $\mathcal{L}(\rho,\sigma)$ is simple, this (nonzero) complex number is not a negative real number. Hence its argument is contained in $(-\pi,\pi)$, and we obtain the first assertion.

To prove the second assertion, pick a large positive real number, h, such that the horizontal line $\Im z = h$ is disjoint from $\mathcal{L}(\rho,\sigma)$. Let M be the submanifold of the Euclidean annulus $\mathbb{C}/\mathbb{Z}$ which is obtained as the image of the region bounded by the horizontal line and $\mathcal{L}(\rho,\sigma)$. Then the desired formula is obtained by applying the Gauss-Bonnet theorem to M.

Corollary 4.2.13. *Under Assumption 4.2.4 (σ-Simple), if $\mathcal{L}(\rho,\sigma)$ is convex to the above at some vertex, then it is convex to the below at some other vertex, and vice versa. In particular, if $\mathcal{L}(\rho,\sigma)$ is not flat (i.e., convex to the above or below) at every vertex, then one of the following holds after a shift of indices.*

1. *$\mathcal{L}(\rho,\sigma)$ is convex to the above at $c(\rho(P_0))$ and convex to the below at $c(\rho(P_1))$ and $c(\rho(P_2))$.*
2. *$\mathcal{L}(\rho,\sigma)$ is convex to the below at $c(\rho(P_0))$ and convex to the above at $c(\rho(P_1))$ and $c(\rho(P_2))$.*

Proof. By the second identity in Lemma 4.2.12, it follows that if one of $\Im(\psi(\overrightarrow{e}_1)/\psi(\overrightarrow{e}_0))$, $\Im(\psi(\overrightarrow{e}_2)/\psi(\overrightarrow{e}_1))$ and $\Im(\psi(\overrightarrow{e}_0)/\psi(\overrightarrow{e}_2))$ is positive (resp. negative) then some of them is negative (positive). The desired result follows from this fact.

Lemma 4.2.14. *Under Assumption 4.2.4 (σ-Simple), the following identity holds for every integer j.*

$$\begin{aligned}\theta^{\epsilon}(\rho,\sigma;s_{[j]}) &= \frac{1}{2}\left(\pi - \epsilon \arg \frac{\overrightarrow{c}(\rho;P_j,P_{j+1})}{\overrightarrow{c}(\rho;P_{j-1},P_j)} - 2\alpha(\rho,\sigma;s_{[j]})\right)\\ &= \frac{1}{2}\left(\pi - \epsilon \arg \frac{\psi(\overrightarrow{e}_{[j-1]})}{\psi(\overrightarrow{e}_{[j+1]})} - 2\alpha(\rho,\sigma;s_{[j]})\right),\end{aligned}$$

Proof. Throughout the proof we use the following notation.

1. $\arg : \mathbb{C} - \mathbb{R}_{\le 0} \to (-\pi,\pi)$ denotes the argument in $(-\pi,\pi)$ of a complex number which does not belong to $\mathbb{R}_{\le 0} := \{z \in \mathbb{R} \mid z \le 0\}$.
2. Let z_1, z_2 and f be non-zero complex numbers such that z_1/f and z_2/f are not nonnegative real numbers and $\arg(z_1/f)$ and $\arg(f/z_2)$ have the same sign, that is, either they are both nonnegative or they are both nonpositive. Then we define

$$\arg_f \frac{z_2}{z_1} := \arg \frac{f}{z_1} + \arg \frac{z_2}{f} \in (-2\pi, 2\pi).$$

Then $\arg_f(z_2/z_1)$ is the signed length of the arc in the unit circle bounded by $z_1/|z_1|$ and $z_2/|z_2|$ which contains $f/|f|$.

For simplicity, we prove the lemma only when $\epsilon = +$ and $j = 1$. Consider the following nonzero complex numbers (see Fig. 4.6):

$$\begin{aligned} f &= \mathrm{Fix}^+(1) - c(1), \\ v_L &= v^+(0,1) - c(1), \\ v_R &= v^+(1,2) - c(1), \\ w_L &= -\overrightarrow{c}(0,1), \\ w_R &= \overrightarrow{c}(1,2). \end{aligned}$$

By the definition of the side parameter,

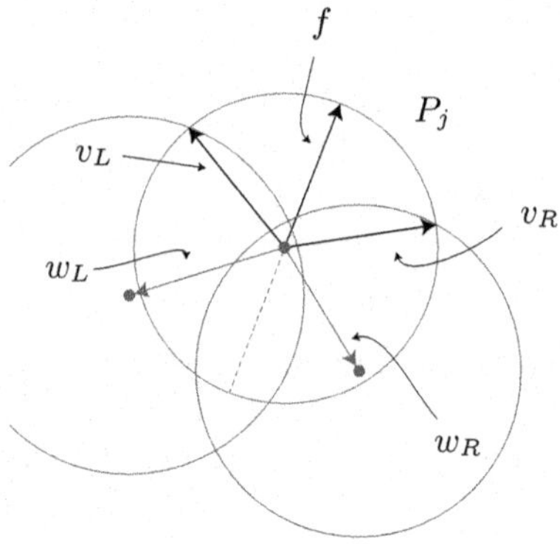

Fig. 4.6. Figure of f, v_L, v_R, w_L and w_R

$$\theta^+(\rho, \sigma; s_1) = \arg \frac{f}{v_R} = \arg \frac{v_L}{f} \in (-\pi, \pi).$$

Hence

$$\arg_f \frac{v_L}{v_R} = \arg \frac{f}{v_R} + \arg \frac{v_L}{f} = 2\theta^+(\rho, \sigma; s_1) \in (-2\pi, 2\pi).$$

By the definition of $\mathrm{Fix}^+(1)$,

$$\arg \frac{f}{w_R} = \arg \frac{w_L}{f} \in (0, \pi).$$

Hence

$$\arg_f \frac{w_L}{w_R} = \arg \frac{f}{w_R} + \arg \frac{w_L}{f} \in (0, 2\pi).$$

Since both $\arg(f/w_R)$ and $\arg(v_R/w_R) = \alpha(\rho,\sigma;s_1)$ belong to $(0,\pi)$, $\arg(f/w_R) - \arg(v_R/w_R)$ belongs to $(-\pi,\pi)$, and hence

$$\theta^+(\rho,\sigma;s_1) = \arg\frac{f}{v_R} = \arg\frac{f}{w_R} - \arg\frac{v_R}{w_R} = \arg\frac{f}{w_R} - \alpha(\rho,\sigma;s_1) \in (-\pi,\pi).$$

Similarly we have

$$\theta^+(\rho,\sigma;s_1) = \arg\frac{v_L}{f} = \arg\frac{w_L}{f} - \arg\frac{w_L}{v_L} = \arg\frac{w_L}{f} - \alpha(\rho,\sigma;s_1) \in (-\pi,\pi).$$

These two equalities imply

$$2\theta^+(\rho,\sigma;s_1) = \arg\frac{w_L}{f} + \arg\frac{f}{w_R} - 2\alpha(\rho,\sigma;s_1) = \arg_f\frac{w_L}{w_R} - 2\alpha(\rho,\sigma;s_1).$$

On the other hand, $-w_L/w_R$ is not a nonpositive real by the assumption that $\mathcal{L}(\rho,\sigma)$ is simple. Therefore we have the well-defined argument $\arg(-w_L/w_R) \in (-\pi,\pi)$. Thus $\pi - \arg(-w_L/w_R)$ is contained in $(0,2\pi)$. This implies

$$\arg_f\frac{w_L}{w_R} = \pi - \arg(-w_L/w_R),$$

because

$$\arg_f\frac{w_L}{w_R} \in (0,2\pi) \quad \text{and} \quad \arg_f\frac{w_L}{w_R} \equiv \pi - \arg(-w_L/w_R) \pmod{2\pi}.$$

Hence

$$2\theta^+(\rho,\sigma;s_1) = \pi - \arg(-w_L/w_R) - 2\alpha(\rho,\sigma;s_1).$$

This is equivalent to the desired identity.

As an immediate corollary to the above lemma, we have the following.

Corollary 4.2.15. *Under Assumption 4.2.4 (σ-Simple), the following holds.*

1. *$\mathcal{L}(\rho,\sigma)$ is convex to the above at $c(\rho(P_j))$, if and only if $\theta^+(\rho,\sigma;s_{[j]}) > \theta^-(\rho,\sigma;s_{[j]})$.*
2. *$\mathcal{L}(\rho,\sigma)$ is convex to the below at $c(\rho(P_j))$, if and only if $\theta^+(\rho,\sigma;s_{[j]}) < \theta^-(\rho,\sigma;s_{[j]})$.*

Proof. The desired result is a consequence of Definition 3.1.8 and the following identity, which follows from Lemma 4.2.14.

$$\theta^-(\rho,\sigma;s_{[j]}) - \theta^+(\rho,\sigma;s_{[j]}) = \arg\frac{\psi(\overrightarrow{e}_{[j-1]})}{\psi(\overrightarrow{e}_{[j+1]})}.$$

We now obtain the following important property concerning the side parameters.

Proposition 4.2.16. *Under Assumption 4.2.4 (σ-Simple), the following hold.*

$$\theta^\epsilon(\rho,\sigma;s_0)+\theta^\epsilon(\rho,\sigma;s_1)+\theta^\epsilon(\rho,\sigma;s_2)=\frac{\pi}{2}.$$

Proof. By Lemmas 4.2.12 and 4.2.14

$$\begin{aligned}\sum_{j=0}^{2}\theta^\epsilon(\rho,\sigma;s_j)&=\frac{1}{2}\sum_{j=0}^{2}\left(\pi-\epsilon\arg\frac{\psi(\overrightarrow{e}_{[j-1]})}{\psi(\overrightarrow{e}_{[j+1]})}-2\alpha(\rho,\sigma;s_{[j]})\right)\\&=\frac{1}{2}\left(3\pi-\epsilon\sum_{j=0}^{2}\arg\frac{\psi(\overrightarrow{e}_{[j-1]})}{\psi(\overrightarrow{e}_{[j+1]})}-2\sum_{j=0}^{2}\alpha(\rho,\sigma;s_{[j]})\right)\\&=\frac{1}{2}(3\pi-0-2\pi)=\frac{\pi}{2}.\end{aligned}$$

Definition 4.2.17. *When all components of $\theta^\epsilon(\rho,\sigma)$ are non-negative, it is regarded as $\pi/2$ times the barycentric coordinate of a point in the triangle σ of the Farey triangulation $\mathcal{D}$. We denote the point by the same symbol $\theta^\epsilon(\rho,\sigma)$. When at most one component of $\theta^\epsilon(\rho,\sigma)$ is zero and the other components are positive, $\theta^\epsilon(\rho,\sigma)$ is also identified with a point in $\mathbb{H}^2$, via its identification with $\mathbb{H}^2\cong|\mathcal{D}|-|\mathcal{D}^{(0)}|$ described in Sect. 1.3.*

The following lemma gives an algebraic characterization of the side parameter, which is used in Chap. 9.

Lemma 4.2.18. *Under Assumption 4.2.4 (σ-Simple), the following identity holds for every integer j.*

$$\phi(s_{[j-1]})+\epsilon i e^{\epsilon i\theta^\epsilon_{[j+1]}}\phi(s_{[j]})-\epsilon i e^{-\epsilon i\theta^\epsilon_{[j]}}\phi(s_{[j+1]})=0,$$

where $\theta^\epsilon_j=\theta^\epsilon(\rho,\sigma;s_j)$ and ϕ is the upward Markoff map inducing ρ. Moreover, each identity is equivalent to the identity

$$\phi(s_0)+\alpha^\epsilon\phi(s_1)+\beta^\epsilon\phi(s_2)=0,$$

where $\alpha^\epsilon=\epsilon i\exp(\epsilon i\theta^\epsilon_2)$ and $\beta^\epsilon=-\epsilon i\exp(-\epsilon i\theta^\epsilon_1)$.

Proof. Since $v^\epsilon(j,j+1)$ and $\mathrm{Fix}^\epsilon(j)$ are contained in $I(j)$, we have

$$|v^\epsilon(j,j+1)-c(j)|=|\,\mathrm{Fix}^\epsilon(j)-c(j)|.$$

Hence, by the definition of the side parameter, we have

$$\begin{aligned}v^\epsilon(j,j+1)-c(j)&=e^{-\epsilon i\theta^\epsilon_{[j]}}(\mathrm{Fix}^\epsilon(j)-c(j))\\&=e^{-\epsilon i\theta^\epsilon_{[j]}}\left(\frac{\epsilon i}{\phi(s_{[j]})}\right)\end{aligned}$$

Thus we have

$$v^\epsilon(j,j+1) = c(j) + \frac{\epsilon i}{\phi(s_{[j]})} e^{-\epsilon i \theta^\epsilon_{[j]}}.$$

Similarly, we also have

$$v^\epsilon(j,j+1) = c(j+1) + \frac{\epsilon i}{\phi(s_{[j+1]})} e^{\epsilon i \theta^\epsilon_{[j+1]}}.$$

Hence,

$$c(j) + \frac{\epsilon i}{\phi(s_{[j]})} e^{-\epsilon i \theta^\epsilon_{[j]}} = c(j+1) + \frac{\epsilon i}{\phi(s_{[j+1]})} e^{\epsilon i \theta^\epsilon_{[j+1]}}.$$

So, by using Lemma 2.4.4(2.1), we have

$$\frac{\phi(s_{[j-1]})}{\phi(s_{[j]})\phi(s_{[j+1]})} = c(j+1) - c(j) = \frac{\epsilon i}{\phi(s_{[j]})} e^{-\epsilon i \theta^\epsilon_{[j]}} - \frac{\epsilon i}{\phi(s_{[j+1]})} e^{\epsilon i \theta^\epsilon_{[j+1]}}.$$

This is equivalent to the following identity:

$$\phi(s_{[j-1]}) = \epsilon i e^{-\epsilon i \theta^\epsilon_{[j]}} \phi(s_{[j+1]}) - \epsilon i e^{\epsilon i \theta^\epsilon_{[j+1]}} \phi(s_{[j]}).$$

Thus we obtain the first identity. The second identity is obtained by putting $j = 1$, and we can easily check that the identity for any integer j is equivalent to this identity.

The above lemma motivates the following definition.

Definition 4.2.19. *(1) Let $\sigma = \langle s_0, s_1, s_2 \rangle$ be a triangle of $\mathcal{D}$ and $\nu = (\theta_0, \theta_1, \theta_2)$ be a point in $\sigma \cap \mathbb{H}^2$. Then $\zeta^\epsilon_{\nu,\sigma}\colon \Phi \to \mathbb{C}$ denotes the map defined by:*

$$\zeta^\epsilon_{\nu,\sigma}(\phi) = \phi(s_0) + \alpha^\epsilon \phi(s_1) + \beta^\epsilon \phi(s_2),$$

where $\alpha^\epsilon = \epsilon i \exp(\epsilon i \theta_2)$ and $\beta^\epsilon = -\epsilon i \exp(-\epsilon i \theta_1)$.

(2) Let σ^ϵ be a triangle of $\mathcal{D}$ and ν^ϵ be a point in $\sigma^\epsilon \cap \mathbb{H}^2$ for each $\epsilon \in \{-,+\}$. Then a Markoff map ϕ is called an algebraic root for $((\nu^-, \sigma^-), (\nu^+, \sigma^+))$ *if $\zeta^\epsilon_{\nu,\sigma^\epsilon}(\phi) = 0$ for each $\epsilon \in \{-,+\}$.*

(3) For a given $\boldsymbol{\nu} = (\nu^-, \nu^+) \in \mathbb{H}^2 \times \mathbb{H}^2$, a Markoff map ϕ is called an algebraic root for $\boldsymbol{\nu}$ *if it is an* algebraic root for $((\nu^-, \sigma^-(\boldsymbol{\nu})), (\nu^+, \sigma^+(\boldsymbol{\nu})))$, *where $\sigma^\epsilon(\boldsymbol{\nu})$ is the ϵ-terminal triangle of $\Sigma(\boldsymbol{\nu})$ (Definition 3.3.3).*

Remark 4.2.20. Suppose a Markoff map ϕ realizes $\boldsymbol{\nu} = (\nu^-, \nu^+) \in \mathbb{H}^2 \times \mathbb{H}^2$, namely, ϕ satisfies Assumption 4.2.4 (σ-Simple) for the two triangles $\sigma^-(\boldsymbol{\nu})$ and $\sigma^+(\boldsymbol{\nu})$ and $\theta^\epsilon(\rho, \sigma^\epsilon(\boldsymbol{\nu}))$ is identified with ν^ϵ for each $\epsilon \in \{-,+\}$. Then ϕ is an algebraic root for $\boldsymbol{\nu}$. However, the converse does not hold, that is, not every algebraic root for $\boldsymbol{\nu}$ does not necessarily realize $\boldsymbol{\nu}$, because it does not necessarily satisfy Assumption 4.2.4 (σ-Simple) for the two triangles $\sigma^-(\boldsymbol{\nu})$ and $\sigma^+(\boldsymbol{\nu})$.

4.3 ϵ-terminal triangles

Lemma 4.3.1. *Under Assumption 4.2.4 (σ-Simple), the following conditions are equivalent for each $j \in \mathbb{Z}$ and $\epsilon \in \{-,+\}$ (see Figs. 4.7 and 4.8):*

1. $\theta^\epsilon(\rho,\sigma;s_{[j]}) > 0$ *(resp.* $\theta^\epsilon(\rho,\sigma;s_{[j]}) = 0$*).*
2. $|\phi(s_{[j+1]})+\epsilon i\phi(s_{[j-1]})| > |\phi(s_{[j]})|$ *(resp.* $|\phi(s_{[j+1]})+\epsilon i\phi(s_{[j-1]})| = |\phi(s_{[j]})|$*).*
3. $\mathrm{Fix}^\epsilon_\sigma(\rho(P_j)) \notin D(\rho(P_{j-1}))$ *(resp.* $\mathrm{Fix}^\epsilon_\sigma(\rho(P_j)) \in I(\rho(P_{j-1}))$*).*
4. $\mathrm{Fix}^\epsilon_\sigma(\rho(P_j)) \notin D(\rho(P_{j+1}))$ *(resp.* $\mathrm{Fix}^\epsilon_\sigma(\rho(P_j)) \in I(\rho(P_{j+1}))$*).*
5. $v^\epsilon(\rho;P_j,P_{j+1}) \notin D(\rho(P_{j-1}))$ *(resp.* $v^\epsilon(\rho;P_{j-1},P_j) = \mathrm{Fix}^\epsilon_\sigma(\rho(P_j))$*).*
6. $v^\epsilon(\rho;P_{j-1},P_j) \notin D(\rho(P_{j+1}))$ *(resp.* $v^\epsilon(\rho;P_j,P_{j+1}) = \mathrm{Fix}^\epsilon_\sigma(\rho(P_j))$*).*

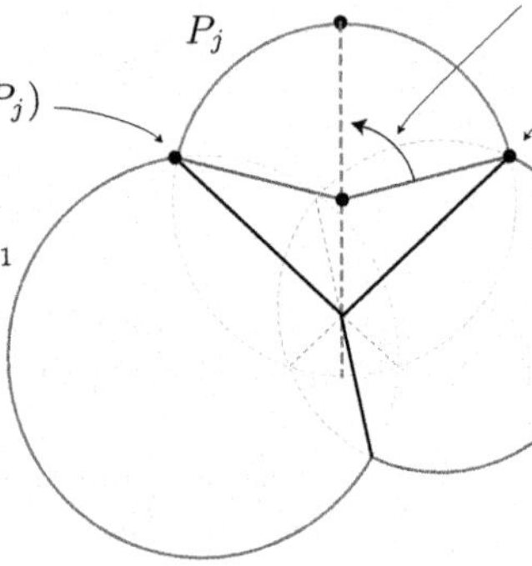

Fig. 4.7.

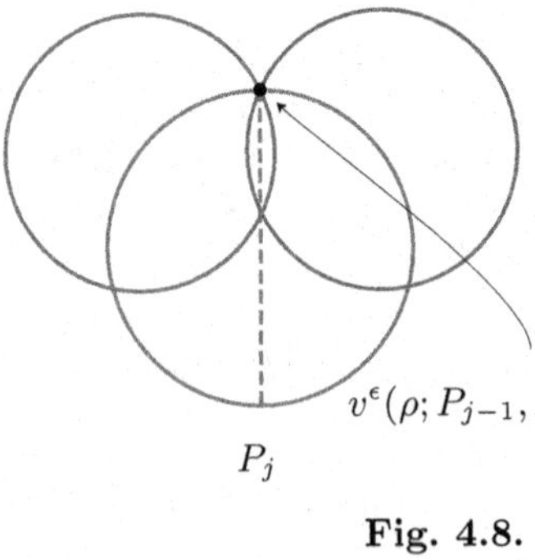

Fig. 4.8.

Proof. For simplicity, we prove the assertion for the case $\epsilon = +$ and $j = 1$. We first show that Condition 1 is equivalent to Conditions 4 and 6. Set $v^\pm = v^\pm(1,2)$, $I_L = I(1) \cap H_L(\overrightarrow{v^-v^+})$ and $I_R = I(1) \cap H_R(\overrightarrow{v^-v^+})$ (cf. Definition 3.1.6).

Claim 4.3.2. $I_R = I(1) \cap D(2)$.

Proof. Since $I(1) \cap I(2) = \{v^-, v^+\}$, $I(1) \cap D(2)$ is equal to either I_L or I_R. To show that it is actually equal to I_R, let x_L and x_R, respectively, be the intersection points of I_L and I_R with the line containing $\overrightarrow{c}(1,2)$. By using the fact that both $c(2)$ and x_R lie on the right hand side of the oriented line $\overrightarrow{f}(1) = \overrightarrow{\mathrm{Fix}^-(1)\,\mathrm{Fix}^+(1)}$, we see $d(c(2), x_R) < d(c(2), x_L)$. Since precisely one of x_L and x_R is contained in $D(2)$, we see $x_R \in \operatorname{int} D(2)$. Hence we obtain the desired result.

We apply a rotational coordinate change of $\mathbb{C}$ so that $\arg(\overrightarrow{f}(1)) = \frac{\pi}{2}$. Then we have the following claim.

Claim 4.3.3. $\Im(v^-) < \Im(v^+)$.

Proof. By Definitions 3.1.3 (Upward) and 4.2.5(1), we have

$$0 < \arg \frac{\overrightarrow{f}(1)}{\overrightarrow{c}(1,2)} < \pi.$$

Since $\arg(\overrightarrow{f}(1)) = \frac{\pi}{2}$ in the new coordinate, we see $\arg(\overrightarrow{c}(1,2)) \in (-\frac{\pi}{2}, \frac{\pi}{2})$. On the other hand, $\arg(\overrightarrow{v^-v^+}) = \arg(\overrightarrow{c}(1,2)) + \frac{\pi}{2}$ by Definition 4.2.5(2). Hence $\arg(\overrightarrow{v^-v^+}) \in (0, \pi)$ and therefore $\Im(v^-) < \Im(v^+)$.

Let $p : \mathbb{R} \to I(1)$ be the covering projection defined by $p(t) = c(1) + re^{it}$ where r is the radius of $I(1)$. By the normalization $\arg(\overrightarrow{f}(1)) = \frac{\pi}{2}$, we have $\mathrm{Fix}^+(1) = p(\frac{\pi}{2})$. Set $\theta^+(1) = \theta^+(\rho, \sigma; s_1)$ and $\tilde{v}^+ = \frac{\pi}{2} - \theta^+(1)$. Then $v^+ = p(\tilde{v}^+)$. Let $\tilde{v}^-$ be the point of $\mathbb{R}$ such that $\tilde{v}^- < \tilde{v}^+$ and $(\tilde{v}^-, \tilde{v}^+) \cap p^{-1}(v^-) = \emptyset$. Then $I_R = p([\tilde{v}^-, \tilde{v}^+])$. Moreover, by Claim 4.3.3, we see the following.

1. If $\theta^+(1) > 0$, then $-\frac{3\pi}{2} < -\frac{3\pi}{2} + \theta^+(1) < \tilde{v}^- < \tilde{v}^+ = \frac{\pi}{2} - \theta^+(1) < \frac{\pi}{2}$.
2. If $\theta^+(1) < 0$, then $-\frac{3\pi}{2} - \theta^+(1) < \tilde{v}^- < \frac{\pi}{2} + \theta^+(1) < \frac{\pi}{2} < \frac{\pi}{2} - \theta^+(1) = \tilde{v}^+$.

In fact, Claim 4.3.3 implies that $-\frac{3\pi}{2} + \theta^+(1) < \tilde{v}^- < \frac{\pi}{2} - \theta^+(1)$ or $-\frac{3\pi}{2} - \theta^+(1) < \tilde{v}^- < \frac{\pi}{2} + \theta^+(1)$ according as $\theta^+(1) > 0$ or $\theta^+(1) < 0$. By using these observations and Claim 4.3.2, we obtain the following claim.

Claim 4.3.4. 1. If $\theta^+(1) > 0$, then neither $\mathrm{Fix}^+(1)$ nor $v^+(0,1)$ belong to $I_R = I(1) \cap D(2)$.
2. If $\theta^+(1) < 0$, then both $\mathrm{Fix}^+(1)$ and $v^+(0,1)$ belong to the interior of $I_R = I(1) \cap D(2)$.
3. If $\theta^+(1) = 0$, then $\mathrm{Fix}^+(1) = v^+(0,1) \in I(1) \cap I(2)$.

Proof. (1) Suppose $\theta^+(1) > 0$. Then $\frac{\pi}{2} + 2k\pi$ and $\frac{\pi}{2} + \theta^+(1) + 2k\pi$ do not belong to $[\tilde{v}^-, \tilde{v}^+]$ for every $k \in \mathbb{Z}$ by the preceding observation. Since $\mathrm{Fix}^+(1) = p(\frac{\pi}{2})$ and $v^+(0,1) = p(\frac{\pi}{2} + \theta^+(1))$, we obtain the desired result.

(2) Suppose $\theta^+(1) < 0$. Then both $\frac{\pi}{2}$ and $\frac{\pi}{2} + \theta^+(1)$ are contained in $(\tilde{v}^-, \tilde{v}^+)$ by the preceding observation. So we obtain the conclusion.

(3) follows from the definition of $\theta^+(1) = \theta^+(\rho, \sigma; s_1)$.

The above claim implies that the Condition 1 is equivalent to the Conditions 4 and 6.

By a parallel argument, we can also see that Condition 1 is equivalent to the Conditions 3 and 5.

Finally, we show that Condition 2 is equivalent to Condition 3. Recall that the radius of $I(0)$ is $1/|\phi(s_0)|$, $c(1)-c(0)=\phi(s_1)/(\phi(s_0)\phi(s_1))$, and that $\mathrm{Fix}^+(1)=c(1)+i/\phi(s_1)$ (see Lemma 2.4.4). Hence, $\mathrm{Fix}^+(1)\notin D(0)$ if and only if

$$\left|\frac{\phi(s_1)}{\phi(s_0)\phi(s_1)}+\frac{i}{\phi(s_1)}\right|>\left|\frac{1}{\phi(s_0)}\right|.$$

This is equivalent to the inequality in Condition 2. Similarly, we can see that the condition $\mathrm{Fix}^+(1)\in I(0)$ is equivalent to the equality in the parenthesis in Condition 2.

This completes the proof of Lemma 4.3.1

Lemma 4.3.5. *Under Assumption 4.2.4 (σ-Simple), $I(\rho(P_j))\subset D(\rho(P_{j-1}))\cup D(\rho(P_{j+1}))$ if and only if $\theta^\epsilon(\rho,\sigma;s_{[j]})\le 0$ for both $\epsilon=-$ and $+$. Moreover, the following hold.*

(1) If $\theta^\epsilon(\rho,\sigma;s_{[j]})\le 0$ and $\theta^{-\epsilon}(\rho,\sigma;s_{[j]})<0$, then

$$Ih(\rho(P_j))\cap(Eh(\rho(P_{j-1}))\cap Eh(\rho(P_{j+1})))=\emptyset.$$

(2) If $\theta^\epsilon(\rho,\sigma;s_{[j]})=0$ and $\theta^{-\epsilon}(\rho,\sigma;s_{[j]})<0$, then

$$\begin{aligned}I(\rho(P_j))\cap(E(\rho(P_{j-1}))\cap E(\rho(P_{j+1})))&=\{v^\epsilon(\rho;P_{j-1},P_j)\}=\{v^\epsilon(\rho;P_j,P_{j+1})\}\\&=\{\mathrm{Fix}^\epsilon_\sigma(\rho(P_j))\},\end{aligned}$$

(3) If $\theta^-(\rho,\sigma;s_{[j]})=\theta^+(\rho,\sigma;s_{[j]})=0$, then

$$Ih(\rho(P_j))\cap(Eh(\rho(P_{j-1}))\cap Eh(\rho(P_{j+1})))=\mathrm{Axis}(\rho(P_j)).$$

Proof. We may assume $j=1$ without loss of generality. Suppose $I(1)\subset D(0)\cup D(2)$. Then $\mathrm{Fix}^\pm(1)\subset D(0)\cup D(2)$. Hence $\theta^\pm(\rho,\sigma;s_1)\le 0$ by Lemma 4.3.1.

To prove the converse, let ℓ be the oriented line in $\mathbb{C}$ containing $\mathrm{proj}(\mathrm{Axis}(1))$ oriented so that ℓ is upward with respect to $\mathcal{L}(\rho,\sigma)$ (cf. Definition 3.1.3). Let $D_L(1)$ (resp. $D_R(1)$) be the closure of the component of $D(1)-\ell$ which lies on the left (resp. right) of ℓ. Now suppose $\theta^\pm(\rho,\sigma;s_1)\le 0$. Then, by Lemma 4.3.1, both $D(0)$ and $D(2)$ contain $\mathrm{Fix}^\pm(1)$ and hence contain the diameter, $\mathrm{proj}(\mathrm{Axis}(1))$, of $D(1)$. This implies $D_L(1)\subset D(0)$ and $D_R(1)\subset D(2)$, because the center $c(0)$ of $D(0)$ lies on the left of ℓ and the center $c(2)$ of $D(2)$ lies on the right of ℓ. Hence $I(1)\subset D(0)\cup D(2)$. This completes the main assertion of the lemma.

The remaining assertions are consequences of the following fact, which follows from the above argument and Lemma 4.3.1: If $\theta^\epsilon(\rho,\sigma;s_1)\le 0$ for both $\epsilon=-$ and $+$, then

$$I(1) \cap \partial (D(0) \cap D(2)) \subset \{\mathrm{Fix}^-(1), \mathrm{Fix}^+(1)\}.$$

Moreover, $\mathrm{Fix}^\epsilon(1)$ is contained in the set on the left hand side if and only if $\theta^\epsilon(\rho, \sigma; s_1) = 0$.

Lemma 4.3.6. *Under Assumption 4.2.4 (σ-Simple), $Ih(\rho(P_{j-1})) \cap Ih(\rho(P_j)) \cap Ih(\rho(P_{j+1}))$ is a singleton if and only if $\theta^\epsilon(\rho, \sigma; s_{[j]}) > 0$ and $\theta^{-\epsilon}(\rho, \sigma; s_{[j]}) < 0$ for some $\epsilon \in \{-, +\}$. Moreover, in this case, the following hold.*

1. $Ih(\rho(P_{j-1})) \cap Ih(\rho(P_j)) \cap Ih(\rho(P_{j+1})) \in \mathrm{Axis}(\rho(P_j))$.
2. $\mathrm{Fix}_\sigma^\epsilon(\rho(P_j)) \notin D(\rho(P_{j-1})) \cap D(\rho(P_{j+1}))$.
3. $\mathrm{Fix}_\sigma^{-\epsilon}(\rho(P_j)) \in \mathrm{int}(D(\rho(P_{j-1})) \cup D(\rho(P_{j+1})))$.

Proof. We may assume $j = 1$ without loss of generality. Since ρ satisfies the triangle inequality at σ by the assumption, we have $Ih(0) \cap Ih(1) \neq \emptyset$ and $Ih(1) \cap Ih(2) \neq \emptyset$ by Lemma 4.2.1. Since $\rho(P_1)$ interchanges $Ih(0) \cap Ih(1)$ and $Ih(1) \cap Ih(2)$ (see Lemma 4.1.3(2) (Chain rule)), $Ih(0) \cap Ih(1) \cap Ih(2)$ is a singleton if and only if $\mathrm{Axis}(1)$ intersects $Ih(0) \cap Ih(1)$ transversely and non-orthogonally. On the other hand, we can see that if $\mathrm{Axis}(1)$ intersects $Ih(0) \cap Ih(1)$ orthogonally, then $\mathcal{L}(\rho, \sigma)$ is not simple, which contradicts Assumption 4.2.4 (σ-Simple) (cf. Lemma 4.8.6). Hence $Ih(0) \cap Ih(1) \cap Ih(2)$ is a singleton if and only if $\mathrm{Axis}(1)$ intersects $Ih(0) \cap Ih(1)$ transversely. The latter condition is satisfied if and only if one of $\mathrm{Fix}^\pm(1)$ is contained in $\mathrm{int}\, D(0)$ and the other is contained in $\mathrm{int}\, E(0)$. By Lemma 4.3.1, this is equivalent to the condition that one of $\theta^\pm(\rho, \sigma; s_1)$ is positive and the other is negative. Thus we have obtained the main assertion and the assertion 1. The remaining assertions follow from Lemma 4.3.1.

Notation 4.3.7. Under Assumption 4.2.4 (σ-Simple), suppose $\theta^\epsilon(\rho, \sigma; s_{[j]}) \geq 0$. Then $e^\epsilon(\rho, \sigma; P_j)$, $e^\epsilon(j)$ for short, denotes the component of

$$I(\rho(P_j)) - \mathrm{int}(D(\rho(P_{j-1})) \cup D(\rho(P_{j+1}))) = I(\rho(P_j)) \cap (E(\rho(P_{j-1})) \cap E(\rho(P_{j+1})))$$

containing $\mathrm{Fix}_\sigma^\epsilon(\rho(P_j))$. Note that $e^\epsilon(\rho, \sigma; P_j)$ is a (possibly degenerate) circular arc of angle $2\theta^\epsilon(\rho, \sigma; s_{[j]})$. Thus:

1. If $\theta^\epsilon(\rho, \sigma; s_{[j]}) > 0$, then $e^\epsilon(\rho, \sigma; P_j)$ is a non-degenerate circular arc, and

$$\partial e^\epsilon(\rho, \sigma; P_j) = \{v^\epsilon(\rho; P_{j-1}, P_j), v^\epsilon(\rho; P_j, P_{j+1})\}.$$

2. If $\theta^\epsilon(\rho, \sigma; s_{[j]}) = 0$, then $e^\epsilon(\rho, \sigma; P_j)$ is a singleton consisting of the point:

$$v^\epsilon(\rho; P_{j-1}, P_j) = v^\epsilon(\rho; P_j, P_{j+1}) = \mathrm{Fix}_\sigma^\epsilon(\rho(P_j)).$$

Definition 4.3.8 (ϵ-Terminal triangle). *Under Assumption 4.2.4 (σ-Simple), σ is called an* ϵ-terminal triangle *of ρ if the following conditions are satisfied:*

1. *All components of* $\theta^\epsilon(\rho,\sigma)$ *are non-negative, and at most one component of* $\theta^\epsilon(\rho,\sigma)$ *can be 0. In particular,* $e^\epsilon(\rho,\sigma;P_j)$ *and* $v^\epsilon(\rho;P_j,P_{j+1})$ *are defined for every* $j \in \mathbb{Z}$.
2. *For each* $j \in \{0,1,2\}$, *if* $\theta^\epsilon(\rho,\sigma;s_j)=0$, *then* $\theta^{-\epsilon}(\rho,\sigma;s_j) \geq 0$.
3. *Let* $E^\epsilon(\rho,\sigma)$ *be the component of* $E(\rho,\sigma)$ *containing the region* $\{z \in \mathbb{C} \mid \epsilon\Im(z) \geq L\}$ *for sufficiently large positive number* L. *Then* $\mathrm{fr}\, E^\epsilon(\rho,\sigma) = \cup_j e^\epsilon(\rho,\sigma;P_j)$ *and it is homeomorphic to* $\mathbb{R}$.
4. *For each* $j \in \mathbb{Z}$, *some neighborhood of* $v^\epsilon(\rho;P_j,P_{j+1})$ *in* $\overline{Ih}(\rho(P_j)) \cap \overline{Ih}(\rho(P_{j+1}))$ *is contained in* $\overline{Eh}(\rho,\sigma)$.

When σ *is an* ϵ*-terminal triangle, we often identify* $\theta^\epsilon(\rho,\sigma)$ *with a point of* $\mathbb{H}^2$ *as in Definition 4.2.17.*

Remark 4.3.9. We describe relations among conditions in Definition 4.3.8.

(1) By Lemma 4.3.5(2), the second condition is a consequence of the last condition. But we include this condition because it is useful in the proof of the key Lemma 4.6.2 and because it is useful in determining whether a given representation is quasifuchsian (see Proposition 6.7.1).

(2) The first two conditions do not imply the third condition. Put $(a_0,a_1,a_2)=(-0.02+0.06\mathrm{i}, 0.26-0.41\mathrm{i}, 0.76+0.35\mathrm{i})$, and let ϕ be the upward Markoff map having (a_0,a_1,a_2) as the complex probability at σ, i.e., ϕ is the Markoff map determined by $(\phi(s_0),\phi(s_1),\phi(s_2)) = (1/\sqrt{a_1a_2}, 1/\sqrt{a_2a_0}, 1/\sqrt{a_0a_1})$. Then $\mathcal{L}(\rho,\sigma)$ is simple and all components of $\theta^+(\rho,\sigma)$ are positive. However, the point $v^+(\rho;P_0,P_1)$ is contained in the interior of $D(\rho(P_3))$. One can check this by putting $z_1=a_2=0.76+0.35i$ and $z_2=a_2+a_0=0.74+0.41i$ in OPTi [78] (see Fig. 4.9).

(3) The first three conditions do not necessarily imply the last condition. In fact, by perturbing the complex probability in (1), we can find an example which satisfies the first three conditions, such that $v^+(\rho;P_0,P_1)$ is contained in $I(\rho(P_3))$. Such an example is obtained by putting $z_1=0.74073902362583166+0.33679617661012978i$ and $z_2=0.74+0.41i$ in OPTi. In this example, we see that $Ih(\rho(P_0)) \cap Ih(\rho(P_1))$ is contained in $\mathrm{int}(Dh(\rho(P_0)) \cup Dh(\rho(P_3)))$. So, the last condition is not satisfied (see Fig. 4.10).

(4) In the third condition, though it is natural to expect that if $\mathrm{fr}\, E^\epsilon(\rho,\sigma) = \cup_j e^\epsilon(\rho,\sigma;P_j)$ then it is homeomorphic to $\mathbb{R}$. But we have not been able to exclude the possibility that $\cup_j e^\epsilon(\rho,\sigma;P_j)$ has a self-tangency.

4.4 Basic properties of ϵ-terminal triangles

Lemma 4.4.1. *Suppose* σ *is an* ϵ*-terminal triangle of* ρ. *Then the following holds for every integer* j *and* k.

1. $D(\rho(P_j)) \cap \mathrm{fr}\, E^\epsilon(\rho,\sigma) = I(\rho(P_j)) \cap \mathrm{fr}\, E^\epsilon(\rho,\sigma)$.
2. $e^\epsilon(\rho,\sigma;P_j) \cap D(\rho(P_k)) = e^\epsilon(\rho,\sigma;P_j) \cap I(\rho(P_k))$.
3. $v^\epsilon(\rho;P_j,P_{j+1}) \cap D(\rho(P_k)) = v^\epsilon(\rho;P_j,P_{j+1}) \cap I(\rho(P_k))$.

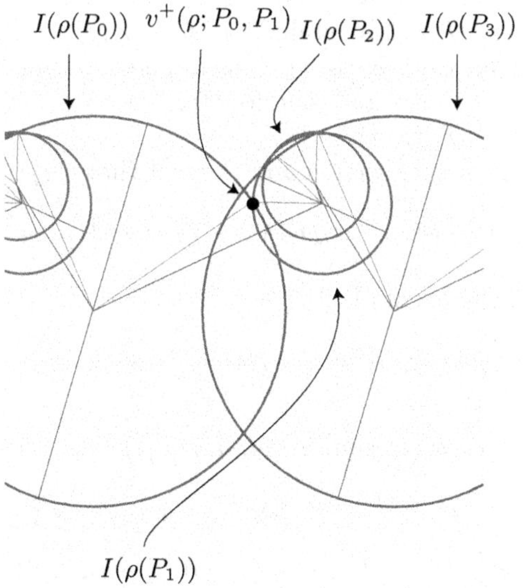

Fig. 4.9. Though all components of $\theta^\epsilon(\rho,\sigma)$ are positive, part of $\operatorname{int} e^\epsilon(\rho,\sigma;P_1)$ is covered by $D(\rho(P_3))$.

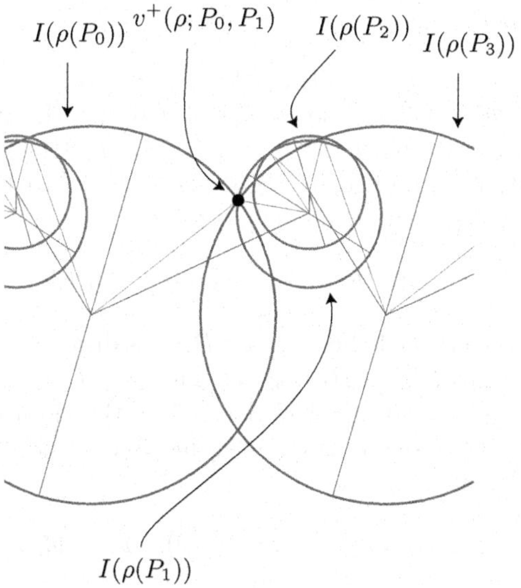

Fig. 4.10. Though $\operatorname{fr} E^\epsilon(\rho,\sigma) = \cup_j e^\epsilon(\rho,\sigma;P_j)$, $\overline{Ih}(\rho(P_0)) \cap \overline{Ih}(\rho(P_1))$ is covered by $\operatorname{int} Dh(\rho(P_3))$, and hence the last condition is not satisfied.

Proof. (1) Since $\operatorname{int} D(j)$ is contained in $\mathbb{C} - E(j) \subset \mathbb{C} - E(\rho,\sigma) \subset \mathbb{C} - E^\epsilon(\rho,\sigma)$, it is disjoint from $E^\epsilon(\rho,\sigma)$. This implies that $\operatorname{int} D(j) \cap \operatorname{fr} E^\epsilon(\rho,\sigma) = \emptyset$, because $E^\epsilon(\rho,\sigma)$ is closed and therefore $\operatorname{fr} E^\epsilon(\rho,\sigma) \subset E^\epsilon(\rho,\sigma)$. Hence $D(j) \cap \operatorname{fr} E^\epsilon(\rho,\sigma) = I(j) \cap \operatorname{fr} E^\epsilon(\rho,\sigma)$.

(2) By using the first assertion and the fact that $e^\epsilon(j) \cap \operatorname{fr} E^\epsilon(\rho,\sigma)$, we see

$$e^\epsilon(j) \cap D(k) = e^\epsilon(j) \cap \operatorname{fr} E^\epsilon(\rho,\sigma) \cap D(k) = e^\epsilon(j) \cap \operatorname{fr} E^\epsilon(\rho,\sigma) \cap I(k) = e^\epsilon(j) \cap I(k).$$

(3) Parallel to the proof of (2).

Lemma 4.4.2. *Suppose σ is an ϵ-terminal triangle of ρ. Then the following holds for every $j \in \mathbb{Z}$.*

1. *$D(\rho(P_j)) \cap \operatorname{fr} E^\epsilon(\rho,\sigma)$ is contained in $I(\rho(P_j)) \cap \mathcal{A}_j$, where $\mathcal{A}_j$ is the vertical strip*
$$\mathcal{A}_j := \{z \in \mathbb{C} \mid |\Re(z - c(\rho(P_j)))| \le 1/2\}.$$
2. *Suppose that $I(\rho(P_j)) \cap I(\rho(P_{j+3})) \neq \emptyset$. Then $I(\rho(P_j)) - (D(\rho(P_{j-3})) \cup D(\rho(P_{j+3}))) = I(\rho(P_j)) \cap \operatorname{int} \mathcal{A}_j$ consists of two open arcs. Moreover, $I(\rho(P_j)) \cap \operatorname{fr} E^\epsilon(\rho,\sigma)$ lies in the closure, a_j^ϵ, of the component of $I(\rho(P_j)) - (D(\rho(P_{j-3})) \cup D(\rho(P_{j+3})))$ which lies on the ϵ-side of the horizontal line $\Im(z) = \Im(c(\rho(P_j)))$.*

Proof. We may assume $j = 0$ without loss of generality.

(1) Recall that $I(\pm 3)$ is the image of $I(0)$ by the parallel translation $\rho(K)^{\pm 1} : z \mapsto z \pm 1$. Hence $D(0) \cap (E(-3) \cap E(3)) \subset D(0) \cap \mathcal{A}_0$. Since $D(0) \cap \operatorname{fr} E^\epsilon(\rho,\sigma)$ is contained in $D(0) \cap (E(-3) \cap E(3))$, it is contained in $D(0) \cap \mathcal{A}_0$, Hence, by Lemma 4.4.1(1), we have

$$D(0) \cap \operatorname{fr} E^\epsilon(\rho,\sigma) \subset D(0) \cap \operatorname{fr} E^\epsilon(\rho,\sigma) \cap \mathcal{A}_0 \subset I(0) \cap \operatorname{fr} E^\epsilon(\rho,\sigma) \cap \mathcal{A}_0 \subset I(0) \cap \mathcal{A}_0.$$

This implies the assertion 1, because $\operatorname{int} D(0)$ is disjoint from $E^\epsilon(\rho,\sigma)$.

(2) Assume that $I(0) \cap I(3) \neq \emptyset$. Then $\cup_m D(3m)$ is connected and $\mathbb{C} - \cup_m D(3m)$ has two components. Let $\tilde{E}^\epsilon$ be the closure of its component which lies on the ϵ-side of the horizontal line $\Im(z) = \Im c(0)$. Then $E^\epsilon(\rho,\sigma)$ is contained in $\tilde{E}^\epsilon$. Thus

$$I(0) \cap \operatorname{fr} E^\epsilon(\rho,\sigma) \subset I(0) \cap \tilde{E}^\epsilon \subset a_0^\epsilon.$$

Lemma 4.4.3. *$\mathbb{C} - \operatorname{fr} E^\epsilon(\rho,\sigma)$ consists of two components, $\tilde{E}^\epsilon$ and $\tilde{E}^{-\epsilon}$, where $\tilde{E}^\epsilon = \operatorname{int} E^\epsilon(\rho,\sigma)$ and $\tilde{E}^{-\epsilon} \supset \cup_j \operatorname{int} D(\rho(P_j))$.*

Proof. Since $\operatorname{fr} E^\epsilon(\rho,\sigma)$ is homeomorphic to $\mathbb{R}$ and is invariant under the parallel translation $\rho(K)$, the complementary region $\mathbb{C} - \operatorname{fr} E^\epsilon(\rho,\sigma)$ consists of two components. Since $\operatorname{int} E^\epsilon(\rho,\sigma) \cap \operatorname{fr} E^\epsilon(\rho,\sigma) = \emptyset$, $\operatorname{int} E^\epsilon(\rho,\sigma)$ is contained in a component of $\mathbb{C} - \operatorname{fr} E^\epsilon(\rho,\sigma)$, which we denote by $\tilde{E}^\epsilon$. Since $\operatorname{fr} \tilde{E}^\epsilon = \operatorname{fr} E^\epsilon(\rho,\sigma)$, we have $\tilde{E}^\epsilon = \operatorname{int} E^\epsilon(\rho,\sigma)$. Since the set of isometric circles $\{I(j) \mid j \in \mathbb{Z}\}$ is locally finite in $\mathbb{C}$, we see $\operatorname{fr} E(\rho,\sigma) = \operatorname{fr} \{\cup_j D(j)\}$ and

$\cup_j \operatorname{int} D(j) = \operatorname{int}\{\cup_j D(j)\}$. Thus the subset $\operatorname{fr} E^{\epsilon}(\rho,\sigma)$ of $\operatorname{fr} E(\rho,\sigma)$ is disjoint from $\cup_j \operatorname{int} D(j)$. Moreover, the union $\cup_j \operatorname{int} D(j)$ is connected, because ρ satisfies the triangle inequality at σ (cf. Lemma 4.2.1 and Definition 4.2.2). Hence $\cup_j \operatorname{int} D(j)$ is contained in one of the components of $\mathbb{C} - \operatorname{fr} E^{\epsilon}(\rho,\sigma)$. Since $\cup_j \operatorname{int} D(j)$ is disjoint from $\operatorname{int} E^{\epsilon}(\rho,\sigma) = \tilde{E}^{\epsilon}$, it is contained in $\tilde{E}^{-\epsilon}$.

Proposition 4.4.4. *Suppose σ is an ϵ-terminal triangle of ρ. Then for every $j \in \mathbb{Z}$, we have:*

$$D(\rho(P_j)) \cap \operatorname{fr} E^{\epsilon}(\rho,\sigma) = e^{\epsilon}(\rho,\sigma;P_j).$$

Proof. We may assume $j = 0$ without loss of generality. By the definition of $e^{\epsilon}(0)$, we have $D(0) \cap \operatorname{fr} E^{\epsilon}(\rho,\sigma) \supset e^{\epsilon}(0)$. Thus we show that $D(0) \cap \operatorname{fr} E^{\epsilon}(\rho,\sigma) \subset e^{\epsilon}(0)$. Since $\operatorname{fr} E^{\epsilon}(\rho,\sigma)$ is the union of the sets $\operatorname{int} e^{\epsilon}(k)$ with $\theta^{\epsilon}(\rho,\sigma;s_{[k]}) > 0$ and the singletons $v^{\epsilon}(k,k+1)$, the desired inclusion is reduced to the following Claims 4.4.5–4.4.9.

Claim 4.4.5. Let k be an integer such that $\theta^{\epsilon}(\rho,\sigma;s_{[k]}) > 0$. Then

$$D(0) \cap \operatorname{int} e^{\epsilon}(k) = \begin{cases} \operatorname{int} e^{\epsilon}(0), & \text{if } k = 0, \\ \emptyset, & \text{otherwise.} \end{cases}$$

Claim 4.4.6. 1. If $\theta^{\epsilon}(\rho,\sigma;s_1) > 0$, then $D(0) \cap v^{\epsilon}(1,2) = \emptyset$. If $\theta^{\epsilon}(\rho,\sigma;s_1) = 0$, then $v^{\epsilon}(1,2) \subset e^{\epsilon}(0)$.
2. If $\theta^{\epsilon}(\rho,\sigma;s_2) > 0$, then $D(0) \cap v^{\epsilon}(-2,-1) = \emptyset$. If $\theta^{\epsilon}(\rho,\sigma;s_2) = 0$, then $v^{\epsilon}(-2,-1) \subset e^{\epsilon}(0)$.

Claim 4.4.7. For any integer l with $l \notin \{-1,0\}$, the following holds.

$$D(0) \cap v^{\epsilon}(3l+1, 3l+2) = \emptyset.$$

Claim 4.4.8. For any integer l with $l \neq 0$, the following holds.

$$D(0) \cap v^{\epsilon}(3l-1, 3l) = \emptyset.$$

Claim 4.4.9. For any integer l with $l \neq 0$, the following holds.

$$D(0) \cap v^{\epsilon}(3l, 3l+1) = \emptyset.$$

In what follows, we prove the claims in the following order: Claim 4.4.5, Claim 4.4.6, Claim 4.4.8 for the case when $l = 1$, Claim 4.4.9 for the case when $l = -1$, Claims 4.4.8 and 4.4.9 for the general case, and Claim 4.4.7.

Proof (Proof of Claim 4.4.5). If $k = 0$, then the claim is obvious. So we may assume $k \neq 0$. By Lemma 4.4.1(2), $D(0) \cap \operatorname{int} e^{\epsilon}(k) = I(0) \cap \operatorname{int} e^{\epsilon}(k)$. Suppose to the contrary that this set is non-empty, and let w be the intersection point. Then, by the above identity, $I(0)$ and $I(k)$ $(\supset e^{\epsilon}(k))$ has the common tangent line at w. Since $\mathcal{L}(\rho,\sigma)$ is simple by assumption and hence $I(0) \neq I(k)$, we have the following three possibilities:

1. $I(0)$ and $I(k)$ lie on the same side of ℓ and $\operatorname{int} D(0) \supset I(k) - \{w\}$.
2. $I(0)$ and $I(k)$ lie on the same side of ℓ and $\operatorname{int} D(k) \supset I(0) - \{w\}$.
3. $I(0)$ and $I(k)$ lie on the different sides of ℓ.

The possibility 1 cannot happen, because it implies that $e^\epsilon(k) = e^\epsilon(k) \cap E(0) = \{w\}$, which contradicts the assumption that $\theta^\epsilon(\rho, \sigma; s_{[k]}) > 0$. Suppose that the possibility 2 happens. Then $e^\epsilon(0) = e^\epsilon(0) \cap E(k) = \{w\}$. Thus $e^\epsilon(0)$ intersects the open arc $\operatorname{int} e^\epsilon(k)$. This contradicts the assumption that $\operatorname{fr} E^\epsilon(\rho, \sigma) = \cup_j e^\epsilon(j)$ is homeomorphic to $\mathbb{R}$ (cf. Definition 4.3.8(3)). Finally, suppose that the possibility 3 happens. Since the set of isometric circles $\{I(j) \mid j \in \mathbb{Z}\}$ is locally finite in $\mathbb{C}$ and since $\operatorname{fr} E^\epsilon(\rho, \sigma) = \cup_j e^\epsilon(j) \cong \mathbb{R}$, there is a disk, B, centered at the point of tangency such that $B \cap \operatorname{fr} E^\epsilon(\rho, \sigma)$ is a properly embedded arc in B. Then one of the two components of $B - \operatorname{fr} E^\epsilon(\rho, \sigma)$ has a nontrivial intersection with $\operatorname{int} D(0)$ and the other has a nontrivial intersection with $\operatorname{int} D(k)$. This contradicts the fact that one of the two sides of $\operatorname{fr} E^\epsilon(\rho, \sigma)$ is equal to $E^\epsilon(\rho, \sigma)$.

Proof (Proof of Claim 4.4.6). We prove only the first assertion, because the second one is proved similarly. If $\theta^\epsilon(\rho, \sigma; s_1) = 0$, then the assertion is obvious. So assume that $\theta^\epsilon(\rho, \sigma; s_1) > 0$. Suppose to the contrary that $D(0) \cap v^\epsilon(1, 2) \neq \emptyset$. Then $v^\epsilon(1, 2)$ is contained in $I(0)$ by Lemma 4.4.1(2). Thus $I(0)$ contains the two (distinct) boundary points $v^\epsilon(0, 1)$ and $v^\epsilon(1, 2)$ of the non-degenerate circular arc $e^\epsilon(1)$. Since $e^\epsilon(1)$ is contained in $I(1)$ and since $I(0) \neq I(1)$, this implies that $I(0) \cap I(1)$ consists of the two points $v^\epsilon(0, 1)$ and $v^\epsilon(1, 2)$. Since $\rho(P_1)$ maps $Ih(1)$ to itself and interchanges $v^\epsilon(0, 1)$ and $v^\epsilon(1, 2)$ (see Lemma 4.2.7), this in tern implies that $\rho(P_1)$ induces an involution of the complete geodesic $Ih(0) \cap Ih(1)$. Therefore $\mathcal{L}(\rho, \sigma)$ is folded at $c(1)$ by Lemma 4.8.6, a contradiction.

Proof (Proof of Claim 4.4.8 when $l = 1$). Suppose to the contrary that $D(0) \cap v^\epsilon(2, 3) \neq \emptyset$. If $\theta^\epsilon(\rho, \sigma; s_2) = 0$, then $v^\epsilon(2, 3) = v^\epsilon(1, 2)$ and hence we have $D(0) \cap v^\epsilon(1, 2) \neq \emptyset$. This contradicts Claim 4.4.6(1), because the condition $\theta^\epsilon(\rho, \sigma; s_2) = 0$ implies $\theta^\epsilon(\rho, \sigma; s_1) > 0$ by Definition 4.3.8(1). So we may assume $\theta^\epsilon(\rho, \sigma; s_2) > 0$. By Lemma 4.4.1(3), $v^\epsilon(2, 3)$ is contained in $I(0)$. Since $v^\epsilon(2, 3) \in I(3)$ by definition, it follows that $v^\epsilon(2, 3) \in I(0) \cap Ih(3)$. This in tern implies

$$\begin{aligned} v^\epsilon(-1, 0) &= \rho(K)^{-1}(v^\epsilon(2, 3)) \\ &\subset \rho(K)^{-1}(I(0) \cap I(3)) \\ &= I(-3) \cap I(0). \end{aligned}$$

Hence, by Lemma 4.4.2(2), $v^\epsilon(-1, 0)$ and $v^\epsilon(2, 3)$, respectively, are the left and right boundary points of the closure of the component of $I(0) - D(-3) \cup D(3)$ which lies on the ϵ-side of the horizontal line $\Im(z) = \Im(c(0))$.

Let ℓ be the line tangent to $I(\rho(P_2))$ at $v^\epsilon(2, 3)$. Since $\theta^\epsilon(\rho, \sigma; s_2) > 0$, $e^\epsilon(2)$ is a non-degenerate arc of $I(2)$, which contains $v^\epsilon(2, 3)$ as an endpoint. Since

$e^\epsilon(2) \subset \mathrm{fr}\, E^\epsilon(\rho,\sigma) \subset E(0) \cap E(3)$, this implies that a germ of ℓ at $v^\epsilon(2,3)$ intersects $E(0) \cap E(3)$. In particular, ℓ is not horizontal, and the left/right hand sides of ℓ are well-defined. The circle $I(2)$ is entirely contained on the left or right hand side of ℓ.

Suppose first that $I(2)$ is contained on the right hand side of ℓ. Consider the half spaces

$$H^\epsilon = \{z \in \mathbb{C} \,|\, \epsilon\Im(z) \geq \epsilon\Im(v^\epsilon(2,3))\}$$
$$H^{-\epsilon} = \{z \in \mathbb{C} \,|\, \epsilon\Im(z) \leq \epsilon\Im(v^\epsilon(2,3))\}.$$

Then the horizontal line $\partial H^\epsilon = \partial H^{-\epsilon}$ is contained in $\cup_m D(3m) \subset \cup_j D(j)$, and hence $\mathrm{int}\, H^{-\epsilon} \cap E^\epsilon(\rho,\sigma) = \emptyset$. Since

$$\theta^\epsilon(\rho,\sigma;s_2) = \epsilon \arg \frac{\mathrm{Fix}^\epsilon(2) - c(2)}{v^\epsilon(2,3) - c(2)} \in (0, \pi/2],$$

a germ of $e^\epsilon(2)$ at $v^\epsilon(2,3)$ lies in $H^{-\epsilon}$ and intersects with $\mathrm{int}\, H^{-\epsilon}$ (see Fig. 4.11) This contradicts the fact that $e^\epsilon(2) \subset \mathrm{fr}\, E^\epsilon(\rho,\sigma)$.

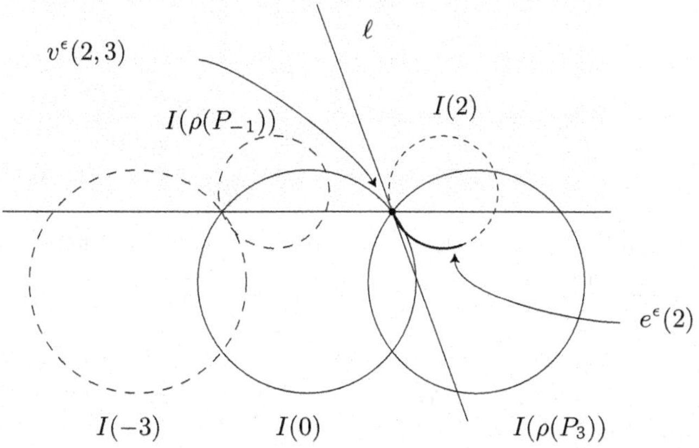

Fig. 4.11. The figure is for the case where $I(2)$ lies in the right hand side of ℓ and $\epsilon = +$.

Suppose finally that $I(2)$ is contained on the left hand side of ℓ. Since ρ satisfies the triangle inequality at σ (cf. Definition 4.2.2), both $I(2)$ and ℓ intersect $I(3)$ transversely. This also implies that $I(0)$ and $I(3)$ intersect transversely. (Otherwise, ℓ is tangent to both $I(0)$ and $I(3)$ at $v^\epsilon(2,3)$, because a germ of ℓ at $v^\epsilon(2,3)$ intersects $E(0) \cap E(3)$.) Since $v^\epsilon(2,3)$ is also a transversal intersection point of $I(0)$ and $I(3)$, we also denote it by $v^\epsilon(0,3)$. Let $v^{-\epsilon}(2,3)$ (resp. $v^{-\epsilon}(0,3)$)) be the point of $I(2) \cap I(3)$ (resp. $I(0) \cap I(3)$) different from $v^\epsilon(2,3) = v^\epsilon(0,3)$. If $v^{-\epsilon}(2,3)$ is contained in $I(3) \cap \mathrm{int}\, D(0)$, then it follows that $Ih(2) \cap Ih(3) \subset \mathrm{int}\, Dh(0)$. This contradicts Condition 4 of Definition

4.3.8 (ϵ-Terminal triangle). Hence $v^{-\epsilon}(2,3)$ must be contained in $I(3) \cap E(0)$. Let D_2' be the disk bounded by the circle which is tangent to ℓ at $v^{\epsilon}(2,3)$ and passes through the point $v^{-\epsilon}(0,3)$ (see Fig. 4.12). Then $D_2' \subset D(2)$, because:

1. Both $I(2)$ and $\partial D_2'$ are tangent to ℓ at $v^{\epsilon}(2,3)$.
2. The point $v^{-\epsilon}(0,3)$ of $\partial D_2'$ is contained in $D(2)$.

Moreover, $I(0) - D(3) \subset D_2'$, because:

1. $I(0)$ and $\partial D_2'$ share the two points $v^{\pm}(0,3)$.
2. The tangent line ℓ to $\partial D_2'$ at $v^{\epsilon}(2,3)$ intersects $E(0) \cap E(3)$ near $v^{\epsilon}(2,3)$.

Hence $I(0) - D(3) \subset D_2' \subset D(2)$. Since $I(0) \neq I(2)$, this implies $I(0) - D(3) \subset \operatorname{int} D(2)$, and hence $v^{\epsilon}(0,1) \subset \operatorname{int} D(2)$. This contradicts the fact that $v^{\epsilon}(0,1) \subset \operatorname{fr} E^{\epsilon}(\rho,\sigma)$.

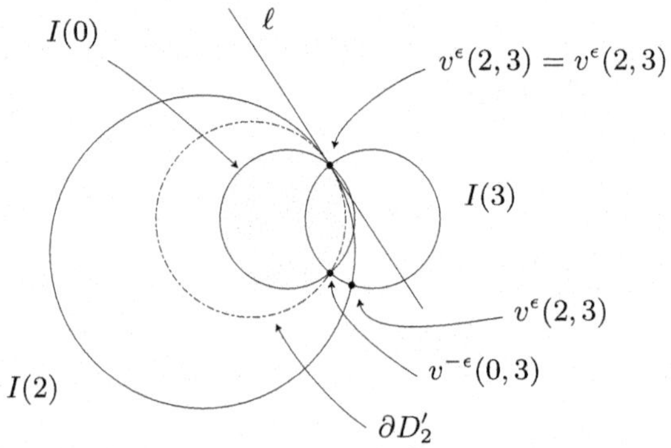

Fig. 4.12. The figure is for the case where $I(2)$ lies in the left hand side of ℓ, $v^{\epsilon}(2,3) \subset E(0)$ and $\epsilon = +$.

Proof (Proof of Claim 4.4.9 when $l = -1$). Parallel to the proof of Claim 4.4.8 when $l = 1$.

Proof (Proof of Claims 4.4.8 and 4.4.9 for the general case). Since $v^{\epsilon}(3l-1,3l)$ and $v^{\epsilon}(3l,3l+1)$ are contained in $I(3l) \cap \operatorname{fr} E^{\epsilon}(\rho,\sigma)$, they are contained in the vertical strip $\mathcal{A}_{3l}$ by Lemma 4.4.2(2). If $|l| \geq 2$, then $\mathcal{A}_{3l} = \rho(K)^l(\mathcal{A}_0)$ is disjoint from $\mathcal{A}_0$, and hence $v^{\epsilon}(3l-1,3l)$ and $v^{\epsilon}(3l,3l+1)$ are disjoint from $I(0) \cap \mathcal{A}_0 \supset D(0) \cap \operatorname{fr} E^{\epsilon}(\rho,\sigma)$. Hence the claims hold in this case. So we may assume $l = \pm 1$. We prove the claims only for the case when $l = 1$, because a similar argument works for the case when $l = -1$. Since Claim 4.4.8 for the case when $l = 1$ is already proved, it suffices to show that $D(0) \cap v^{\epsilon}(3,4) = \emptyset$. Suppose this is not the case, namely, $v^{\epsilon}(3,4)$ is contained in $D(0)$. Then it follows that $v^{\epsilon}(3,4)$ is contained in the vertical line $\mathcal{A}_0 \cap \mathcal{A}_3$, because $v^{\epsilon}(3,4)$ is also contained in $\mathcal{A}_3$ as observed in the above. Hence $v^{\epsilon}(0,1) = \rho(K)^{-1}(v^{\epsilon}(3,4))$ is contained in the vertical line $\mathcal{A}_{-3} \cap \mathcal{A}_0$. By Lemma 4.4.2(2), both $v^\epsilon(0,1)$ and $\mathrm{Fix}^\epsilon(j)$ are contained in the closure, a_j^ϵ, of the component $I(0) \cap \mathrm{int}\, \mathcal{A}_0$ which lies on the ϵ-side of the horizontal line $\Re(z) = \Re(c(0))$. The above observation implies that $v^\epsilon(0,1)$ forms the left boundary point of the arc a_j^ϵ. Since

$$\theta^\epsilon(\rho, \sigma; s_0) = \epsilon \arg \frac{\mathrm{Fix}^\epsilon(0) - c(0)}{v^\epsilon(0,1) - c(0)} \in [0, \pi/2],$$

the above observation implies that $\mathrm{Fix}^\epsilon(0)$ is also equal to the left boundary point of a_j^ϵ and therefore $\theta^\epsilon(\rho, \sigma; s_0) = 0$. Hence $v^\epsilon(-1,0) = v^\epsilon(0,1) \in D(-3))$, and so $v^\epsilon(2,3)) \in D(\rho(P_0))$. This contradicts Claim 4.4.8 for the case when $l = 1$, which we have already proved.

Proof (Proof of Claim 4.4.7). We prove the claim for $l > 0$. (A similar argument works for the case when $l < 0$.) Suppose to the contrary that $D(0) \cap v^\epsilon(3l+1, 3l+2) \neq \emptyset$ for some $l > 0$. Set $w_1 = v^\epsilon(3l+1, 3l+2)$, $w_2 = v^\epsilon(3l+2, 3l+3)$. Then $e^\epsilon(3l+2)$ is a circular arc joining the point $w_1 \in I(0)$ with the point $w_2 \in I(3l+3)$. Set $w_2' = \rho(K)^{-(l+1)}(w_2)$. Then w_2' is contained in $I(0) \cap \mathrm{fr}\, E^\epsilon(\rho, \sigma)$, because

$$\begin{aligned} w_2 &\in I(3l+3) \cap \mathrm{fr}\, E^\epsilon(\rho, \sigma) \\ &= \rho(K)^{l+1}(I(0)) \cap \mathrm{fr}\, E^\epsilon(\rho, \sigma) \\ &= \rho(K)^{l+1}(I(0) \cap \mathrm{fr}\, E^\epsilon(\rho, \sigma)). \end{aligned}$$

Thus both w_1 and w_2' are contained in $I(0) \cap \mathrm{fr}\, E^\epsilon(\rho, \sigma)$ and hence in $I(0) \cap \mathcal{A}_0$ by Lemma 4.4.2(1). Thus there is a simple arc, τ_0, joining w_2' with w_1 such that $\mathrm{int}\, \tau_0 \subset \mathrm{int}\, D(\rho(P_0)) \cap \mathcal{A}_0$. Consider the union $\tau := \tau_0 \cup e^\epsilon(3l+2)$. Since $\mathrm{int}\, \tau_0 \subset \mathrm{int}\, D(0)$ and $e^\epsilon(3l+2) \subset \mathrm{fr}\, E^\epsilon(\rho, \sigma)$, we see that τ_0 and $e^\epsilon(3l+2)$ intersects only at w_1. Hence τ is a simple arc in $\mathbb{C}$ with endpoints w_2' and w_2. Then τ projects onto a closed curve in the open annulus $\mathbb{C}/\langle \rho(K) \rangle$. Since the closed curve represents $l+1$ times the generator of the first integral homology group of the open annulus and since $l + 1 \geq 2$ by assumption, it cannot be simple.

On the other hand, since $\mathrm{int}\, \tau_0$ is contained in the vertical strip $\mathcal{A}_0$, which forms a fundamental domain for the action of $\langle \rho(K) \rangle$ on $\mathbb{C}$, the image of $\mathrm{int}\, \tau_0$ in $\mathbb{C}/\langle \rho(K) \rangle$ is simple. Similarly, we see, by using Lemma 4.4.2(1), that the image of $\mathrm{int}\, e^\epsilon(3l+2)$ in $\mathbb{C}/\langle \rho(K) \rangle$ is also simple. Moreover the image of $\mathrm{int}\, \tau_0$ is disjoint from the image of $e^\epsilon(3l+2)$, because $\mathrm{int}\, \tau_0 \subset \mathrm{int}\, D(0)$ and $e^\epsilon(3l+2) \subset \mathrm{fr}\, E^\epsilon(\rho, \sigma)$. Since the image of τ is a non-simple closed curve, the above observations imply that the image of w_1 is equal to the image of w_2. Hence, there is an integer l' such that $w_2 = \rho(K)^{l'}(w_1)$, namely,

$$v^\epsilon(3l+2, 3l+3) = \rho(K)^{l'} v^\epsilon(3l+1, 3l+2) = v^\epsilon(3(l+l')+1, 3(l+l')+2).$$

By Condition 3 in Definition 4.3.8 (ϵ-Terminal triangle), this happens only when $\theta^\epsilon(\rho, \sigma; s_2) = 0$ and $l' = 0$. Then $v^\epsilon(3l+2, 3l+3)$ is equal to

$v^{\epsilon}(3l+1, 3l+2)$, and hence is contained in $D(0)$ by the primary assumption. Since $l > 0$, this contradicts Claim 4.4.8, which we have already proved.

We have proved Claims 4.4.5–4.4.9. Thus the proof of Proposition 4.4.4 is complete.

4.5 Relation between side parameters at adjacent triangles

In this section, we study relation between side parameters at adjacent triangles. To explain the results, we need the following definition.

Definition 4.5.1. *Let ρ be a type-preserving representation, and let σ_1 and σ_2 be mutually adjacent triangles of $\mathcal{D}$.*

(1) $\mathcal{L}(\rho, \{\sigma_1, \sigma_2\})$ is said to be simple *if $\mathcal{L}(\rho, (\sigma_1, \sigma_2))$ or $\mathcal{L}(\rho, (\sigma_2, \sigma_1))$ is simple (cf. Definition 3.2.2).*

(2) $\mathcal{L}(\rho, \{\sigma_1, \sigma_2\})$ is said to be flat *if both $\mathcal{L}(\rho, \sigma_1)$ and $\mathcal{L}(\rho, \sigma_2)$ are simple and $|\mathcal{L}(\rho, \sigma_1)| = |\mathcal{L}(\rho, \sigma_2)|$.*

The following lemma shows that the property that ϕ is upward at a triangle is inherited by adjacent triangles:

Lemma 4.5.2. *Under Notation 2.1.14 (Adjacent triangles) and Assumption 2.4.6 (σ-NonZero), suppose that $\phi^{-1}(0) \cap (\sigma \cup \sigma')^{(0)} = \emptyset$ and that $\mathcal{L}(\rho, \{\sigma, \sigma'\})$ is simple or flat. Then ϕ is upward at σ if and only if ϕ is upward at σ'. Moreover, we have:*

$$\mathrm{Fix}^{\epsilon}_{\sigma}(\rho(P_0)) = \mathrm{Fix}^{\epsilon}_{\sigma'}(\rho(P'_0)), \qquad \mathrm{Fix}^{\epsilon}_{\sigma}(\rho(P_2)) = \mathrm{Fix}^{\epsilon}_{\sigma'}(\rho(P'_1)).$$

Proof. We prove the lemma only when $\mathcal{L}(\rho, \{\sigma, \sigma'\})$ is simple, because the proof for the case when $\mathcal{L}(\rho, \{\sigma, \sigma'\})$ is flat is essentially the same. Suppose that ϕ is upward at σ. We first show that $i/\phi(s'_1) = i/\phi(s_2)$ is upward at $c(\rho(P'_1)) = c(\rho(P_2))$ with respect to $\mathcal{L}(\rho, \sigma')$. To this end, let ℓ be the oriented line in $\mathbb{C}$ containing $\pi(\mathrm{Axis}(\rho(P'_1))) = \pi(\mathrm{Axis}(\rho(P_2)))$, and orient ℓ so that it is upward with respect to $\mathcal{L}(\rho, \sigma)$. Note that ℓ bisects the angle $\angle(c(\rho(P'_0)), c(\rho(P'_1)), c(\rho(P'_2)))$ (see Proposition 2.4.4(2.1)). Moreover, by Lemma 3.1.2,

$$\angle(c(\rho(P_1)), c(\rho(P_2)), c(\rho(P_0))) = \angle(c(\rho(P'_3)), c(\rho(P'_1)), c(\rho(P'_2))).$$

On the other hand, since $\mathcal{L}(\rho, \{\sigma, \sigma'\})$ is simple, the triangle $\Delta(c(\rho(P_1)), c(\rho(P_2)), c(\rho(P_0)))$ intersects the triangle $\Delta(c(\rho(P'_3)), c(\rho(P'_1)), c(\rho(P'_2)))$ only at the common vertex $c(\rho(P_2)) = c(\rho(P'_1))$. Hence $c(\rho(P'_0)) = c(\rho(P_0))$ must lie on the left of the oriented line ℓ, and $c(\rho(P'_2))$ must lie on the right of ℓ (see Fig. 4.13). Hence $i/\phi(s'_1) = i/\phi(s_2)$ is upward at $c(\rho(P'_1))$ with respect to

$\mathcal{L}(\rho, \sigma')$. By a parallel argument, we can also see that $i/\phi(s_0') = i/\phi(s_0)$ is upward at $c(\rho(P_0')) = c(\rho(P_0))$ with respect to $\mathcal{L}(\rho, \sigma')$. These imply that $i/\phi(s_2')$ is also upward at $c(\rho(P_2'))$ with respect to $\mathcal{L}(\rho, \sigma')$, by the last assertion of Lemma 3.1.4. Hence ϕ is upward at σ'. Since the argument is symmetric, we obtain the first assertion of the lemma.

To see the second assertion, let ϕ be the Markoff map inducing ρ which is upward at σ and σ'. Then

$$\mathrm{Fix}^{\epsilon}_{\sigma}(\rho(P_0)) = c(\rho(P_0)) + \frac{\epsilon i}{\phi(s_0)} = c(\rho(P_0')) + \frac{\epsilon i}{\phi(s_0')} = \mathrm{Fix}^{\epsilon}_{\sigma'}(\rho(P_0')).$$

The remaining identity is proved similarly.

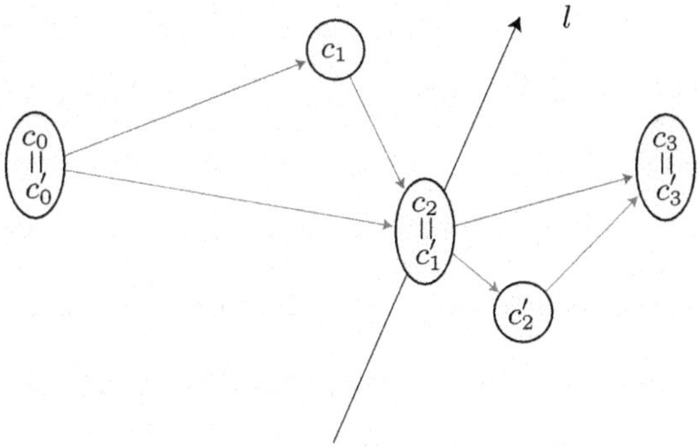

Fig. 4.13. Figure for Lemma 4.5.2

Lemma 4.5.3. *Under Notation 2.1.14 (Adjacent triangles) and Assumption 4.2.4 (σ-Simple), suppose $\phi^{-1}(0) \cap (\sigma \cup \sigma')^{(0)} = \emptyset$, $\mathcal{L}(\rho, \{\sigma, \sigma'\})$ is simple or flat, and ρ satisfies the triangle inequality at σ and σ'. Then the following hold.*

(1) $\theta^{\epsilon}(\rho, \sigma'; s_2')$ is positive, 0, or negative according as $\theta^{\epsilon}(\rho, \sigma; s_1)$ is negative, 0, or positive.

(2) Suppose $\theta^{\epsilon}(\rho, \sigma; s_1) = 0$, or equivalently $\theta^{\epsilon}(\rho, \sigma'; s_2') = 0$. Then $\theta^{\epsilon}(\rho, \sigma; s_0) = \theta^{\epsilon}(\rho, \sigma'; s_0')$ and $\theta^{\epsilon}(\rho, \sigma; s_2) = \theta^{\epsilon}(\rho, \sigma'; s_1')$. In particular, when these two numbers are positive, $\theta^{\epsilon}(\rho, \sigma)$ and $\theta^{\epsilon}(\rho, \sigma')$ determine the same point in $\operatorname{int} \tau \subset \mathbb{H}^2$ (cf. Definition 4.2.17).

Proof. (1) By Lemmas 4.5.2, the Markoff map ϕ is also upward at σ'. Hence we have the following by Lemma 4.3.1:

1. $\theta^{\epsilon}(\rho, \sigma; s_1) < 0$ if and only if $|\phi(s_2) + \epsilon i \phi(s_0)| < |\phi(s_1)|$.
2. $\theta^{\epsilon}(\rho, \sigma'; s_2') > 0$ if and only if $|\phi(s_0') + \epsilon i \phi(s_1')| > |\phi(s_2')|$.

Put $(x,y,z)=(\phi(s_0),\phi(s_1),\phi(s_2))$. Then $(\phi(s_0'),\phi(s_1'),\phi(s_2'))=(x,z,xz-y)$, and hence we have

$$\begin{aligned}|\phi(s_2)+\epsilon i\phi(s_0)||\phi(s_0')+\epsilon i\phi(s_1')| &= |z+\epsilon ix|\cdot|x+\epsilon iz|\\ &= |x^2+z^2|\\ &= |xyz-y^2|\\ &= |\phi(s_1)\phi(s_2')|.\end{aligned}$$

By this identity and the above facts, we obtain the first assertion.

(2) Since $\theta^\epsilon(\rho,\sigma;s_1)=0$, we have $v^\epsilon(\rho;P_0,P_1)=v^\epsilon(\rho;P_1,P_2)=\mathrm{Fix}_\sigma^\epsilon(\rho(P_1))$. By using the facts that $P_0'=P_0$ and $P_1'=P_2$, we see that this point is also equal to $v^\epsilon(\rho;P_0',P_1')\in I(\rho(P_0'))\cap I(\rho(P_1'))$. Moreover, $\mathrm{Fix}_\sigma^\epsilon(\rho(P_0))=\mathrm{Fix}_{\sigma'}^\epsilon(\rho(P_0'))$ by Lemma 4.5.2. Hence $\theta^\epsilon(\rho,\sigma;s_0)=\theta^\epsilon(\rho,\sigma';s_0')$. Similarly, we have $\theta^\epsilon(\rho,\sigma;s_2)=\theta^\epsilon(\rho,\sigma';s_1')$. Thus we have obtained the desired result.

Lemma 4.5.4. *Under Assumption 4.2.4 (σ-Simple), suppose $\theta^\epsilon(\rho,\sigma)=(+,0,+)$ and $\theta^{-\epsilon}(\rho,\sigma;s_1)>0$. Then the following hold under Notation 2.1.14 (Adjacent triangles):*

1. *$\mathcal{L}(\rho,\sigma')$ lies on the ϵ-side of $\mathcal{L}(\rho,\sigma)$, $\mathcal{L}(\rho,\{\sigma,\sigma'\})$ is simple, and ρ satisfies the triangle inequality at σ'. (In particular, $\theta^\epsilon(\rho,\sigma')$ is well-defined.)*
2. *$\theta^\epsilon(\rho,\sigma';s_2')=0$. Moreover, $\theta^\epsilon(\rho,\sigma)$ and $\theta^\epsilon(\rho,\sigma')$ determine the same point of $\mathrm{int}\,\tau$.*
3. *$D(\rho(P_2'))\subset D(\rho(P_2))\cup D(\rho(P_3))$ and $D(\rho(P_2'))\cap\partial(D(\rho(P_2))\cup D(\rho(P_3)))=\{v^\epsilon(\rho;P_2,P_3)\}$.*
4. *$Dh(\rho(P_2'))\cap(Eh(\rho(P_2))\cap Eh(\rho(P_3)))=\emptyset$.*

In the above lemma, $\theta^\epsilon(\rho,\sigma)=(+,0,+)$ means

$$\theta^\epsilon(\rho,\sigma;s_0)>0,\quad \theta^\epsilon(\rho,\sigma;s_1)=0,\quad \theta^\epsilon(\rho,\sigma;s_2)>0.$$

To prove this lemma, we need the following lemma, whose proof is referred to Sect. 4.7.

Lemma 4.5.5. *Under Assumption 4.2.4 (σ-Simple), suppose that all components of $\theta^\epsilon(\rho,\sigma)$ are non-negative. Then any two of $\{\Delta_k^\epsilon(\rho,\sigma)\,|\,k\in\mathbb{Z}\}$ (see Notation 4.2.10) intersect only at a common edge or a common vertex.*

Proof (Proof of Lemma 4.5.4). Recall Notation 4.2.10, and set $\alpha_j=\alpha(\rho,\sigma;s_j)$. By Lemma 4.2.11, the angles of $\Delta_0^\epsilon(\rho,\sigma)$ at the vertices $c(\rho(P_0))$ and $c(\rho(P_1))$ are equal to α_0 and α_1, respectively, and the angles of $\Delta_1^\epsilon(\rho,\sigma)$ at the vertices $c(\rho(P_1))$ and $c(\rho(P_2))$ are equal to α_1 and α_2, respectively (see Fig. 4.5). By the assumption and Sublemma 4.5.5, $\Delta_0^\epsilon(\rho,\sigma)$ and $\Delta_1^\epsilon(\rho,\sigma)$ intersect only in the common edge $\overline{c(\rho(P_1))v^\epsilon(\rho;P_0,P_1)}=\overline{c(\rho(P_1))v^\epsilon(\rho;P_1,P_2)}$, and hence $\Delta_0^\epsilon(\rho,\sigma)\cup\Delta_1^\epsilon(\rho,\sigma)$ forms a quadrangle. Note that the (inner) angles of the quadrangle are α_0, α_2, $2\alpha_1$ and $\alpha_0+\alpha_1=\pi-\alpha_2$. Since $\theta^\epsilon(\rho,\sigma;s_1)=0$ and $\theta^{-\epsilon}(\rho,\sigma;s_1)>0$, Lemma 4.2.11(3) implies:

$$2\alpha_1 = \pi - \theta^{\epsilon}(\rho, \sigma; s_1) - \theta^{-\epsilon}(\rho, \sigma; s_1) < \pi.$$

Thus every corner of the quadrangle has angle $< \pi$, and hence the quadrangle is convex. Hence the edge $\overline{c(\rho(P_0))c(\rho(P_2))}$ of $\mathcal{L}(\rho, \sigma')$ is contained in $\Delta_0^{\epsilon}(\rho, \sigma) \cup \Delta_1^{\epsilon}(\rho, \sigma)$, and therefore it lies on the ϵ-side of $\mathcal{L}(\rho, \sigma)$ (see Fig. 4.14). By Lemma 3.1.2, we have the following inequalities among angles.

$$\angle(c(\rho(P_3')), c(\rho(P_1')), c(\rho(P_2'))) = \angle(c(\rho(P_0)), c(\rho(P_2)), c(\rho(P_1))) < \alpha_2,$$
$$\angle(c(\rho(P_2')), c(\rho(P_3')), c(\rho(P_1'))) = \angle(c(\rho(P_1)), c(\rho(P_0)), c(\rho(P_2))) < \alpha_0.$$

Hence the vertex $c(\rho(P_2'))$ lies in the interior of $\Delta_2^{\epsilon}(\rho, \sigma)$. Therefore the edges $\overline{c(\rho(P_1'))c(\rho(P_2'))}$ and $\overline{c(\rho(P_2'))c(\rho(P_3'))}$ of $\mathcal{L}(\rho, \sigma')$ lie in $\Delta_2^{\epsilon}(\rho, \sigma)$. Since $\Delta_2^{\epsilon}(\rho, \sigma)$ and $\Delta_0^{\epsilon}(\rho, \sigma) \cup \Delta_1^{\epsilon}(\rho, \sigma)$ intersect only at the common vertex $c(\rho(P_2))$ by Sublemma 4.5.5, we see that $\mathcal{L}(\rho, \sigma')$ lies on the ϵ-side of $\mathcal{L}(\rho, \sigma)$ and that $\mathcal{L}(\rho, \{\sigma, \sigma'\})$ is simple. Since $\theta^{\epsilon}(\rho, \sigma; s_1) = 0$, $I(\rho(P_0')) = I(\rho(P_0))$ and $I(\rho(P_1')) = I(\rho(P_2))$ intersect at $v^{\epsilon}(\rho; P_0, P_1) = v^{\epsilon}(\rho; P_1, P_2)$. Since $\mathcal{L}(\rho, \sigma')$ is simple, this intersection point must be a transversal intersection point by Lemma 4.8.6. Hence ρ satisfies the triangle inequality at σ' by Lemma 4.2.1. This completes the proof of the first assertion.

The second assertion follows from the assumption $\theta^{\epsilon}(\rho, \sigma; s_1) = 0$ and Lemma 4.5.3.

To see the third assertion, we note that $\theta^{\epsilon}(\rho, \sigma'; s_2') = 0$ and $\theta^{-\epsilon}(\rho, \sigma'; s_2') < 0$, where the latter inequality follows from the assumption $\theta^{-\epsilon}(\rho, \sigma; s_1) > 0$ and Lemma 4.5.3 (1). Hence we obtain the desired result by Lemma 4.3.5(2).

The last assertion is a consequence of the third assertion.

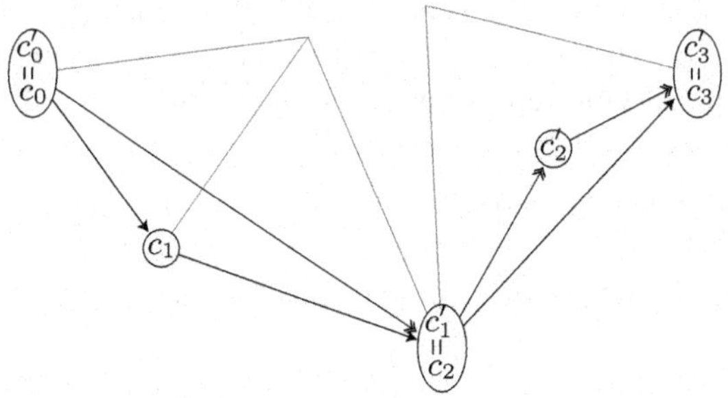

Fig. 4.14. $\mathcal{L}(\rho, \sigma)$ and $\Delta_k^{\epsilon}(\rho, \sigma)$

By a similar argument, we obtain the following lemma.

Lemma 4.5.6. *Under Assumption 4.2.4 (σ-Simple), suppose $\theta^{\epsilon}(\rho, \sigma) = (+, 0, +)$ and $\theta^{-\epsilon}(\rho, \sigma; s_1) = 0$. Then the following hold under Notation 2.1.14 (Adjacent triangles):*

1. *$\mathcal{L}(\rho, \{\sigma, \sigma'\})$ is flat, and ρ satisfies the triangle inequality at σ'. (In particular, $\theta^\epsilon(\rho, \sigma')$ is well-defined.)*
2. *$\theta^\epsilon(\rho, \sigma'; s_2') = 0$. Moreover, $\theta^\epsilon(\rho, \sigma)$ and $\theta^\epsilon(\rho, \sigma')$ determine the same point of $\operatorname{int} \tau$.*
3. *$I(\rho(P_2')) \subset D(\rho(P_2)) \cup D(\rho(P_3))$ and*

$$I(\rho(P_2')) \cap \partial(D(\rho(P_2)) \cup D(\rho(P_3))) = \{v^-(\rho; P_2, P_3), v^+(\rho; P_2, P_3)\}.$$

4. *$Dh(\rho(P_2')) \cap (Eh(\rho(P_2)) \cap Eh(\rho(P_3))) = Ih(\rho(P_2')) \cap (Eh(\rho(P_2)) \cap Eh(\rho(P_3)))$ is a geodesic with endpoints $v^\pm(\rho; P_2, P_3)$.*

By combining Proposition 4.4.4 with Lemmas 4.5.4 and 4.5.6, we obtain the following analogy of Proposition 4.4.4.

Proposition 4.5.7. *Under Notation 2.1.14 (Adjacent triangles) and Assumption 2.4.6 (σ-NonZero), let $\sigma = \langle s_0, s_1, s_2 \rangle$ be an ϵ-terminal triangle of ρ and assume that $\theta^\epsilon(\rho, \sigma; s_1) = 0$. Then the following hold.*

1. *$\theta^\epsilon(\rho, \sigma')$ is well-defined, and $\theta^\epsilon(\rho, \sigma'; s_2') = 0$. Moreover, $\theta^\epsilon(\rho, \sigma)$ and $\theta^\epsilon(\rho, \sigma')$ determine the same point of $\operatorname{int} \tau$.*
2. *$\operatorname{fr} E^\epsilon(\rho, \sigma) = \operatorname{fr} E^\epsilon(\rho, \sigma')$ and*

$$e^\epsilon(\rho, \sigma'; P_{3j}') = e^\epsilon(\rho, \sigma; P_{3j}),$$
$$e^\epsilon(\rho, \sigma'; P_{3j+1}') = e^\epsilon(\rho, \sigma; P_{3j+2}),$$
$$e^\epsilon(\rho, \sigma'; P_{3j+2}') = v^\epsilon(\rho; P_{3j+2}, P_{3j+3}).$$

3. *For every $j \in \mathbb{Z}$, $D(\rho(P_j')) \cap \operatorname{fr} E^\epsilon(\rho, \sigma') = e^\epsilon(\rho, \sigma'; P_j')$.*

4.6 Transition of terminal triangles

We are now ready to prove the following key Lemmas 4.6.1, 4.6.2 and 4.6.7, which describes how the terminal triangles changes according to small deformation of type-preserving representations. These lemmas hold the key to the proof of Proposition 6.2.1 (Openness in $\mathcal{X}$).

Lemma 4.6.1. *Under Assumption 2.4.6 (σ-NonZero), let σ be an ϵ-terminal triangle of ρ and assume that $\theta^\epsilon(\rho, \sigma) \in \operatorname{int} \sigma$. Then there is a neighborhood U of ρ in $\mathcal{X}$, such that for any element ρ' of U, σ is an ϵ-terminal triangle of ρ'.*

Proof. This is almost obvious and all arguments necessary for the proof are contained in the proof of the next Lemma 4.6.2. So we omit the proof.

Lemma 4.6.2. *Under Notation 2.1.14 (Adjacent triangles) and Assumption 2.4.6 (σ-NonZero), let $\sigma = \langle s_0, s_1, s_2 \rangle$ be an ϵ-terminal triangle of ρ and assume that $\theta^\epsilon(\rho, \sigma; s_1) = 0$ and $\theta^{-\epsilon}(\rho, \sigma; s_1) > 0$. (In particular, $\theta^\epsilon(\rho, \sigma) \in \operatorname{int} \tau$, where $\tau = \langle s_0, s_2 \rangle$.) For notational convenience, we set $\sigma^* := \sigma'$. Then there is a neighborhood U of ρ in $\mathcal{X}$ such that the following conditions hold for every $\rho' \in U$ (see Fig. 4.15):*

1. *If* $\theta^\epsilon(\rho', \sigma; s_1) \geq 0$*, then* σ *is an* ϵ*-terminal triangle of* ρ'*.*
2. *If* $\theta^\epsilon(\rho', \sigma; s_1) < 0$*, then* σ^* *is an* ϵ*-terminal triangle of* ρ'*. Moreover,* $D(\rho'(P_1)) \cap E^\epsilon(\rho', \sigma^*) = \emptyset$*, and both* $Ih(\rho'(P_0)) \cap Ih(\rho'(P_1)) \cap Ih(\rho'(P_2))$ *and* $Ih(\rho'(P_1')) \cap Ih(\rho'(P_2')) \cap Ih(\rho'(P_3'))$ *are singletons.*

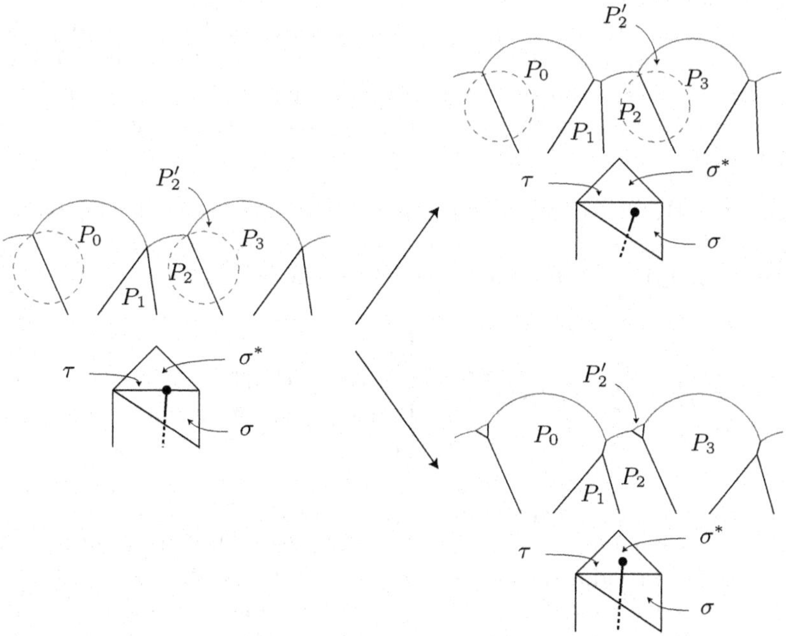

Fig. 4.15. Transition of terminal triangles

Proof. By the assumption and Lemmas 4.5.4, $\mathcal{L}(\rho, \{\sigma, \sigma^*\})$ is simple and ρ satisfies the triangle inequality at σ and σ^*. This implies that there is a neighborhood U_1 of ρ in $\mathcal{X}$ such that for every $\rho' \in U_1$, $\mathcal{L}(\rho', \{\sigma, \sigma^*\})$ is simple, and ρ' satisfies the triangle inequality at σ and σ^*. In particular, $\theta^\epsilon(\rho', \sigma)$ and $\theta^\epsilon(\rho', \sigma^*)$ are defined. Since $\theta^\epsilon(\rho, \sigma) = (+, 0, +)$, we have $\theta^\epsilon(\rho, \sigma^*) = (+, +, 0)$ by Lemma 4.5.3. So we can find a neighborhood U_2 of ρ in U_1 such that every $\rho' \in U_2$ satisfies the following conditions:

1. The components of the side parameter $\theta^\epsilon(\rho', \sigma)$, except possibly $\theta^\epsilon(\rho', \sigma; s_1)$, are positive.
2. $\theta^{-\epsilon}(\rho', \sigma; s_1) > 0$.
3. The components of the side parameter $\theta^\epsilon(\rho', \sigma^*)$ are positive, except possibly $\theta^\epsilon(\rho', \sigma^*; s_2')$.

Here the second condition follows from the assumption that $\theta^{-\epsilon}(\rho, \sigma; s_1) > 0$. Thus, by using Lemma 4.5.3, we see that one of the following holds for each $\rho' \in U_2$.

1. $\theta^\epsilon(\rho', \sigma; s_1) > 0$ and hence all components of $\theta^\epsilon(\rho', \sigma)$ are positive.
2. $\theta^\epsilon(\rho', \sigma; s_1) < 0$ and hence all components of $\theta^\epsilon(\rho', \sigma^*)$ are positive.
3. $\theta^\epsilon(\rho', \sigma; s_1) = 0$ and hence all components of $\theta^\epsilon(\rho', \sigma)$ are non-negative

In the following, we show, through case-by-case argument, that every $\rho' \in U_2$ satisfies the conclusion of Lemma 4.6.2 provided that U_2 is chosen to be small enough in advance.

Case 1. $\theta^\epsilon(\rho', \sigma; s_1) > 0$ and hence all components of $\theta^\epsilon(\rho', \sigma)$ are positive. Then the first two conditions for σ to be an ϵ-terminal triangle (Definition 4.3.8) are satisfied. In particular, $e^\epsilon(\rho', \sigma; P_j)$ and $v^\epsilon(\rho'; P_j, P_{j+1})$ are defined for every $j \in \mathbb{Z}$. To show that the remaining conditions are satisfied, we need the following claim.

Claim 4.6.3. By choosing U_2 small enough in advance, the following hold. For each $j \in \mathbb{Z}$, $e^\epsilon(\rho', \sigma; P_j)$ is a non-degenerate circular arc, and satisfies the following condition (cf. the upper-right figure in Fig. 4.15).

$$e^\epsilon(\rho', \sigma; P_j) \cap D(\rho'(P_k)) = \begin{cases} e^\epsilon(\rho', \sigma; P_j) & \text{if } k = j, \\ v^\epsilon(\rho'; P_j, P_{j+1}) & \text{if } k = j+1, \\ v^\epsilon(\rho'; P_{j-1}, P_j) & \text{if } k = j-1, \\ \emptyset & \text{otherwise.} \end{cases}$$

Proof. Since all components of $\theta^\epsilon(\rho', \sigma)$ are positive by the assumption, each $e^\epsilon(\rho', \sigma; P_j)$ is a non-degenerate circular arc. Thus we have the first assertion.

To see the second assertion, note that, by virtue of Proposition 4.4.4 and the assumption $\theta^\epsilon(\rho, \sigma; s_1) = 0$, $D(\rho(P_k))$ has nonempty intersection with $e^\epsilon(\rho, \sigma; P_j)$ if and only if one of the following conditions holds.

1. $k = j$ or $j \pm 1$.
2. $k = j + 2$ and $j \equiv 0 \pmod 3$.
3. $k = j - 2$ and $j \equiv 2 \pmod 3$.

Suppose k does not satisfy any of the above conditions. Then $D(\rho(P_k))$ is disjoint from $e^\epsilon(\rho, \sigma; P_j)$, and hence we may assume, by choosing U_2 sufficiently small in advance, that $D(\rho'(P_k))$ is disjoint from $e^\epsilon(\rho', \sigma; P_j)$. Since the Euclidean distance between $D(\rho(P_k))$ and $e^\epsilon(\rho, \sigma; P_j)$ is large when $|k-j|$ is large, and since we have the parallel translation $\rho(K)$, we can choose U_2 in advance, so that the same conclusion holds for every such k.

Suppose $k = j$. Then it is obvious that $e^\epsilon(\rho', \sigma; P_j) \cap D(\rho'(P_j)) = e^\epsilon(\rho', \sigma; P_j)$.

Suppose $k = j \pm 1$. Since $I(\rho'(P_j))$ intersect $I(\rho'(P_{j+1}))$ transversely and since $e^\epsilon(\rho, \sigma; P_j) \cap I(\rho(P_{j+1})) = v^\epsilon(\rho; P_j, P_{j+1})$, we see $e^\epsilon(\rho', \sigma; P_j) \cap I(\rho(P_{j+1})) = v^\epsilon(\rho'; P_j, P_{j+1})$ by choosing U_2 small enough in advance.

Suppose $j \equiv 0 \pmod 3$ and $k = j + 2$. Then we may assume $j = 0$ and $k = 2$. By Proposition 4.4.4, $e^\epsilon(\rho, \sigma; P_0) \cap D(\rho(P_2))$ is equal to the degenerate arc $e^\epsilon(\rho, \sigma; P_1) = v^\epsilon(\rho; P_0, P_1) = v^\epsilon(\rho; P_1, P_2)$. By the preceding argument

$v^\epsilon(\rho'; P_0, P_1) \notin D(\rho'(P_2))$. Moreover, by Lemma 4.2.1, $I(\rho(P_0)) = I(\rho(P_0'))$ and $I(\rho(P_2)) = I(\rho(P_1'))$ intersect transversely. Hence we can choose U_2 small enough in advance so that $e^\epsilon(\rho', \sigma; P_0) \cap D(\rho'(P_2)) = \emptyset$ (see Fig. 4.16). In fact this is guaranteed by the following observations.

1. Since $I(\rho(P_0))$ and $I(\rho(P_2))$ intersect transversely and since $v^\epsilon(\rho'; P_0, P_1) \notin D(\rho'(P_2))$, we may assume a small neighborhood of $v^\epsilon(\rho'; P_0, P_1)$ in $e^\epsilon(\rho', \sigma; P_0)$ is disjoint from $D(\rho'(P_2))$, by choosing U_2 small enough in advance.
2. Since the complement of a small neighborhood of $v^\epsilon(\rho; P_0, P_1)$ in $e^\epsilon(\rho, \sigma; P_0)$ is disjoint from $D(\rho(P_2))$, we may assume the complement of the above small neighborhood of $v^\epsilon(\rho'; P_0, P_1)$ in $e^\epsilon(\rho', \sigma; P_0)$ is also disjoint from $D(\rho'(P_2))$, by choosing U_2 small enough in advance.

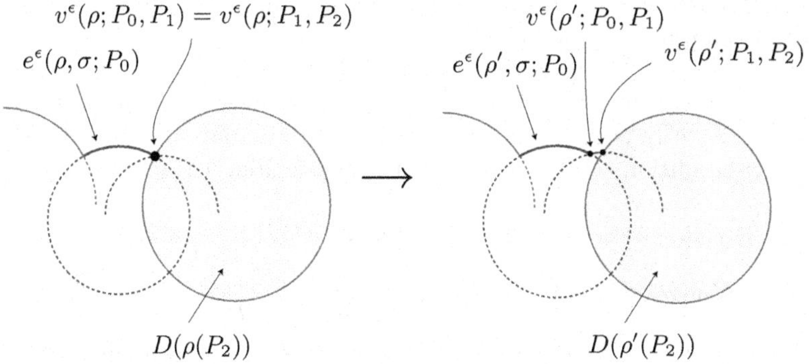

Fig. 4.16. Deformation from $\theta^\epsilon(\rho, \sigma; P_1) = 0$ to $\theta^\epsilon(\rho', \sigma; P_1) > 0$

Suppose $j \equiv 2 \pmod 3$ and $k = j - 2$. Then by an argument similar to the above, we see $e^\epsilon(\rho', \sigma; P_j)$ is disjoint from $D(\rho'(P_k))$.

This completes the proof of the claim.

Claim 4.6.4. Let U_2 be as in Claim 4.6.3. Then $\mathrm{fr}\, E^\epsilon(\rho', \sigma) = \cup_j e^\epsilon(\rho', \sigma; P_j) \cong \mathbb{R}$.

Proof. By Claim 4.6.3, $\cup_j e^\epsilon(\rho', \sigma; P_j)$ is homeomorphic to $\mathbb{R}$ and is contained in $\mathrm{fr}\, E(\rho', \sigma)$. Since it is invariant under the parallel translation $\rho(K)$, the complement $\mathbb{C} - \cup_j e^\epsilon(\rho', \sigma; P_j)$ consists of two components. Let $\tilde{E}^+$ (resp. $\tilde{E}^-$) be the component of $\mathbb{C} - \cup_j e^\epsilon(\rho', \sigma; P_j)$ which contains the region $\{z \in \mathbb{C} \,|\, \Im(z) \geq L\}$ (resp. $\{z \in \mathbb{C} \,|\, -\Im(z) \geq L\}$) for sufficiently large number L. Since $\mathrm{int}\, E^\epsilon(\rho', \sigma)$ is disjoint from $\cup_j e^\epsilon(\rho', \sigma; P_j) \subset \mathrm{fr}\, E(\rho', \sigma)$ and contains the region $\{z \in \mathbb{C} \,|\, \epsilon\Im(z) \geq L\}$ for sufficiently large number L, it is contained in $\tilde{E}^\epsilon$.

On the other hand $\cup_j \mathrm{int}\, D(\rho'(P_j))$ is disjoint from $\cup_j e^\epsilon(\rho', \sigma; P_j) \subset \mathrm{fr}\, E(\rho', \sigma)$ and hence it is contained in $\tilde{E}^-$ or $\tilde{E}^+$. But $\cup_j \mathrm{int}\, D(\rho'(P_j))$

contains $\mathcal{L}(\rho',\sigma)$, which in tern is contained in $\tilde{E}^{-\epsilon}$. Thus $\cup_j \operatorname{int} D(\rho'(P_j))$ is contained in $\tilde{E}^{-\epsilon}$. Now every point in $\tilde{E}^{\epsilon}$ is joined to the region $\{z \in \mathbb{C} \mid \epsilon\Im(z) \geq L\}$ by a path in $\tilde{E}^{\epsilon}$. The path is disjoint from $\cup_j \operatorname{int} D(\rho'(P_j))$ by the above observation. Thus we have $\tilde{E}^{\epsilon} \subset \operatorname{int} E^{\epsilon}(\rho',\sigma)$. Hence $\tilde{E}^{\epsilon} = \operatorname{int} E^{\epsilon}(\rho',\sigma)$, and therefore $\operatorname{fr} E^{\epsilon}(\rho',\sigma) = \operatorname{fr} \tilde{E}^{\epsilon} = \cup_j e^{\epsilon}(\rho',\sigma;P_j)$.

By Claim 4.6.4, the pair (ρ',σ) satisfies the third condition in Definition 4.3.8.

By Claim 4.6.3, $\overline{Dh}(\rho'(P_k))$ intersects (a small neighborhood of) $v^{\epsilon}(\rho';P_j, P_{j+1})$ if and only if $k = j$ or $j+1$. So, the last condition in Definition 4.3.8 is also satisfied. Hence σ is an ϵ-terminal triangle of ρ'.

Case 2. $\theta^{\epsilon}(\rho',\sigma;s_1) < 0$ and hence all components of $\theta^{\epsilon}(\rho',\sigma^*)$ are positive. Then the first two conditions for σ^* to be an ϵ-terminal triangle (Definition 4.3.8) are satisfied. In particular, $e^{\epsilon}(\rho',\sigma^*;P'_j)$ and $v^{\epsilon}(\rho';P'_j,P'_{j+1})$ are defined for every $j \in \mathbb{Z}$. To show that the remaining conditions are satisfied, we need the following claim.

Claim 4.6.5. By choosing U_2 small enough in advance, the following hold. For each $j \in \mathbb{Z}$, $e^{\epsilon}(\rho',\sigma;P'_j)$ is a non-degenerate circular arc, and satisfies the following condition (cf. the lower-right figure in Fig. 4.15).

$$e^{\epsilon}(\rho',\sigma^*;P'_j) \cap D(\rho'(P'_k)) = \begin{cases} e^{\epsilon}(\rho',\sigma^*;P'_j) & \text{if } k = j, \\ v^{\epsilon}(\rho';P'_j,P'_{j+1}) & \text{if } k = j+1, \\ v^{\epsilon}(\rho';P'_{j-1},P'_j) & \text{if } k = j-1, \\ \emptyset & \text{otherwise.} \end{cases}$$

Proof. By Proposition 4.5.7, $D(\rho(P'_k))$ has nonempty intersection with $e^{\epsilon}(\rho,\sigma;P_j)$ if and only if one of the following conditions holds.

1. $k = j$ or $j \pm 1$.
2. $k = j+2$ and $j \equiv 1 \pmod 3$.
3. $k = j-2$ and $j \equiv 3 \pmod 3$.

As in the proof of Claim 4.6.3, we may assume the conclusion holds for every k which does not satisfy any of the above conditions. For each k satisfying one of the above conditions, the intersection is controlled by the side parameter as shown in the proof of Claim 4.6.3. Thus we obtain the claim.

By using Claim 4.6.5 instead of Claim 4.6.3 in the proof of Claim 4.6.4, we see that $\operatorname{fr} E^{\epsilon}(\rho',\sigma^*) = \cup_j e^{\epsilon}(\rho',\sigma^*;P'_j) \cong \mathbb{R}$ whenever $\rho' \in U_2$, provided that U_2 is chosen to be small enough in advance. This implies that the pair (ρ',σ^*) satisfies the third condition in Definition 4.3.8. By Claim 4.6.5 and by an argument parallel to that in Case 1, the last condition in Definition 4.3.8 is also satisfied. Hence σ^* is an ϵ-terminal triangle of ρ'.

Next, we show that $D(\rho'(P_1)) \cap E^{\epsilon}(\rho',\sigma^*) = \emptyset$. Note that $E^{\epsilon}(\rho',\sigma^*)$ ($\rho' \in U_2$) moves continuously with respect to the Chabauty topology. Since

$D(\rho(P_1)) \cap E^\epsilon(\rho, \sigma^*) = \{\mathrm{Fix}^\epsilon_\sigma(\rho(P_1))\}$ and since $I(\rho(P_j))$ $(j = 0, 1, 2)$ intersect transversely at $\mathrm{Fix}^\epsilon_\sigma(\rho(P_1))$, we can find, by using the above continuity, a round disk V with center $\mathrm{Fix}^\epsilon_\sigma(\rho(P_1))$ such that the following conditions are satisfied for all $\rho' \in U_2$.

1. $D(\rho'(P_1)) \cap E^\epsilon(\rho', \sigma^*) \subset V$.
2. $D(\rho'(P_j)) \cap V$ is a bigon, i.e., a convex region bounded by an arc in $I(\rho'(P_j))$ and an arc in ∂D, for $j = 0, 1, 2$.
3. $(I(\rho'(P_j)) \cap I(\rho'(P_{j+1}))) \cap V = \{v^\epsilon(\rho'; P_j, P_{j+1})\}$ for $j = 0, 1$.

Since $\theta^\epsilon(\rho', \sigma; s_1) < 0$ by the assumption, we see by Lemma 4.3.1 that $v^\epsilon(\rho'; P_0, P_1) \in \mathrm{int}\, D(\rho'(P_2))$ and $v^\epsilon(\rho'; P_1, P_2) \in \mathrm{int}\, D(\rho'(P_0))$. These imply that the $D(\rho'(P_1)) \cap V$ is contained in the interior of $(D(\rho'(P_0)) \cap D(\rho'(P_2))) \cap V$ in V. Hence we have $D(\rho'(P_1)) \cap E^\epsilon(\rho', \sigma^*) = \emptyset$.

Since $\theta^\epsilon(\rho', \sigma; s_1) < 0$ and $\theta^{-\epsilon}(\rho', \sigma; s_1) > 0$, $Ih(\rho'(P_0)) \cap Ih(\rho'(P_1)) \cap Ih(\rho'(P_2))$ is a singleton by Lemma 4.3.6. Hence $Ih(\rho'(P'_1)) \cap Ih(\rho'(P'_2)) \cap Ih(\rho'(P'_3))$ is also a singleton by Lemma 4.1.3 (Chain rule).

Case 3. $\theta^\epsilon(\rho', \sigma; s_1) = 0$ and hence all components of $\theta^\epsilon(\rho', \sigma)$ are non-negative. Then the first two conditions for σ to be an ϵ-terminal triangle are satisfied, because $\theta^{-\epsilon}(\rho', \sigma; s_1) > 0$. Moreover, we have the following analogy of Claim 4.6.3.

Claim 4.6.6. By choosing U_2 small enough in advance, the following hold. For each $j \in \mathbb{Z}$, $e^\epsilon(\rho', \sigma; P_j)$ is a degenerate or non-degenerate circular arc according as $j \equiv 1 \pmod 3$ or not, and satisfies the following condition.

$$e^\epsilon(\rho', \sigma; P_j) \cap D(\rho'(P_k)) = \begin{cases} e^\epsilon(\rho', \sigma; P_j) & \text{if } k = j, \\ v^\epsilon(\rho'; P_j, P_{j+1}) & \text{if } k = j+1, \\ v^\epsilon(\rho'; P_{j-1}, P_j) & \text{if } k = j-1, \\ e^\epsilon(\rho', \sigma; P_{j+1}) & \text{if } k = j+2 \text{ and } j \equiv 0 \pmod 3) \\ e^\epsilon(\rho', \sigma; P_{j-1}) & \text{if } k = j-2 \text{ and } j \equiv 0 \pmod 3) \\ \emptyset & \text{otherwise.} \end{cases}$$

Proof. The proof is parallel to that of Claim 4.6.3. The following are the only difference.

- Suppose $k = j + 1$ and $j \equiv 1 \pmod 3$, say $k = 1$ and $j = 0$, then $\theta^\epsilon(\rho', \sigma; s_1) = 0$ and hence $e^\epsilon(\rho', \sigma; P_1)$ is a singleton. So it is obvious that $e^\epsilon(\rho', \sigma; P_1) \cap D(\rho'(P_0))$ is equal to the degenerate arc $e^\epsilon(\rho', \sigma; P_1)$.
- Suppose $k = j + 2$ and $j \equiv 0 \pmod 3$, say $k = 2$ and $j = 0$. Then $e^\epsilon(\rho', \sigma; P_1)$ is a degenerate arc and is equal to $v^\epsilon(\rho'; P_0, P_1) = v^\epsilon(\rho'; P_1, P_2)$. Thus $e^\epsilon(\rho', \sigma; P_1)$ is contained in the intersection $e^\epsilon(\rho', \sigma; P_0) \cap D(\rho'(P_2))$. On the other hand, since $I(\rho'(P_0))$ and $I(\rho'(P_2))$ intersect transversely and since $e^\epsilon(\rho, \sigma; P_0) \cap D(\rho(P_2))$ is equal to the singleton $e^\epsilon(\rho, \sigma; P_1)$, we see that $e^\epsilon(\rho', \sigma; P_0) \cap D(\rho'(P_2)) = e^\epsilon(\rho', \sigma; P_1)$ by choosing U_2 small enough in advance.

- Suppose $k = j - 2$ and $j \equiv 0 \pmod 3$. Then we can apply the above argument to this case.

By using Claim 4.6.6 instead of Claim 4.6.3 in the proof of Claim 4.6.4, we see that $\mathrm{fr}\, E^{\epsilon}(\rho', \sigma) = \cup_j e^{\epsilon}(\rho', \sigma; P_j) \cong \mathbb{R}$ whenever $\rho' \in U_2$, provided that U_2 is chosen to be small enough in advance. This implies that the pair (ρ', σ) satisfies the third condition in Definition 4.3.8. By Claim 4.6.6 and the fact that $\theta^{-\epsilon}(\rho', \sigma; s_1) > 0$, the last condition in Definition 4.3.8 is also satisfied. Hence σ is an ϵ-terminal triangle of ρ'.

Thus we have proved that if we choose U_2 small enough in advance, then every $\rho' \in U_2$ satisfies the conclusion of Lemma 4.6.2. This completes the proof of Lemma 4.6.2.

Finally, we prove the following analogy to Lemmas 4.6.1 and 4.6.2.

Lemma 4.6.7. *Under Notation 2.1.14 (Adjacent triangles) and Assumption 2.4.6 (σ-NonZero), let $\sigma = \langle s_0, s_1, s_2 \rangle$ be an ϵ-terminal triangle of ρ and assume that $\theta^{\epsilon}(\rho, \sigma; s_1) = \theta^{-\epsilon}(\rho, \sigma; s_1) = 0$. For notational convenience, we set $\sigma^* := \sigma'$. Then there is a neighborhood U of ρ in $\mathcal{X}$ such that the following conditions hold for every $\rho' \in U$:*

1. *If $\theta^{\epsilon}(\rho', \sigma; s_1) > 0$, then σ is an ϵ-terminal triangle of ρ'. Moreover $D(\rho'(P_2'))$ is disjoint from $E^{\epsilon}(\rho', \sigma)$.*
2. *If $\theta^{\epsilon}(\rho', \sigma; s_1) < 0$, then σ^* is an ϵ-terminal triangle of ρ'. Moreover $D(\rho'(P_1))$ is disjoint from $E^{\epsilon}(\rho', \sigma^*)$.*
3. *If $\theta^{\epsilon}(\rho', \sigma; s_1) = 0$ and $\theta^{-\epsilon}(\rho', \sigma; s_1) > 0$, then σ is an ϵ-terminal triangle of ρ'.*
4. *If $\theta^{\epsilon}(\rho', \sigma; s_1) = 0$ and $\theta^{-\epsilon}(\rho', \sigma; s_1) < 0$, then σ^* is an ϵ-terminal triangle of ρ'.*
5. *If $\theta^{\epsilon}(\rho', \sigma; s_1) = 0$ and $\theta^{-\epsilon}(\rho', \sigma; s_1) = 0$, then both σ and σ^* are ϵ-terminal triangles of ρ'.*

Proof. The proof of Lemma 4.6.7 is parallel to that of Lemma 4.6.2. The only difference are the following.

- $\mathcal{L}(\rho, \{\sigma, \sigma^*\})$ is not simple but is flat (Definition 4.5.1(2)). Thus for each nearby representation ρ', $\mathcal{L}(\rho', \{\sigma, \sigma^*\})$ is either simple or flat. However, the results in Sect. 4.5, which were used in the proof of Lemma 4.6.2, are proved under this weaker condition.
- The condition $\theta^{-\epsilon}(\rho, \sigma; s_1) > 0$ in Lemma 4.6.2 was used only at the following two points.
  1. In the treatment of ρ' such that $\theta^{\epsilon}(\rho, \sigma; s_1) < 0$, it was used to prove that both $Ih(\rho'(P_0)) \cap Ih(\rho'(P_1)) \cap Ih(\rho'(P_2))$ and $Ih(\rho'(P_1')) \cap Ih(\rho'(P_2')) \cap Ih(\rho'(P_3'))$ are singletons. However, Lemma 4.6.7 does not contain corresponding assertion. (Such a problem is studied in Sect. 7.3.)

2. In the treatment of ρ' such that $\theta^\epsilon(\rho', \sigma; s_1) = 0$. However, in Lemma 4.6.7, the corresponding case is divided into three cases according as $\theta^{-\epsilon}(\rho', \sigma; s_1)$ is > 0, $= 0$ or < 0. If $\theta^\epsilon(\rho', \sigma; s_1) = 0$ and $\theta^{-\epsilon}(\rho', \sigma; s_1) \le 0$, then $\theta^\epsilon(\rho', \sigma^*; s_2') = 0$ and $\theta^{-\epsilon}(\rho', \sigma^*; s_2') \ge 0$ by Lemma 4.5.3. Thus these cases are essentially equivalent to Case 3 in the proof of Lemma 4.6.2, where σ may be replaced with σ^*, except when $\theta^\epsilon(\rho', \sigma; s_1) = \theta^{-\epsilon}(\rho', \sigma; s_1) = 0$. In this exceptional case, (ρ, τ) is an isosceles representation (see the forthcoming Definition 5.2.2) by virtue of the forthcoming Proposition 5.2.3. Thus the conclusion also holds by the forthcoming Proposition 5.2.8.

4.7 Proof of Lemma 4.5.5

Throughout this section, we use Notation 4.2.6. We shall reduce the proof of Lemma 4.5.5 to the generic case, where all components of $\theta^\epsilon(\rho, \sigma)$ are positive for some $\epsilon \in \{-, +\}$ and $\mathcal{L}(\rho, \sigma)$ is not flat at every vertex. By Corollary 4.2.13, we may assume without loss of generality that $\mathcal{L}(\rho, \sigma)$ is convex to the above at $c(0)$ and convex to the below at $c(1)$ and $c(2)$, because the argument for the other case is parallel. The following lemma describes the shape of $E(\rho, \sigma)$ and $\mathcal{L}(\rho, \sigma)$ in this generic case, and it holds the key to the proof of Lemma 4.5.5.

Lemma 4.7.1. *Under Assumption 4.2.4 (σ-Simple), suppose that all components of $\theta^\epsilon(\rho, \sigma)$ are positive for some $\epsilon \in \{-, +\}$ and that $\mathcal{L}(\rho, \sigma)$ is convex to the above at $c(0)$ and convex to the below at $c(1)$ and $c(2)$. Apply a rotational coordinate change of $\mathbb{C}$ so that the oriented line segment $\overrightarrow{f}(0)$ is parallel to the imaginary axis, i.e.,* $\arg(\overrightarrow{f}(0)) = \frac{\pi}{2}$.

(1) If $\epsilon = -$, then the following inequalities hold in the new coordinate (see Fig. 4.17).
  - *(i)* $\Re(c(0)) < \Re(v^-(0,1)) < \Re(c(1)) < \Re(c(2)) < \Re(v^-(2,3)) < \Re(c(3))$.
  - *(ii)* $\Re(c(0)) < \Re(v^-(0,1)) < \Re(v^-(1,2)) < \Re(v^-(2,3)) < \Re(c(3))$.
  - *(iii)* $0 < \arg(\overrightarrow{f}(1)) \le \pi/2 \le \arg(\overrightarrow{f}(2)) < \pi$.
  - *(iv)* $\arg(\overrightarrow{c}(0,1)) \in (-\pi/2, 0)$, $\arg(\overrightarrow{c}(1,2)) \in (-\pi/2, \pi/2)$, $\arg(\overrightarrow{c}(2,3)) \in (0, \pi/2)$.

(2) If $\epsilon = +$, then the following inequalities hold in the new coordinate (see Fig. 4.18).
  - *(i)* $\Re(c(0)) < \Re(c(1)) < \Re(v^+(1,2)) < \Re(c(2)) < \Re(c(3))$.
  - *(ii)* $\Re(c(0)) < \Re(v^+(0,1)) < \Re(v^+(1,2)) < \Re(v^+(2,3)) < \Re(c(3))$.
  - *(iii)* $0 < \arg(\overrightarrow{f}(1)) \le \pi/2 \le \arg(\overrightarrow{f}(2)) < \pi$.
  - *(iv)* $\arg(\overrightarrow{c}(0,1)) \in (-\pi/2, 0)$, $\arg(\overrightarrow{c}(1,2)) \in (-\pi/2, \pi/2)$, $\arg(\overrightarrow{c}(2,3)) \in (0, \pi/2)$.

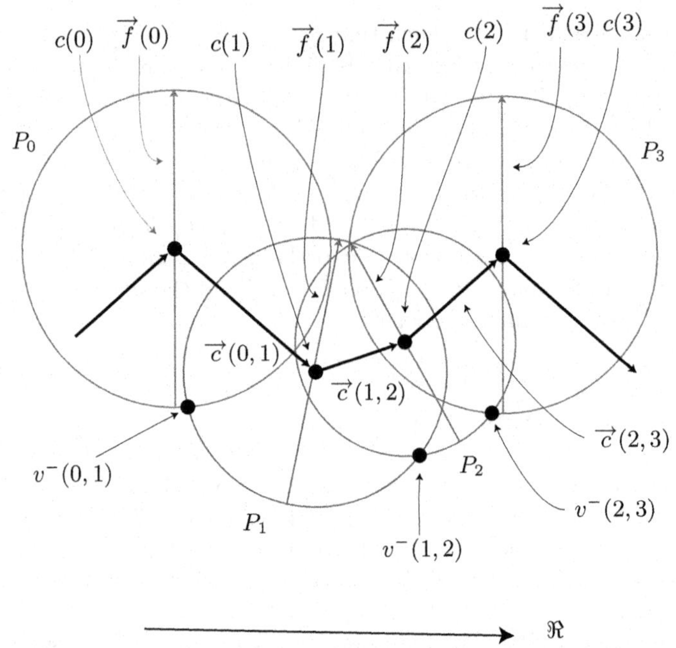

Fig. 4.17. Isometric circles after a coordinate change (a)

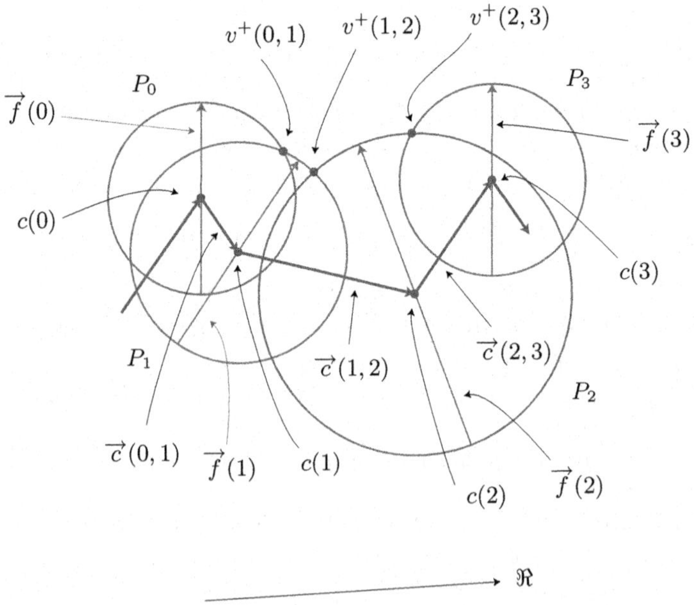

Fig. 4.18. Isometric circles after a coordinate change (b)

Remark 4.7.2. Though one might expect the stronger inequality

$$\Re(c(0)) < \Re(v^+(0,1)) < \Re(c(1)) < \Re(v^+(1,2)) \\ < \Re(c(2)) < \Re(v^+(2,3)) < \Re(c(3)),$$

it does not necessarily hold. In fact, it can happen that $\Re(c(1)) > \Re(v^+(1,2))$ or $\Re(c(2)) > \Re(v^+(2,3))$. Such a phenomena can be observed in the example in Remark 4.3.9(2).

Proof. We begin by the following observation.

Claim 4.7.3. Under the setting of Lemma 4.7.1, the following hold.

(1) $c(j-1) \in \mathrm{int}(H_L(\overrightarrow{f}(j)))$ and $c(j+1) \in \mathrm{int}(H_R(\overrightarrow{f}(j)))$.
(2) $v^+(j,j+1) \in \mathrm{int}(H_L(\overrightarrow{c}(j,j+1)))$ and $v^-(j,j+1) \in \mathrm{int}(H_R(\overrightarrow{c}(j,j+1)))$.
(3) $v^\epsilon(j-1,j) \in \mathrm{int}(H_L(\overrightarrow{f}(j)))$ and $v^\epsilon(j,j+1) \in \mathrm{int}(H_R(\overrightarrow{f}(j)))$.
(4) $\mathrm{Fix}_\sigma^+(\rho(P_j)) \in \mathrm{int}(H_L(\overrightarrow{c}(j-1,j))) \cap \mathrm{int}(H_L(\overrightarrow{c}(j,j+1)))$ and $\mathrm{Fix}_\sigma^-(\rho(P_j)) \in \mathrm{int}(H_R(\overrightarrow{c}(j-1,j))) \cap \mathrm{int}(H_R(\overrightarrow{c}(j,j+1)))$.
(5) $-\arg(\overrightarrow{c}(-1,0)) = \arg(\overrightarrow{c}(0,1)) \in (-\pi/2, 0)$.

Proof. By Definitions 3.1.3 (Upward) and 4.2.5(1), we have

$$0 < \arg \frac{-\overrightarrow{c}(j-1,j)}{\overrightarrow{f}(j)} = \arg \frac{\overrightarrow{f}(j)}{\overrightarrow{c}(j,j+1)} < \pi.$$

This implies the assertion (1).

The assertions (2) and (4) follow from Definition 4.2.5.

By the assumption that all components of $\theta^\epsilon(\rho,\sigma)$ are positive, we have

$$\arg\left(\frac{v^\epsilon(j-1,j) - c(j)}{\overrightarrow{f}(j)}\right) = \arg\left(\frac{\overrightarrow{f}(j)}{v^\epsilon(j,j+1) - c(j)}\right) = \theta^\epsilon(\rho,\sigma; s_{[j]}) > 0.$$

(cf. Lemma 4.2.8.) This implies the assertion (3).

By Definition 3.1.3 and the assumption that $\overrightarrow{f}(0)$ is parallel to the imaginary axis (in the new coordinate), we have

$$-\arg(\overrightarrow{c}(-1,0)) = \arg(\overrightarrow{c}(0,1)) \in (-\pi/2, \pi/2).$$

Since $\mathcal{L}(\rho,\sigma)$ is convex to the above at $c(0)$ by the assumption, this implies that these angles lie in $(-\pi/2, 0)$. Thus we obtain the assertion (5).

Claim 4.7.4. $\Re(c(0)) < \Re(c(1)) < \Re(c(2)) < \Re(c(3))$.

Proof. By Claim 4.7.3(5), we have $\Re(c(-1)) < \Re(c(0)) < \Re(c(1))$. Since $c(3) - c(2) = c(0) - c(-1)$, we also have $\Re(c(2)) < \Re(c(3))$.

Next, we prove $\Re(c(0)) < \Re(c(3))$. Suppose to the contrary that $\Re(c(3)) \le \Re(c(0))$. Then we have $\Re(c(2)) < \Re(c(3)) \le \Re(c(0))$ and therefore $c(2) \in$

$H_L(\overrightarrow{f}(0))$. Moreover, since $\mathcal{L}(\rho,\sigma)$ is convex to the below at $c(1)$ by the assumption, we have $c(2) \in \mathrm{int}(H_L(\overrightarrow{c}(0,1)))$. Thus $c(2) \in \mathrm{int}(H_L(\overrightarrow{c}(0,1))) \cap H_L(\overrightarrow{f}(0))$. Since $\arg(\overrightarrow{c}(0,1)) \in (-\pi/2, 0)$ by Claim 4.7.3(5) and since $\overrightarrow{f}(0)$ is parallel to the imaginary axis, this implies $\Im(c(2)) > \Im(c(1))$ (see Fig. 4.19). On the other hand, we can also see $c(1) \in \mathrm{int}(H_L(\overrightarrow{c}(2,3))) \cap H_R(\overrightarrow{f}(3))$, through a parallel argument, by using the fact that $\Re(c(2)) < \Re(c(3)) \leq \Re(c(0)) < \Re(c(1))$ and the assumption that $\mathcal{L}(\rho,\sigma)$ is convex to the below at $c(2)$. Since $\arg(\overrightarrow{c}(2,3)) \in (0, \pi/2)$ by Claim 4.7.3(5) and since $\overrightarrow{f}(3)$ is parallel to the imaginary axis, this implies $\Im(c(1)) > \Im(c(2))$, a contradiction. Hence $\Re(c(0)) < \Re(c(3))$.

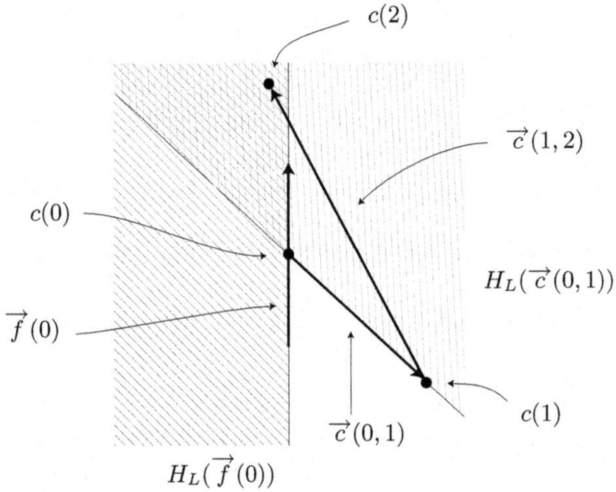

Fig. 4.19. $\Im(c(2)) > \Im(c(0))$

Finally, we prove $\Re(c(1)) < \Re(c(2))$. Suppose to the contrary that $\Re(c(2)) \leq \Re(c(1))$. Then $c(2) \in \mathrm{int}(H_L(\overrightarrow{c}(0,1))) \cap \{z \in \mathbb{C} \,|\, \Re(z) \leq \Re(c(1))\}$, because $\mathcal{L}(\rho,\sigma)$ is convex to the below at $c(1)$. Since $\arg(\overrightarrow{c}(0,1)) \in (-\pi/2, 0)$, this implies $\Im(c(2)) > \Im(c(1))$. However, we can also conclude $\Im(c(1)) > \Im(c(2))$ by using $\arg(\overrightarrow{c}(2,3)) \in (0,\pi/2)$ and the assumption that $\mathcal{L}(\rho,\sigma)$ is convex to the below at $c(2)$. This is a contradiction. Hence we have $\Re(c(1)) < \Re(c(2))$.

Claim 4.7.5. $-\pi/2 < \arg(\overrightarrow{c}(1,2)) < \pi/2$ and $0 < \arg(\overrightarrow{f}(1)) < \pi/2 < \arg(\overrightarrow{f}(2)) < \pi$.

Proof. Since $\Re(c(1)) < \Re(c(2))$ by Claim 4.7.4, it follows that $-\pi/2 < \arg(\overrightarrow{c}(1,2)) < \pi/2$. Since $\mathcal{L}(\rho,\sigma)$ is convex to the below at $c(2)$, we have $\arg(\overrightarrow{c}(1,2)) < \arg(\overrightarrow{c}(2,3)) = -\arg(\overrightarrow{c}(0,1))$. Thus $\arg(\overrightarrow{c}(0,1)) + \arg(\overrightarrow{c}(1,2)) < 0$. Hence

$$\arg(\overrightarrow{f}(1)) = \frac{1}{2}\{\arg(\overrightarrow{c}(0,1)) + \arg(\overrightarrow{c}(1,2))\} + \frac{\pi}{2} \in (0, \pi/2).$$

By a parallel argument, we can also prove that $\arg(\overrightarrow{f}(2)) \in (\pi/2, \pi)$.

Claim 4.7.6. $\Re(c(0)) < \Re(v^\epsilon(0,1)) < \Re(v^\epsilon(1,2)) < \Re(v^\epsilon(2,3)) < \Re(c(3))$.

Proof. By Claim 4.7.3(3), we have $\Re(c(0)) < \Re(v^\epsilon(0,1))$. To prove $\Re(v^\epsilon(0,1)) < \Re(v^\epsilon(1,2))$, note that

$$v^\epsilon(0,1) = c(1) + \epsilon r e^{i(\alpha+\epsilon\beta)}, \quad v^\epsilon(1,2) = c(1) + \epsilon r e^{i(\alpha-\epsilon\beta)},$$

where $\alpha = \arg(\overrightarrow{f}(1))$, $\beta = \theta^\epsilon(\rho,\sigma;s_1)$ and $r = r(\rho(P_1))$. Thus

$$\begin{aligned}\Re(v^\epsilon(1,2)) - \Re(v^\epsilon(0,1)) &= \Re(re^{i(\alpha-\beta)} - re^{i(\alpha+\beta)}) \\ &= r(\cos(\alpha-\beta) - \cos(\alpha+\beta)) = 2r\sin\alpha\sin\beta.\end{aligned}$$

Since $\alpha \in (0,\pi/2)$ by Claim 4.7.5 and since $\beta \in (0,\pi/2)$ by the assumption that all components of $\theta^\epsilon(\rho,\sigma)$ are positive, the above number is positive. Hence $\Re(c(0)) < \Re(v^\epsilon(0,1)) < \Re(v^\epsilon(1,2))$. The remaining inequalities are proved by a parallel argument.

Claim 4.7.7. If $\epsilon = -$, then $\Re(v^-(0,1)) < \Re(c(1))$ and $\Re(c(2)) < \Re(v^-(2,3))$.

Proof. By the assertions (2) and (3) of Claim 4.7.3, $v^-(0,1)$ is contained in the open region $\operatorname{int} H_R(\overrightarrow{c}(0,1)) \cap \operatorname{int} H_L(\overrightarrow{f}(1)) \cap \operatorname{int} H_R(\overrightarrow{f}(0))$. By Claims 4.7.3(5) and 4.7.5, the open region is contained in the open strip $\{z \in \mathbb{C} \mid \Re(c(0)) < \Re(z) < \Re(c(1))\}$ (see Fig. 4.20). Hence we have $\Re(v^-(0,1)) < \Re(c(1))$. By a parallel argument, we also have $\Re(c(2)) < \Re(v^-(2,3))$.

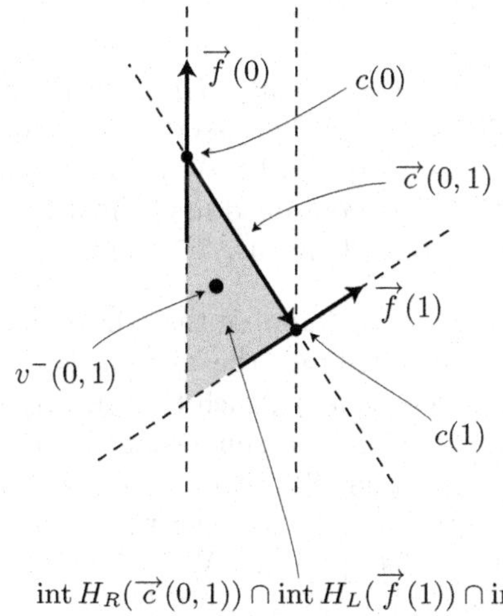

Fig. 4.20. $\Re(c(0)) < \Re(v^-(0,1)) < \Re(c(1))$

Claim 4.7.8. If $\epsilon = +$, then $\Re(c(1)) < \Re(v^+(1,2)) < \Re(c(2))$.

Proof. By the assertions (2) and (3) of Claim 4.7.3, $v^+(1,2)$ is contained in the open region

$$\mathrm{int}(H_L(\overrightarrow{c}(1,2))) \cap \mathrm{int}(H_R(\overrightarrow{f}(1))) \cap \mathrm{int}(H_L(\overrightarrow{f}(2))).$$

By Claim 4.7.5, this open region is contained in the following region (see Fig. 4.21).

$$H_L(\overrightarrow{c}(1,2)) \cap \{z \in \mathbb{C} \mid \Re(c(1)) \le \Re(z) \le \Re(c(2))\}$$

Hence we obtain the desired result.

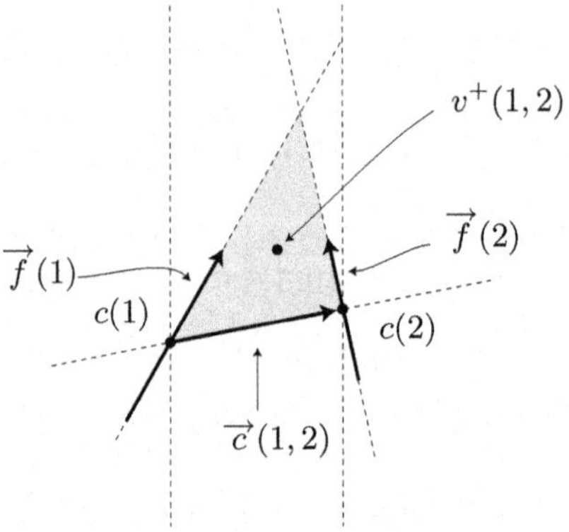

Fig. 4.21. $\Re(c(1)) < \Re(v^+(1,2)) < \Re(c(2))$

This completes the proof of Lemma 4.7.1. In fact, the assertion (1-i) follows from Claims 4.7.4, 4.7.6 and 4.7.7. The assertion (2-i) follows from Claims 4.7.4 and 4.7.8. The assertions (1-ii) and (2-ii) are equivalent to Claim 4.7.6. The assertion (1-iii) and (2-iii) are contained in Claim 4.7.5. Finally, the assertions (1-iv) and (2-iv) follow from Claims 4.7.3(5) and 4.7.5.

Proof (Proof of Lemma 4.5.5). Suppose that the lemma does not hold. Then there exists a pair (ρ, σ) satisfying the assumption of Lemma 4.5.5 and a couple of distinct integers j_1 and j_2, such that $\mathrm{int}(\Delta^{\epsilon}_{j_1}) \cap \mathrm{int}(\Delta^{\epsilon}_{j_2}) \neq \emptyset$. Since this is an open condition, we may assume that all components of $\theta^{\epsilon}(\rho, \sigma)$ are positive and that $\mathcal{L}(\rho, \sigma)$ is not flat at every vertex. According to Corollary 4.2.13, we may assume that $\mathcal{L}(\rho, \sigma)$ satisfies the assumptions of Lemma 4.7.1.

By Lemma 4.2.11(1), $\mathrm{int}(\Delta^{\epsilon}_0) \cap \mathrm{int}(\Delta^{\epsilon}_1) = \mathrm{int}(\Delta^{\epsilon}_1) \cap \mathrm{int}(\Delta^{\epsilon}_2) = \emptyset$. In what follows, we see that $\mathrm{int}(\Delta^{\epsilon}_0) \cap \mathrm{int}(\Delta^{\epsilon}_2) = \emptyset$. If $\epsilon = -$, then, by Lemma 4.7.1, the vertices $c(0)$, $c(1)$ and $v^-(0,1)$ of Δ^-_0 are contained in $\{z \in \mathbb{C} \mid \Re(z) \le \Re(c(1))\}$, and the vertices $c(2)$, $c(3)$ and $v^-(2,3)$ of Δ^-_2 are contained in $\{z \in \mathbb{C} \mid \Re(z) \ge \Re(c(2))\}$ in the coordinate in Lemma 4.7.1. Since these two regions are disjoint by Lemma 4.7.1(1), we have $\text{int}(\Delta_0^-) \cap \text{int}(\Delta_2^-) = \emptyset$. If $\epsilon = +$, then, by Lemma 4.7.1, the vertices $c(0)$, $c(1)$ and $v^\epsilon(0,1)$ of Δ_0^ϵ are contained in $\{z \in \mathbb{C} \,|\, \Re(z) \leq \Re(v^+(1,2))\}$, and the vertices $c(2)$, $c(3)$ and $v^+(2,3)$ of Δ_2^+ are contained in $\{z \in \mathbb{C} \,|\, \Re(z) \geq \Re(v^+(1,2))\}$ in the coordinate in Lemma 4.7.1. Hence $\text{int}(\Delta_0^+) \cap \text{int}(\Delta_2^+) = \emptyset$. Thus we have proved that the three successive triangles Δ_0^ϵ, Δ_1^ϵ and Δ_2^ϵ have mutually disjoint interiors.

On the other hand, $\text{int}(\Delta_0^\epsilon)$, $\text{int}(\Delta_1^\epsilon)$ and $\text{int}(\Delta_2^\epsilon)$ are contained in the open strip $\{z \in \mathbb{C} \,|\, \Re(c(0)) < \Re(z) < \Re(c(3))\}$, because the vertices of these triangles are contained in the closure of the strip by Lemma 4.7.1. Since the Euclidean transformation $\rho(K)$ maps Δ_j^ϵ to Δ_{j+3}^ϵ, and since the images of the above open strip by the powers of $\rho(K)$ are disjoint, we see that the triangles Δ_j^ϵ $(j \in \mathbb{Z})$ have mutually disjoint interiors. This completes the proof of Lemma 4.5.5.

4.8 Representations which are weakly simple at σ

In this section, we collect basic facts concerning representations ρ such that $\mathcal{L}(\rho, \sigma)$ is weakly simple but is not simple (Definition 3.1.1). To this end we introduce the following definition.

Definition 4.8.1. *Under Assumption 2.4.6 (σ-NonZero):*

1. *$\mathcal{L}(\rho,\sigma)$ is said to be* folded *at $c(\rho(P_j))$ if $\psi(\overrightarrow{e}_{[j-1]}) = c(\rho(P_{j+1})) - c(\rho(P_j))$ is a real negative multiple of $\psi(\overrightarrow{e}_{[j+1]}) = c(\rho(P_j)) - c(\rho(P_{j-1}))$.*
2. *$\mathcal{L}(\rho,\sigma)$ is said to be* singly folded *if $\mathcal{L}(\rho,\sigma)$ is weakly simple and there is a unique element $k \in \{0,1,2\}$ such that $\mathcal{L}(\rho,\sigma)$ is folded at $c(\rho(P_j))$ if and only if $[j] = k$.*
3. *$\mathcal{L}(\rho,\sigma)$ is said to be* doubly folded *if $\mathcal{L}(\rho,\sigma)$ is weakly simple and there are two elements k_1 and $k_2 \in \{0,1,2\}$ such that $\mathcal{L}(\rho,\sigma)$ is folded at $c(\rho(P_j))$ if and only if $[j] = k_1$ or k_2.*

Lemma 4.8.2. *Under Assumption 2.4.6 (σ-NonZero), suppose that $\mathcal{L}(\rho,\sigma)$ is weakly simple but is not simple. Then $\mathcal{L}(\rho,\sigma)$ is either singly folded or doubly folded. Moreover the following hold.*

1. *If $\mathcal{L}(\rho,\sigma)$ is singly folded, then there is a unique Markoff map inducing ρ which is upward at σ.*
2. *If $\mathcal{L}(\rho,\sigma)$ is doubly folded, then there are precisely two Markoff maps inducing ρ which are upward at σ.*

Proof. Since $\mathcal{L}(\rho,\sigma)$ is weakly simple, there is a sequence $\{\rho_n\}$ of type-preserving representations converging to ρ, such that $\mathcal{L}(\rho_n,\sigma)$ is simple. Following Notation 4.2.6, let $\overrightarrow{c_n}(j, j+1)$ be the oriented line segment $\overrightarrow{c}(\rho_n; P_j, P_{j+1})$ and let $c_n(j, j+1)$ be the edge of $\mathcal{L}(\rho_n,\sigma)$ with endpoints $c_n(j)$ and $c_n(j+1)$. The assumption that $\mathcal{L}(\rho,\sigma)$ is not simple implies the following claim.

Claim 4.8.3. There are integers k and j such that $k \neq j, j+1$ and that the vertex $c(k)$ is contained in the edge $c(j, j+1)$.

Proof. Since $\mathcal{L}(\rho, \sigma)$ is not simple, one of the following condition holds.

1. $c(j-1, j) \cap c(j, j+1)$ is strictly larger than $\{c(j)\}$ for some j.
2. $c(j, j+1) \cap c(k, k+1) \neq \emptyset$ for some integers j and k such that $|j-k| \geq 2$.

If the first condition holds, then $\mathcal{L}(\rho, \sigma)$ is folded at $c(j)$ and we have either $c(j-1) \in c(j, j+1)$ or $c(j+1) \in c(j-1, j)$. Suppose the second conditions holds. Suppose further that $\operatorname{int} c(j, j+1) \cap \operatorname{int} c(k, k+1) \neq \emptyset$. Since $\mathcal{L}(\rho, \sigma)$ is weakly simple, this intersection cannot be a transversal intersection. Hence we see that $c(j, j+1)$ and $c(k, k+1)$ are contained in a single line and the boundary of one of the edges is contained in the other, which implies the desired result. Suppose $\operatorname{int} c(j, j+1) \cap \operatorname{int} c(k, k+1) = \emptyset$. Since $c(j, j+1) \cap c(k, k+1) \neq \emptyset$, we again come to the same conclusion.

Claim 4.8.4. $\mathcal{L}(\rho, \sigma)$ is folded at some vertex.

Proof. By Claim 4.8.3, some vertex $c(k)$ is contained in the edge $c(0, 1)$ for some $k \neq 0, 1$, after a shift of indices. We may assume $k > 1$, because the case $k < 0$ is treated by a parallel argument. We also assume that k is the minimal integer greater than 1 such that $c(k) \in c(0, 1)$.

Case 1. $k = 3j$ for some $j \in \mathbb{N}$. Since $c(3j) = c(0) + j$, $c(0, 1)$ is horizontal (i.e., parallel to the real axis), and $\Re(c(0)) < \Re(c(3j)) \leq \Re(c(1))$. Moreover we have $j = 1$ and $k = 3$ by the minimality of k. Suppose $\Im(c(2)) = \Im(c(0))$. Then $c(2)$ and $c(3)$ lie in the horizontal line containing $c(0, 1)$. Hence $\mathcal{L}(\rho, \sigma)$ is folded at $c(1)$ or $c(2)$ according as $\Re(c(2)) < \Re(c(1))$ or $\Re(c(2)) > \Re(c(1))$. Suppose $\Im(c(2)) > \Im(c(0))$ $(= \Im(c(1)))$. If $c(3) = c(1)$ then $\mathcal{L}(\rho, \sigma)$ is folded at $c(2)$. So we may assume $c(3) \in c(0, 1) - c(1)$ and hence $\Re(c(3)) < \Re(c(1))$. Thus $\Im(c_n(2)) > \Im(c_n(0)), \Im(c_n(1))$ and $\Re(c_n(3)) < \Re(c_n(1))$ for all sufficiently large n. This implies that $\arg(\overrightarrow{c_n}(2, 3)/\overrightarrow{c_n}(1, 2)) \in (0, \pi)$. Hence $\mathcal{L}(\rho_n, \sigma)$ is convex to the below at $c_n(2)$. On the other hand, since $\Im(c_n(2)) > \Im(c_n(0))$, $\Im(c_n(1))$, Lemma 3.1.9 implies that $\mathcal{L}(\rho_n, \sigma)$ is convex to the above at $c_n(2)$. This is a contradiction. Hence $c(3) = c(1)$ and $\mathcal{L}(\rho, \sigma)$ is folded at $c(2)$. If $\Im(c(2)) < \Im(c(0))$, then we obtain the same conclusion by a parallel argument. Thus the desired result holds in Case 1.

Case 2. $k = 3j + 1$ for some $j \in \mathbb{N}$. Since $c(3j + 1) = c(1) + j$, $c(0, 1)$ is horizontal, and $\Re(c(1)) < \Re(c(3j + 1)) \leq \Re(c(0))$. Moreover we have $j = 1$ and $k = 4$ by the minimality of k. Suppose $\Im(c(2)) = \Im(c(0))$. Then $c(2)$ and $c(3)$ lie in the horizontal line containing $c(0, 1)$. Hence $\mathcal{L}(\rho, \sigma)$ is folded at $c(1)$ or $c(2)$ according as $\Re(c(2)) > \Re(c(1))$ or $\Re(c(2)) < \Re(c(1))$. Suppose $\Im(c(2)) \neq \Im(c(0))$. Let ℓ_0 and ℓ_2, respectively, be the horizontal line containing $c(0)$ and $c(2)$. Then $c(4)$ and $c(3) = c(0)+1$ lie in ℓ_0 in this order, and $c(2)$ and $c(5) = c(2)+3$ lie in ℓ_2 in this order. Since $\ell_0 \neq \ell_2$, this implies that $\operatorname{int} c(2, 3)$ and $\operatorname{int} c(4, 5)$ intersect transversely. This contradicts the assumption that $\mathcal{L}(\rho, \sigma)$ is weakly simple. Thus the desired result holds in Case 2.

Case 3. $k = 3j + 2$ for some $j \in \mathbb{N} \cup \{0\}$. If $j = 0$, then $\mathcal{L}(\rho, \sigma)$ is folded at $c(1)$. Suppose $j > 0$. Suppose further that $c(0, 1)$ is horizontal. Then all vertices of $\mathcal{L}(\rho, \sigma)$ lies in a single horizontal line. Hence, by using the assumption $c(3j + 2) \in c(0, 1)$, we see that $\mathcal{L}(\rho, \sigma)$ is folded at some vertex. Suppose that $c(0, 1)$ is not horizontal. Then the line ℓ containing $c(3, 4)$ separates $c(0, 1)$ from $c(3j + 3) = c(0) + (j + 1)$. Since $c(3j + 2) \in c(0, 1)$, $\operatorname{int} c(3j + 2, 3j + 3)$ intersects ℓ transversely. Since $c(3j + 3) = c(0) + j + 1$ and $c(3, 4) = c(0, 1) + 1$, the intersection point is contained in $\operatorname{int} c(3, 4)$. Hence $\operatorname{int} c(3, 4)$ and $\operatorname{int} c(3j + 2, 3j + 3)$ intersect transversely, a contradiction. Thus the desired result holds in Case 3.

Since $\rho(K)(c(j+3)) = c(j)+1$, $\mathcal{L}(\rho, \sigma)$ cannot be folded at three successive vertices. Hence $\mathcal{L}(\rho, \sigma)$ is either singly folded or doubly folded.

To see the remaining assertion, let ϕ be a Markoff map inducing ρ. Suppose first that $\mathcal{L}(\rho, \sigma)$ is singly folded. We may assume $\mathcal{L}(\rho, \sigma)$ is folded at $c(0)$. Then we can see that ϕ is upward at σ if and only if the angle

$$\arg \frac{i/\phi(s_{[j]})}{\vec{c}(j, j+1)} = \arg \frac{-\vec{c}(j-1, j)}{i/\phi(s_{[j]})}$$

is equal to π if $j = 0$ and belongs to $(0, \pi)$ if $j = 1$ or 2. Hence there is a unique Markoff map inducing ρ which is upward at σ. Suppose next that $\mathcal{L}(\rho, \sigma)$ is doubly folded. We may assume $\mathcal{L}(\rho, \sigma)$ is folded at $c(1)$ and $c(2)$. Then ϕ is upward if and only if the above angle is equal to π for $j = 1$ or 2. The angle for $j = 0$ is equal to π or $-\pi$ and both possibilities actually happen. Hence there are precisely two Markoff maps inducing ρ which are upward at σ. This completes the proof of Lemma 4.8.2.

The following Lemmas 4.8.5 ad 4.8.6 describe characteristic properties of those (ρ, σ)'s such that $\mathcal{L}(\rho, \sigma)$ is folded.

Lemma 4.8.5. *Under Assumption 2.4.6 (σ-NonZero), suppose that $\mathcal{L}(\rho, \sigma)$ is folded at $c(\rho(P_1))$. Then $\overline{Ih}(\rho(P_0)) \cap \overline{Ih}(\rho(P_1)) = \overline{Ih}(\rho(P_1)) \cap \overline{Ih}(\rho(P_2))$, and $\rho(P_1)$ preserves this set. Moreover, if $\overline{Ih}(\rho(P_0)) \cap \overline{Ih}(\rho(P_1))$ is a complete geodesic, then $\rho(P_1)$ interchanges the endpoints of the complete geodesic.*

Proof. Since $\mathcal{L}(\rho, \sigma)$ is folded at $c(1)$, the centers $c(j)$ ($j \in \{0, 1, 2\}$) lie in the line, ℓ, containing $\operatorname{proj}(\overline{\operatorname{Axis}(1)})$ by Lemma 2.4.4(2.1). Suppose first that $\overline{Ih}(0) \cap \overline{Ih}(1)$ is a complete geodesic. Then $\operatorname{proj}(\overline{\operatorname{Axis}(1)})$ is orthogonal to $\operatorname{proj}(\overline{Ih}(0) \cap \overline{Ih}(1))$ by the above observation. Thus $\overline{\operatorname{Axis}}(1)$ is orthogonal to $\overline{Ih}(0) \cap \overline{Ih}(1)$ and hence $\rho(P_1)$ maps $\overline{Ih}(0) \cap \overline{Ih}(1)$ to itself by interchanging the two endpoints. Since $\rho(P_1)$ maps $\overline{Ih}(0) \cap \overline{Ih}(1)$ to $\overline{Ih}(1) \cap \overline{Ih}(2)$ by Lemma 4.1.3 (Chain rule), we have $\overline{Ih}(0) \cap \overline{Ih}(1) = \overline{Ih}(1) \cap \overline{Ih}(2)$, and obtain the conclusion. Next, suppose that $\overline{Ih}(0) \cap \overline{Ih}(1)$ is a singleton, then we see that the singleton is contained in $\operatorname{proj}(\overline{\operatorname{Axis}(1)})$ and hence in $\overline{\operatorname{Axis}}(1)$. Thus we obtain the conclusion as in the first case. Finally, if $\overline{Ih}(0) \cap \overline{Ih}(1)$ is neither a complete geodesic nor a singleton, then it is the empty set and the conclusion obviously holds.

Lemma 4.8.6. *Under Assumption 2.4.6 (σ-NonZero), $\mathcal{L}(\rho, \sigma)$ is folded at $c(\rho(P_1))$, if one of the following conditions holds.*

1. *$\overline{Ih}(\rho(P_0)) \cap \overline{Ih}(\rho(P_1)) = \overline{Ih}(\rho(P_1)) \cap \overline{Ih}(\rho(P_2))$, and it is a complete geodesic, and $\rho(P_1)$ interchanges the two endpoints of the geodesic.*
2. *$\overline{Ih}(\rho(P_0)) \cap \overline{Ih}(\rho(P_1)) = \overline{Ih}(\rho(P_1)) \cap \overline{Ih}(\rho(P_2))$, and it is a singleton.*

Proof. Suppose Condition 1 holds. Then the assumption that $\overline{Ih}(0) \cap \overline{Ih}(1) = \overline{Ih}(1) \cap \overline{Ih}(2)$ is a complete geodesic implies that the centers $c(\rho(P_j))$ ($j \in \{0, 1, 2\}$) are contained in the perpendicular bisector, ℓ, of the line segment $\mathrm{proj}(\overline{Ih}(0) \cap \overline{Ih}(1))$. The condition that $\rho(P_1)$ interchanges the two endpoints of the geodesic implies that $\mathrm{proj}(\overline{\mathrm{Axis}}(1))$ is orthogonal to $\mathrm{proj}(\overline{Ih}(0) \cap \overline{Ih}(1))$ and therefore contained in ℓ. Hence by using Lemma 2.4.4(2.1) we see that $\mathcal{L}(\rho, \sigma)$ is folded at $c(1)$.

Suppose Condition 2 holds. Then by a similar argument we again see that the centers $c(\rho(P_j))$ ($j \in \{0, 1, 2\}$) are contained in a single line, ℓ, and that $\mathrm{proj}(\overline{\mathrm{Axis}}(1))$ is contained in ℓ. Hence we obtain the conclusion by using Lemma 2.4.4(2.1).

At the end of this section, we prove the following Lemma 4.8.7, which implies the useful Lemma 8.9.3 that in tern is used in Sect. 8.9.

Lemma 4.8.7. *Under Notation 2.1.14 (Adjacent triangles), let Σ be the chain (σ, σ') or (σ', σ). Let ρ be an element of $\mathcal{X}$, satisfying the following conditions.*

1. *$\phi^{-1}(0) \cap \Sigma^{(0)} = \emptyset$.*
2. *$\mathcal{L}(\rho, \sigma)$ is folded at $c(\rho(P_1))$.*
3. *$\mathcal{L}(\rho, \sigma)$ is not doubly folded.*

Then there is a neighborhood U of ρ in $\mathcal{X}$ such that $\mathcal{L}(\rho', \Sigma)$ is not weakly simple for any element ρ' of U.

Proof. Let (a_0, a_1, a_2) and (a_0', a_1', a_2'), respectively, be the complex probability of ρ at σ and σ', which are well-defined by the first assumption. By the second assumption, $a_0 = \lambda a_2$ for some negative real number λ. By Lemma 2.4.1(3), this implies

$$a_0' = \frac{\lambda}{1+\lambda} a_1, \quad a_1' = \frac{1}{1+\lambda} a_1, \quad a_2' = (1+\lambda) a_2.$$

In particular, $\lambda \neq -1$ and $a_0' = \lambda a_1'$. Thus we have

$$c(\rho(P_2)) - c(\rho(P_1)) = \lambda(c(\rho(P_1)) - c(\rho(P_0))),$$
$$c(\rho(P_2')) - c(\rho(P_1')) = \lambda(c(\rho(P_3')) - c(\rho(P_2'))).$$

Suppose first that $-1 < \lambda < 0$. Then the above identity implies that $c(\rho(P_2))$ lies in the interior of the edge $\overline{c(\rho(P_0))c(\rho(P_1))}$ and that $c(\rho(P_1'))$ lies in the interior of the edge $\overline{c(\rho(P_2'))c(\rho(P_3'))}$. Hence the interiors of the edges $\overline{c(\rho(P_0))c(\rho(P_1))}$ and $\overline{c(\rho(P_2'))c(\rho(P_3'))}$ have a common point $c(\rho(P_2)) = c(\rho(P_1'))$. These two edges are not parallel by the third condition, and hence their interiors intersect transversely. Therefore, for any representation ρ' which is sufficiently close to ρ, $\mathcal{L}(\rho', \Sigma)$ is weakly singular.

Next, suppose that $\lambda < -1$. Then the identity implies $c(\rho(P_0)) \in \operatorname{int} \overline{c(\rho(P_1))c(\rho(P_2))}$ and $c(\rho(P_3')) \in \operatorname{int} \overline{c(\rho(P_1'))c(\rho(P_2'))}$. Hence the interiors of the edges $\overline{c(\rho(P_1))c(\rho(P_2))}$ and $\overline{c(\rho(P_{-2}'))c(\rho(P_{-1}'))}$ intersect transversely at the point $c(\rho(P_0)) = c(\rho(P_0'))$. Hence we also obtain the desired result in this case.

5

Special examples

In this chapter, we give a detailed study of special examples.

In Sect. 5.1, we study the *real* representations, i.e., the type-preserving $PSL(2,\mathbb{R})$ representations. We give a direct proof to the well-known fact that every such representation is fuchsian, by explicitly constructing the Ford domains (Proposition 5.1.3). This gives the starting point to the proof of the main Theorem 1.2.2. We also show that the restriction of the side parameter map $\boldsymbol{\nu} : \mathcal{QF} \to \mathbb{H}^2 \times \mathbb{H}^2$ to the subspace $\mathcal{F}$ of fuchsian representations is a bijection (actually a homeomorphism) to the diagonal set (Proposition 5.1.5). This proves that Theorem 1.3.2 is valid for the subspace $\mathcal{F}$.

In Sect. 5.2, we study the *isosceles* representations, i.e., those representations ρ such that $\mathcal{L}(\rho, \sigma)$ together with a horizontal line forms periodic pattern of isosceles triangles (see Definition 5.2.2). We give a complete characterization of those isosceles representations which are quasifuchsian, and determine the Ford domains (Proposition 5.2.8). These representations are important by the following reasons.

1. Various geometric quantities, including the Ford domain, the bending laminations, the convex core volume, and the width of the limit sets, of these representations can be explicitly calculated (cf. [5, Example 3.3], [57]).
2. The isosceles representations play special roles in the proof of the main theorems, because their Ford domains have "hidden isometric hemispheres" (see Definition 7.1.1 and Lemma 7.1.6). In fact we treat these representations separately in the proof Proposition 6.2.1 (Openness) in Sect. 7.3.
3. We can easily see how Jorgensen's theory can be extended to the outside of $\mathcal{QF}$ for isosceles representations (Remark 5.2.9).

Proposition 5.2.12, which describes the side parameters of those representations near to isosceles representations, plays an important role in the proof of the main theorems (cf. Proof of Proposition 8.3.2).

In Sect. 5.3. we study the groups generated by two parabolic transformations. After describing their relation with the Markoff maps which vanish at some vertex (Lemma 5.3.2), we describe basic properties of these groups. We also calculate the degrees of the polynomials which arise from a Markoff map ϕ such that $\phi(1/0) = 0$ and $\phi(0/1) = x$ (Lemmas 5.3.12 and 5.3.14). These results are used in the proof of Proposition 9.1.13 in Chap. 9.

In Sect. 5.4, we recall a result of Bowditch [17] concerning the "imaginary representations" which are induced by Markoff maps which take value in $\mathbb{R} \cup i\mathbb{R}$ (Proposition 5.4.3).

In Sect. 5.5, we characterize those type-preserving representations which send a non-peripheral element of $\pi_1(T)$ to a parabolic or elliptic transformation (Lemmas 5.5.1, 5.5.3 and 5.5.4). We also present a basic Lemma 5.5.6, which is used in Sect. 8.7 in Chap. 8.

5.1 Real representations

Definition 5.1.1 (Real representation). *A type-preserving representation $\rho : \pi_1(\mathcal{O}) \to PSL(2,\mathbb{C})$ is called a* real *representation if it is conjugate to a representation into $PSL(2,\mathbb{R})$.*

By using the identities 2.4 and 2.5, we can easily obtain the following characterization of the real representations.

Lemma 5.1.2. *For a type-preserving representation ρ, the following conditions are equivalent.*

1. *ρ is a real representation.*
2. *The Markoff map $\phi = \phi_\rho$ takes only real numbers.*
3. *The complex probability map $\psi = \psi_\rho$ takes only positive real numbers.*

As is noted by [17, Proposition 4.11]), every real representation ρ is fuchsian, i.e., discrete and faithful. The following proposition describes the Ford domains of real representations.

Proposition 5.1.3. *Let $\rho : \pi_1(\mathcal{O}) \to PSL(2,\mathbb{R})$ be a type-preserving real representation. Then it is faithful and discrete. Moreover, there is a unique triangle or edge, $\delta(\rho)$, of $\mathcal{D}$ such that the Ford domain $Ph(\rho)$ is equal to*

$$\begin{aligned} &Eh(\rho, \delta(\rho)) \\ &\quad := \bigcap \{Eh(\rho(P)) \mid P \text{ is an elliptic generator with slope } s(P) \in \delta(\rho)^{(0)}\} \end{aligned}$$

and has the structure as illustrated in Fig. 5.1. Furthermore, $\delta(\rho)$ is characterized by the following properties:

1. *Suppose $\delta(\rho)$ is a triangle $\sigma = \langle s_0, s_1, s_2 \rangle$ of $\mathcal{D}$. Then $|\phi_\rho(s_i)| \le |\phi_\rho(s_i')|$ for each $i \in \{0,1,2\}$, where s_i' is the vertex of $\mathcal{D}$ opposite to s_i with respect to the edge of σ which does not contain s_i.*

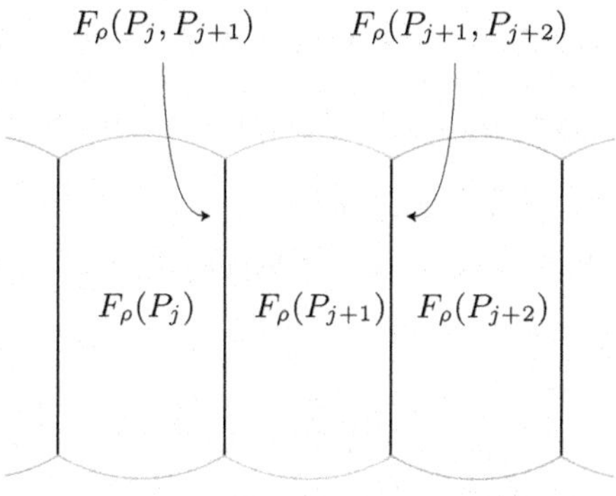

Fig. 5.1. $Eh(\rho, \sigma)$ for a real representation

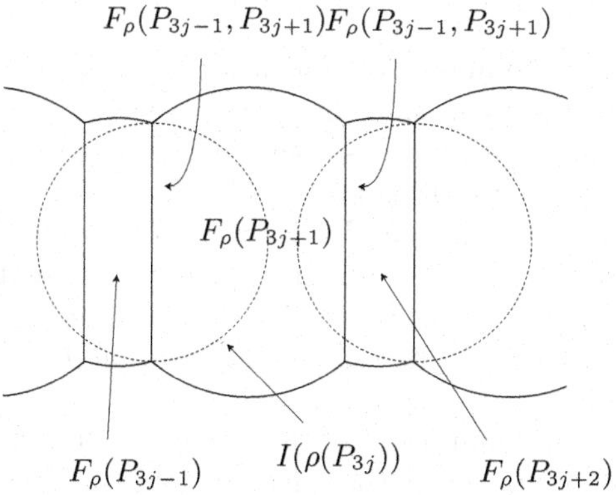

Fig. 5.2. $Eh(\rho, \sigma)$ for a "thin" real representation

2. *Suppose $\delta(\rho)$ is an edge, say τ. Then the two triangles having τ as an edge satisfies the condition in 1.*

Proof. By [17, Lemma 3.17], we may assume ϕ_ρ takes only positive real values. Moreover, we have $\phi_\rho(s) > 2$ for every $s \in \mathcal{D}^{(0)}$ by [17, Proposition 3.18]. Hence, by [17, Corollary 3.7], there is a triangle $\sigma = \langle s_0, s_1, s_2 \rangle$ of $\mathcal{D}$ such that $0 < \phi_\rho(s_i) \le \phi_\rho(s_i')$ for each $i \in \{0, 1, 2\}$, where s_i' is the vertex of $\mathcal{D}$ opposite to s_i with respect to the edge of σ which does not contain s_i. Set $(x, y, z) = (\phi_\rho(s_0), \phi_\rho(s_1), \phi_\rho(s_2))$. Then we have

$$x \le yz - x, \quad y \le zx - y, \quad z \le xy - z.$$

This implies

$$|x| \le |y \pm iz|, \quad |y| \le |z \pm ix|, \quad |z| \le |x \pm iy|.$$

In fact, $x \leq yz - x$ implies $x^2 \leq xyz - x^2 = y^2 + z^2$ and therefore $|x| \leq |y \pm iz|$. Hence, by Lemma 4.3.1, $\theta^+(\rho, \sigma; s_j) = \theta^-(\rho, \sigma; s_j) \geq 0$ for every $j \in \{0, 1, 2\}$. Let $\{P_j\}$ be the sequence of elliptic generators associated with σ, and set

$$\begin{aligned} Eh(\rho, \sigma) &:= \cap_{j \in \mathbb{Z}} Eh(\rho(P_j)), \\ F_\rho(P_j) &:= Ih(\rho(P_j)) \cap (Eh(\rho(P_{j-1})) \cap Eh(\rho(P_{j+1}))), \\ F_\rho(P_j, P_{j+1}) &:= Ih(\rho(P_j)) \cap Ih(\rho(P_{j+1})) \end{aligned}$$

Suppose first that $\theta^+(\rho, \sigma; s_j) = \theta^-(\rho, \sigma; s_j) > 0$ for every $j \in \{0, 1, 2\}$. Then $F_\rho(P_j, P_{j+1})$ is an edge of $Eh(\rho, \sigma)$ and $F_\rho(P_j)$ is a face of $Eh(\rho, \sigma)$ which is bounded by the two edges $F_\rho(P_{j-1}, P_j)$ and $F_\rho(P_j, P_{j+1})$ (see Fig. 5.1(1)). The involution $\rho(P_j)$ preserves the face $F_\rho(P_j)$ and interchanges the edges $F_\rho(P_{j-1}, P_j)$ and $F_\rho(P_j, P_{j+1})$. Thus the set

$$\{F_\rho(P_{j-1}, P_j), F_\rho(P_j, P_{j+1}), F_\rho(P_{j+1}, P_{j+2})\}$$

forms an "edge cycle modulo $\rho(K)$" for each integer j. The sum of the dihedral angles of $Eh(\rho, \sigma)$ along the edges in the edge cycle is equal to 2π by Lemma 4.1.3 (chain rule). Hence we can see that $Eh(\rho, \sigma)$ together with the gluing isometries $\{\rho(P_j)\}$ satisfies the conditions for the "Poincare's theorem on fundamental polyhedra modulo $\rho(K)$". (See Sect. 6.4 for details.) From this fact, we see that ρ is discrete and faithful and $Eh(\rho, \sigma)$ is a fundamental domain of $\rho(\pi_1(\mathcal{O}))$ modulo $\rho(K)$. Since $Eh(\rho, \sigma)$ is obtained as the intersection of the exteriors of isometric hemispheres, this implies that $Eh(\rho, \sigma)$ is equal to the Ford domain $Ph(\rho)$.

Suppose that some component of the side parameter vanishes, say $\theta^+(\rho, \sigma; s_0) = \theta^-(\rho, \sigma; s_0) = 0$. Then the edges $F_\rho(P_{3j-1}, P_{3j})$ and $F_\rho(P_{3j}, P_{3j+1})$ become an identical edge, which we denote by $F_\rho(P_{3j-1}, P_{3j+1})$ (see Fig. 5.2). Moreover, the face $F_\rho(P_{3j})$ degenerates to the edge $F_\rho(P_{3j-1}, P_{3j+1})$, and $Eh(\rho, \sigma)$ becomes identical with

$$\begin{aligned} Eh(\rho, \tau) &:= \bigcap \{Eh(\rho(P)) \mid P \text{ is an elliptic generator with slope } s(P) \in \tau^{(0)}\} \\ &= \bigcap \{Eh(\rho(P_j)) \mid j \not\equiv 0 \pmod 3\}, \end{aligned}$$

where $\tau = \langle s_1, s_2 \rangle$ (cf. Fig. 5.5). Then $\partial Eh(\rho, \tau)$ consists of the faces $F_\rho(P_{3j-1})$, $F_\rho(P_{3j+1})$ and the edges $F_\rho(P_{3j-1}, P_{3j+1})$, $F_\rho(P_{3j+1}, P_{3j+2})$ where j runs over $\mathbb{Z}$. The involution $\rho(P_{3j-1})$ preserves the face $F_\rho(P_{3j-1})$ and interchanges $F_\rho(P_{3j-2}, P_{3j-1})$ with $F_\rho(P_{3j-1}, P_{3j+1})$, whereas the involution $\rho(P_{3j+1})$ preserves the face $F_\rho(P_{3j+1})$ and interchanges $F_\rho(P_{3j-1}, P_{3j+1})$ with $F_\rho(P_{3j+1}, P_{3j+2})$. Thus the set

$$\{F_\rho(P_{3j-1}, P_{3j+1}), F_\rho(P_{3j+1}, P_{3j+2})\}$$

forms an "edge cycle modulo $\rho(K)$" for each integer j. On the other hand, the dihedral angle of $Eh(\rho, \tau)$ along $F_\rho(P_{3j-1}, P_{3j+1})$ is equal to $\theta_{3j-1} + \theta_{3j} - \pi$, where θ_j be the dihedral angle of $Eh(\rho(P_j)) \cap Eh(\rho(P_{j+1}))$. Hence the sum of the dihedral angles of $Eh(\rho, \tau)$ along the edges in the edge cycle is equal to π by Lemma 4.1.3 (chain rule). Hence we can see that $Eh(\rho, \tau)$ together with the gluing isometries satisfies the conditions for the "Poincare's theorem on fundamental polyhedra modulo $\rho(K)$". (See Sect. 6.4 for details.) Here the above angle sum π corresponds to the fact that $F_\rho(P_{3j-1}, P_{3j+1})$ is equal to the axis of the involution $\rho(P_{3j})$. From this fact, we see that ρ is discrete and faithful, and $Eh(\rho, \tau)$ is equal to the Ford domain $Ph(\rho)$ as in the previous case. Moreover, by the argument for the previous case, the condition $\theta^+(\rho, \sigma; s_0) = \theta^-(\rho, \sigma; s_0) = 0$ implies $x = yz - x$. Hence the condition in 1 is also satisfied for the triangle sharing the edge τ with σ. This completes the proof of Proposition 5.1.3.

Remark 5.1.4. Set $\boldsymbol{\nu} = (\theta^-(\rho, \sigma), \theta^+(\rho, \sigma))$ and consider the labeled representation $\boldsymbol{\rho} = (\rho, \boldsymbol{\nu})$. Then, in the generic case when $\delta(\rho) = \sigma$, the convex polyhedra $Eh(\rho, \tau)$, $F_\rho(P_j)$ and $F_\rho(P_j, P_{j+1})$ introduced in the proof, are equal to $Eh(\boldsymbol{\rho})$, $F_{\boldsymbol{\rho}}(P_j)$, $F_{\boldsymbol{\rho}}(P_j, P_{j+1})$ respectively (see Definitions 3.4.3 and 3.4.6). Similar identifications are also valid for the non-generic case, as explained in the forthcoming Remark 5.2.10.

Proposition 5.1.5. *For any point $\nu \in \mathbb{H}^2$, there is a type-preserving real representation $\rho : \pi_1(\mathcal{O}) \to PSL(2, \mathbb{R})$ which realizes (ν, ν), that is, $Ph(\rho) = Eh(\rho, \sigma)$ and $\theta^{\pm}(\rho, \sigma) = \nu$, where σ is a triangle of $\mathcal{D}$ containing ν. Moreover, the element ρ in $\mathcal{X}$ is uniquely determined by ν.*

Proof. Set $\nu = (\theta_0, \theta_1, \theta_2) \in \sigma$. For each positive real h, we can find points c_j ($j \in \{0, 1, 2, 3\}$), v_{01}, v_{12} and v_{23} in the complex plane which satisfy the following conditions (see Fig. 5.3).

1. $c_j \in \mathbb{R}$ and $c_0 < c_1 < c_2 < c_3$.
2. v_{01}, v_{12} and v_{23} are contained in the horizontal line $\Im(z) = h$.
3. $\arg(v_{01} - c_0) = \frac{\pi}{2} - \theta_0$, $\arg(v_{01} - c_1) = \frac{\pi}{2} + \theta_1$, $\arg(v_{12} - c_1) = \frac{\pi}{2} - \theta_1$, $\arg(v_{12} - c_2) = \frac{\pi}{2} + \theta_2$, $\arg(v_{23} - c_2) = \frac{\pi}{2} - \theta_2$, $\arg(v_{23} - c_3) = \frac{\pi}{2} + \theta_0$.

By choosing a suitable h, we may assume $c_3 - c_0 = 1$. Put $(a_0, a_1, a_2) = (c_2 - c_1, c_3 - c_2, c_1 - c_0)$. Then $a_0 + a_1 + a_2 = 1$, and therefore there is a complex probability map ψ such that $(\psi(\vec{e}_0), \psi(\vec{e}_1), \psi(\vec{e}_2)) = (a_0, a_1, a_2)$, where $(\vec{e}_0, \vec{e}_1, \vec{e}_2)$ is dual to the triangle $\sigma = \langle s_0, s_1, s_2 \rangle$ (cf. Sect. 2.4). Let ρ be the type-preserving representation determined by ψ. Then ρ is (equivalent to) a real representation, because the complex probability (a_0, a_1, a_2) consists of real positive numbers. To show that ρ satisfies the desired properties, we normalize ρ so that $c(\rho(P_0)) = c_0$. Observe that the triangles $\Delta(c_0, c_1, v_{01})$, $\Delta(v_{12}, c_1, c_2)$ and $\Delta(c_3, v_{23}, c_2)$ are mutually similar, by comparing the angles. By comparing the edge lengths, we see that these triangles are similar to a triangle $\Delta(v_0, v_1, v_2)$ such that $|v_2 - v_1| = \sqrt{|a_0|}$, $|v_0 - v_2| = \sqrt{|a_1|}$ and $|v_1 - v_0| = \sqrt{|a_2|}$, which in tern is similar to a triangle $\Delta(w_0, w_1, w_2)$ such that $|w_2 - w_1| = |\phi_\rho(s_0)|$, $|w_0 - w_2| = |\phi_\rho(s_1)|$ and $|w_1 - w_0| = |\phi_\rho(s_2)|$.

Thus, by Lemma 4.2.11(2), the above three triangles are identical with the triangles $\Delta_0^+(\rho,\sigma)$, $\Delta_1^+(\rho,\sigma)$ and $\Delta_2^+(\rho,\sigma)$ in Notation 4.2.10(1), respectively. Thus we have $\theta^{\pm}(\rho,\sigma)=\nu$ by Lemma 4.2.11(1). The above fact also implies that $v_{j,j+1}\in I(\rho(P_j))\cap I(\rho(P_{j+1}))$ for each $j\in\{0,1,2\}$. Moreover, we see that $Eh(\rho,\sigma)$ has the structure as that in the proof of Proposition 5.1.3. Hence we have $Ph(\rho)=Eh(\rho,\sigma)$. The uniqueness of ρ follows from the fact that the configuration in Fig. 5.3 is uniquely determined by ν (cf. Lemma 5.2.12).

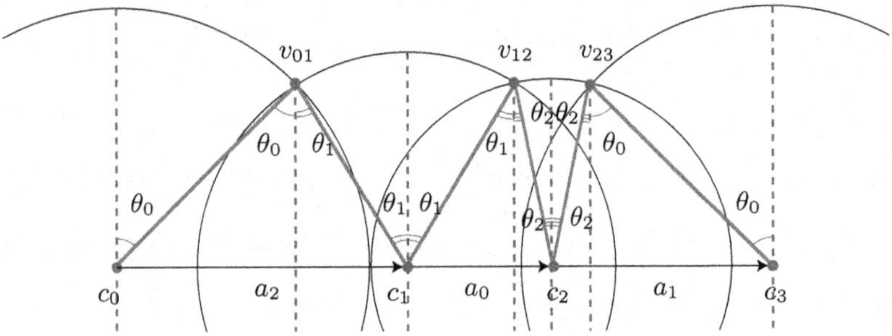

Fig. 5.3. angles

5.2 Isosceles representations and thin labels

By using Lemma 3.1.2, we can easily prove the following lemma.

Lemma 5.2.1. *Let $\rho:\pi_1(\mathcal{O})\to PSL(2,\mathbb{C})$ be a type-preserving representation such that $\phi_\rho^{-1}(0)\cap(\sigma\cup\sigma')^{(0)}=\emptyset$ under Notation 2.1.14 (Adjacent triangles). Then the following conditions are equivalent.*

1. *The complex probability (a_0,a_1,a_2) of ρ at σ satisfies the following conditions.*
$$a_2=ta_0 \quad \text{for some } t>0, \quad a_0+a_2=\bar{a}_1.$$
That is, the point c_1 lies in the interior of the edge $\overline{c_0c_2}$ and the triangle $\Delta(c_0,c_2,c_3)$ is a (possibly degenerate) isosceles triangle with base $\overline{c_0c_3}$, where $c_j=c(\rho(P_j))$ (see Fig. 5.4).
2. *The complex probability (a_0',a_1',a_2') of ρ at σ' satisfies the following conditions.*
$$a_1'=ta_0' \quad \text{for some } t>0, \quad a_0'+a_1'=\bar{a}_2'.$$
That is, the point c_2' lies in the interior of the edge $\overline{c_1'c_3'}$ and the triangle $\Delta(c_0',c_1',c_3')$ is a (possibly degenerate) isosceles triangle with base $\overline{c_0'c_3'}$, where $c_j'=c(\rho(P_j'))$.

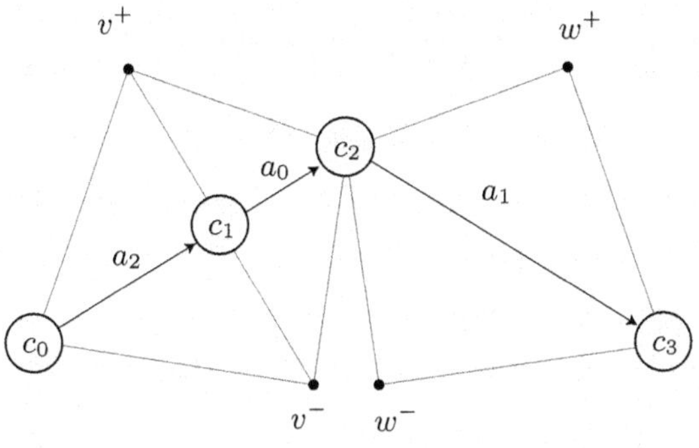

Fig. 5.4. Complex probabilities of an isosceles representation

Definition 5.2.2 (Isosceles representation). *Let $\rho : \pi_1(\mathcal{O}) \to PSL(2,\mathbb{C})$ be a type-preserving representation. Under Notation 2.1.14 (Adjacent triangles), the pair (ρ,τ), where $\tau = \sigma \cap \sigma'$, is called an* isosceles *representation if the mutually equivalent conditions in Lemma 5.2.1 are satisfied.*

We have the following characterization of isosceles representations.

Proposition 5.2.3. *Let $\rho : \pi_1(\mathcal{O}) \to PSL(2,\mathbb{C})$ be a type-preserving representation such that $\phi_\rho^{-1}(0) \cap (\sigma \cup \sigma')^{(0)} = \emptyset$ under Notation 2.1.14 (Adjacent triangles). Then the following conditions are equivalent.*

1. *(ρ,τ) is an isosceles representation, where $\tau = \sigma \cap \sigma'$.*
2. *The side parameters $\theta^{\pm}(\rho,\sigma)$ are defined and $\theta^-(\rho,\sigma;s_1) = \theta^+(\rho,\sigma;s_1) = 0$.*
3. *The side parameters $\theta^{\pm}(\rho,\sigma')$ are defined and $\theta^-(\rho,\sigma';s_2') = \theta^+(\rho,\sigma';s_2') = 0$.*
4. *$Ih(\rho(P_0)) \cap Ih(\rho(P_2)) = \mathrm{Axis}(\rho(P_1))$.*
5. *$Ih(\rho(P_1')) \cap Ih(\rho(P_3')) = \mathrm{Axis}(\rho(P_2'))$.*

Proof. (1) $\to$ (2). Suppose (ρ,τ) is an isosceles representation. Let v^+ and v^- be the points in $\mathbb{C}$ satisfying the following conditions (see Fig. 5.4).

1. v^+ and v^- lie on the perpendicular to the edge $\overline{c_0c_2}$ passing through c_1.
2. v^+ (resp. v^-) lies on the left (resp. right) to the directed line $\overrightarrow{c_0c_2}$
3. $\angle(c_0, v^\epsilon, c_2)$ is the right angle for each $\epsilon \in \{-,+\}$.

We show that the triangles $\Delta(c_0,c_1,v^\epsilon)$ and $\Delta(c_1,c_2,v^\epsilon)$ coincide with the triangles $\Delta_0^\epsilon(\rho,\sigma)$ and $\Delta_1^\epsilon(\rho,\sigma)$ in Notation 4.2.10, respectively. To this end, let w^ϵ be the image of v^ϵ by the reflection in the perpendicular bisector of the edge $\overline{c_0c_3}$, which is equal to the perpendicular to $\overline{c_0c_3}$ passing through c_2. Then we can see, by using the fact that $\angle(c_0,v^\epsilon,c_2) = \pi/2$, that the right triangles $\Delta(c_0,c_1,v^\epsilon)$, $\Delta(v^\epsilon,c_1,c_2)$ and $\Delta(c_3,w^\epsilon,c_2)$ are mutually similar. By using this fact we have the following claim.

Claim 5.2.4. Set $(x, y, z) = (\phi_\rho(s_0), \phi_\rho(s_1), \phi_\rho(s_2))$. Then we have:

$$|v^\epsilon - c_0| = \frac{1}{|x|}, \quad |v^\epsilon - c_1| = \frac{1}{|y|}, \quad |v^\epsilon - c_2| = |w^\epsilon - c_2| = \frac{1}{|z|}.$$

Proof. We prove only the last equality, because the remaining equalities can be proved similarly.

$$\begin{aligned}|v^\epsilon - c_2|^2 &= |v^\epsilon - c_2| \cdot |w^\epsilon - c_2| \\ &= |c_1 - c_2| \cdot |c_3 - c_2| \text{ because } \Delta(v^\epsilon, c_1, c_2) \text{ is similar to } \Delta(c_3, w^\epsilon, c_2) \\ &= \left|\frac{x}{yz}\right| \cdot \left|\frac{y}{zx}\right| = \frac{1}{|z^2|}.\end{aligned}$$

The above claim implies $v^\epsilon \in I(\rho(P_0)) \cap I(\rho(P_1)) \cap I(\rho(P_2))$, $w^\epsilon \in I(\rho(P_2))$ and that the triangles $\Delta(c_0, c_1, v^\epsilon)$, $\Delta(v^\epsilon, c_1, c_2)$ and $\Delta(c_3, w^\epsilon, c_2)$ are similar to the triangle $\Delta(\rho, \sigma)$ in Notation 4.2.10(2). Hence the triangles $\Delta(c_0, c_1, v^\epsilon)$ and $\Delta(c_1, c_2, v^\epsilon)$ coincide with the triangles $\Delta_0^\epsilon(\rho, \sigma)$ and $\Delta_1^\epsilon(\rho, \sigma)$ in Notation 4.2.10, respectively. Since $\mathcal{L}(\rho, \sigma)$ is obviously simple, this implies that $\theta^\pm(\rho, \sigma)$ are defined. Hence we obtain $\theta^-(\rho, \sigma; s_1) = \theta^+(\rho, \sigma; s_1) = 0$ by Lemma 4.2.11(1).

(2) $\to$ (4). Though this is a consequence of Lemma 4.3.5(3), we give a direct proof. If $\theta^-(\rho, \sigma; s_1) = \theta^+(\rho, \sigma; s_1) = 0$, then $v^\pm(\rho; P_0, P_1) = v^\pm(\rho; P_1, P_2) = \mathrm{Fix}_\sigma^\pm(\rho(P_1))$ by the definition of $\theta^\pm(\rho, \sigma; s_1)$. Hence $\mathrm{Axis}(\rho(P_1)) = Ih(\rho(P_0)) \cap Ih(\rho(P_2))$.

(4) $\to$ (1). Suppose $\mathrm{Axis}(\rho(P_1)) = Ih(\rho(P_0)) \cap Ih(\rho(P_2))$. Then c_0, c_1 and c_2 lie on the perpendicular bisector ℓ of the open interval $\pi(\mathrm{Axis}(\rho(P_1)))$. Since $\pi(\mathrm{Axis}(\rho(P_1)))$ bisects the angle $\angle(c_0, c_1, c_2)$ (see Proposition 2.4.4(2.1)), c_0, c_1 and c_2 lie on ℓ in this order. Thus we have $a_2 = ta_0$ for some $t > 0$. Let v^+ (resp. v^-) be the point of $\mathrm{Fix}(\rho(P_1)) = I(\rho(P_0)) \cap I(\rho(P_1)) \cap I(\rho(P_2))$ which lies on the left (resp. right) to the directed line $\overrightarrow{c_0 c_2}$. Then $\angle(c_0, c_1, v^\epsilon) = \angle(v^\epsilon, c_1, c_2) = \pi/2$ and $\Delta(c_0, c_1, v^\epsilon)$ is similar to $\Delta(v^\epsilon, c_1, c_2)$. Thus $\Delta(c_0, v^\epsilon, c_2)$ is also similar to $\Delta(v^\epsilon, c_1, c_2)$. Let γ be the reflection in the line containing $\pi(\mathrm{Axis}(\rho(P_2)))$, and set $w^\epsilon = \gamma(v^\epsilon)$. Since $v^\epsilon \in I(\rho(P_2))$, w^ϵ is equal to the image of v^ϵ by $\rho(P_2)$, and hence $w^\epsilon \in I(\rho(P_2)) \cap I(\rho(P_3))$ by Lemma 4.1.3(2) (Chain rule). Since $\Delta(\gamma(c_0), w^\epsilon, c_2) = \gamma(\Delta(c_0, v^\epsilon, c_2))$ is similar to $\Delta(v^\epsilon, c_1, c_2)$, we have $c_3 = \gamma(c_0)$. Since $c_3 - c_0 = 1 \neq 0$, $\Delta(c_0, c_2, c_3)$ is an isosceles triangle with base $\overline{c_0 c_3}$. Moreover, c_1 lies in the interior of the edge $\overline{c_0 c_2}$. Hence (ρ, τ) is an isosceles representation.

Thus we have proved that the conditions (1), (2) and (4) are mutually equivalent. By a parallel argument we can see that (1), (3) and (5) are mutually equivalent.

Corollary 5.2.5. *Let (ρ, τ) an isosceles representation. Then, under Notation 2.1.14 (Adjacent triangles), we have the following.*

$$Ih(\rho(P_1)) \cap (Eh(\rho(P_0)) \cap Eh(\rho(P_2))) = \mathrm{Axis}(\rho(P_1))$$
$$Ih(\rho(P_2')) \cap (Eh(\rho(P_1')) \cap Eh(\rho(P_3'))) = \mathrm{Axis}(\rho(P_2')).$$

We also present the following two characterizations of isosceles representations.

Lemma 5.2.6. *Under Assumption 2.4.6 (σ-NonZero), suppose that the intersection $Ih(\rho(P_{j-1})) \cap Ih(\rho(P_j)) \cap Ih(\rho(P_{j+1}))$ is non-empty and is not a singleton. Then one of the following holds.*

1. *(ρ, τ) is an isosceles representation, where $\tau = \langle s_{[j-1]}, s_{[j+1]} \rangle$.*
2. *$\mathcal{L}(\rho, \sigma)$ is folded at $c(\rho(P_j))$ (cf. Definition 4.8.1).*

Proof. Note that both $Ih(\rho(P_{j-1})) \cap Ih(\rho(P_j))$ and $Ih(\rho(P_j)) \cap Ih(\rho(P_{j+1}))$ are geodesics by Condition 2.4.6 (σ-NonZero). Suppose that $Ih(\rho(P_{j-1})) \cap Ih(\rho(P_j)) \cap Ih(\rho(P_{j+1}))$ is nonempty and is not a singleton. Then it follows that the intersection is equal to the geodesic $Ih(\rho(P_{j-1})) \cap Ih(\rho(P_j)) = Ih(\rho(P_j)) \cap Ih(\rho(P_{j+1}))$, and the centers $c(\rho(P_{j-1}))$, $c(\rho(P_j))$ and $c(\rho(P_{j+1}))$ lie on the perpendicular bisector ℓ of the open interval $\mathrm{proj}(Ih(\rho(P_{j-1})) \cap Ih(\rho(P_j)))$. Since $\rho(P_j)$ interchanges $Ih(\rho(P_{j-1})) \cap Ih(\rho(P_j))$ and $Ih(\rho(P_j)) \cap Ih(\rho(P_{j+1}))$, one of the following holds:

1. $\mathrm{Axis}(\rho(P_j)) = Ih(\rho(P_{j-1})) \cap Ih(\rho(P_j))$.
2. $\mathrm{proj}(\mathrm{Axis}(\rho(P_j)))$ is contained in ℓ.

If the first condition is satisfied, then (ρ, τ), with $\tau = \langle s_{[j-1]}, s_{[j+1]} \rangle$, is an isosceles representation by Proposition 5.2.3. If the second condition holds, then we see that $\mathcal{L}(\rho, \sigma)$ is folded at $c(\rho(P_j))$ by Lemma 4.8.6. This completes the proof.

Lemma 5.2.7. *Let $\rho : \pi_1(\mathcal{O}) \to PSL(2, \mathbb{C})$ be a type-preserving representation which is not a real representation. Then, under Notation 2.1.14 (Adjacent triangles), (ρ, τ) is an isosceles representation if and only if both $\phi_\rho(s_0)$ and $\phi_\rho(s_2)$ are non-zero real numbers.*

Proof. Set $(x, y, z) = (\phi_\rho(s_0), \phi_\rho(s_1), \phi_\rho(s_2))$, and let (a_0, a_1, a_2) be the complex probability of ρ at σ (if it is defined). Recall the identities 2.4 and 2.5.

$$a_0 = \frac{x}{yz}, \quad a_1 = \frac{y}{zx}, \quad a_2 = \frac{z}{xy}, \qquad x^2 = \frac{1}{a_1 a_2}, \quad y^2 = \frac{1}{a_2 a_0}, \quad z^2 = \frac{1}{a_0 a_1}.$$

Suppose (ρ, τ) is an isosceles representation. Then $\arg a_0 \equiv \arg a_2 \equiv -\arg a_1 \pmod{2\pi}$, and hence x and z are non-zero real numbers. Suppose conversely that x and z are non-zero real numbers. Then $y \neq 0$ by the Markoff identity, and therefore the complex probability of ρ at σ is defined. Since $a_1 a_2 = 1/x^2$ is a positive real, we have $\arg a_1 + \arg a_2 \equiv 0 \pmod{2\pi}$. Similarly, we have $\arg a_0 + \arg a_1 \equiv 0 \pmod{2\pi}$. Thus $\arg a_2 \equiv -\arg a_1 \equiv \arg a_0 \pmod{2\pi}$ and hence $a_2 = t a_0$ for some $t > 0$. On the other hand, since ρ is not real, $\Im y \neq 0$ and therefore $\Im a_1 \neq 0$. By using the facts that $\arg(a_0 + a_2) = -\arg a_1$, $\Im(a_0 + a_1 + a_2) = 0$ and $\Im a_1 \neq 0$, we see $a_0 + a_2 = \bar{a}_1$. Hence (ρ, τ) is an isosceles representation.

The following proposition determines when an isosceles representation is quasifuchsian.

Proposition 5.2.8. *Let (ρ, τ) be an isosceles representation, and assume that $\Im(c(\rho(P_0))) \le \Im(c(\rho(P_2)))$ under Notation 2.1.14 (Adjacent triangles). Then the following hold.*

(1) $\rho(KP_0)$ (resp. $\rho(KP_2)$) is purely-hyperbolic, parabolic or elliptic according as $\theta^-(\rho,\sigma;s_2)$ (resp. $\theta^+(\rho,\sigma;s_0)$) is positive, 0 or negative.

(2) Suppose $\theta^-(\rho,\sigma;s_2) > 0$ and $\theta^+(\rho,\sigma;s_0) > 0$. Then ρ is a quasifuchsian representation and its Ford domain $Ph(\rho)$ is equal to the following polyhedron (see Fig. 5.5):

$$\begin{aligned} Eh(\rho,\tau) :&= \bigcap \{Eh(\rho(P)) \mid P \text{ is an elliptic generator with slope } s(P) \in \tau^{(0)}\} \\ &= \bigcap \{Eh(\rho(P_j)) \mid j \not\equiv 1 \pmod 3\} \\ &= \bigcap \{Eh(\rho(P'_j)) \mid j \not\equiv 2 \pmod 3\}. \end{aligned}$$

Moreover $\theta^\epsilon(\rho,\sigma) = (+,0,+)$ and $\theta^\epsilon(\rho,\sigma') = (+,+,0)$, and they determine the same point of $\operatorname{int}\tau$, for each $\epsilon \in \{-,+\}$.

Remark 5.2.9. If one of $\rho(KP_0)$ and $\rho(KP_2)$ is parabolic and the other is purely hyperbolic, then ρ gives a cusp group. If both of them are parabolic, then ρ gives the simplest double cusp group. If one of them becomes elliptic, then ρ is not discrete anymore in general. But it can be regarded as the holonomy representation of a hyperbolic cone manifold whose 'Ford domain' is given by the formula in the second assertion. This observation is the starting point to the forthcoming paper [11], whose main results had been announced in [9].

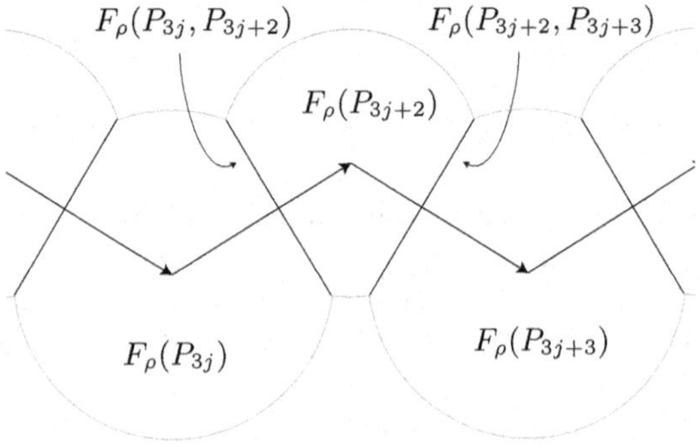

Fig. 5.5. Ford domain of an isosceles representation

Proof. (1) Recall that $\operatorname{tr}(\rho(KP_j)) = \phi_\rho(s_j)$ up to sign. Since $\phi_\rho(s_0)$ is real by Proposition 5.2.7, $\rho(KP_0)$ is purely-hyperbolic, parabolic or elliptic according as $|\phi_\rho(s_0)|$ is greater than 2, equal to 2, or less than 2. Since $r(\rho(P_0)) = r(\rho(P_3)) = 1/|\phi_\rho(s_0)|$ and $c(\rho(P_3)) - c(\rho(P_0)) = 1$, the latter conditions are equivalent to the conditions that $I(\rho(P_0)) \cap I(\rho(P_3))$ is empty, a singleton, or two points, respectively. To study the intersection of these two isometric circles, consider the quadrangles

$$\Diamond_0^L := \Delta_{-1}^-(\rho,\sigma) \cup \Delta_{-1}^+(\rho,\sigma), \quad \Diamond_0^R := \Delta_0^-(\rho,\sigma) \cup \Delta_0^+(\rho,\sigma) \cup \Delta_1^-(\rho,\sigma) \cup \Delta_1^+(\rho,\sigma).$$

Let $\Diamond_3^L$ and $\Diamond_3^R$, respectively, be the images of $\Diamond_0^L$, $\Diamond_0^R$ by the translation $z \mapsto z+1$.

Case 1. $\theta^-(\rho,\sigma;s_2) > 0$. Let x_0^R be the intersection of $I(\rho(P_0))$ with the horizontal ray emanating from $c(\rho(P_0))$ which contains $c(\rho(P_3))$. Similarly, let x_3^L be the intersection of $I(\rho(P_3))$ with the horizontal ray emanating from $c(\rho(P_3))$ which contains $c(\rho(P_0))$. Then $I(\rho(P_0)) \cap I(\rho(P_3)) = \emptyset$ if and only if the line segments $[c(\rho(P_0)), x_0^R]$ and $[c(\rho(P_3)), x_3^L]$ intersect. To show that these two line segments are disjoint, note that

$$\theta^-(\rho,\sigma;s_0) = (\pi/2) - (\theta^-(\rho,\sigma;s_1) + \theta^-(\rho,\sigma;s_2)) = (\pi/2) - \theta^-(\rho,\sigma;s_2) < \pi/2.$$

This implies that $[c(\rho(P_0)), x_0^R]$ is contained in $\Diamond_0^R$. Similarly, $[c(\rho(P_3)), x_3^L]$ is contained in $\Diamond_3^L$. On the other hand, since $\mathcal{L}(\rho,\sigma)$ is convex to the above at $c(\rho(P_2))$, we have $\theta^+(\rho,\sigma;s_2) > \theta^-(\rho,\sigma;s_2) > 0$ by Corollary 4.2.15. This implies $\Delta_1^\epsilon(\rho,\sigma) \cap \Delta_2^\epsilon(\rho,\sigma) = \{c(\rho(P_2))\}$ for each $\epsilon \in \{-,+\}$ and hence $\Diamond_0^R \cap \Diamond_3^L = \{c(\rho(P_2))\}$. Hence $[c(\rho(P_0)), x_0^R]$ and $[c(\rho(P_3)), x_3^L]$ are disjoint. Thus we have $I_0^R \cap I_3^L = \emptyset$. Hence $I(\rho(P_0)) \cap I(\rho(P_3)) = \emptyset$. and therefore $\rho(P_0)$ is purely-hyperbolic.

Case 2. $\theta^-(\rho,\sigma;s_2) = 0$. Then we have $\theta^-(\rho,\sigma;s_0) = \pi/2$ and hence $[c(\rho(P_0)), x_0^R]$ and $[c(\rho(P_3)), x_3^L]$ are contained in (the boundary of) $\Diamond_0^R$ and $\Diamond_3^L$, respectively. On the other hand, $\theta^+(\rho,\sigma;s_2) > \theta^-(\rho,\sigma;s_2) = 0$ as in Case 1. This implies $\Delta_1^\epsilon(\rho,\sigma) \cap \Delta_2^\epsilon(\rho,\sigma) = \overline{c_2 v^-}$, where $v^- := v^-(\rho;P_1,P_2) = v^-(\rho;P_2,P_3)$. Hence $\Diamond_0^R \cap \Diamond_3^L = \overline{c_2 v^-}$. Thus we have $[c(\rho(P_0)), x_0^R] \cap [c(\rho(P_3)), x_3^L] = \{v^-\}$. This implies $I(\rho(P_0)) \cap I(\rho(P_3)) = \{v^-\}$. Hence $\rho(P_0)$ is parabolic.

Case 3. $\theta^-(\rho,\sigma;s_2) < 0$. Then we have $\theta^+(\rho,\sigma;s_2) = \pi - \theta^-(\rho,\sigma;s_2) - 2\alpha(\rho,\sigma;s_2) > \pi - 2\alpha(\rho,\sigma;s_2) > 0$ (cf. Notation 4.2.10(2)). Since (ρ,τ) is thin, we have $\alpha(\rho,\sigma;s_1) = \pi/2$ and hence $\alpha(\rho,\sigma;s_2) < \pi/2$. Hence $Ih(\rho(P_1)) \cap Ih(\rho(P_2)) \cap Ih(\rho(P_3))$ is a singleton by Lemma 4.3.6. This singleton lies in $\operatorname{Axis}(\rho(P_2))$ by Lemma 4.1.3(4) (Chain rule). On the other hand, $Ih(\rho(P_1)) \cap Ih(\rho(P_2)) = \operatorname{Axis}(\rho(P_1))$ by Corollary 5.2.5. Hence $\operatorname{Axis}(\rho(P_1)) \cap \operatorname{Axis}(\rho(P_2)) \neq \emptyset$, and therefore $\rho(KP_0) = \rho(P_2)\rho(P_1)$ is an elliptic transformation.

Thus we have proved the assertion for $\rho(P_0)$. We can prove the assertion for $\rho(P_2)$ by a parallel argument.

(2) Suppose $\theta^-(\rho,\sigma;s_2) > 0$ and $\theta^+(\rho,\sigma;s_0) > 0$. Since $\mathcal{L}(\rho,\sigma)$ is convex to the above at $c(\rho(P_2))$ and convex to the below at $c(\rho(P_0))$, we have

$\theta^-(\rho, \sigma; s_0) > \theta^+(\rho, \sigma; s_0) > 0$ and $\theta^+(\rho, \sigma; s_2) > \theta^-(\rho, \sigma; s_2) > 0$ by Corollary 4.2.15. As in the latter half of the proof of Proposition 5.1.3, we set:

$$\begin{aligned}
Eh(\rho, \tau) &= \cap\{Eh(\rho(P_j)) \mid j \not\equiv 1 \pmod 3\},\\
F_\rho(P_{3j}) &= Ih(\rho(P_{3j})) \cap (Eh(\rho(P_{3j-1})) \cap Eh(\rho(P_{3j+2}))),\\
F_\rho(P_{3j+2}) &= Ih(\rho(P_{3j+2})) \cap (Eh(\rho(P_{3j})) \cap Eh(\rho(P_{3j+3}))),\\
F_\rho(P_{3j}, P_{3j+2}) &= Ih(\rho(P_{3j})) \cap Ih(\rho(P_{3j+2})),\\
F_\rho(P_{3j-1}, P_{3j}) &= Ih(\rho(P_{3j-1})) \cap Ih(\rho(P_{3j})).
\end{aligned}$$

By Corollary 5.2.5, we have

$$\begin{aligned}
F_\rho(P_{3j}, P_{3j+2}) &= Ih(\rho(P_{3j})) \cap Ih(\rho(P_{3j+1}))\\
&= Ih(\rho(P_{3j+1})) \cap Ih(\rho(P_{3j+2})) = \mathrm{Axis}(\rho(P_{3j+1})).
\end{aligned}$$

By using the fact that all components of $\theta^\pm(\rho, \sigma)$ are non-negative, we see the following:

1. The geodesics $F_\rho(P_{3j-1}, P_{3j})$ and $F_\rho(P_{3j}, P_{3j+2})$ are disjoint, and $F_\rho(P_{3j})$ is a 2-dimensional convex polyhedron bounded by these two geodesics.
2. The geodesics $F_\rho(P_{3j}, P_{3j+2})$ and $F_\rho(P_{3j+2}, P_{3j+3})$ are disjoint, and $F_\rho(P_{3j+2})$ is a 2-dimensional convex polyhedron bounded by these two geodesics.

Thus $\partial Eh(\rho, \tau)$ consists of the faces $F_\rho(P_{3j})$, $F_\rho(P_{3j+2})$ and the edges $F_\rho(P_{3j-1}, P_{3j})$, $F_\rho(P_{3j}, P_{3j+2})$ where j runs over $\mathbb{Z}$. The involution $\rho(P_{3j})$ preserves the face $F_\rho(P_{3j})$ and interchanges $F_\rho(P_{3j-1}, P_{3j})$ with $F_\rho(P_{3j}, P_{3j+2})$, whereas the involution $\rho(P_{3j+2})$ preserves the face $F_\rho(P_{3j+2})$ and interchanges $F_\rho(P_{3j}, P_{3j+2})$ with $F_\rho(P_{3j+2}, P_{3j+3})$. Thus the set

$$\{F_\rho(P_{3j-1}, P_{3j}), F_\rho(P_{3j}, P_{3j+2})\}$$

forms an "edge cycle modulo $\rho(K)$" for each integer j. On the other hand, the dihedral angle of $E(\rho, \tau)$ along $F_\rho(P_{3j}, P_{3j+2})$ is equal to $\theta_{3j} + \theta_{3j+1} - \pi$, where θ_j be the dihedral angle of $Eh(\rho(P_j)) \cap Eh(\rho(P_{j+1}))$. Hence the sum of the dihedral angles of $Eh(\rho, \tau)$ along the edges in the edge cycle is equal to π by Lemma 4.1.3 (chain rule). Hence we can see that $Eh(\rho, \tau)$ together with the gluing isometries satisfies the conditions for the "Poincare's theorem on fundamental polyhedra modulo $\rho(K)$". (See Sect. 6.4 for details.) From this fact, we see that ρ is discrete and faithful, and $Eh(\rho, \tau)$ is equal to the Ford domain $Ph(\rho)$.

Moreover, by Lemma 4.5.3(2), $\theta^\epsilon(\rho, \sigma)$ and $\theta^\epsilon(\rho, \sigma')$, determine the same point of $\mathrm{int}\,\tau$, for each $\epsilon \in \{-, +\}$. This completes the proof of Proposition 5.2.8.

Remark 5.2.10. Set $\boldsymbol{\nu} = (\theta^-(\rho, \sigma), \theta^+(\rho, \sigma))$ and $\boldsymbol{\rho} = (\rho, \boldsymbol{\nu})$. Then the convex polyhedra $Eh(\rho, \tau)$, $F_\rho(P_j)$, $F_\rho(P_{3j-1}, P_{3j})$ and $F_\rho(P_{3j}, P_{3j+2})$, introduced in the proof, are equal to $Eh(\boldsymbol{\rho})$, $F_{\boldsymbol{\rho}}(P_j)$, $F_{\boldsymbol{\rho}}(P_{3j-1}, P_{3j})$ and $F_{\boldsymbol{\rho}}(P_{3j}, P_{3j+2})$, respectively (cf. Definitions 3.4.3 and 3.4.6).

Next, we prove the following proposition, which shows that every thin label is realized as the side parameter of a quasifuchsian representation.

Proposition 5.2.11. *Let* $\boldsymbol{\nu} = (\nu^-, \nu^+)$ *be a thin label,* τ *the edge of* $\mathcal{D}$ *containing* $\nu^\pm$*, and* σ *a triangle of* $\mathcal{D}$ *containing* τ*. Then there is a unique element* $\rho \in \mathcal{QF}$ *which realizes* $\boldsymbol{\nu}$*, that is, the following hold.*

1. (ρ, τ) *is an isosceles representation, and the Ford domain* $Ph(\rho)$ *is equal to* $Eh(\rho, \tau)$.
2. $(\theta^-(\rho, \sigma), \theta^+(\rho, \sigma))$ *is well-defined and represents the point* $\boldsymbol{\nu}$.

To prove this proposition, we prepare the following lemma.

Lemma 5.2.12. *Let* σ *be a triangle of* $\mathcal{D}$ *and let* $\nu^\pm$ *be points in* $\sigma \cap \mathbb{H}^2$*. Then there is a unique (non-trivial) Markoff map* ϕ *which is an algebraic root for* $((\nu^-, \sigma), (\nu^+, \sigma))$ *(Definition 4.2.19).*

Proof. Let ϕ be a non-trivial Markoff map with $(x, y, z) = (\phi(s_0), \phi(s_1), \phi(s_2))$, where $\sigma = \langle s_0, s_1, s_2 \rangle$. Then ϕ is an algebraic root for $((\nu^-, \sigma), (\nu^+, \sigma))$ if and only if (x, y, z) is a root of the following system of equations.

$$x^2 + y^2 + z^2 = xyz, \quad x + \alpha^\epsilon y + \beta^\epsilon z = 0 \quad (\epsilon \in \{-, +\}).$$

Here $\alpha^\epsilon = \epsilon i \exp(\epsilon i \theta_2^\epsilon)$ and $\beta^\epsilon = -\epsilon i \exp(-\epsilon i \theta_1^\epsilon)$, where $\nu^\epsilon = (\theta_0^\epsilon, \theta_1^\epsilon, \theta_2^\epsilon) \in \sigma^\epsilon$. By the two linear equations, we have

$$y = \frac{(\beta^+ - \beta^-)x}{\alpha^+\beta^- - \alpha^-\beta^+}, \qquad z = \frac{(-\alpha^+ + \alpha^-)x}{\alpha^+\beta^- - \alpha^-\beta^+}.$$

Here $\alpha^+\beta^- - \alpha^-\beta^+ \neq 0$, because $\boldsymbol{\nu} \neq (s_0, s_0)$. By substituting y and z in the Markoff equation with the above, we obtain:

$$(\alpha^+\beta^- - \alpha^-\beta^+)^2 x^2 + (\beta^+ - \beta^-)^2 x^2 + (-\alpha^+ + \alpha^-)^2 x^2 \\ = (\beta^+ - \beta^-)(-\alpha^+ + \alpha^-)x^3.$$

Hence we have either $x = 0$ or

$$x = \frac{(\alpha^+\beta^- - \alpha^-\beta^+)^2 + (\beta^+ - \beta^-)^2 + (-\alpha^+ + \alpha^-)^2}{(\beta^+ - \beta^-)(-\alpha^+ + \alpha^-)}.$$

Here $-\alpha^+ + \alpha^- \neq 0$ and $\beta^+ - \beta^- \neq 0$, because $\boldsymbol{\nu} \neq (s_2, s_2)$ and $\boldsymbol{\nu} \neq (s_1, s_1)$. If $x = 0$ then $y = z = 0$ and hence ϕ is the trivial Markoff map. So x must take the latter value. Hence there is a unique algebraic root for $((\nu^-, \sigma), (\nu^+, \sigma))$.

Proof (Proof of Proposition 5.2.11). We may assume τ is the edge $\langle s_1, s_2 \rangle$ of the triangle $\sigma = \langle s_0, s_1, s_2 \rangle$ in the proof of Lemma 5.2.12. Then $\beta^\epsilon = \epsilon i \alpha^\epsilon$, and hence the Markoff triple (x, y, z) determined by the non-trivial algebraic root for $((\nu^-, \sigma), (\nu^+, \sigma))$ is given by:

$$x = \frac{4i\alpha^+\alpha^-(\alpha^+\alpha^- + 1)}{(\alpha^+ + \alpha^-)(-\alpha^+ + \alpha^-)}, \quad y = \frac{2i(\alpha^+\alpha^- + 1)}{(\alpha^+ - \alpha^-)}, \quad z = \frac{-2(\alpha^+\alpha^- + 1)}{(\alpha^+ + \alpha^-)}.$$

Thus the complex probability (a_0, a_1, a_2) at σ is:

$$a_0 = \frac{\alpha^+\alpha^-}{\alpha^+\alpha^- + 1}, \quad a_1 = \frac{(\alpha^+ + \alpha^-)^2}{4\alpha^+\alpha^-(\alpha^+\alpha^- + 1)}, \quad a_2 = \frac{-(\alpha^+ - \alpha^-)^2}{4\alpha^+\alpha^-(\alpha^+\alpha^- + 1)}.$$

So we have

$$\Re a_0 = \Re\left(\frac{(1+\cos\theta) + i\sin\theta}{2(1+\cos\theta)}\right) = \frac{1}{2}$$

$$\frac{a_2}{a_1} = \frac{-(\alpha^+ - \alpha^-)^2}{(\alpha^+ + \alpha^-)^2} = -\left(\frac{1 - e^{-i(\theta_2^+ + \theta_2^-)}}{1 + e^{-i(\theta_2^+ + \theta_2^-)}}\right)^2 = -\left(\frac{i\sin\frac{\theta_2^+ + \theta_2^-}{2}}{\cos\frac{\theta_2^+ + \theta_2^-}{2}}\right)^2 \in \mathbb{R}_+.$$

Hence (ρ, τ) is an isosceles representation. Since ρ is quasifuchsian, Proposition 5.2.8 implies that the Ford domain $Ph(\rho)$ is equal to $Eh(\tau)$. Moreover, $(\theta^-(\rho,\sigma), \theta^+(\rho,\sigma))$ is well-defined by Proposition 5.2.3. To show that it is equal to $\boldsymbol{\nu}$, we have only to show that ϕ is upward at σ, because ϕ is an algebraic root for $\boldsymbol{\nu}$. To this end, note that

$$\frac{i/x}{a_2} = \frac{i/x}{z/(xy)} = \frac{iy}{x} = \frac{\alpha^+ - \alpha^-}{\alpha^+ + \alpha^-} = \frac{1 - e^{-i(\theta_2^+ + \theta_2^-)}}{1 + e^{-i(\theta_2^+ + \theta_2^-)}} = \frac{i\sin\frac{\theta_2^+ + \theta_2^-}{2}}{\cos\frac{\theta_2^+ + \theta_2^-}{2}},$$

$$\frac{i/y}{a_0} = \frac{i/y}{x/(yz)} = \frac{iz}{x} = \frac{\alpha^+ - \alpha^-}{2\alpha^+\alpha^-} = \frac{1}{2}\left(\frac{1}{\alpha^-} - \frac{1}{\alpha^+}\right)$$
$$= \frac{1}{2}\left(e^{i(\theta_2^- + \frac{\pi}{2})} - e^{-i(\theta_2^+ + \frac{\pi}{2})}\right)$$

$$\frac{i/y}{a_1} = \frac{i/z}{y/(zx)} = \frac{ix}{y} = \frac{-2i(\alpha^+\alpha^-)}{\alpha^+ + \alpha^-} = \frac{2}{-e^{i\theta_2^+} + e^{-i\theta_2^-}}$$

Hence we see that $\arg(\frac{i/x}{a_2}) = \pi/2$ and that both $\arg(\frac{i/y}{a_0})$ and $\arg(\frac{i/y}{a_1})$ belongs to $(0, \pi)$ by using the fact that $\theta_2^{\pm} \in (0, \pi/2)$. So ϕ is upward at σ and hence $(\theta^-(\rho,\sigma), \theta^+(\rho,\sigma)) = \boldsymbol{\nu}$. This completes the proof of Proposition 5.2.11.

Next, we prove the following proposition, which is used in the proof of Proposition 8.3.2.

Proposition 5.2.13. *Under Notation 2.1.14 (Adjacent triangles), there is a neighborhood, U, of* $\operatorname{int}\tau \times \operatorname{int}\tau$ *in* $(\sigma' \cap \mathbb{H}^2) \times (\sigma \cap \mathbb{H}^2)$ *and a continuous map* $U \ni \boldsymbol{\nu} \mapsto \phi_{\boldsymbol{\nu}} \in \Phi$ *which have the following properties.*

1. *$\phi_{\boldsymbol{\nu}}$ is an algebraic root for $((\nu^-, \sigma'), (\nu^+, \sigma))$, where $(\nu^-, \nu^+) = \boldsymbol{\nu}$.*
2. *The representation induced by $\phi_{\boldsymbol{\nu}}$ is quasifuchsian.*
3. *The representation induced by any other algebraic root for $((\nu^-, \sigma'),$ $(\nu^+, \sigma))$ is not quasifuchsian.*

To prove this proposition, we need the following lemma.

Lemma 5.2.14. *Under Notation 2.1.14 (Adjacent triangles), let* $\boldsymbol{\nu} = (\nu^-, \nu^+)$ *be a point of* $(\sigma' \cap \mathbb{H}^2) \times (\sigma \cap \mathbb{H}^2)$. *Then the following hold.*

1. *There are at most two non-trivial Markoff maps, counted with multiplicity, which are algebraic roots for* $((\nu^-, \sigma'), (\nu^+, \sigma))$.
2. *If* $\nu^\pm \in \operatorname{int} \tau$, *then there is a unique non-trivial Markoff map, counted with multiplicity, which is an algebraic root for* $((\nu^-, \sigma'), (\nu^+, \sigma))$. *Moreover it is equal to the unique non-trivial algebraic root for* $((\nu^-, \sigma), (\nu^+, \sigma))$ *given in Lemma 5.2.12.*

Proof. Set $\nu^- = (\theta_0^-, \theta_1^-, \theta_2^-) \in \sigma'$ and $\nu^+ = (\theta_0^+, \theta_1^+, \theta_2^+) \in \sigma$. Then a Markoff map ϕ is an algebraic root for $((\nu^-, \sigma'), (\nu^+, \sigma))$ if and only if the value $(x, y, z, w) = (\phi(s_0), \phi(s_1), \phi(s_2), \phi(s_2'))$ is a root of the following system of equations:

$$x^2 + y^2 + z^2 = xyz, \quad y + w = xz, \quad x + \alpha^+ y + \beta^+ z = 0, \quad x + \alpha^- z + \beta^- w = 0$$

Here $\alpha^\epsilon = \epsilon i \exp(\epsilon i \theta_2^\epsilon)$ and $\beta^\epsilon = -\epsilon i \exp(-\epsilon i \theta_1^\epsilon)$. Note that the last two equations are equivalent to the equations

$$y = -\overline{\alpha^+} x - \gamma^+ z, \quad w = -\overline{\beta^-} x - \gamma^- z,$$

where $\gamma^\epsilon = i \exp(i \theta_0^\epsilon)$. The second equation and these two equations imply

$$xz = -(\overline{\alpha^+} + \overline{\beta^-}) x - (\gamma^+ + \gamma^-) z, \quad \text{and hence} \quad z = \frac{-(\overline{\alpha^+} + \overline{\beta^-}) x}{x + (\gamma^+ + \gamma^-)}.$$

By substituting y in the Markoff equation with the linear combination of x and z, we obtain

$$(1 + (\alpha^+)^{-2}) x^2 + (1 + (\gamma^+)^2) z^2 + xz(\overline{\alpha^+} x + \gamma^+ z + 2\overline{\alpha^+} \gamma^+) = 0.$$

By substituting z with $\frac{-(\overline{\alpha^+} + \overline{\beta^-}) x}{x + (\gamma^+ + \gamma^-)}$, this equation turns into the equation

$$\left(\frac{x}{\alpha^+ \beta^- (x + (\gamma^+ + \gamma^-))} \right)^2 f(x) = 0,$$

where $f(x)$ is a quadratic polynomial in x. By putting $x = X - (\gamma^+ + \gamma^-)$, $f(x)$ turns into the quadratic polynomial

$$g(X) := \alpha^+ \beta^- (\alpha^+ \beta^- - 1) X^2 + (\alpha^+ + \beta^-)(\alpha^+ \gamma^+ + \beta^- \gamma^-) X \\ + (\alpha^+ + \beta^-)^2 (1 - \gamma^+ \gamma^-).$$

The fact that $\boldsymbol{\nu} \neq (s_2', s_1)$ implies that the coefficient of X^2 is non-zero, and the fact that $\boldsymbol{\nu} \notin \mathit{Diagonal}(\tau^{(0)})$ implies that the constant term is non-zero. Thus the above equation is equivalent to the equation $x^2 f(x) = 0$, and

$f(x) = 0$ has two roots counted with multiplicity. Thus the equation has at most two non-zero roots. Hence we obtain the first assertion.

To see the second assertion, assume that $\nu^{\pm} \in \operatorname{int} \tau$. Then we see

$$f(x) = \alpha^+\beta^-(\alpha^+\beta^- - 1)x^2 - 2i\alpha^+\beta^-(\alpha^+ + \beta^-)x.$$

Hence it has a unique non-zero root

$$x = \frac{2i(\alpha^+ + \beta^-)}{(\alpha^+\beta^- - 1)}.$$

By noting that the coordinate of the point ν^- in σ is $(\theta_0^-, 0, \theta_1^-)$, we see that this is equal to the non-zero root in Lemma 5.2.12. Thus we obtain the second assertion.

Proof (Proof of Proposition 5.2.14). Let $\varphi_j : (\sigma' \cap \mathbb{H}^2) \times (\sigma \cap \mathbb{H}^2) \to \Phi$ $(j = 0, 1)$ be continuous maps such that $\varphi_1(\boldsymbol{\nu})$ and $\varphi_2(\boldsymbol{\nu})$ correspond to the roots of the quadratic polynomial $f(x)$ for $\boldsymbol{\nu}$ in the proof of Lemma 5.2.14. We assume that if $\boldsymbol{\nu}$ is thin then $\varphi_2(\boldsymbol{\nu})$ is the trivial Markoff map. Then the type-preserving representation ρ induced by $\varphi_1(\boldsymbol{\nu})$ determines an isosceles representation (ρ, τ) and ρ is quasifuchsian whenever $\boldsymbol{\nu}$ is thin (Proposition 5.2.11). Since $\mathcal{QF}$ is open in $\mathcal{X}$, there is a neighborhood U of $\operatorname{int} \tau \times \operatorname{int} \tau$ in $(\sigma' \cap \mathbb{H}^2) \times (\sigma \cap \mathbb{H}^2)$ such that the representation induced by $\varphi_1(\boldsymbol{\nu})$ is quasifuchsian for every $\boldsymbol{\nu} \in U$. Since $\varphi_2(\boldsymbol{\nu})$ is the trivial Markoff map whenever $\boldsymbol{\nu}$ is thin, we can choose U so that the representation induced by $\varphi_2(\boldsymbol{\nu})$ is not quasifuchsian for every $\boldsymbol{\nu} \in U$. Since $\varphi_1(\boldsymbol{\nu})$ and $\varphi_2(\boldsymbol{\nu})$ are the only (possibly) non-trivial algebraic roots for $\boldsymbol{\nu} \in (\sigma' \cap \mathbb{H}^2) \times (\sigma \cap \mathbb{H}^2)$, we obtain the desired conclusion, by setting $\phi_{\boldsymbol{\nu}} = \varphi_1(\boldsymbol{\nu})$.

At the end of this section, we point out the following corollary, which gives another characterization of isosceles representations.

Corollary 5.2.15. *Under the assumption of Proposition 5.2.3, the following conditions are equivalent.*

1. *(ρ, τ) is an isosceles representation, where $\tau = \sigma \cap \sigma'$.*
2. *The side parameters $\theta^{\pm}(\rho, \sigma)$ and $\theta^{\pm}(\rho, \sigma')$ are defined, and $\theta^-(\rho, \sigma'; s_2') = \theta^+(\rho, \sigma; s_1) = 0$.*
3. *The side parameters $\theta^{\pm}(\rho, \sigma)$ and $\theta^{\pm}(\rho, \sigma')$ are defined, and $\theta^-(\rho, \sigma; s_1) = \theta^+(\rho, \sigma'; s_2') = 0$*

Proof. By Proposition 5.2.3, the first condition implies the remaining conditions. Suppose the second condition is satisfied. Then, by Lemma 5.2.14, ρ is induced by the Markoff map which is the non-trivial algebraic root for $((\nu^-, \sigma), (\nu^+, \sigma))$, where ν^- and ν^+, respectively, are the points of τ determined by $\theta^-(\rho, \sigma')$ and $\theta^+(\rho, \sigma)$. Hence, by Proposition 5.2.11, (ρ, τ) is an isosceles representation. Similarly, the third condition implies the first. This completes the proof.

5.3 Groups generated by two parabolic transformations

In this section, we study those type-preserving representations induced by Markoff maps which vanish at some vertex of $\mathcal{D}$. To describe geometric meaning of these representations, let (B^3, t) be the pair of a 3-ball B^3 and a pair t of arcs properly embedded in B^3 as illustrated in Fig. 5.6, which is called a *trivial tangle*. A disk D properly embedded in B^3 which separate the two components of t is called a *meridian disk* of the trivial tangle. We identify

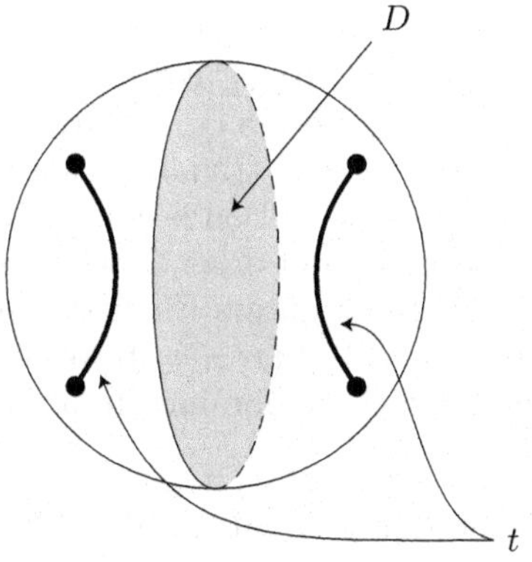

Fig. 5.6. Trivial tangle

$\partial B^3 - \partial t$ with the 4-times punctured sphere $S = (\mathbb{R}^2 - \mathbb{Z}^2)/G$ introduced in Sect. 2.1, so that the boundary of a meridian disk is identified with the essential simple loop $\tilde{\alpha}_\infty$ of slope ∞. Then $\pi_1(B^3 - t)$ is identified with the quotient group $\pi_1(S)/\langle\langle\tilde{\alpha}_\infty\rangle\rangle$. Here we identify an (oriented) essential simple loop with the (conjugacy class of an) element of the fundamental group represented by the loop. Note that $\tilde{\alpha}_\infty = \alpha_\infty^2 \in \pi_1(\mathcal{O})$, where α_∞ is the essential simple loop in $\mathcal{O}$ (or in T) of slope ∞. Let $\{P_j\}$ be the sequence of elliptic generators associated with the triangle $\langle 0, 1, \infty\rangle$ of $\mathcal{D}$. (Thus $(P_0, P_1, P_2) = (P_0, Q_0, R_0)$ in the presentation 2.3 by Convention 2.1.5.) Then, by Definition 2.1.3 and Proposition 2.1.2, we see $\alpha_\infty = K^{-1}P_2 = P_0P_1$ and hence $\tilde{\alpha}_\infty = (P_0P_1)^2$. On the other hand,

$$K_1K_2 = K^{P_0}K^{P_1} = P_0(P_2P_1P_0)P_0 \cdot P_1(P_2P_1P_0)P_1 = (P_0P_1)^2 = \tilde{\alpha}_\infty$$
$$K_0K_3 = KK^{P_2} = (P_2P_1P_0) \cdot P_2(P_2P_1P_0)P_2 = P_2(P_1P_0)^2P_2 = P_2\tilde{\alpha}_\infty^{-1}P_2^{-1}.$$

Hence we obtain the following lemma.

Lemma 5.3.1. *Denote the image in $\pi_1(B^3 - t)$ of the generator K_i in the presentation 2.2 of $\pi_1(S)$ by the same symbol K_i ($i \in \{0, 1, 2, 3\}$). Then $\pi_1(B^3 - t)$ is the free group freely generated by K_0 and K_1. Moreover $K_2 = K_1^{-1}$ and $K_3 = K_0^{-1}$ in $\pi_1(B^3 - t)$.*

The following lemma describes the meaning of those Markoff maps which vanish at some vertex of $\mathcal{D}$.

Lemma 5.3.2. *Let $\rho : \pi_1(S) \to PSL(2,\mathbb{C})$ be a type-preserving representation and ϕ a Markoff map inducing ρ. Then ρ descends to a representation of $\pi_1(B^3 - t) = \langle K_0, K_1 \rangle$, if and only if $\phi(\infty) = 0$. Moreover if this condition is satisfied, then*

$$\rho(K_0) = \begin{pmatrix} 1 & 1 \\ 0 & 1 \end{pmatrix}, \qquad \rho(K_1) = \begin{pmatrix} 1 & 0 \\ \omega & 1 \end{pmatrix},$$

where $\omega = -x^2$ with $x = \phi(0)$.

Proof. Since $\pi_1(B^3 - t) = \pi_1(S)/\langle\langle \tilde{\alpha}_\infty \rangle\rangle$ and since $\tilde{\alpha}_\infty = \alpha_\infty^2$, ρ descends to a representation of $\pi_1(B^3 - t) = \langle K_0, K_1 \rangle$, if and only if $\rho(\alpha_\infty)$ is trivial or of order 2. However, $\rho(\alpha_\infty)$ cannot be trivial, because $\pi_1(T) = \langle \alpha_\infty, \alpha_0 \rangle$ and ρ is irreducible. Hence the above condition is equivalent to the condition that $\rho(\alpha_\infty)$ is an elliptic transformation of order 2, which in tern is equivalent to the condition $\phi(\infty) = 0$. Hence we obtain the first assertion. By the Markoff identity, we see $(\phi(0), \phi(1), \phi(\infty)) = (x, \pm ix, 0)$ for some $x \in \mathbb{C} - \{0\}$. Thus the formula for $\rho(K_0)$ and $\rho(K_1)$ are obtained from Lemma 2.3.7(3).

By the above lemma, the image of $\rho : \pi_1(B^3 - t) \to PSL(2,\mathbb{C})$ induced by a type-preserving representation is equal to the group

$$G_\omega = \left\langle \begin{pmatrix} 1 & 1 \\ 0 & 1 \end{pmatrix}, \begin{pmatrix} 1 & 0 \\ \omega & 1 \end{pmatrix} \right\rangle,$$

for some $\omega \in \mathbb{C} - \{0\}$. Discreteness problem for the non-elementary two-parabolic groups G_ω has been studied extensively by various authors, including [20, 48, 1, 45]. In particular, the following facts have been proved.

Proposition 5.3.3. *(1)* (Brenner [20]) *If $|\omega| \geq 4$, then G_ω is discrete and free.*

(2) (Knapp [48]) *If ω is real, then G_ω is discrete if and only if ω or $-\omega$ belongs to*

$$[4, \infty) \cup \{4\cos^2(\pi/n) \mid n \in \mathbb{N}\}.$$

The following characterization of the discrete two-parabolic groups of co-finite volume is proved by Adams by using the orbifold uniformization theorem.

Proposition 5.3.4 (Adams [1]). *G_ω is discrete and* vol$(\mathbb{H}^3/G_\omega)$ *is finite, if and only if G_ω is isomorphic to the link group, $\pi_1(S^3 - K)$, of a hyperbolic 2-bridge link K.*

R. Riley devoted massive effort to identify and to plot those complex numbers corresponding to the 2-bridge link groups and produced a mysterious output [69] (Fig. 0.2a). It also outlines the following domain, which is now called the Riley slice of the Schottky space (Fig. 0.2b).

Definition 5.3.5 (Riley slice). *The* Riley slice $\mathcal{R}$ *of the Schottky space is the subspace of* $\mathbb{C}$ *consisting of those complex numbers* ω *such that* G_ω *is discrete and free and that the quotient* $\Omega(G_\omega)/G_\omega$ *of the domain of discontinuity is homeomorphic to the 4-times punctured sphere* S.

Keen-Series[45] (see also Komori-Series[49]) introduces the pleating coordinates on the Riley slice $\mathcal{R}$ and produced a beautiful picture of $\mathcal{R}$. In the forthcoming paper [11], we add to their picture those complex numbers ω corresponding to the 2-bridge link groups, by extending Jorgensen's theory to the outside of quasifuchsian space (cf. Fig. 0.2b and the announcement [9]).

In the remainder of this section, we describe the Ford domains of the Kleinian groups $\rho(\pi_1(\mathcal{O}))$ which are finite extensions of those groups in Proposition 5.3.3. We begin with the following elementary facts.

Lemma 5.3.6. *Let* ϕ *be a nontrivial Markoff map with* $\phi(\infty) = 0$. *Then the following hold.*

1. *There is a sign* $\epsilon \in \{-, +\}$ *and a complex number* $x \in \mathbb{C} - \{0\}$, *such that* $\phi(k) = (\epsilon i)^k x$ *for every integer* k.
2. *Let* $\{P_j\}$ *be the sequence of elliptic generators associated with* $\sigma = \langle 0, 1, \infty \rangle$. *Then the isometric hemisphere* $Ih(\rho(P_j))$ *is defined if and only if* $j \not\equiv 2 \pmod 3$, *and* $\rho(P_{3j+2})$ *is the* π*-rotation around the vertical geodesic above the point* $c(\rho(P_{3j})) + (1/2)$. *Moreover,* $Ih(\rho(P_{3j})) = Ih(\rho(P_{3j+1}))$, *and* $\rho(P_{3j+1}P_{3j})$ *is the* π*-rotation around the vertical geodesic above the point* $c(\rho(P_{3j}))$.
3. *Let* $\sigma' = \langle s_0', s_1', s_2' \rangle$ *be a triangle of* $\mathcal{D}$ *such that* $\langle s_0', s_2' \rangle$ *is adjacent to* ∞, *i.e.,* $\langle s_0', s_2' \rangle = \langle k+1, k \rangle$ *for some* $k \in \mathbb{Z}$, *and* $s_1' \neq \infty$. *Then the complex probability of* ρ *at* σ' *is equal to* $(a, 1, -a)$ *for some* $a \in \mathbb{C} - \{0\}$. *In particular,* $\mathcal{L}(\rho, \sigma')$ *is folded at* $c(\rho(P_1'))$, *and* $Ih(\rho(P_{3j}')) = Ih(\rho(P_{3j+2}'))$, *where* $\{P_j'\}$ *is the sequence of elliptic generators associated with* σ'.

Proof. (1) As in the proof of Lemma 5.3.2, we see by Markoff identity that $(\phi(0), \phi(1), \phi(\infty)) = (x, \epsilon i x, 0)$ for some $x \in \mathbb{C} - \{0\}$ and a sign $\epsilon \in \{-, +\}$. Since $\langle k, k+1, \infty \rangle$ is a triangle of $\mathcal{D}$, we see by Proposition 2.3.4(2) that $\phi(k-1) + \phi(k+1) = \phi(k)\phi(\infty) = 0$. Hence we obtain the desired formula.

(2) This is equivalent to Proposition 2.4.4(2).

(3) By (1) we see $(\phi(s_0'), \phi(s_2')) = (x, \epsilon i x)$ for some $\epsilon \in \{-, +\}$. Thus we have $\phi(s_1') = \phi(s_0')\phi(s_2') - \phi(\infty) = \epsilon i x^2$ by Proposition 2.3.4(2). Hence the complex probability (a_0', a_1', a_2') of ρ at σ' is equal to $(-1/x^2, 1, 1/x^2)$. Thus we obtain the first assertion, which implies the remaining assertions.

Proposition 5.3.3, together with the above lemma, motivates the following definition.

Definition 5.3.7. *(1) (Brenner representation) A type-preserving representation* ρ *is called a* Brenner representation *of slope* $s \in \hat{\mathbb{Q}}$ *if* $\phi(s) = 0$ *and* $|\phi(s')| \geq 2$ *for some (or every) vertex* s' *adjacent to* s.

(2) (Knapp representation) A type-preserving representation ρ is called a Knapp representation *of slope $s \in \hat{\mathbb{Q}}$ and of degree $n \geq 3$ if $\phi(s) = 0$ and $\phi(s') = \pm 2\cos(\pi/n)$ or $\pm 2i\cos(\pi/n)$ for some vertex s' adjacent to s.*

Then we have the following propositions.

Proposition 5.3.8. *Let ρ be a Brenner representation of slope s. Pick a triangle $\sigma = \langle s_0, s_1, s_2 \rangle$ of $\mathcal{D}$ with $s_2 = s$, and let $\{P_j\}$ be the sequence of elliptic generators associated with σ. Then ρ is discrete, and the Ford domain of $\rho(\pi_1(\mathcal{O}))$ is equal to the common exterior $\cap_j Eh(\rho(P_{3j}))$ of mutually disjoint isometric hemispheres $\{Ih(\rho(P_{3j}))\}$ (see Fig. 5.7). Moreover, we have the following group presentations:*

$$\rho(\pi_1(\mathcal{O})) \cong \langle P_0, P_1, P_2 \mid P_0^2 = P_1^2 = P_2^2 = 1, (P_0P_1)^2 = 1 \rangle,$$
$$\rho(\pi_1(S)) \cong \langle K_0, K_1 \rangle.$$

Proof. By Lemma 5.3.6 and the fact that $r(\rho(P_{3j})) = 1/|\phi(s_0)| \leq 1/2$, we see that the isometric hemispheres $\{Ih(\rho(P_{3j}))\}$ are mutually disjoint. Consider the fundamental domain

$$R_0 = \{(z,t) \in \mathbb{H}^3 \mid |\Re(z - c(\rho(P_0)))| \leq 1/2, \quad \Im(z - c(\rho(P_0))) \geq 0\}$$

of the infinite dihedral Kleinian group $\langle \rho(K), \rho(P_0P_1) \rangle$, and set $R := R_0 \cap Eh(\rho(P_0))$. Then we can check that the polyhedron R, together with the face pairing isometries $\rho(K)$, $\rho(P_1P_0)$, $\rho(P_0)$ and $\rho(P_1)$, satisfies the conditions of Poincare's theorem on fundamental polyhedra (see Fig. 5.8). In the extreme case where $\phi(s_0)$ (resp. $\phi(s_1)$) is real, $\rho(P_1)$ (resp. $\rho(P_0)$) does not belong to the set of the face pairing isometries. This implies the discreteness of ρ and the desired group presentation of $\rho(\pi_1(\mathcal{O}))$. This also implies that $\cap_j Eh(\rho(P_{3j}))$ is a fundamental domain of $\rho(\pi_1(\mathcal{O}))$ modulo the stabilizer $\langle \rho(K), \rho(P_0P_1) \rangle$ of ∞. Hence we see that $\cap_j Eh(\rho(P_{3j}))$ is the Ford domain of $\rho(\pi_1(\mathcal{O}))$. The group presentation of $\rho(\pi_1(S))$ is obtained through the Reidemeister-Schreier method by using the following isomorphism.

$$\rho(\pi_1(\mathcal{O}))/\rho(\pi_1(S)) \cong \langle P_0 \mid P_0^2 = 1 \rangle \oplus \langle P_1 \mid P_1^2 = 1 \rangle, \quad \text{where } P_2 = P_0P_1.$$

Proposition 5.3.9. *Let ρ be a Knapp representation of slope s and of order n. Pick a triangle $\sigma = \langle s_0, s_1, s_2 \rangle$ of $\mathcal{D}$ with $s_2 = s$ and $\phi(s_0) = \pm 2\cos(\pi/n)$, and let $\{P_j\}$ be the sequence of elliptic generators associated with σ. Then ρ is discrete, and the Ford domain of $\rho(\pi_1(\mathcal{O}))$ is equal to $\cap_j Eh(\rho(P_{3j}))$ (see Fig. 5.9). Moreover, we have the following group presentations:*

$$\rho(\pi_1(\mathcal{O})) \cong \langle P_0, P_1, P_2 \mid P_0^2 = P_1^2 = P_2^2 = 1, (P_0P_1)^2 = 1, (P_1P_2)^n = 1 \rangle,$$
$$\langle \rho(P_1), \rho(K) \rangle \cong \langle P_0, K \mid P_0^2 = (KP_0)^n = 1 \rangle \cong \pi_1(O(2,n,\infty)),$$
$$\rho(\pi_1(S)) \cong \begin{cases} \pi_1(O(2,n,\infty)) & \textit{if } n\textit{:odd}, \\ \pi_1(O(n/2,\infty,\infty)) & \textit{if } n\textit{:even}. \end{cases}$$

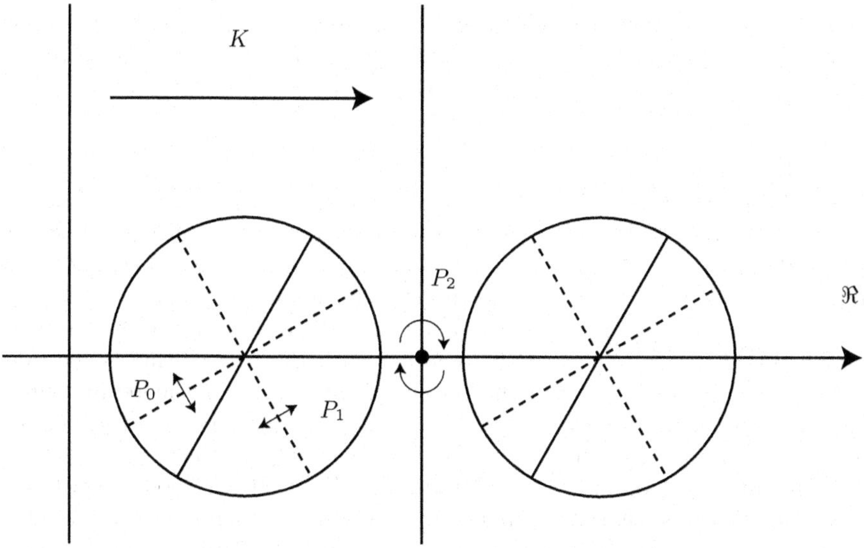

Fig. 5.7. Ford domain of a Brenner representation

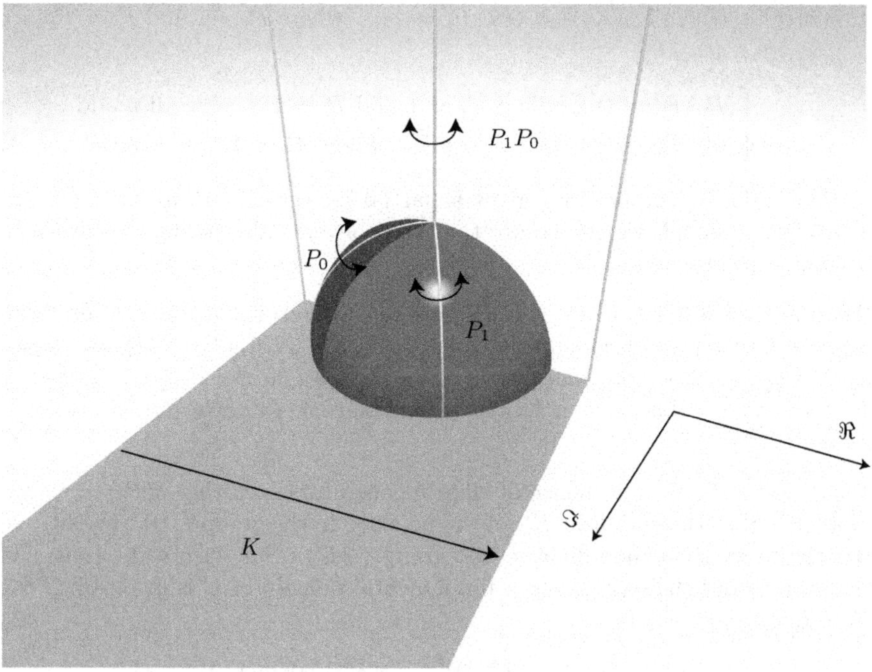

Fig. 5.8. Fundamental domain of a Brenner representation

Here $O(p,q,r)$ denotes the 2-dimensional hyperbolic orbifold with underlying space S^2 and with cone points of indices p, q, r, where a cone point of index ∞ is regarded as a puncture; in particular, $\pi_1(O(p,q,r))$ is the triangle group $\langle x,y,z \,|\, x^p = y^q = z^r = xyz = 1\rangle$ of type (p,q,r).

Proof. By Lemma 5.3.6 and the fact $r(\rho(P_{3j})) = 1/|\phi(s_0)| = 1/(2\cos(\pi/n))$, we see that the dihedral angle of the polyhedron $Eh(\rho(P_0)) \cap Eh(\rho(P_3))$ along the edge $Ih(\rho(P_0)) \cap Ih(\rho(P_3))$ is equal to $2\pi/n$. Let $f := c(\rho(P_0)) + (1/2)$ be the fixed point in $\mathbb{C}$ of the π-rotation $\rho(P_2)$ (cf. Lemma 5.3.6(2)). Let R_0 be the fundamental domain of the infinite dihedral Kleinian group $\langle \rho(K), \rho(P_0P_1)\rangle$ introduced in the proof of Proposition 5.3.8. Then we can check that the intersection $R := R_0 \cap Eh(\rho(P_0))$, together with the face pairing isometries $\rho(K)$, $\rho(P_1P_0)$, $\rho(P_0)$, satisfies the conditions of Poincare's theorem on fundamental polyhedra (see Fig. 5.10). This implies the discreteness of ρ and the desired group presentation of $\rho(\pi_1(\mathcal{O}))$. This also implies that $\cap_j Eh(\rho(P_{3j}))$ is a fundamental domain of $\rho(\pi_1(\mathcal{O}))$ modulo the stabilizer $\langle \rho(K), \rho(P_0P_1)\rangle$ of ∞. Hence we see that $\cap_j Eh(\rho(P_{3j}))$ is the Ford domain of $\rho(\pi_1(\mathcal{O}))$. The remaining group presentations are obtained through the Reidemeister-Schreier method by using the following isomorphisms.

$$\rho(\pi_1(\mathcal{O}))/\langle\langle \rho(P_1), \rho(K)\rangle\rangle \cong \langle P_1 \,|\, P_1^2 = 1\rangle, \quad \text{where } P_0 = 1 \text{ and } P_2 = P_1.$$

$$\rho(\pi_1(\mathcal{O}))/\rho(\pi_1(S)) \cong \begin{cases} \langle P_1 \,|\, P_1^2 = 1\rangle, \quad \text{where } P_0 = 1 \text{ and } P_2 = P_1 & \text{if } n\text{:odd,} \\ \langle P_0 \,|\, P_0^2 = 1\rangle \oplus \langle P_1 \,|\, P_1^2 = 1\rangle, \quad \text{where } P_2 = P_0P_1 & \text{if } n\text{:even.} \end{cases}$$

The following proposition gives a partial converse to Propositions 5.3.8 and 5.3.9. (Though we conjecture that the converse also holds, we have not been able to prove it.)

Proposition 5.3.10. *Let ρ be a discrete type-preserving representation, and suppose that the Ford domain $Ph(\rho)$ is equal to $\cap\{Ih(\rho(P)) \,|\, s(P) = s\}$ for some slope s. Then the image of ρ is isomorphic to the image of either a Brenner representation or a Knapp representation. In particular, ρ does not belong to $\overline{\mathcal{QF}}$.*

Proof. Let $\{P_j\}$ be a sequence of elliptic generators such that $s(P_0) = s$. We first show that the stabilizer Γ_∞ of the point ∞ in the group $\Gamma := \rho(\pi_1(\mathcal{O}))$ is strictly larger than the infinite cyclic group $\langle \rho(K)\rangle$. Suppose to the contrary that this is not the case. Then a fundamental domain of Γ is given by $\tilde{R} := \tilde{R}_0 \cap Eh(\rho(P_0))$, where

$$\tilde{R}_0 = \{(z,t) \in \mathbb{H}^3 \,|\, |\Re(z - c(\rho(P_0)))| \le 1/2\}$$

is a fundamental domain of $\langle \rho(K)\rangle$. Then we have the following claim.

Claim 5.3.11. The set of gluing isometries for the fundamental domain $\tilde{R}$ consists of $\rho(K)$ and $\rho(P_0)$, and Γ has one of the following presentations.

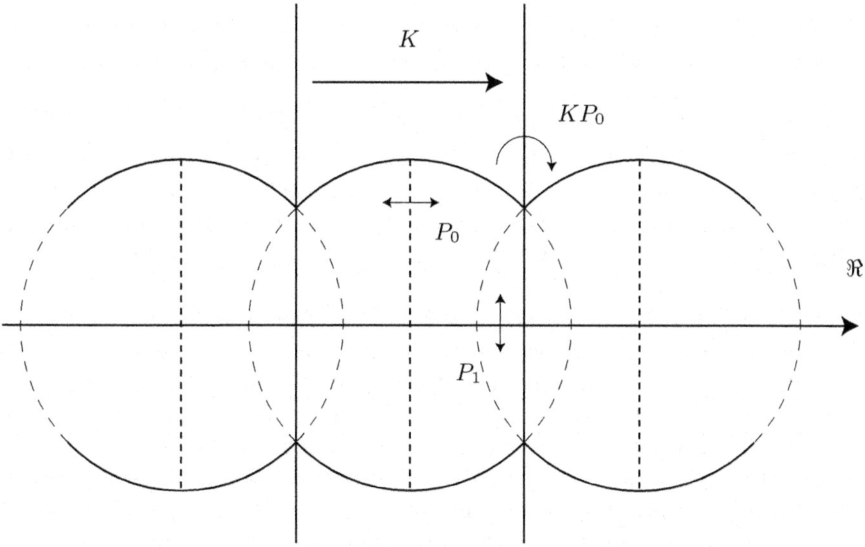

Fig. 5.9. Ford domain of a Knapp representation

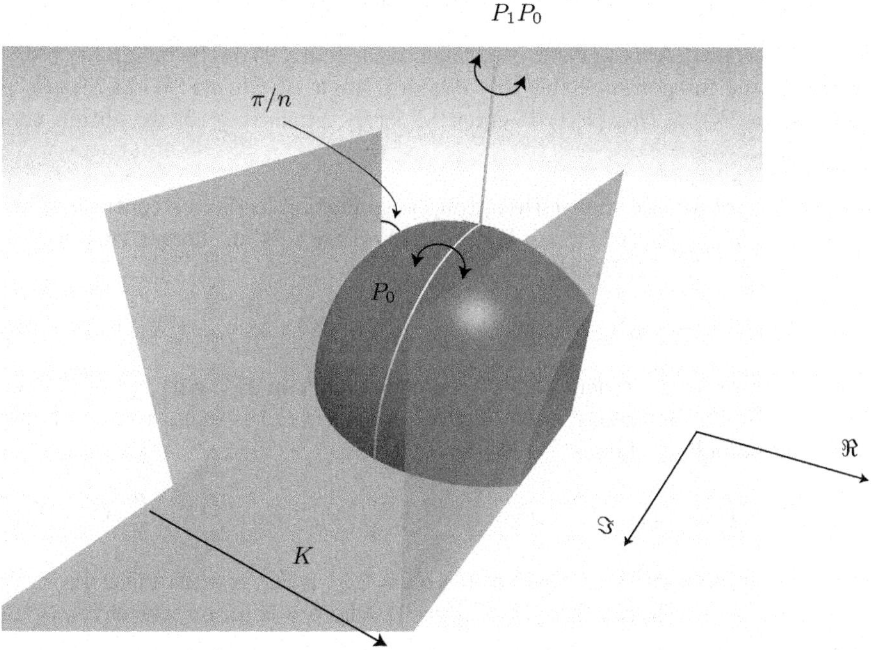

Fig. 5.10. Fundamental domain of a Knapp representation

$\langle P_0, K \mid P_0^2 = (KP_0)^n = 1\rangle$ where n is an integer with $n \geq 3$ or $n = \infty$,
$\langle P_0, K \mid P_0^2 = [K, P_0]^n = 1\rangle$ where n is an integer with $n \geq 2$ or $n = \infty$.

Proof. Suppose $Ih(\rho(P_0)) \cap Ih(\rho(P_3)) = \emptyset$. Then $\partial\tilde{R}$ is the disjoint union of $Ih(\rho(P_0))$ and two vertical planes. The face $Ih(\rho(P_0))$ is paired with itself by $\rho(P_0)$ and the vertical planes are paired with each other by $\rho(K)$. Hence we obtain the first assertion. By applying Poincare's theorem on fundamental polyhedra to this fundamental domain, we have $\Gamma \cong \langle P_0, K \mid P_0^2 = 1\rangle \cong \langle P_0, K \mid P_0^2 = (KP_0)^\infty = 1\rangle \cong \langle P_0, K \mid P_0^2 = [K, P_0]^\infty = 1\rangle$.

Suppose $Ih(\rho(P_0)) \cap Ih(\rho(P_3)) \neq \emptyset$. Then $\partial\tilde{R}$ is the union of the face $Ih(\rho(P_0)) \cap \partial\tilde{R}$ and two vertical faces. If $\mathrm{proj}(\mathrm{Axis}(\rho(P_0)))$ is not parallel to the real axis nor the imaginary axis, then $\rho(P_0)(\mathrm{int}\, Ph(\rho))$ has a non-empty intersection with $\mathrm{int}\, Ph(\rho)$, a contradiction. Hence $\mathrm{proj}(\mathrm{Axis}(\rho(P_0)))$ is parallel to the real axis or the imaginary axis, and therefore the face $Ih(\rho(P_0)) \cap \partial\tilde{R}$ is paired to itself by $\rho(P_0)$. Moreover the vertical faces are paired with each other by $\rho(K)$. Hence we obtain the first assertion. By applying Poincare's theorem on fundamental polyhedra to this fundamental domain, we see that Γ is isomorphic to

$$\langle P_0, K \mid P_0^2 = (KP_0)^n = 1\rangle \quad \text{or} \quad \langle P_0, K \mid P_0^2 = [K, P_0]^{n/2} = 1\rangle,$$

according as $\mathrm{proj}(\mathrm{Axis}(\rho(P_0)))$ is parallel to the real axis or the imaginary axis. Here n is the integer such that the dihedral angle of $Eh(\rho(P_0)) \cap Eh(\rho(P_3))$ along $Ih(\rho(P_0)) \cap Ih(\rho(P_3))$ is equal to $2\pi/n$. Since $n \geq 3$, we obtain the conclusion.

In what follows we show either group presentation leads to a contradiction.

Case 1. $\Gamma \cong \langle P_0, K \mid P_0^2 = (KP_0)^n = 1\rangle$ where n is an integer with $n \geq 3$ or $n = \infty$. Set $X = KP_0 \in \Gamma$. Then

$$\Gamma \cong \langle P_0 \mid P_0^2\rangle * \langle X \mid X^n = 1\rangle \text{ and } H_1(\Gamma) \cong \langle [P_0] \mid 2[P_0] = 0\rangle \oplus \langle [X] \mid n[X] = 0\rangle.$$

Hence each order 2 element of Γ is conjugate to P_0 or $K^{\frac{n}{2}}$ with n even. This implies that the homology class of $\rho(K) = \rho(P_2P_1P_0)$ is equal to one of the following homology classes.

$$[P_0]+[P_0]+[P_0] = [P_0],\ [P_0]+[P_0]+\frac{n}{2}[X] = \frac{n}{2}([X]),\ [P_0]+\frac{n}{2}[X]+\frac{n}{2}[X] = [P_0].$$

Since $n \geq 3$, none of them is equal to $[K] = [P_0] + [X]$, a contradiction.

Case 2. $\Gamma \cong \langle P_0, K \mid P_0^2 = [K, P_0]^n = 1\rangle$ where n is an integer with $n \geq 2$ or $n = \infty$. Then

$$H_1(\Gamma) \cong \langle [P_0] \mid 2[P_0] = 0\rangle \oplus \langle [K]\rangle.$$

Thus the homology classes of $\rho(P_j)$ is either $[P_0]$ or 0 for every integer j. In particular the homology class of $\rho(K) = \rho(P_2P_1P_0)$ is a (possibly trivial) torsion element. This contradicts the fact that $[K]$ is of infinite order.

Thus we have proved that Γ_∞ is strictly larger than $\langle \rho(K) \rangle$.

Next we show that Γ_∞ is an infinite dihedral group. To this end, let T be an element of Γ_∞ which does not belong to $\langle \rho(K) \rangle$. After composing a power of $\rho(K)$ we may assume T preserves $Ih(\rho(P_0))$. Since T is an Euclidean isometry, this implies that T is a rotation about the vertical geodesic above $c(\rho(P_0))$. Since T preserves $Ph(\rho) = \cap_j Eh(\rho(P_{3j}))$, the angle of rotation of T must be equal to π. Thus Γ_∞ is the infinite dihedral group $\langle \rho(K), T \rangle$.

Let R_0 and R be as in the proofs of Propositions 5.3.8 and 5.3.9. Then R is a fundamental domain of Γ and we can see that the set of gluing isometries consist of T, $\rho(K)$, $\rho(P_0)$ and $\rho(P_0)T$. (In the special case when the proj(Axis($\rho(P_0)$)) or proj(Axis($\rho(P_0)T$)) is parallel to the real line, $\rho(P_0)$ or T accordingly does not give a face pairing.) By applying Poincare's theorem on fundamental polyhedra, we see

$$\Gamma \cong \langle P_0, K, T \,|\, P_0^2 = T^2 = (P_0T)^2 = (TK)^2 = (KP_0)^n = 1 \rangle.$$

Set $Q = TP_0$ and $R = KP_0Q$. Then we see

$$\Gamma \cong \langle P_0, Q, R \,|\, P_0^2 = Q^2 = R^2 = 1, (P_0Q)^2 = 1, (QR)^n = 1 \rangle.$$

This completes the proof.

At the end of this section, we prove two lemmas for the Markoff maps which vanish at ∞. They are analogies of [44, Proposition 3.1] for the Markoff maps which takes the value 2 at ∞, and it is used in Sect. 9.1.

Lemma 5.3.12. *Let Φ_0 be the space of Markoff maps, including the trivial one, such that*

$$(\phi(1/0), \phi(0/1), \phi(1/1)) = (0, x, ix)$$

for some $x \in \mathbb{C}$, and identify Φ_0 with the complex plane by the correspondence $\phi \mapsto x = \phi(0/1)$. For each $r \in \hat{\mathbb{Q}}$, let $V_r : \Phi_0 \to \mathbb{C}$ be the function defined by $V_r(\phi) = \phi(r)$. Then V_r is a polynomial in the variable x, which we denote by $V[r]$, and satisfies the following conditions.

1. *$V[r+1](x) = V[r](ix)$ and $V[-r](x) = \overline{V[r]}(x)$. Here $\overline{V[r]}(x)$ denotes the polynomial obtained from $V[r](x)$ by converting each coefficient into its complex conjugate.*
2. *For each rational number $q/p \in [0, 1/2]$,*

$$V[q/p](x) = i^q(x^p - c_{q/p}x^{p-2} + (\text{lower terms})),$$

for some integer $c_{q/p}$. Here $c_{q/p} = 0$ if $q/p = 0/1$ or $1/2$. If $q/p \in (0, 1/2)$, then $c_{q/p} > 0$.

Proof. (1) is proved by induction on the depth of r, i.e., the edge path distance between ∞ and r in $\mathcal{D}$, by using the identity in Proposition 2.3.4(2).

(2) Since $V[0/1] = ix$ and $V[1/2] = ix^2$, the assertion is valid for $0/1$ and $1/0$. We show that the assertion is valid for $(q_1 + q_2)/(p_1 + p_2) \in (0, 1/2)$ by

assuming that it is valid for q_1/p_1, q_2/p_2 and $(q_2-q_1)/(p_2-p_1)$, where p_1, p_2, q_1 and q_2 are integers such that $1 \leq p_1 \leq p_2$, $0 \leq q_1 \leq q_2$ and $\begin{vmatrix} q_1 & q_1 \\ p_1 & p_2 \end{vmatrix} = \pm 1$. By Proposition 2.3.4(2),

$$\begin{aligned}
&V[(q_1+q_2)/(p_1+p_2)] \\
&= V[q_1/p_1]V[q_2/p_2] - V[(q_2-q_1)/(p_2-p_1)] \\
&= i^{q_1}(x^{p_1} - c_{q_1/p_1}x^{p_1-2} + (\text{lower terms})) \\
&\qquad\qquad \times i^{q_2}(x^{p_2} - c_{q_2/p_2}x^{p_2-2} + (\text{lower terms})) \\
&\qquad - i^{q_2-q_1}(x^{p_2-p_1} + (\text{lower terms})) \\
&= i^{q_1+q_2}(x^{p_1+p_2} - (c_{q_1/p_1} + c_{q_2/p_2})x^{p_1+p_2-2} \\
&\qquad - (-1)^{q_1}x^{p_2-p_1} + (\text{lower terms})).
\end{aligned}$$

If $p_1 \geq 2$, then $p_1 + p_2 - 2 > p_2 - p_1$ and hence the assertion holds, and we have $c_{(q_1+q_2)/(p_1+p_2)} = c_{q_1/p_1} + c_{q_2/p_2}$.

If $p_1 = 1$, then $q_1 = 0$ and $p_1 + p_2 - 2 = p_2 - p_1$. Thus the coefficient of $x^{p_1+p_1-2}$ is equal to $-i^{q_1+q_2}(c_{q_1/p_1} + c_{q_2/p_2} + 1) = -i^{q_1+q_2}(c_{q_2/p_2} + 1)$. Hence the assertion also holds, and we have $c_{(q_1+q_2)/(p_1+p_2)} = c_{q_1/p_1} + c_{q_2/p_2} + 1 = c_{q_2/p_2} + 1$.

Remark 5.3.13. It is easy to see that there is an integral polynomial $F[q/p]$ such that $V[q/p](x)$ is equal to $i^q xF[q/p](x^2)$ or $i^q x^2 F[q/p](x^2)$ according as p is odd or even. This polynomial is essentially equal to the polynomial f introduced in [71, Sect. II.5], which gives the hyperbolicity equation for the complement of the 2-bridge link $K(q/p)$.

Lemma 5.3.14. *For each integer $n \geq 2$, we have*

$$V[n/(2n+1)](x) = i^n(x^{2n+1} - x^{2n-1} - (n-1)x^{2n-3} + (\text{lower terms})),$$

where $V[n/(2n+1)](x)$ is as in Lemma 5.3.12.

Proof. This is proved by induction on n. As noted in Remark 5.3.13, the corresponding polynomial $F[n/(2n+1)]$ gives the hyperbolicity equation for the twist knot $K(n/(2n+1))$ and detailed study of the polynomial has been made in [37] and [36].

5.4 Imaginary representations

Definition 5.4.1. *(Imaginary representation) A type-preserving representation ρ is called an* imaginary representation *if it is induced by an imaginary Markoff map ϕ, i.e., a non-real Markoff map whose image is contained in $\mathbb{R} \cup i\mathbb{R}$ (see [17, Sect. 5]).*

By using the identities 2.4 and 2.5, we can easily obtain the following characterization of the imaginary representations.

Lemma 5.4.2. *Let ρ be a type-preserving representation of $\pi_1(\mathcal{O})$ and ψ the complex probability map corresponding to ρ. Then the following conditions are equivalent.*

1. *ρ is an imaginary representation.*
2. *The values of ϕ at the vertices of some triangle of $\mathcal{D}$ are contained in $\mathbb{R} \cup i\mathbb{R}$, but not in $\mathbb{R}$.*
3. *The value (a_0, a_1, a_2) of ψ at some triangle of $\mathcal{D}_\phi$ consists of real numbers, such that at least one of which is negative.*
4. *The value (a_0, a_1, a_2) of ψ at any triangle of $\mathcal{D}_\phi$ consists of real numbers, such that at least one of which is negative.*

In particular, if ρ is doubly folded at some triangle, then ρ is an imaginary representation.

By using a result of Bowditch [17], we have the following proposition.

Proposition 5.4.3. *Let ρ be an imaginary representation. Then ρ is discrete if and only if it is either a Brenner representation or a Knapp representation. In particular, no representation in $\overline{\mathcal{QF}}$ is imaginary.*

Proof. By [17, Proposition 5.1], an imaginary representation ρ can be discrete only when the corresponding Markoff map ϕ vanish at some vertex, say s. Let s' be a vertex adjacent to s and put $x = \phi(s')$ and $\omega = -x^2$. Then the two-parabolic group G_ω is a finite index subgroup of $\rho(\pi_1(\mathcal{O}))$ by Lemma 5.3.2, and therefore ρ is discrete if and only if G_ω is discrete. On the other hand, ω is real, because ρ is imaginary and x belongs to $\mathbb{R} \cup i\mathbb{R}$. Hence G_ω is discrete if and only ω or $-\omega$ belongs to

$$[4, \infty) \cup \{4\cos^2(\pi/n) \mid n \in \mathbb{N}\}.$$

by Proposition 5.3.3(2). This is equivalent to the condition that ρ is either a Brenner representation or a Knapp representation of slope s. The last assertion follows from the fact that Brenner representations and Knapp representations are not faithful.

5.5 Representations with accidental parabolic/elliptic transformations

In this section we study those representations which send a generator of $\pi_1(T)$ to a parabolic or elliptic transformation. The results in this section is used in Sects. 6.7 and 8.7.

Lemma 5.5.1. *Under Assumption 2.4.6 (NonZero), suppose that $\mathcal{L}(\rho,\sigma)$ is not folded at $c(\rho(P_1))$ nor $c(\rho(P_2))$. Then the following conditions are equivalent:*

(1) $\rho(KP_0)$ *is an elliptic transformation.*
(2) $\mathrm{Axis}(\rho(P_1)) \cap \mathrm{Axis}(\rho(P_2)) \neq \emptyset$.
(3) $\cap_{j=0}^{3} Ih(\rho(P_j)) \neq \emptyset$.

Moreover, if the above mutually equivalent conditions are satisfied, then $\cap_{j=0}^{3} Ih(\rho(P_j)) = \mathrm{Axis}(\rho(P_1)) \cap \mathrm{Axis}(\rho(P_2))$ *and it is a singleton contained in the axis of the elliptic transformation $\rho(KP_0)$ (see Fig. 5.11).*

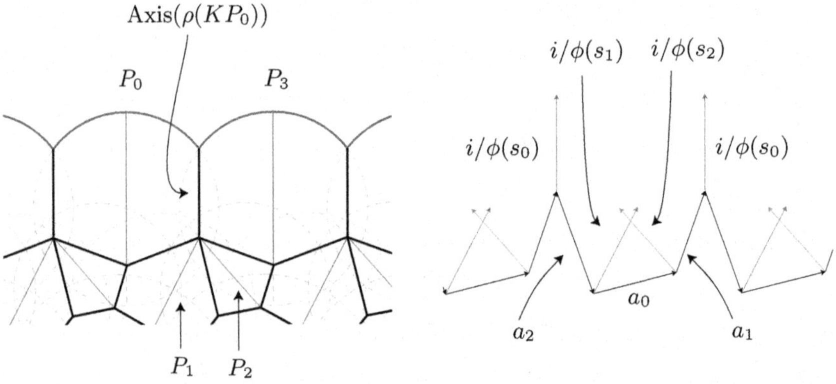

Fig. 5.11. Frontier of elliptic type: this figure is for $\epsilon = +$.

Proof. (1)⇒(2) This follows from the general fact that $\rho(KP_0) = \rho(P_2P_1)$ is elliptic, parabolic or loxodromic according as $\mathrm{Axis}(\rho(P_1))$ and $\mathrm{Axis}(\rho(P_2))$ share a common point, share a single common endpoint, or $\overline{\mathrm{Axis}(\rho(P_1))}$ and $\overline{\mathrm{Axis}(\rho(P_2))}$ are disjoint.

(2)⇒(3) Suppose the axes of $\rho(P_1)$ and $\rho(P_2)$ have a common point, say x. Since $\mathrm{Axis}(\rho(P_j)) \subset Ih(\rho(P_j))$, we see $x \in Ih(\rho(P_1)) \cap Ih(\rho(P_2))$. By using Lemma 4.1.3(2) (chain rule) as in the above, we see that $x \in Ih(\rho(P_0)) \cap Ih(\rho(P_1))$ and $x \in Ih(\rho(P_2)) \cap Ih(\rho(P_3))$. Hence $\cap_{j=0}^{3} Ih(\rho(P_j))$ contains x, and therefore it is non-empty.

(3)⇒(1). Suppose that $\cap_{j=0}^{3} Ih(\rho(P_j)) \neq \emptyset$. Then both $\cap_{j=0}^{2} Ih(\rho(P_j))$ and $\cap_{j=1}^{3} Ih(\rho(P_j))$ are non-empty. Since $\rho(P_1)$ interchanges $Ih(\rho(P_0)) \cap Ih(\rho(P_1))$ and $Ih(\rho(P_1)) \cap Ih(\rho(P_2))$ by Lemma 4.1.3(2) (Chain rule), $\cap_{j=0}^{2} Ih(\rho(P_j))$ is preserved by $\rho(P_1)$. Similarly $\cap_{j=1}^{3} Ih(\rho(P_j))$ is preserved by $\rho(P_2)$. Thus if both of them are singletons, then they are fixed by $\rho(P_1)$ and $\rho(P_2)$, respectively. Moreover both of them must be equal to the non-empty set $\cap_{j=0}^{3} Ih(\rho(P_j))$. Hence $\rho(KP_0) = \rho(P_2P_1)$ fixes the singleton $\cap_{j=0}^{3} Ih(\rho(P_j))$ in $\mathbb{H}^3$, and hence $\rho(KP_0)$ is elliptic. Suppose that either $\cap_{j=0}^{2} Ih(\rho(P_j))$ or

$\cap_{j=1}^3 Ih(\rho(P_j))$ is not a singleton. For simplicity, we assume that $\cap_{j=0}^2 Ih(\rho(P_j))$ is not a singleton. (The other case can be treated similarly.) Then it must be a complete geodesic. Thus by Lemma 5.2.6 and the assumption that $\mathcal{L}(\rho,\sigma)$ is not folded at $c(\rho(P_1))$, (ρ,τ) is an isosceles representation, where $\tau = \langle s_0, s_2 \rangle$. So $\cap_{j=0}^2 Ih(\rho(P_j)) = \mathrm{Axis}(\rho(P_1))$ by Proposition 5.2.3. Moreover $Ih(\rho(P_2)) \cap Ih(\rho(P_3)) = Ih(\rho(P_1')) \cap Ih(\rho(P_3')) = \mathrm{Axis}(\rho(P_2'))$ by Proposition 5.2.3, and therefore it is a geodesic, where P_j' are as in the proposition. Hence

$$\begin{aligned}\cap_{j=1}^3 Ih(\rho(P_j)) &= (Ih(\rho(P_1)) \cap Ih(\rho(P_2))) \cap (Ih(\rho(P_2)) \cap Ih(\rho(P_3)))\\ &= \mathrm{Axis}(\rho(P_1)) \cap \mathrm{Axis}(\rho(P_2')).\end{aligned}$$

Since (ρ,τ) is an isosceles representation, this implies that $\cap_{j=1}^3 Ih(\rho(P_j))$ is a singleton. Thus it is fixed by $\rho(P_2)$ by the previous argument. Since it is also fixed by $\rho(P_1)$, because it is contained in $\mathrm{Axis}(\rho(P_1))$. Hence $\rho(KP_0) = \rho(P_2P_1)$ fixes the singleton, and hence $\rho(KP_0)$ is elliptic.

Thus we have proved that the three conditions are equivalent. The remaining assertion is already proved in the above argument.

Remark 5.5.2. In the above lemma, we cannot drop the condition that $\mathcal{L}(\rho,\sigma)$ is not folded at any vertex. In fact we can easily find a representation ρ such that $\mathcal{L}(\rho,\sigma)$ is folded at $c(\rho(P_1))$, $\cap_{j=0}^3 Ih(\rho(P_j)) \neq \emptyset$ but that $\rho(KP_0)$ is not elliptic. There exists also a representation ρ such that $\mathcal{L}(\rho,\sigma)$ is doubly folded, $\rho(KP_0)$ is elliptic and that $\cap_{j=0}^3 Ih(\rho(P_j))$ is a complete geodesic.

By a similar argument, we can also prove the following lemma:

Lemma 5.5.3. *Under Assumption 2.4.6 (NonZero), suppose that $\mathcal{L}(\rho,\sigma)$ is not folded at $c(\rho(P_1))$ nor $c(\rho(P_2))$. Then the following conditions are equivalent:*

(1) $\rho(KP_0)$ is a parabolic transformation.
(2) $\overline{\mathrm{Axis}}(\rho(P_1))$ and $\overline{\mathrm{Axis}}(\rho(P_2))$ share a unique common endpoint.
(3) The four isometric circles $I(\rho(P_j))$ $(0 \le j \le 4)$ have a unique common point.

Moreover, if the above mutually equivalent conditions are satisfied, then the parabolic fixed point of $\rho(KP_0)$ is equal to $c(\rho(P_0)) + (1/2)$. Furthermore, this point is equal to $\cap_{j=0}^3 I(\rho(P_j)) = \overline{\mathrm{Axis}}(\rho(P_1)) \cap \overline{\mathrm{Axis}}(\rho(P_2))$ (see Fig. 5.12).

Proof. All assertions, except the location of the parabolic fixed point, are proved by arguments parallel to the proof of Lemma 5.5.1. To prove the remaining assertion, suppose $\rho(KP_0)$ is parabolic. Then $\phi(s_0) = \pm 2$ and $\rho(P_0)$ is the π-rotation around the geodesic with endpoints $c(\rho(P_0)) \pm (i/2)$ by Proposition 2.4.4(1.1). Hence

$$\rho(KP_0)(c(\rho(P_0)) + (1/2)) = \rho(K)(c(\rho(P_0)) - (1/2)) = c(\rho(P_0)) + (1/2).$$

Hence the parabolic fixed point of $\rho(KP_0)$ is equal to $c(\rho(P_0)) + (1/2)$.

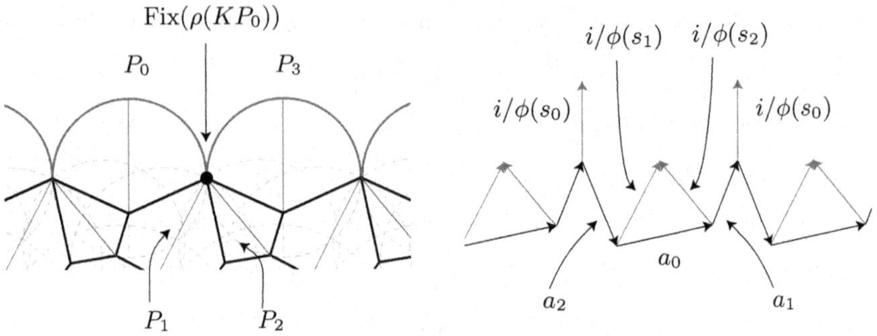

Fig. 5.12. Frontier of parabolic type: this figure is for $\epsilon = +$.

In the remainder of this section, we present further properties of representations which send KP_0 to a parabolic transformation.

Lemma 5.5.4. *Under Assumption 4.2.4 (σ-Simple), assume that $\rho(KP_0)$ is parabolic. Then the following hold (see Fig. 5.12).*

(1) $\arg a_1 = -\arg a_2 \in (-\pi/2, 0) \cup (0, \pi/2)$, *where (a_0, a_1, a_2) is the complex probability of ρ at σ.*

(2) For the parabolic fixed point, $\mathrm{Fix}(\rho(KP_0))$, *of $\rho(KP_0)$, the following hold.*

$$\begin{aligned}\mathrm{Fix}(\rho(KP_0)) &= c(\rho(P_0)) + (1/2) \\ &= \mathrm{Fix}^+_\sigma(\rho(P_1)) = \mathrm{Fix}^+_\sigma(\rho(P_2)) \\ &= v^+(\rho; P_0, P_1) = v^+(\rho; P_1, P_2) = v^+(\rho; P_2, P_3),\end{aligned}$$

where $\epsilon = -$ or $+$ according as $\arg a_1$ *is contained in $(-\pi/2, 0)$ or $(0, \pi/2)$.*

(3) $\theta^\epsilon(\rho, \sigma) = (\pi/2, 0, 0)$, where ϵ is as in the above.

Conversely, if $\theta^\epsilon(\rho, \sigma) = (\pi/2, 0, 0)$ for some $\epsilon \in \{-, +\}$, then $\rho(KP_0)$ is parabolic.

Proof. Under Assumption 4.2.4 (σ-Simple), suppose $\rho(KP_0)$ is parabolic. Then $a_1 a_2 = 1/\phi(s_0)^2 = 1/4$, we have $\arg a_1 \equiv -\arg a_2 \pmod{2\pi}$. We can easily see that if this argument belongs to either $[-\pi, -\pi/2]$ or $[\pi/2, \pi]$ then $\mathcal{L}(\rho, \sigma)$ is not simple. Hence the argument belongs to $(-\pi/2, \pi/2)$. We show that the argument is not equal to 0. Suppose this is not the case. Then a_1, a_2 and $a_0 = 1 - a_1 - a_2$ are non-zero real numbers. Since ρ is not an imaginary representation, a_0, a_1 and a_2 are positive real numbers by Lemma 5.4.2. Hence $a_1 + a_2 \geq 2\sqrt{a_1 a_2} = 2/|\phi(s_0)| = 1$, and hence $a_0 = 1 - a_1 - a_2 \leq 0$. Since $a_0 \neq 0$, we have $a_0 < 0$, a contradiction. Hence we obtain the first assertion, $\arg a_1 = -\arg a_2 \in (-\pi/2, 0) \cup (0, \pi/2)$.

To prove the second and third assertion, set $f = c(\rho(P_0)) + (1/2) = c(\rho(P_3)) - (1/2)$. Then Lemma 5.5.3 says that f is equal to the parabolic fixed point of $\rho(KP_0)$ and that it is the unique common fixed point of $\rho(P_1)$

and $\rho(P_2)$. Suppose first that $\arg a_1 = -\arg a_2 \in (0, \pi/2)$. Then $\Im(c(\rho(P_1)))$ and $\Im(c(\rho(P_2)))$ are smaller that $\Im(f)$, and hence

$$\arg \frac{c(\rho(P_0)) - c(\rho(P_1))}{f - c(\rho(P_1))} \in (0, \pi) \quad \text{and} \quad \arg \frac{f - c(\rho(P_2))}{c(\rho(P_3)) - c(\rho(P_2))} \in (0, \pi).$$

Thus we have $f = \mathrm{Fix}^+_\sigma(\rho(P_1)) = \mathrm{Fix}^+_\sigma(\rho(P_2))$ by definition. This implies that the point f is also equal to $v^+(\rho; P_1, P_2)$. By Lemma 4.2.7, this in turn implies $v^+(\rho; P_0, P_1) = v^+(\rho; P_1, P_2) = v^+(\rho; P_2, P_3)$. Thus we have the second assertion. This also implies $\theta^+(\rho, \sigma; s_1) = \theta^+(\rho, \sigma; s_2) = 0$, and therefore we obtain the third assertion $\theta^+(\rho, \sigma) = (\pi/2, 0, 0)$. By a parallel argument we can also show that if $\arg a_1 = -\arg a_2 \in (-\pi/2, 0)$ then $\theta^-(\rho, \sigma) = (\pi/2, 0, 0)$. Thus we obtain the assertion (3).

Conversely, suppose $\theta^\epsilon(\rho, \sigma) = (\pi/2, 0, 0)$ for some $\epsilon \in \{-, +\}$. Then we have

$$\mathrm{Fix}^\epsilon_\sigma(\rho(P_1)) = v^\epsilon(\rho; P_1, P_2) = \mathrm{Fix}^\epsilon_\sigma(\rho(P_2)).$$

Hence $\rho(KP_0) = \rho(P_2P_1)$ is parabolic (cf. Lemma 5.5.1). This completes the proof of Lemma 5.5.4.

We occasionally need to study the sequence of elliptic generators associated with a chain all of whose triangles share a common vertex, which is explained in the following:

Notation 5.5.5 (InfiniteFan). For the sequence of elliptic generators $\{P_j\}$ associated with a triangle $\sigma = \langle s_0, s_1, s_2 \rangle$ of $\mathcal{D}$, the symbols $\{P_j^{(k)}\}_j$ $(k \in \mathbb{Z})$ denote the sequences of elliptic generators obtained from $\{P_j\}$ by the following recursive formula:

$$(P_0^{(0)}, P_1^{(0)}, P_2^{(0)}) = (P_0, P_1, P_2)$$
$$(P_0^{(k+1)}, P_1^{(k+1)}, P_2^{(k+1)}) = (P_0, P_2^{(k)}, P_2^{(k)} P_1^{(k)} P_2^{(k)})$$

We set $\sigma^{(k)} = \langle s(P_0^{(k)}), s(P_1^{(k)}), s(P_2^{(k)}) \rangle$. It should be noted that $(\cdots, \sigma^{(-1)}, \sigma^{(0)}, \sigma^{(1)}, \cdots)$ forms an "infinite fan", or an "bi-infinite chain" sharing s_0 as the common vertex (see Fig. 5.13).

Lemma 5.5.6. *Let ρ be a type-preserving representation such that $\rho(KP_0)$ is parabolic, and let ϕ be a Markoff map inducing ρ. Then the following hold under Notation 5.5.5 (InfiniteFan).*

(1) If $\phi(s(P_1^{(k)})) = 0$ for some k, then ρ is an imaginary representation.

(2) Suppose ρ is not an imaginary representation. Let $(a_0^{(k)}, a_1^{(k)}, a_2^{(k)})$ be the complex probability of ρ at $\sigma^{(k)}$. Then the following hold:

(i) $\arg a_1^{(k)} = -\arg a_2^{(k)} \in (-\pi, 0) \cup (0, \pi)$ for every k.

(ii) The sign of $\arg a_1^{(k)} = -\arg a_2^{(k)}$ does not depend on k. In other words, the sign of $\Im a_j^{(k)}$ does not depend on k for each $j = 1, 2$ (see Fig. 5.13).

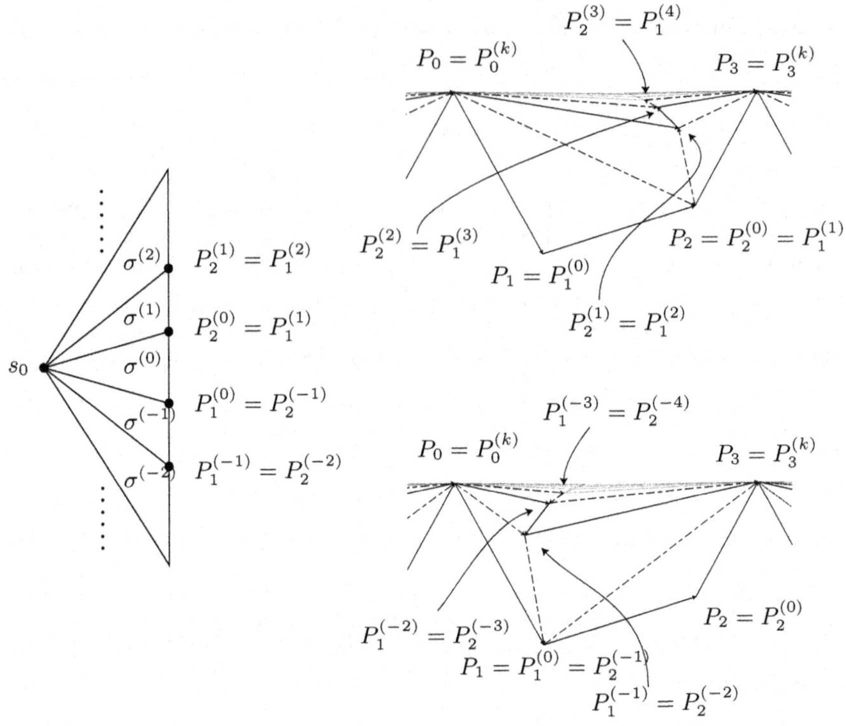

Fig. 5.13. Infinite Fan

Proof. (1) Suppose $\phi(s(P_1^{(k)})) = 0$ for some k. Then

$$\phi(s(P_1^{(k+1)})) = \phi(s(P_2^{(k)})) = \pm i\phi(s(P_0^{(k)})) = \pm i\phi(s_0) \in i\mathbb{R}.$$

Hence ρ is an imaginary representation by Lemma 5.4.2.

(2) We note that the complex probability of ρ at $\sigma^{(k)}$ is well-defined by the first assertion. We see by the proof of Lemma 5.5.4(1) that $\arg a_1^{(k)} = -\arg a_2^{(k)}$. Since ρ is not an imaginary representation, $a_1^{(k)}$ and $a_2^{(k)}$ are not real. Hence $\arg a_1^{(k)} = -\arg a_2^{(k)} \in (-\pi, \pi) - \{0\}$. In particular, $\Im a_1^{(k)}$ and $\Im a_2^{(k)}$ have different signs. On the other hand, since $a_2^{(k+1)} = 1 - a_1^{(k)}$, we have $\Im a_2^{(k+1)} = -\Im a_1^{(k)}$. Thus $\Im a_1^{(k)}$ and $\Im a_1^{(k+1)}$ have the same sign. Thus the signs of $\Im a_1^{(k)}$ and $\Im a_2^{(k)}$ do not depend on k. This completes the proof.

Remark 5.5.7. The above lemma holds under the weaker assumption that $\rho(KP_0)$ is parabolic or elliptic with rotation angle $\neq \pi$. Actually the proof is valid under this weaker assumption with a slight modification.

6

Reformulation of Main Theorem 1.3.5 and outline of the proof

In this chapter, we give a "2-dimensional" reformulation of the Main Theorem 1.3.5 and present a route map of its proof.

In Sect. 6.1, we introduce the definition of "quasifuchsian" labeled representations and "good" labeled representations (Definition 6.1.7). A labeled representation is defined to be quasifuchsian if (i) ρ is quasifuchsian, (ii) the Ford domain $Ph(\rho)$ is as described by Jorgensen, and (iii) $\boldsymbol{\nu}$ is equal to the side parameter of ρ (Definition 6.1.1), whereas a labeled representation is defined to be *good* if it satisfies the three conditions, Nonzero, Frontier and Duality (Definition 6.1.7). Theorem 6.1.7 (Good implies quasifuchsian) claims that if a labeled representation is good, then it is quasifuchsian. Though the condition for a labeled representation to be good is rather complicated, it is not difficult to check if a given labeled representation satisfy the condition. By introducing the space $\mathcal{J}[\mathcal{QF}] \subset \mathcal{QF} \times (\mathbb{H}^2 \times \mathbb{H}^2)$ of all good labeled representations, Jorgensen's result is replaced with the assertion that the projections $\boldsymbol{\mu}_1 : \mathcal{J}[\mathcal{QF}] \to \mathcal{QF}$ and $\boldsymbol{\mu}_2 : \mathcal{J}[\mathcal{QF}] \to \mathbb{H}^2 \times \mathbb{H}^2$ are homeomorphisms (Modified Main Theorem 6.1.11).

In Sect. 6.2, we give a route map for the proof of Modified Main Theorem 6.1.11.

In Sects. 6.3, 6.4 and 6.5, we prove Theorem 6.1.7 (Good implies quasifuchsian) by using Poincare's theorem on fundamental polyhedra. Here we actually need a variation of Poincare's theorem. Since its rigorous proof is rather complicated as is the original Poincare's theorem [28] and since we also need its generalization when we treat hyperbolic cone-manifolds in the forthcoming paper [11], we present only a sketch of the proof. A detailed proof will be included in [11]. However we believe that readers have no fear concerning the validity of the variation of Poincare's theorem. (Such a variation is used in [64] without even mentioning that it is different from the original one.)

In Sect. 6.6, we prove Theorem 6.1.12 which describe the structure of the ideal simplicial complex $\Delta_{\mathbb{E}}(\rho)$ dual to the Ford domain. As a corollary we prove Proposition 6.6.1. To prove the theorem, we present a general recipe for understanding the structure of $\Delta_{\mathbb{E}}(\Gamma)$ from the Ford domain $Ph(\Gamma)$ (Lemma

6.6.2). A variation of the argument for the proof of Lemma 6.6.2 leads to an important Lemma 7.1.3.

In Sect. 6.7, we give a characterization of the chain $\Sigma(\boldsymbol{\nu})$ for a good labeled representation (Propositions 6.7.1 and 6.7.2). They enable us to reconstruct the chain $\Sigma(\boldsymbol{\nu})$ of a good (or quasifuchsian) labeled representation $(\rho, \boldsymbol{\nu})$ from a single triangle in the chain. This fact provides a key step for checking if a given representation is quasifuchsian, and is used in the Bers' slice project in [50].

6.1 Reformulation of Main Theorem 1.3.5

Definition 6.1.1 (Quasifuchsian labeled representation). *A labeled representation $\boldsymbol{\rho} = (\rho, \boldsymbol{\nu})$ is said to be* quasifuchsian *if $\rho \in \mathcal{QF}$ and the weighted spine* Ford$(\rho(\pi_1(T)))$ *of $\bar{M}(\rho(\pi_1(T)))$ is equivalent to* Spine$(\boldsymbol{\nu})$ *(see Definition 1.3.4).*

The first task for the proof of Main Theorem 1.3.5 is to give a characterization of the quasifuchsian labeled representations in terms of the isometric hemispheres of the images of elliptic generators. For this purpose, we recall Definition 3.3.4 and introduce two more definitions.

Definition 6.1.2 (NonZero). *A labeled representation $\boldsymbol{\rho} = (\rho, \boldsymbol{\nu})$ is said to satisfy the condition* NonZero *if $\phi_\rho^{-1}(0) \cap \Sigma(\boldsymbol{\nu})^{(0)} = \emptyset$.*

Definition 6.1.3 (Duality). *Let $\boldsymbol{\rho} = (\rho, \boldsymbol{\nu})$ be a labeled representation satisfying the condition NonZero. Then we say that $\boldsymbol{\rho}$ satisfies the condition* Duality, *if the following conditions are satisfied.*

1. *For each simplex ξ of $\mathcal{L}(\boldsymbol{\nu})$, $F_{\boldsymbol{\rho}}(\xi)$ is contained in $Eh(\boldsymbol{\rho})$, and it is a convex polyhedron of dimension $2 - \dim \xi$ (see Definitions 3.4.3 and 3.4.6).*
2. *If P and P' are mutually distinct vertices of $\mathcal{L}(\boldsymbol{\nu})$, then $Ih(\rho(P)) \neq Ih(\rho(P'))$.*

We shall show that if $\boldsymbol{\rho}$ satisfies the conditions NonZero and Duality, then the collection $\{F_{\boldsymbol{\rho}}(\xi) \,|\, \xi \in \mathcal{L}(\boldsymbol{\nu})^{(\le 2)}\}$ gives the cellular structure of $\partial Eh(\boldsymbol{\rho})$, which is dual to $\mathcal{L}(\boldsymbol{\nu})$ (see Proposition 6.3.1).

Definition 6.1.4 (Frontier). *Let $\boldsymbol{\rho} = (\rho, \boldsymbol{\nu})$ with $\boldsymbol{\nu} = (\nu^-, \nu^+)$ be a labeled representation satisfying the condition NonZero.*

*(1) We say that $\boldsymbol{\rho}$ satisfies the condition ϵ-*Frontier *if the following conditions are satisfied.*

1. *$\sigma^\epsilon(\boldsymbol{\nu})$ (see Definition 3.3.3) is an ϵ-terminal triangle of ρ (see Definition 4.3.8).*
2. *$\nu^\epsilon = \theta^\epsilon(\rho, \sigma^\epsilon(\boldsymbol{\nu}))$ (see Definition 4.2.17).*
3. *$E^\epsilon(\boldsymbol{\rho}) = E^\epsilon(\rho, \sigma^\epsilon(\boldsymbol{\nu}))$ (see Definition 3.4.3).*

(2) We say that $\boldsymbol{\rho} = (\rho, \boldsymbol{\nu})$ *satisfies the condition* Frontier *if it satisfies* ϵ*-Frontier for each* $\epsilon = \pm$.

Remark 6.1.5. If $\boldsymbol{\nu}$ is thin, then $\sigma^\epsilon(\boldsymbol{\nu})$ is not uniquely determined by $\boldsymbol{\nu}$. However, by virtue of Propositions 5.2.3, 5.2.8 and Corollary 5.2.15, if $\boldsymbol{\rho} = (\rho, \boldsymbol{\nu})$ satisfies the condition Frontier for some choice of $\sigma^\epsilon(\boldsymbol{\nu})$, then it also satisfies the condition Frontier for arbitrary choice of $\sigma^\epsilon(\boldsymbol{\nu})$.

Remark 6.1.6. (1) If a labeled representation satisfies the conditions NonZero and Duality, then we can see by an argument in the proof of Proposition 8.3.6 that it also satisfies the first and the third conditions in Definition 6.1.4(1) for each $\epsilon \in \{-, +\}$. However, we include these two conditions in the definition of Frontier, because it is useful in the proof of a key Proposition 6.2.1.

(2) Let $\boldsymbol{\rho} = (\rho, \boldsymbol{\nu})$ be a quasifuchsian labeled representation. Then it follows that $\sigma^\epsilon(\boldsymbol{\nu})$ is an ϵ-terminal triangle of ρ (see Remark 6.1.9 below). However, the converse does not hold; that is, ρ may have an ϵ-terminal triangle which is different from $\sigma^\epsilon(\boldsymbol{\nu})$. Let ϕ be the Markoff map with $(\phi(s_0), \phi(s_1), \phi(s_2)) = (20, 1.9966 + 0.001\mathrm{i}, 19.88056385 - 2.303580328\mathrm{i})$. Then we see that $\rho = \rho_\phi$ is quasifuchsian and σ is a $(+)$-terminal triangle of ρ. However, $\sigma^+(\boldsymbol{\nu}(\rho))$ is different from σ, where $\boldsymbol{\nu}(\rho)$ is the image of ρ by the map $\boldsymbol{\nu}$ in Theorem 1.3.2. In fact, for each elliptic generator, P, of slope s_0, $Ih(\rho(P))$ has the small radius $1/20$ and does not support a face of the Ford domain of ρ. One can check this by putting $z_1 = 0.4972138848 + 0.05736029875\mathrm{i}$ and $z_2 = 0.5021685589 + 0.05793691669\mathrm{i}$ in OPTi. This phenomenon is related to the existence of an exotic component of the linear slice of quasifuchsian punctured torus space [51].

Definition 6.1.7 (Good labeled representation). *A labeled representation* $\boldsymbol{\rho} = (\rho, \boldsymbol{\nu})$ *is said to be* good *if it satisfies the conditions NonZero, Frontier and Duality.*

In Sect. 6.5, we prove the following characterization of quasifuchsian labeled representations.

Theorem 6.1.8 (Good implies quasifuchsian). *A labeled representation* $\boldsymbol{\rho} = (\rho, \boldsymbol{\nu})$ *is quasifuchsian if it is good.*

Remark 6.1.9. The converse to the above theorem also holds. But we do not give a direct proof, because we do not need it for the proof of Main Theorem 1.3.5 and because it is a consequence of the theorem.

To reformulate Main Theorem 1.3.5 by using Theorem 6.1.8, we introduce the following notations.

Definition 6.1.10. *The symbol* $\mathcal{J}[\mathcal{QF}]$ *denotes the subset of* $\mathcal{X} \times (\mathbb{H}^2 \times \mathbb{H}^2)$ *consisting of all good labeled representations. By* $\boldsymbol{\mu}_1 : \mathcal{J}[\mathcal{QF}] \to \mathcal{X}$ *and* $\boldsymbol{\mu}_2 : \mathcal{J}[\mathcal{QF}] \to \mathbb{H}^2 \times \mathbb{H}^2$*, respectively, we denote the projection to the first and the second factors.*

Then by virtue of Theorem 6.1.8, we see that Main Theorem 1.3.5 is a consequence of the following theorem.

Modified Main Theorem 6.1.11. *(1) The projection* $\boldsymbol{\mu}_1 : \mathcal{J}[\mathcal{QF}] \to \mathcal{X}$ *induces a homeomorphism* $\mathcal{J}[\mathcal{QF}] \to \mathcal{QF}$.

(2) The projection $\boldsymbol{\mu}_2 : \mathcal{J}[\mathcal{QF}] \to \mathbb{H}^2 \times \mathbb{H}^2$ *is a homeomorphism.*

Proof (Proof of Main Theorem 1.3.5 by assuming Theorem 6.1.8 and Modified Main Theorem 6.1.11). By Modified Main Theorem 6.1.11(1), we can define a continuous map $\boldsymbol{\nu} : \mathcal{QF} \to \mathbb{H}^2 \times \mathbb{H}^2$ as the composition $\boldsymbol{\mu}_2 \circ \left(\boldsymbol{\mu}_1|_{\mathcal{J}[\mathcal{QF}]}\right)^{-1}$. Let ρ be an element of $\mathcal{QF}$. Then $(\rho, \boldsymbol{\nu}(\rho))$ is a good labeled representation by the definition of the map $\boldsymbol{\nu}$. Thus $(\rho, \boldsymbol{\nu}(\rho))$ is quasifuchsian by Theorem 6.1.8, and hence the weighted spine Ford(ρ) of $M(\rho)$ is equivalent to Spine$(\boldsymbol{\nu}(\rho))$ (see Definition 6.1.1). This implies that the map $\boldsymbol{\nu} : \mathcal{QF} \to \mathbb{H}^2 \times \mathbb{H}^2$ is equal to the map in Main Theorem 1.3.5 (cf. Corollary 6.2.5). By Modified Main Theorem 6.1.11(2) and (3), this continuous map is bijective. Finally, we see by Proposition 6.2.1 that the inverse of $\boldsymbol{\nu}$ is also continuous. Hence we obtain Main Theorem 1.3.5.

A route map of the proof of this theorem is given in Sect. 6.2, and the remaining chapters of this paper are devoted to the proof of this theorem.

In Sect. 6.6, we shall also prove the following theorem, which implies Theorem 1.4.2.

Theorem 6.1.12. *Let* $\boldsymbol{\rho}$ *be a good labeled representation. Then the ideal polyhedral complex* $\Delta_{\mathbb{E}}(\boldsymbol{\rho})$ *is combinatorially equivalent to* Trg$(\boldsymbol{\nu})$.

6.2 Route map of the proof of Modified Main Theorem 6.1.11

To prove Modified Main Theorem 6.1.11(1), we need the following Propositions 6.2.1, 6.2.3, 6.2.4 and 6.2.5. Proposition 6.2.1 is proved in Chap. 7, Propositions 6.2.3 and 6.2.4 are proved in Chap. 8, and Proposition 6.2.5 is proved in Sect. 6.4 of this chapter.

Proposition 6.2.1 (Openness). *For any good labeled representation* $\boldsymbol{\rho}_0 = (\rho_0, \boldsymbol{\nu}_0)$, *there is an open neighborhood* U *of* ρ_0 *in* $\mathcal{X}$ *and a continuous map* $U \ni \rho \mapsto \boldsymbol{\nu}(\rho) \in \mathbb{H}^2 \times \mathbb{H}^2$ *with* $\boldsymbol{\nu}(\rho_0) = \boldsymbol{\nu}_0$, *such that* $(\rho, \boldsymbol{\nu}(\rho))$ *is good for any* $\rho \in U$.

Definition 6.2.2 (SameStratum). *Let* $\{\boldsymbol{\rho}_n\}$ *be a sequence of labeled representations. We say that the sequence satisfies the condition* SameStratum *if all* ν_n^ϵ *(*$n \in \mathbb{N}$*) is contained in a common (open) cell of* $\mathcal{D}$ *for each* $\epsilon \in \{-,+\}$. *In this case, we denote the common chain* $\Sigma(\boldsymbol{\nu}_n)$ *(resp. elliptic generator complex* $\mathcal{L}(\boldsymbol{\nu}_n)$*) by* Σ_0 *(resp.* $\mathcal{L}_0$*).*

Proposition 6.2.3 (SameStratum). *Let* $\{\boldsymbol{\rho}_n\} = \{(\rho_n, \boldsymbol{\nu}_n)\}$ *be a sequence in* $\mathcal{J}[\mathcal{QF}]$ *such that* $\{\rho_n\}$ *converges to* $\rho_\infty \in \mathcal{QF}$*. Then there is a subsequence of* $\{\boldsymbol{\rho}_n\}$*, denoted by the same symbol, satisfying the following conditions:*

(1) $\{\boldsymbol{\rho}_n\}$ *converges to a labeled representation* $\boldsymbol{\rho}_\infty = (\rho_\infty, \boldsymbol{\nu}_\infty)$ *for some* $\boldsymbol{\nu}_\infty \in \mathbb{H}^2 \times \mathbb{H}^2$.
(2) $\{\boldsymbol{\rho}_n\}$ *satisfies the condition SameStratum.*

Proposition 6.2.4 (Closedness). *Let* $\{\boldsymbol{\rho}_n\} = \{(\rho_n, \boldsymbol{\nu}_n)\}$ *be a sequence in* $\mathcal{J}[\mathcal{QF}]$ *satisfying the following conditions:*

(1) $\{\boldsymbol{\rho}_n\}$ *converges to a labeled representation* $\boldsymbol{\rho}_\infty = (\rho_\infty, \boldsymbol{\nu}_\infty) \in \overline{\mathcal{QF}} \times (\mathbb{H}^2 \times \mathbb{H}^2)$.
(2) $\{\boldsymbol{\rho}_n\}$ *satisfies the condition SameStratum.*

Then the limit $\boldsymbol{\rho}_\infty$ *is a good labeled representation and hence belongs to* $\mathcal{J}[\mathcal{QF}]$.

Proposition 6.2.5 (Uniqueness of good label). *For each quasifuchsian representation* $\rho \in \mathcal{QF}$*, there is at most one label* $\boldsymbol{\nu}$ *such that* $(\rho, \boldsymbol{\nu})$ *is good.*

Proof (Proof of Modified Main Theorem 6.1.11(1)). We first show that $\boldsymbol{\mu}_1 : \mathcal{J}[\mathcal{QF}] \to \mathcal{X}$ induces a surjection $\mathcal{J}[\mathcal{QF}] \to \mathcal{QF}$. To this end, let $\mathcal{QF}_0$ be the subspace of $\mathcal{QF}$ consisting of $\rho \in \mathcal{QF}$ such that $(\rho, \boldsymbol{\nu}) \in \mathcal{J}[\mathcal{QF}]$ for some label $\boldsymbol{\nu} \in \mathbb{H}^2 \times \mathbb{H}^2$. It is clear that $\mathcal{QF}_0$ is non-empty by Proposition 5.1.5. We show that $\mathcal{QF}_0$ is open and closed in $\mathcal{QF}$.

The openness of $\mathcal{QF}_0$ follows from Proposition 6.2.1 (Openness).

To prove the closedness, let $\{\boldsymbol{\rho}_n\} = \{(\rho_n, \boldsymbol{\nu}_n)\}$ be a sequence in $\mathcal{J}[\mathcal{QF}]$, such that ρ_n converges to ρ_∞ in $\mathcal{QF}$. By Proposition 6.2.3 (SameStratum), we may assume that $\{\boldsymbol{\rho}_n\}$ satisfies the condition of Proposition 6.2.4 (Closedness). Hence, $(\rho_\infty, \boldsymbol{\nu}_\infty)$ is good by the proposition, where $\boldsymbol{\nu}_\infty = \lim \boldsymbol{\nu}_n$. This proves the closedness of $\mathcal{QF}_0$. Since $\mathcal{QF}$ is connected, we have $\mathcal{QF} = \mathcal{QF}_0$. Hence $\boldsymbol{\mu}_1 : \mathcal{J}[\mathcal{QF}] \to \mathcal{X}$ induces a surjection $\mathcal{J}[\mathcal{QF}] \to \mathcal{QF}$.

By Proposition 6.2.5 (Uniqueness of good label), the above map is also injective, and hence bijective. Since it is a restriction of the projection $\boldsymbol{\mu}_1$, it is continuous. Moreover, its inverse map is also continuous by Proposition 6.2.1. Hence $\boldsymbol{\mu}_1 : \mathcal{J}[\mathcal{QF}] \to \mathcal{X}$ induces a homeomorphism $\mathcal{J}[\mathcal{QF}] \to \mathcal{QF}$.

Modified Main Theorem 6.1.11(2) is a consequence of the following proposition, which is proved in Chap. 9.

Proposition 6.2.6 (Unique realization). *For every label* $\boldsymbol{\nu} \in \mathbb{H}^2 \times \mathbb{H}^2$*, there is a unique quasifuchsian representation* $\rho \in \mathcal{QF}$*, such that* $(\rho, \boldsymbol{\nu})$ *is good.*

Proof (Proof of Modified Main Theorem 6.1.11(2)). By Proposition 6.2.6 (Unique realization), $\boldsymbol{\mu}_2 : \mathcal{J}[\mathcal{QF}] \to \mathbb{H}^2 \times \mathbb{H}^2$ is bijective. Since it is a restriction of the projection, it is continuous. Since $\mathcal{J}[\mathcal{QF}] \cong \mathcal{QF} \cong \mathbb{R}^4$ by

Modified Main Theorem 6.1.11(1), $\boldsymbol{\mu}_2$ is regarded as a bijective continuous map from $\mathbb{R}^4$ to $\mathbb{H}^2 \times \mathbb{H}^2 \cong \mathbb{R}^4$. This together with the invariance of domain implies that $\boldsymbol{\mu}_2$ is a homeomorphism. In fact, the restriction of $\boldsymbol{\mu}_2$ to a closed ball B is a homeomorphism and hence $\boldsymbol{\mu}_2(\operatorname{int} B)$ is a subspace of $\mathbb{H}^2 \times \mathbb{H}^2$ homeomorphic to $\operatorname{int} B$. Thus the invariance of domain implies that $\boldsymbol{\mu}_2(\operatorname{int} B)$ is open in $\mathbb{H}^2 \times \mathbb{H}^2$. Hence $\boldsymbol{\mu}_2$ is an open map and therefore it is a homeomorphism.

Finally, we note that the following convergence theorem holds a key to the proof of Proposition 6.2.6 (Unique realization), which in tern is proved in Sect. 8.2.

Proposition 6.2.7 (Convergence). *Let $\{\boldsymbol{\rho}_n\} = \{(\rho_n, \boldsymbol{\nu}_n)\}$ be a sequence of good labeled representations satisfying the following conditions:*

(1) $\{\boldsymbol{\nu}_n\}$ converges to $\boldsymbol{\nu}_\infty \in \mathbb{H}^2 \times \mathbb{H}^2$.
(2) $\{\boldsymbol{\rho}_n\}$ satisfies the condition SameStratum.

Then $\{\boldsymbol{\rho}_n\}$ has a convergent subsequence.

6.3 The cellular structure of $\partial Eh(\boldsymbol{\rho})$

In this section, we describe the cellular structure of the boundary of the virtual Ford domain $Eh(\boldsymbol{\rho})$ of a good labeled representation $\boldsymbol{\rho}$.

Proposition 6.3.1 (Cellular structure). *Let $\boldsymbol{\rho} = (\rho, \boldsymbol{\nu})$ be a labeled representation satisfying the conditions NonZero and Duality. Then the collection $\{F_{\boldsymbol{\rho}}(\xi) \,|\, \xi \in \mathcal{L}(\boldsymbol{\nu})^{(\leq 2)}\}$ gives the cellular structure of $\partial Eh(\boldsymbol{\rho})$, which is dual to $\mathcal{L}(\boldsymbol{\nu})$. Namely, we have the following:*

1. *For each $\xi \in \mathcal{L}(\boldsymbol{\nu})^{(\leq 2)}$, $F_{\boldsymbol{\rho}}(\xi)$ is a convex polyhedron of dimension $2 - \dim \xi$ contained in $\partial Eh(\boldsymbol{\rho})$.*
2. *For each $\xi \in \mathcal{L}(\boldsymbol{\nu})^{(\leq 2)}$, $\partial F_{\boldsymbol{\rho}}(\xi)$ is the disjoint union of $\{\operatorname{int} F_{\boldsymbol{\rho}}(\xi') \,|\, \xi < \xi' \in \mathcal{L}(\boldsymbol{\nu})^{(\leq 2)}\}$.*
3. *$\partial Eh(\boldsymbol{\rho})$ is the disjoint union of $\{\operatorname{int} F_{\boldsymbol{\rho}}(\xi) \,|\, \xi \in \mathcal{L}(\boldsymbol{\nu})^{(\leq 2)}\}$.*

To prove this proposition, we prepare a few lemmas. Throughout this section, $\boldsymbol{\rho}$ denotes a labeled representation satisfying the conditions NonZero and Duality.

Lemma 6.3.2. *Let $\boldsymbol{\rho} = (\rho, \boldsymbol{\nu})$ be a labeled representation satisfying the conditions NonZero and Duality. Then then following hold.*

(1) For any $\xi \in \mathcal{L}(\boldsymbol{\nu})^{(\leq 2)}$, we have

$$\begin{aligned} F_{\boldsymbol{\rho}}(\xi) &= \left(\bigcap \{Ih(\rho(P)) \,|\, P \in \xi^{(0)}\}\right) \cap Eh(\boldsymbol{\rho}) \\ &= \left(\bigcap \{Ih(\rho(P)) \,|\, P \in \xi^{(0)}\}\right) \cap \partial Eh(\boldsymbol{\rho}) \end{aligned}$$

(2) For any $\xi,\ \xi' \in \mathcal{L}(\boldsymbol{\nu})^{(\leq 2)}$ with $\xi < \xi'$, we have $F_{\boldsymbol{\rho}}(\xi) \supset F_{\boldsymbol{\rho}}(\xi')$.

Proof. (1) Since $\boldsymbol{\rho}$ satisfies Duality, we have $F_{\boldsymbol{\rho}}(\xi) \subset \partial Eh(\boldsymbol{\rho}) \subset Eh(\boldsymbol{\rho})$. Hence, we have

$$\begin{aligned} F_{\boldsymbol{\rho}}(\xi) &= F_{\boldsymbol{\rho}}(\xi) \cap Eh(\boldsymbol{\rho}) \\ &= \left(\left(\bigcap\{Ih(\rho(P)) \,|\, P \in \xi^{(0)}\}\right) \cap Eh(\rho; \mathrm{lk}(\xi, \mathcal{L}(\boldsymbol{\nu})))\right) \cap Eh(\boldsymbol{\rho}) \\ &= \left(\bigcap\{Ih(\rho(P)) \,|\, P \in \xi^{(0)}\}\right) \cap Eh(\boldsymbol{\rho}). \end{aligned}$$

Since $Ih(\rho(P)) \cap Eh(\boldsymbol{\rho}) = Ih(\rho(P)) \cap \partial Eh(\boldsymbol{\rho})$ for any $P \in \mathcal{L}(\boldsymbol{\nu})^{(0)}$, we also have

$$F_{\boldsymbol{\rho}}(\xi) = \left(\bigcap\{Ih(\rho(P)) \,|\, P \in \xi^{(0)}\}\right) \cap \partial Eh(\boldsymbol{\rho}).$$

(2) By (1), we see

$$\begin{aligned} F_{\boldsymbol{\rho}}(\xi) &= \left(\bigcap\{Ih(\rho(P)) \,|\, P \in \xi^{(0)}\}\right) \cap Eh(\boldsymbol{\rho}) \\ &\supset \left(\bigcap\{Ih(\rho(P)) \,|\, P \in (\xi')^{(0)}\}\right) \cap Eh(\boldsymbol{\rho}) \\ &= F_{\boldsymbol{\rho}}(\xi'). \end{aligned}$$

Lemma 6.3.3. *Let $\boldsymbol{\rho} = (\rho, \boldsymbol{\nu})$ be a labeled representation satisfying the conditions NonZero and Duality. Then the following hold.*

(1) For any $\xi \in \mathcal{L}(\boldsymbol{\nu})^{(\le 2)}$ and any $X \in \mathcal{L}(\boldsymbol{\nu})^{(0)} - \xi^{(0)}$, we have $F_{\boldsymbol{\rho}}(\xi) \cap Ih(\rho(X)) \subset \partial F_{\boldsymbol{\rho}}(\xi)$.

(2) For any ξ, $\xi' \in \mathcal{L}(\boldsymbol{\nu})^{(\le 2)}$ with $\xi < \xi'$, we have $\partial F_{\boldsymbol{\rho}}(\xi) \supset F_{\boldsymbol{\rho}}(\xi')$.

Proof. First, we prove (2) by assuming (1). Pick an element $X \in (\xi')^{(0)} - \xi^{(0)}$. Then, by Lemma 6.3.2 (2),

$$F_{\boldsymbol{\rho}}(\xi') = F_{\boldsymbol{\rho}}(\xi) \cap F_{\boldsymbol{\rho}}(\xi') \subset F_{\boldsymbol{\rho}}(\xi) \cap Ih(\rho(X)).$$

Since the last set is contained in $\partial F_{\boldsymbol{\rho}}(\xi)$ by (1), we have $F_{\boldsymbol{\rho}}(\xi') \subset \partial F_{\boldsymbol{\rho}}(\xi)$.

Next, we prove (1). Pick elements $\xi \in \mathcal{L}(\boldsymbol{\nu})^{(\le 2)}$ and $X \in \mathcal{L}(\boldsymbol{\nu})^{(0)} - \xi^{(0)}$.

Case 1. ξ is a 0-simplex $\langle P \rangle$. Suppose contrary that $F_{\boldsymbol{\rho}}(P) \cap Ih(\rho(X))$ contains a point in the interior of the 2-dimensional face $F_{\boldsymbol{\rho}}(P)$. Then since $F_{\boldsymbol{\rho}}(P) \subset \partial Eh(\boldsymbol{\rho})$ we must have $Ih(\rho(X)) = Ih(\rho(P))$. Hence, by the condition Duality, we have $X = P$, a contradiction.

Case 2. ξ is a 1-simplex $\langle P, Q \rangle$. Suppose contrary that $F_{\boldsymbol{\rho}}(\xi) \cap Ih(\rho(X))$ contains a point in the interior of $F_{\boldsymbol{\rho}}(\xi)$. Then $Ih(\rho(X))$ contains an open circular arc of $Ih(\rho(P)) \cap Ih(\rho(Q))$, because $F_{\boldsymbol{\rho}}(\xi)$ is a sub-arc of $Ih(\rho(P)) \cap Ih(\rho(Q))$ which is contained in $Eh(\boldsymbol{\rho})$. Hence, the three isometric hemispheres $Ih(\rho(P))$, $Ih(\rho(Q))$ and $Ih(\rho(X))$ intersect in a common half circle. So, one of the three isometric hemispheres, say $Ih(\rho(Y))$ ($Y \in \{P, Q, X\}$), must be contained in the union of the half balls bounded by the remaining two isometric hemispheres. Since the three isometric hemispheres are mutually distinct by the condition Duality, this implies that $F_{\boldsymbol{\rho}}(Y)$ cannot be a 2-dimensional convex polyhedron contained in $\partial Eh(\boldsymbol{\rho})$, a contradiction.

Case 3. ξ is a 2-simplex $\langle P, Q, R \rangle$. Then we have the following claim.

Claim 6.3.4. There is an open neighborhood U of the vertex $F_{\boldsymbol{\rho}}(\xi)$ in $\partial Eh(\boldsymbol{\rho})$ such that $U \subset \cup\{\text{int}\, F_{\boldsymbol{\rho}}(\xi') \,|\, \xi' \leq \xi\}$.

Proof. Note that the following facts are deduced from the proof of (2) and the conclusions for Cases 1 and 2.

1. The 0-dimensional convex polyhedron $F_{\boldsymbol{\rho}}(\xi)$ is a vertex of each of the 1-dimensional convex polyhedra $F_{\boldsymbol{\rho}}(P,Q)$, $F_{\boldsymbol{\rho}}(Q,R)$ and $F_{\boldsymbol{\rho}}(P,R)$.
2. The 1-dimensional convex polyhedron $F_{\boldsymbol{\rho}}(P,Q)$ is an edge of the 2-dimensional convex polyhedra $F_{\boldsymbol{\rho}}(P)$ and $F_{\boldsymbol{\rho}}(Q)$. Similarly $F_{\boldsymbol{\rho}}(Q,R)$ (resp. $F_{\boldsymbol{\rho}}(P,R)$) is an edge of the 2-dimensional convex polyhedron $F_{\boldsymbol{\rho}}(Q)$ and $F_{\boldsymbol{\rho}}(R)$ (resp. $F_{\boldsymbol{\rho}}(P)$ and $F_{\boldsymbol{\rho}}(R)$).

Let C_P be the convex hull of $\text{int}\, F_{\boldsymbol{\rho}}(P,Q)$, $\text{int}\, F_{\boldsymbol{\rho}}(P,R)$ and $F_{\boldsymbol{\rho}}(\xi)$. Then, by the above facts, we see that C_P is a sector of $F_{\boldsymbol{\rho}}(P)$ contained in the subset $\text{int}\, F_{\boldsymbol{\rho}}(\xi) \cup \text{int}\, F_{\boldsymbol{\rho}}(P,Q) \cup \text{int}\, F_{\boldsymbol{\rho}}(P,R) \cup \text{int}\, F_{\boldsymbol{\rho}}(P)$. Let C_Q and C_R,respectively, be the sectors of $F_{\boldsymbol{\rho}}(Q)$ and $F_{\boldsymbol{\rho}}(R)$, defined in a similar way, and set $U := C_P \cup C_Q \cup C_R$. Then U is contained in $\cup\{\text{int}\, F_{\boldsymbol{\rho}}(\xi') \,|\, \xi' \leq \xi\}$. On the other hand, since $F_{\boldsymbol{\rho}}(\xi) = Ih(\rho(P)) \cap Ih(\rho(Q)) \cap Ih(\rho(R))$ is a 0-dimensional convex polyhedron, the relative position of $Ih(\rho(P))$, $Ih(\rho(Q))$ and $Ih(\rho(R))$ is as illustrated in Fig. 6.1. Hence we see that U is an open disk in $\partial Eh(\boldsymbol{\rho})$ containing the vertex $F_{\boldsymbol{\rho}}(\xi)$. This implies that U is an open neighborhood of $F_{\boldsymbol{\rho}}(\xi)$ in $\partial Eh(\boldsymbol{\rho})$, because $\partial Eh(\boldsymbol{\rho})$ is a 2-dimensional manifold. Hence we have the claim.

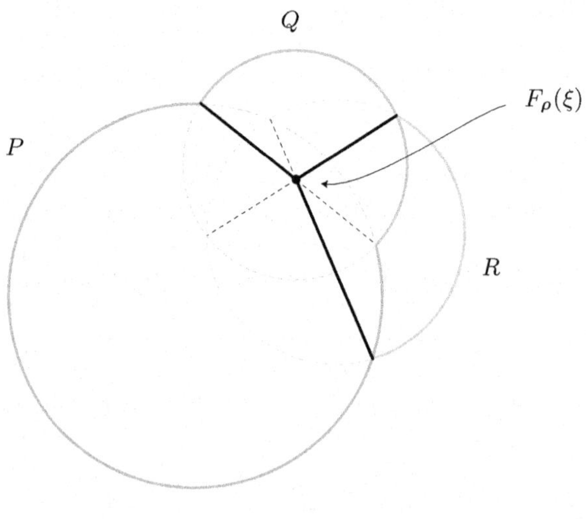

Fig. 6.1. Three isometric circles: $\xi^{(0)} = \{P, Q, R\}$

Suppose contrary that $F_{\boldsymbol{\rho}}(\xi) \cap Ih(\rho(X))$ contains a point in the interior of $F_{\boldsymbol{\rho}}(\xi)$. Then the vertex $F_{\boldsymbol{\rho}}(\xi)$ is contained in $Ih(\rho(X))$, and hence it is contained in the 2-dimensional face $Ih(\rho(X)) \cap Eh(\boldsymbol{\rho}) = F_{\boldsymbol{\rho}}(X)$ (see Lemma 6.3.2 (1)). For any open neighborhood U of $F_{\boldsymbol{\rho}}(\xi)$ in $\partial Eh(\boldsymbol{\rho})$, we have $U \cap F_{\boldsymbol{\rho}}(X) - F_{\boldsymbol{\rho}}(\xi) \neq \emptyset$, because the vertex $F_{\boldsymbol{\rho}}(\xi)$ is contained in the 2-dimensional face $F_{\boldsymbol{\rho}}(X)$. Hence there is a simplex $\xi' < \xi$ such that $\operatorname{int} F_{\boldsymbol{\rho}}(\xi') \cap F_{\boldsymbol{\rho}}(X) \neq \emptyset$ (and hence $\operatorname{int} F_{\boldsymbol{\rho}}(\xi') \cap Ih(\rho(X)) \neq \emptyset$) by the above sublemma. Since $\dim \xi' \leq 1$, this contradicts the conclusion in Case 1 or Case 2.

Lemma 6.3.5. *Let $\boldsymbol{\rho} = (\rho, \boldsymbol{\nu})$ be a labeled representation satisfying the conditions NonZero and Duality. Then for any distinct elements ξ and ξ' of $\mathcal{L}(\boldsymbol{\nu})^{(\leq 2)}$, we have* $\operatorname{int} F_{\boldsymbol{\rho}}(\xi) \cap \operatorname{int} F_{\boldsymbol{\rho}}(\xi') = \emptyset$.

Proof. We may assume that there is a vertex X of ξ' which is not a vertex of ξ. Then $F_{\boldsymbol{\rho}}(\xi) \cap F_{\boldsymbol{\rho}}(\xi') \subset F_{\boldsymbol{\rho}}(\xi) \cap Ih(\rho(X)) \subset \partial F_{\boldsymbol{\rho}}(\xi)$ by Lemma 6.3.3 (1). Hence

$$\operatorname{int} F_{\boldsymbol{\rho}}(\xi) \cap \operatorname{int} F_{\boldsymbol{\rho}}(\xi') \subset \operatorname{int} F_{\boldsymbol{\rho}}(\xi) \cap (F_{\boldsymbol{\rho}}(\xi) \cap F_{\boldsymbol{\rho}}(\xi')) \subset \operatorname{int} F_{\boldsymbol{\rho}}(\xi) \cap \partial F_{\boldsymbol{\rho}}(\xi) = \emptyset,$$

and we obtain the conclusion.

Lemma 6.3.6. *Let $\boldsymbol{\rho} = (\rho, \boldsymbol{\nu})$ be a labeled representation satisfying the conditions NonZero and Duality. Then for any $\xi \in \mathcal{L}(\boldsymbol{\nu})^{(\leq 2)}$, we have*

$$\partial F_{\boldsymbol{\rho}}(\xi) = \bigcup \{F_{\boldsymbol{\rho}}(\xi') \mid \xi < \xi' \in \mathcal{L}(\boldsymbol{\nu})^{(\leq 2)}, \ \dim \xi' = \dim \xi + 1\}.$$

Proof. We first show that

$$\partial F_{\boldsymbol{\rho}}(\xi) = \bigcup \{F_{\boldsymbol{\rho}}(\xi) \cap Ih(\rho(X)) \mid X \in \operatorname{lk}(\xi, \mathcal{L}(\boldsymbol{\nu}))^{(0)}\}.$$

By Lemma 6.3.3 (1), the left hand side contains the right hand side. Suppose contrary that the left hand side is not contained in the right hand side. Then there is a point $x \in \partial F_{\boldsymbol{\rho}}(\xi)$ such that $x \notin Ih(\rho(X))$ for any $X \in \operatorname{lk}(\xi, \mathcal{L}(\boldsymbol{\nu}))^{(0)}$. Then x lies in the open set $\cap\{\operatorname{int} Eh(\rho(X)) \mid X \in \operatorname{lk}(\xi, \mathcal{L}(\boldsymbol{\nu}))^{(0)}\} = \operatorname{int} Eh(\rho; \operatorname{lk}(\xi, \mathcal{L}(\boldsymbol{\nu})))$. By the definition of $F_{\boldsymbol{\rho}}(\xi)$, this implies that $x \in \operatorname{int} F_{\boldsymbol{\rho}}(\xi)$, a contradiction. Thus we have the identity.

On the other hand, for each $X \in \operatorname{lk}(\xi, \mathcal{L}(\boldsymbol{\nu}))^{(0)}$, we have $F_{\boldsymbol{\rho}}(\xi) \cap Ih(\rho(X)) = F_{\boldsymbol{\rho}}(\xi')$, where ξ' is the simplex of $\mathcal{L}(\boldsymbol{\nu})$ spanned by ξ and X. Because Lemma 6.3.2 (1) implies

$$\begin{aligned} F_{\boldsymbol{\rho}}(\xi) \cap Ih(\rho(X)) &= \left(\left(\bigcap \{Ih(\rho(P)) \mid P \in \xi^{(0)}\}\right) \cap Eh(\boldsymbol{\rho})\right) \cap Ih(\rho(X)) \\ &= \left(\bigcap \{Ih(\rho(P)) \mid P \in (\xi')^{(0)}\}\right) \cap Eh(\boldsymbol{\rho}) \\ &= F_{\boldsymbol{\rho}}(\xi'). \end{aligned}$$

Hence we obtain the desired result.

Proof (Proof of Proposition 6.3.1). The first assertion directly follows from the condition Duality.

Next, we prove the second assertion. If $\dim \xi = 2$, the assertion is obvious. If $\dim \xi = 1$, the assertion follows from Lemma 6.3.6. The assertion for the case $\dim \xi = 0$ follows from this observation and Lemmas 6.3.5 and 6.3.6.

Finally, we prove the third assertion. Since $\{Ih(\rho(X)) \,|\, X \in \mathcal{L}(\boldsymbol{\nu})^{(0)}\}$ is locally finite, we have $\partial Eh(\boldsymbol{\rho}) = \cup\{Ih(\rho(X)) \cap Eh(\rho) \,|\, X \in \mathcal{L}(\nu)^{(0)}\}$. By Lemma 6.3.2 (1), this implies $\partial Eh(\boldsymbol{\rho}) = \cup\{F_{\boldsymbol{\rho}}(X) \,|\, X \in \mathcal{L}(\boldsymbol{\nu})^{(0)}\}$. Hence we obtain the third conclusion by the second conclusion and Lemma 6.3.5.

Corollary 6.3.7. *Let $\boldsymbol{\rho}$ be a labeled representation satisfying the conditions NonZero and Duality, and let ξ_1 and ξ_2 be simplices of $\mathcal{L}(\boldsymbol{\nu})$. Then the following conditions are equivalent.*

1. *$F_{\boldsymbol{\rho}}(\xi_1) \cap F_{\boldsymbol{\rho}}(\xi_2) \neq \emptyset$.*
2. *There is a simplex ξ' of $\mathcal{L}(\boldsymbol{\nu})$ which contains both ξ_1 and ξ_2 as faces.*

Proof. By Proposition 6.3.1, $F_{\boldsymbol{\rho}}(\xi_i)$ is the disjoint union of $\{\operatorname{int} F_{\boldsymbol{\rho}}(\xi') \,|\, \xi_i \le \xi' \in \mathcal{L}(\boldsymbol{\nu})^{(\le 2)}\}$. Since $\partial Eh(\boldsymbol{\rho})$ is the disjoint union of $\{\operatorname{int} F_{\boldsymbol{\rho}}(\xi) \,|\, \xi \in \mathcal{L}(\boldsymbol{\nu})^{(\le 2)}\}$, this implies the desired result.

6.4 Applying Poincare's theorem on fundamental polyhedra

In this section, we prove the following key proposition for the proof of Theorem 6.1.8

Proposition 6.4.1. *Let $\boldsymbol{\rho} = (\rho, \boldsymbol{\nu})$ be a good labeled representation. Then ρ is discrete and the Ford domain $Ph(\rho)$ is equal to the virtual Ford domain $Eh(\boldsymbol{\rho})$.*

First, we note the following consequence of Lemma 4.1.3 (Chain rule).

Lemma 6.4.2. *Let $\boldsymbol{\rho} = (\rho, \boldsymbol{\nu})$ be a labeled representation satisfying the conditions NonZero and Duality. Then we have the following.*

(1) For each 2-dimensional face $F_{\boldsymbol{\rho}}(P)$, the involution $\rho(P)$ maps $F_{\boldsymbol{\rho}}(P)$ onto itself.

(2) Suppose a face $F_{\boldsymbol{\rho}}(P)$ contains an edge $F_{\boldsymbol{\rho}}(\xi)$ ($\xi = \langle P_0, P_1\rangle$), that is, $P = P_0$ or P_1. Then the involution $\rho(P)$ maps $F_{\boldsymbol{\rho}}(\xi)$ onto $F_{\boldsymbol{\rho}}(\xi')$ where $\xi' = \langle P_{-1}, P_0\rangle$ or $\langle P_1, P_2\rangle$ according as $P = P_0$ or P_1.

(3) Suppose a face $F_{\boldsymbol{\rho}}(P)$ contains a vertex $F_{\boldsymbol{\rho}}(\xi)$ ($\xi = \langle P_0, P_1, P_2\rangle$), that is, $P = P_0$, P_1 or P_2. Then the involution $\rho(P)$ maps $F_{\boldsymbol{\rho}}(\xi)$ to $F_{\boldsymbol{\rho}}(\xi')$ where $\xi' = \langle P_{-1}, P_1^{P_0}, P_0\rangle$, $\langle P_0, P_1, P_2\rangle$ or $\langle P_2, P_1^{P_2}, P_3\rangle$ according as $P = P_0$, P_1 or P_2. In particular, $\rho(P_1)$ fixes $F_{\boldsymbol{\rho}}(\xi)$.

The above lemma (together with Proposition 6.3.1) implies that the family of the involutions $\{\rho(P) \,|\, P \in \mathcal{L}(\boldsymbol{\nu})^{(0)}\}$ determines a "gluing data" for the polyhedron $Eh(\boldsymbol{\rho})$. The key ingredient of the proof of Proposition 6.4.1 is to show that $Eh(\boldsymbol{\rho})$ together with the gluing data satisfies the conditions of (a variation of) Poincare's theorem on fundamental polyhedra. We note that $Eh(\boldsymbol{\rho})$ is a fundamental domain only modulo the action of the infinite cyclic Kleinian group $\langle \rho(K) \rangle$, i.e., an actual fundamental domain is the intersection of $Eh(\boldsymbol{\rho})$ with a fundamental domain for $\langle \rho(K) \rangle$. So, if we were to apply the usual version of Poincare's theorem on fundamental polyhedra (e.g. [28] and [53]), then we need to work on an artificial polyhedron and an artificial cellular structure. To avoid this troublesome phenomena, we work with the quotient of $Eh(\boldsymbol{\rho})$ by $\langle \rho(K) \rangle$.

Throughout the remainder of this section, we assume that $\boldsymbol{\rho} = (\rho, \boldsymbol{\nu})$ is a thick labeled representation which satisfies the conditions NonZero, Frontier and Duality. To make the above mentioned setting clear, we introduce the following definition.

Definition 6.4.3. *(1)* $\mathrm{Cusp}(K)$ *denotes the hyperbolic manifold* $\mathbb{H}^3/\langle \rho(K) \rangle$. $\overline{\mathrm{Cusp}}(K)$ *and* $\partial \mathrm{Cusp}(K)$ *denote* $\overline{\mathbb{H}}^3/\langle \rho(K) \rangle$ *and* $\mathbb{C}/\langle \rho(K) \rangle$, *respectively.*

(2) Let $q_K : \overline{\mathbb{H}}^3 \to \overline{\mathrm{Cusp}}(K)$ *be the projection, and set*

$$\begin{aligned} Eh_K(\boldsymbol{\rho}) &:= q_K(Eh(\boldsymbol{\rho})) \subset \mathrm{Cusp}(K), \\ E_K(\boldsymbol{\rho}) &:= q_K(E(\boldsymbol{\rho})) \quad \subset \partial\overline{\mathrm{Cusp}}(K), \\ F_{K,\boldsymbol{\rho}}(\xi) &:= q_K(F_{\boldsymbol{\rho}}(\xi)) \quad (\xi \in (\mathcal{L}(\boldsymbol{\nu})/\langle K \rangle)^{(\le 2)}). \end{aligned}$$

Lemma 6.4.4. *Let* $\boldsymbol{\rho} = (\rho, \boldsymbol{\nu})$ *be a good labeled representation. Then for each* $\xi \in \cup_{i=0}^{2} \mathcal{L}(\boldsymbol{\nu})^{(i)}$, *the restriction of the projection* q_K *to* $F_{\boldsymbol{\rho}}(\xi)$ *is injective.*

Proof. Suppose that the restriction of q_K to $F_{\boldsymbol{\rho}}(\xi)$ is not injective. Then $F_{\boldsymbol{\rho}}(\xi) \cap \rho(K^n)(F_{\boldsymbol{\rho}}(\xi)) \neq \emptyset$ for some nonzero integer n. By Corollary 6.3.7, this implies that ξ and $K^n(\xi)$ is contained in a simplex of $\mathcal{L}(\boldsymbol{\nu})$. But this is impossible, because $n \neq 0$. Hence we obtain the desired result.

Thus $F_{K,\boldsymbol{\rho}}(\xi)$ can be regarded as a "polyhedron" in $\mathrm{Cusp}(K)$, and we obtain a map

$$F_{K,\boldsymbol{\rho}} : (\mathcal{L}(\boldsymbol{\nu})/\langle K \rangle)^{(\le 2)} \to \{\text{polyhedron in } \mathrm{Cusp}(K)\}.$$

By Proposition 6.3.1, the images $\{F_{K,\boldsymbol{\rho}}(\xi)\}$ determine a "cellular structure" of the boundary of $Eh_K(\boldsymbol{\rho})$.

On the other hand, by Lemma 6.4.2, for each 2-dimensional face $F_{K,\boldsymbol{\rho}}(P)$ of $\partial Eh_K(\boldsymbol{\rho})$ $\rho(P)$ induces an isometric involution on $F_{K,\boldsymbol{\rho}}(P)$. These involutions determine the "gluing data" for the polyhedron $Eh_K(\boldsymbol{\rho})$ in $\mathrm{Cusp}(K)$.

By Lemma 6.4.2 (2), we can see that each "edge cycle" of the gluing data is of the following type:

- The cycle $\{F_{K,\boldsymbol{\rho}}(P_0, P_1), F_{K,\boldsymbol{\rho}}(P_1, P_2), F_{K,\boldsymbol{\rho}}(P_2, P_3)\}$ of length 3, where $\{P_j\}$ is the sequence of elliptic generators associate with a triangle of $\Sigma(\boldsymbol{\nu})$. The sum of the dihedral angles in the cycle is equal to 2π by Lemma 4.1.3 (Chain rule).

Let $\sim$ be the equivalence relation on $Eh_K(\boldsymbol{\rho})$ induced by the gluing data, and let $M_{\mathcal{O}}(\boldsymbol{\rho})$ denote the quotient pseudometric space of the hyperbolic manifold $Eh_K(\boldsymbol{\rho})$. Then, by the above facts, we can see that the pseudometric of $M_{\mathcal{O}}(\boldsymbol{\rho})$ is actually a metric, which is locally isometric to $\mathbb{H}^3$ or to the quotient of $\mathbb{H}^3$ by an isometric involution. Moreover, we can prove that $M_{\mathcal{O}}(\boldsymbol{\rho})$ is complete by using the condition Frontier and the fact that the gluing data has "consistent horoballs", that is, we can associate each ideal vertex v of an edge of $\partial Eh_K(\boldsymbol{\rho})$ a horoball, H_v, with center v so that the following conditions are satisfied (cf. [28, Theorem 6.3]).

1. H_v does not intersect the faces of $\partial Eh_K(\boldsymbol{\rho})$ whose closure in $\overline{\mathrm{Cusp}}(K)$ does not contain v.
2. For each 2-dimensional face $F_{K,\boldsymbol{\rho}}(P)$ whose closure contains v, the involution, h, of $F_{K,\boldsymbol{\rho}}(P)$ induced by $\rho(P)$ maps $H_v \cap F_{K,\boldsymbol{\rho}}(P)$ to $H_{h(v)} \cap F_{K,\boldsymbol{\rho}}(P)$.

(For the details of the above arguments, please see the subsequent paper [11], in which we describe constructions of hyperbolic cone manifolds from polyhedra in $\overline{\mathrm{Cusp}}(K)$ by gluing their faces.)

Thus we see that $M_{\mathcal{O}}(\boldsymbol{\rho})$ is a hyperbolic orbifold and that $Eh(\boldsymbol{\rho})$ is a fundamental domain of the discrete group $\rho(\pi_1(\mathcal{O}))$ modulo $\langle \rho(K) \rangle$. Since $Eh(\boldsymbol{\rho})$ is bounded by isometric hemispheres of elements of $\rho(\pi_1(\mathcal{O}))$, the Ford domain $Ph(\rho(\pi_1(\mathcal{O})))$ is contained in $Eh(\boldsymbol{\rho})$. This implies $Ph(\rho(\pi_1(\mathcal{O}))) = Eh(\boldsymbol{\rho})$, since both of them are fundamental domains of $\rho(\pi_1(\mathcal{O}))$ modulo $\langle \rho(K) \rangle$. Hence we have $Ph(\rho(\pi_1(T)) = Ph(\rho(\pi_1(\mathcal{O}))) = Eh(\boldsymbol{\rho})$ by Lemma 2.2.8. This completes the proof of Proposition 6.4.1.

Proof (Proof of Proposition 6.2.5 (Uniqueness of good label)). Assume that two labeled representations $(\rho, \boldsymbol{\nu}_1)$ and $(\rho, \boldsymbol{\nu}_2)$ are good for a quasifuchsian representation $\rho \in \mathcal{QF}$ and two labels $\boldsymbol{\nu}_j = (\nu_j^-, \nu_j^+)$ $(j = 1, 2)$. Then by Proposition 6.4.1 and Lemma 2.5.4(2-iii), both $\Sigma(\boldsymbol{\nu}_1)^{(0)}$ and $\Sigma(\boldsymbol{\nu}_2)^{(0)}$ consist of the slopes of the elliptic generators P such that $Ih(\rho(P))$ supports a 2-dimensional face of the Ford domain of $Ph(\rho)$. Hence we have $\Sigma(\boldsymbol{\nu}_1)^{(0)} = \Sigma(\boldsymbol{\nu}_2)^{(0)}$ and therefore $\Sigma(\boldsymbol{\nu}_1) = \Sigma(\boldsymbol{\nu}_2)$. Let σ^ϵ be the ϵ-terminal triangle of $\Sigma(\boldsymbol{\nu}_1) = \Sigma(\boldsymbol{\nu}_2)$. Then we have $\nu_1^\epsilon = \theta^\epsilon(\rho, \sigma^\epsilon) = \nu_2^\epsilon$. Hence we have $\boldsymbol{\nu}_1 = \boldsymbol{\nu}_2$.

6.5 Proof of Theorem 6.1.8 (Good implies quasifuchsian)

In this section, we complete the proof of Theorem 6.1.8. Let $\boldsymbol{\rho} = (\rho, \boldsymbol{\nu})$ be a good labeled representation. Then, by Proposition 6.4.1, ρ is discrete and the Ford domain $Ph(\rho)$ is equal to the virtual Ford domain $Eh(\boldsymbol{\rho})$. By using this fact, we show that the Ford complex $\mathrm{Ford}(\rho) = \mathrm{Ford}(\rho(T))$ of $M(\rho) = M(\rho(\pi_1(T))$ is equivalent to $\mathrm{Spine}(\boldsymbol{\nu}) = \mathrm{Spine}(\delta^-(\boldsymbol{\nu}), \delta^+(\boldsymbol{\nu}))$.

To this end, we give detailed explanation of the relation between the Ford complex and the Ford domain, which are sketched at the end of Sect. 1.2 (see Fig. 1.2). Consider the foliation of each face $F_{\boldsymbol{\rho}}(P)$ of $\partial Eh(\boldsymbol{\rho})$ by geodesic segments orthogonal to $F_\rho(P) \cap \mathrm{Axis}(\rho(P))$. We say that a leaf of the foliation is *inner* if it is a (possibly degenerate) geodesic segment with endpoints in $\partial F_\rho(P)$. If a leaf is not inner, then it is a bi-infinite geodesic which joins two points of $\overline{F}_\rho(P) \cap \mathrm{fr}\, E(\rho)$. By joining the inner leaves, we obtain a foliation, $\{L_t\}_{t\in(-1,1)}$, of an open submanifold of $\partial Eh(\rho)$ by periodic piecewise geodesic lines, because the following observation shows that L_t does not branch at vertices of $\partial Eh(\rho)$:

- For each vertex $F_\rho(P, Q, R)$ of $\partial Eh(\rho)$, the following hold (see Fig. 6.2).
  1. The vertex $F_\rho(P, Q, R)$ is the intersection of the three faces $F_\rho(P)$, $F_\rho(Q)$ and $F_\rho(R)$.
  2. The vertex $F_\rho(P, Q, R)$ is contained in $\mathrm{Axis}(\rho(Q))$ but is contained in neither $\mathrm{Axis}(\rho(P))$ nor $\mathrm{Axis}(\rho(R))$. Hence each of $F_\rho(P)$ and $F_\rho(R)$ contains a unique inner leaf having the vertex as an endpoint, and the leaf in $F_\rho(Q)$ containing the vertex is the singleton $\{F_\rho(P, Q, R)\}$.

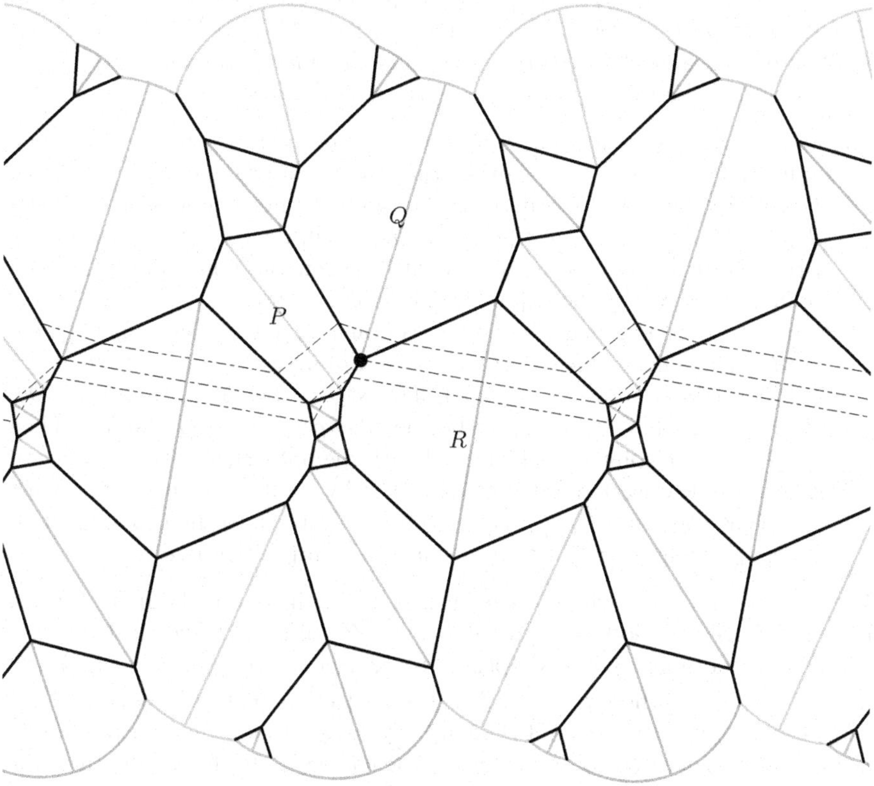

Fig. 6.2. $F_\rho(P, Q, R)$: the dotted broken lines are the leaves of the foliation $\{L_t\}$

Let $L_{\pm 1}$ be the piecewise geodesic line in $\partial \overline{Eh}(\boldsymbol{\rho})$ obtained as the limits $\lim_{t\to\pm 1} L_t$ (in the space of closed subsets of $\overline{\mathbb{H}}^3$ with Chabauty topology). Consider the piecewise linear (affine) lines $\pi(L_{\pm 1})$ in $\mathbb{C}$, where $\pi : \mathbb{H}^3 \to \mathbb{C}$ is the projection. Then the interior of the region in $\mathbb{C}$ bounded by these two lines is disjoint from $E(\boldsymbol{\rho})$ and $\mathrm{fr}\, E^-(\boldsymbol{\rho})$ (resp. $\mathrm{fr}\, E^+(\boldsymbol{\rho})$) lies below $\pi(L_{-1})$ (resp. above $\pi(L_{+1})$). By using this observation, we have the following lemma.

Lemma 6.5.1. *Let $\boldsymbol{\rho} = (\rho, \boldsymbol{\nu})$ be a good labeled representation. Then $E(\boldsymbol{\rho}) = E^-(\boldsymbol{\rho}) \sqcup E^+(\boldsymbol{\rho})$ and $\Omega(\rho)/\rho(\pi_1(T))$ forms a pair of punctured tori. In particular, ρ is quasifuchsian.*

Proof. The first assertion follows from the preceding observation. Since $Ph(\rho) = Eh(\boldsymbol{\rho})$ by Proposition 6.4.1, this implies $P(\rho) = E(\boldsymbol{\rho}) = E^-(\boldsymbol{\rho}) \sqcup E^+(\boldsymbol{\rho})$ (cf. Lemma 3.4.9). By using the fact that $\sigma^{\epsilon}(\boldsymbol{\nu})$ is an ϵ-terminal triangle of ρ (see Definition 4.3.8), we see that the image of $E^{\epsilon}(\boldsymbol{\rho})$ ($\subset \Omega(\rho)$) in $\Omega(\rho)/\rho(\pi_1(T))$ is a once-punctured torus. Hence we obtain the conclusion.

Let $S_t := \pi^{-1}(\pi(L_t)) \cap Eh(\boldsymbol{\rho})$ be the piecewise totally geodesic plane in $Eh(\boldsymbol{\rho})$ lying above L_t, let T_t be the image of S_t in $M(\rho) \cong T \times (-1, 1)$. Then T_t is isotopic to a level punctured torus, and the image, C_t, of $L_t = \partial S_t$ in T_t is a spine of T_t. Moreover, we can observe the following.

1. Suppose L_t is generic, i.e., it does not contain a vertex of $\partial Eh(\boldsymbol{\rho})$. Then by using Lemma 6.4.2, we can see that there is a triangle σ of $\Sigma(\boldsymbol{\nu})$ such that L_t is obtained by joining non-degenerate inner leaves of the faces $\{F_{\boldsymbol{\rho}}(P_j)\}$, where $\{P_j\}$ is the sequence of elliptic generators associated with σ. In this case C_t is identified with the generic spine, $\mathrm{spine}(\sigma)$ (when we identify T_t with a level punctured torus $T \times \{t\}$).
2. Suppose L_t is not generic, i.e., it contains a vertex, say $F_{\boldsymbol{\rho}}(P, Q, R)$, of $\partial Eh(\boldsymbol{\rho})$. Then, since (P, Q, R) is a 2-simplex of $\mathcal{L}(\boldsymbol{\nu})$, both $\sigma := \langle s(P), s(Q), s(R)\rangle$ and $\sigma' := \langle s(P), s(R), s(Q^R)\rangle$ belong to $\Sigma(\boldsymbol{\nu})$ (see Definitions 3.2.3 and 3.3.5). Let τ be the edge $\sigma \cap \sigma' = \langle s(P), s(R)\rangle$. Then we see that the spine C_t is identified with the non-generic spine, $\mathrm{spine}(\tau)$. Moreover, after changing the parameter t by $-t$ if necessary, we may assume that $C_{t-\eta}$ and $C_{t+\eta}$, respectively, are isotopic to the generic spines, $\mathrm{spine}(\sigma)$ and $\mathrm{spine}(\sigma')$, for every small enough positive real η.

We explain the reason why C_t is identified with $\mathrm{spine}(\sigma)$ in the first observation. Let $\tilde{\gamma}_j$ be the geodesic segment $L_t \cap F_{\boldsymbol{\rho}}(P_j)$ and $\tilde{\beta}_j$ the vertical geodesic ray joining the point $\tilde{\gamma}_j \cap \mathrm{Axis}(\rho(P_j))$ to ∞. Set $\gamma_j = p(\tilde{\gamma}_j) = p(\tilde{\gamma}_{j+3})$ and $\beta_j = p(\tilde{\beta}_j \cup \tilde{\beta}_{j+3})$, where $p : S_t \to T_t$ is the projection. Then C_t consists of the three edges γ_j ($j = 0, 1, 2$) and $\beta_0 \cup \beta_1 \cup \beta_2$ gives a topological ideal triangulation of the punctured torus T_t dual to the spine C_t. By the definition of the slopes of elliptic generators (Definition 2.1.3), we can see that the slope of the arc β_j is equal to the slope $s(P_j) = s(P_{j+3})$. Hence $\beta_0 \cup \beta_1 \cup \beta_2$ is identified with the topological ideal triangulation $\mathrm{trg}(\sigma)$ and hence C_t is identified with spine(σ) (see Sect. 1.2). The corresponding assertion in the second observation can be seen similarly.

After modifying the continuous family $\{(T_t, C_t)\}_{t\in(-1,1)}$ near $t = \pm 1$, keeping the above two conditions, we can extend it to a continuous family $\{(T_t, C_t)\}_{t\in[-1,1]}$ such that $\{T_t\}_{t\in[-1,1]}$ gives a foliation of $\bar{M}(\rho) \cong T \times [-1,1]$ and that $\mathrm{Ford}(\rho) = \cup_{t\in[-1,1]} C_t$. In particular, $C_{\epsilon 1}$ is equal to the image of $\mathrm{fr}\, E^{\epsilon}(\boldsymbol{\rho})$ in $T_{\epsilon 1} = \partial^{\epsilon}\bar{M}(\rho)$, and hence it is identified with spine($\delta^{\epsilon}(\boldsymbol{\nu})$). This can be done as follows. Let N^+ be the submanifold of $\bar{M}(\rho)$ bounded by $T_{t_0^+}$ and $\partial^+\bar{M}(\rho)$ for a real number $t_0^+ < 1$ sufficiently close to 1. Then we can see that $(N^+, N^+ \cap \mathrm{Ford}(\rho))$ is homeomorphic to $(T_{t_0^+}, C_{t_0^+}) \times [t_0^+, 1]$ or its quotient by the equivalence relation which collapses (an edge of $C_{t_0^+}) \times 1$ to a point, according as ν^+ is generic or non-generic (i.e., according as $\delta^+(\boldsymbol{\nu})$ is a triangle or an edge). Similar assertion holds near $t = -1$. By using these facts, we obtain the desired family $\{(T_t, C_t)\}_{t\in[-1,1]}$.

We note that $C_{\pm 1}$ is identified with spine($\delta^{\pm}(\boldsymbol{\nu})$) and that $\{C_t\}_{t\in[-1,1]}$ gives a sequence of elementary transformations relating spine($\delta^-(\boldsymbol{\nu})$) to spine($\delta^+(\boldsymbol{\nu})$). Moreover, the number of Whitehead transformations (= transformations from generic spines to generic spines via a single non-generic spine) that occur when t moves in the open interval $(-1,1)$ is equal to the number of the vertices of $\partial Eh(\boldsymbol{\rho})$ modulo the equivalence relation generated by the gluing data, which in tern is equal to the number of triangles in $\Sigma(\boldsymbol{\nu})$ minus 1. Hence $\{C_t\}_{t\in[-1,1]}$ realizes the canonical sequence of elementary transformations

$$\mathrm{spine}(\delta^-(\boldsymbol{\nu})) = \mathrm{spine}(\delta_0) \to \mathrm{spine}(\delta_1) \to \cdots \to \mathrm{spine}(\delta_m) = \mathrm{spine}(\delta^+(\boldsymbol{\nu}))$$

introduced in Sect. 1.2. Thus $\mathrm{Ford}(\rho)$ is isotopic to $\mathrm{Spine}(\delta^-(\boldsymbol{\nu}), \delta^+(\boldsymbol{\nu}))$. Furthermore, since $\boldsymbol{\rho}$ satisfies the condition Frontier, it follows that $\nu^{\epsilon} = \theta^{\epsilon}(\rho, \sigma^{\epsilon}(\boldsymbol{\nu}))$. This completes the proof of Theorem 6.1.8.

6.6 Structure of the complex $\Delta_{\mathbb{E}}$ and the proof of Theorem 6.1.12

In this section, we prove Theorem 6.1.12 and the following proposition.

Proposition 6.6.1. *Let $\boldsymbol{\rho} = (\rho, \boldsymbol{\nu})$ be a quasifuchsian labeled representation. Then $\mathcal{L}(\rho, \boldsymbol{\nu})$ is simple.*

To this end, we describe a method for obtaining the complex $\Delta_{\mathbb{E}}(\Gamma)$ from the Ford domain in the general setting where Γ is a non-elementary Kleinian group such that Γ_∞ contains parabolic transformations. Let $\sim$ be the equivalence relation on $Ph(\Gamma)$ such that $M(\Gamma) \cong Ph(\Gamma)/\sim$, namely two points in $Ph(\Gamma)$ are equivalent if and only if they project to the same point of $M(\Gamma)$. For each $p \in \partial Ph(\Gamma)$, let $[p]$ be the $\sim$-equivalence class of p, namely

$$[p] = \{x \in \overline{Ph}(\Gamma) \mid x = A(p) \text{ for some } A \in \Gamma\}.$$

Lemma 6.6.2. *Let e be a cell of $\partial Ph(\Gamma)$ and p a point in the (relative) interior of e. Then the following hold.*

1. *$[p]/\Gamma_\infty$ is a finite set.*
2. *Let $\{p = p_1, p_2, \cdots, p_n\} \subset [p]$ be a representative for $[p]/\Gamma_\infty$, and let $A_i \in \Gamma$ be an element such that $A_i(p) = p_i$. Then the ideal polyhedron $\widetilde{\Delta}_e$ dual to e is spanned by the ideal vertices*

$$\{A(\infty) \,|\, A \in \bigsqcup_{i=1}^{n} \Gamma_p A_i^{-1}\},$$

where Γ_p is the stabilizer of p with respect to the action of Γ.

Proof. (1) This follows from Lemma A.1.10.

(2) Recall that the ideal vertices V_e of the ideal polyhedron $\widetilde{\Delta}_e$ is given by

$$V_e = \{A(\infty) \,|\, d(p, A(H_\infty)) = d(p, \Gamma_\infty H_\infty), \quad A \in \Gamma\}.$$

On the other hand we have the following equivalence:

$$\begin{aligned}
&d(p, A(H_\infty)) = d(p, \Gamma_\infty H_\infty),\\
&\Leftrightarrow d(A^{-1}(p), H_\infty) = d(A^{-1}(p), \Gamma_\infty H_\infty),\\
&\Leftrightarrow A^{-1}(p) \in Ph(\Gamma),\\
&\Leftrightarrow A^{-1}(p) \in [p],\\
&\Leftrightarrow A^{-1}(p) = BA_i(p) \text{ for some } i \in \{1, \cdots, n\} \text{ and } B \in \Gamma_\infty,\\
&\Leftrightarrow A^{-1} = BA_iC \text{ for some } i \in \{1, \cdots, n\},\ B \in \Gamma_\infty \text{ and } C \in \Gamma_p,\\
&\Leftrightarrow A \in \textstyle\bigsqcup_{i=1}^{n} \Gamma_p A_i^{-1} \Gamma_\infty.
\end{aligned}$$

Since every element of a double coset $\Gamma_p A_i^{-1} \Gamma_\infty$ maps ∞ to Hence we obtain the desired result.

We apply the above lemma to the setting of Theorem 6.1.12. We note that, by virtue of Proposition 2.2.8(3), we may work with $\Gamma := \rho(\pi_1(\mathcal{O}))$ instead of $\rho(\pi_1(T))$.

Lemma 6.6.3. *Let $\boldsymbol{\rho} = (\rho, \boldsymbol{\nu})$ be a thick good labeled representation. Then for each cell e of $\partial Ph(\rho) = \partial Eh(\boldsymbol{\rho})$, the ideal polyhedron $\widetilde{\Delta}_e$ dual to e is described as follows:*

1. *If e is a face $F_{\boldsymbol{\rho}}(P)$, then $\widetilde{\Delta}_e$ is the ideal edge spanned by $\{\infty, \rho(P)(\infty)\}$.*
2. *If e is an edge $F_{\boldsymbol{\rho}}(P, Q)$, then $\widetilde{\Delta}_e$ is the ideal triangle spanned by*

$$\{\infty, \rho(P)(\infty), \rho(Q)(\infty)\}.$$

3. *If e is a vertex $F_{\boldsymbol{\rho}}(P, Q, R)$, then $\widetilde{\Delta}_e$ is the ideal tetrahedron spanned by*

$$\{\infty, \rho(P)(\infty), \rho(Q)(\infty), \rho(R)(\infty)\}.$$

Proof. We prove only (3), because the remaining assertion can be proved similarly. Let p be the vertex $F_{\boldsymbol{\rho}}(P, Q, R)$. Then the $(\sim)$-equivalence class $[p]$ modulo $\Gamma_\infty = \langle \rho(K) \rangle$ is equal to $\{p, \rho(P)(p)\}$ (see Sect. 6.4). Since p lies on the axis of $\rho(Q)$, the image of p in the hyperbolic orbifold $\mathbb{H}^3/\Gamma$ lies in a cone axis of cone angle π. Thus we have $\Gamma_p = \langle \rho(Q) \rangle$. Hence, by Lemma 6.6.2, the ideal polyhedron $\widetilde{\Delta}_e$ is spanned by the image of ∞ by the elements in the set

$$\rho\left\{\{1, Q\} \cup \{1, Q\}P^{-1}\right\} = \rho\{1, P, Q, QP\}.$$

Since $\rho(QP)(\infty) = \rho(RRQP)(\infty) = \rho(R)(\infty)$, we obtain the desired result.

Lemma 6.6.4. *Let $\boldsymbol{\rho} = (\rho, \boldsymbol{\nu})$ be a thin good labeled representation such that $\nu^\pm \in \operatorname{int} \tau$ under Notation 2.1.14 (Adjacent triangles). Then, modulo the action of $\langle \rho(K) \rangle$, $\partial Ph(\rho) = \partial Eh(\boldsymbol{\rho})$ consists of the two 2-dimensional convex polyhedra $F_{\boldsymbol{\rho}}(P_0)$ and $F_{\boldsymbol{\rho}}(P_2)$, together with the two 1-dimensional convex polyhedra $F_{\boldsymbol{\rho}}(P_0, P_2)$ and $F_{\boldsymbol{\rho}}(P_2, P_3)$. Moreover the ideal polyhedron $\widetilde{\Delta}_e$ dual to a cell e of $\partial Ph(\rho)$ is described as follows:*

1. *If e is the face $F_{\boldsymbol{\rho}}(P_0)$ (resp. $F_{\boldsymbol{\rho}}(P_2)$), then $\widetilde{\Delta}_e$ is the ideal edge spanned by $\{\infty, \rho(P_0)(\infty)\}$ (resp. $\{\infty, \rho(P_2)(\infty)\}$).*
2. *If e is the edge $F_{\boldsymbol{\rho}}(P_0, P_2)$ (resp. $F_{\boldsymbol{\rho}}(P_2, P_3) = F_{\boldsymbol{\rho}}(P_1', P_3')$), then $\widetilde{\Delta}_e$ is the ideal quadrangle spanned by $\{\infty, \rho(P_0)(\infty), \rho(P_1)(\infty), \rho(P_2)(\infty)\}$ (resp. $\{\infty, \rho(P_1')(\infty), \rho(P_2')(\infty), \rho(P_3')(\infty)\}$).*

Proof. By the definition of a good labeled representation and the definition of $\mathcal{L}(\boldsymbol{\nu})$ for the thin label $\boldsymbol{\nu}$ (Definition 3.2.7), we obtain the classification of the cells of $\partial Ph(\rho) = \partial Eh(\boldsymbol{\rho})$. Since (1) is obtained easily, we prove only (2) for the case when $e = F_{\boldsymbol{\rho}}(P_0, P_2)$. Let p be a point of $F_{\boldsymbol{\rho}}(P_0, P_2)$. Then the $(\sim)$-equivalence class $[p]$ modulo $\Gamma_\infty = \langle \rho(K) \rangle$ is equal to $\{p, \rho(P_2)(p)\}$ (see Sect. 6.4). On the other hand, we see, by the forthcoming Proposition 6.7.3, that (ρ, τ) is an isosceles representation. So $F_{\boldsymbol{\rho}}(P_0, P_2) = Ih(\rho(P_0)) \cap Ih(\rho(P_2))$ is identical with $\operatorname{Axis}(\rho(P_1))$ by Proposition 5.2.3. Thus we see $\Gamma_p = \langle \rho(P_1) \rangle$ as in the proof of Lemma 6.6.3. Hence, by Lemma 6.6.2, the ideal polyhedron $\widetilde{\Delta}_e$ is spanned by the image of ∞ by the elements in the set

$$\rho\left\{\{1, P_1\} \cup \{1, P_1\}P_2^{-1}\right\} = \rho\{1, P_1, P_2, P_1P_2\}.$$

Since $\rho(P_1P_2)(\infty) = \rho(P_0P_0P_1P_2)(\infty) = \rho(P_0)(\infty)$, $\widetilde{\Delta}_e$ is spanned by

$$\{\infty, \rho(P_0)(\infty), \rho(P_1)(\infty), \rho(P_2)(\infty)\}.$$

Since (ρ, τ) is an isosceles representation, the convex hull of these four points is an ideal quadrangle.

As a corollary to Lemmas 6.6.3 and 6.6.4, we obtain the following.

Corollary 6.6.5. *For every good labeled representation $\boldsymbol{\rho} = (\rho, \boldsymbol{\nu})$, $\mathcal{L}(\boldsymbol{\rho})$ is equal to the projection to $\mathbb{C}$ of the cross section of the ideal polyhedral complex $\Delta_{\mathbb{E}}(\rho)$ with the horizontal horosphere ∂H_∞.*

Proof. We first note that any ideal polyhedron of $\widetilde{\Delta}_{\mathbb{E}}(\rho)$ which has ∞ as an ideal vertex is the geometric dual to a cell of $\partial Ph(\rho) \subset \widetilde{\mathrm{Ford}}(\rho)$. Thus the cross section of $\widetilde{\Delta}_{\mathbb{E}}(\rho)$ with the horizontal horosphere ∂H_∞ consists of the cross sections of the ideal polyhedra which appear in Lemmas 6.6.3 or 6.6.4, according as $\boldsymbol{\nu}$ is thick or thin.

Suppose $\boldsymbol{\nu}$ is thick. Then (the image by the projection to $\mathbb{C}$ of) the cross section of $\widetilde{\Delta}_e$ with ∂H_∞ is equal to the Euclidean simplex $\langle \rho(P)(\infty)\rangle$, $\langle \rho(P)(\infty), \rho(Q)(\infty)\rangle$ or $\langle \rho(P)(\infty), \rho(Q)(\infty), \rho(R)(\infty)\rangle$ according as e is $F_{\boldsymbol{\rho}}(P)$, $F_{\boldsymbol{\rho}}(P,Q)$ or $F_{\boldsymbol{\rho}}(P,Q,R)$. Hence these (images of the) cross sections constitute $\mathcal{L}(\boldsymbol{\rho})$.

Suppose $\boldsymbol{\nu}$ is thin. Then (the image by the projection to $\mathbb{C}$ of) the cross section of $\widetilde{\Delta}_e$ with ∂H_∞ is equal to the Euclidean simplex $\langle \rho(P_0)(\infty)\rangle$, $\langle \rho(P_2)(\infty)\rangle$, $\langle \rho(P_0)(\infty), \rho(P_2)(\infty)\rangle$ or $\langle \rho(P_2)(\infty), \rho(P_3)(\infty)\rangle$ according as e is $F_{\boldsymbol{\rho}}(P_0)$, $F_{\boldsymbol{\rho}}(P_2)$, $F_{\boldsymbol{\rho}}(P_0,P_2)$ or $F_{\boldsymbol{\rho}}(P_2,P_3)$. Hence, again, these (images of the) cross sections constitute $\mathcal{L}(\boldsymbol{\rho})$. This completes the proof of the corollary.

Proof (Proof of Proposition 6.6.1). By Corollary 6.6.5, $\mathcal{L}(\boldsymbol{\rho})$ is identified with the cross section of the ideal polyhedral complex $\widetilde{\Delta}_{\mathbb{E}}(\rho)$ with the horizontal horosphere ∂H_∞. Hence $\mathcal{L}(\rho,\boldsymbol{\nu})$ satisfies the first two conditions in Definition 3.2.2 (Simple). So we have only to show that the last condition is satisfied. If the length m of the chain $\Sigma(\boldsymbol{\nu})$ is equal to 1, then we have nothing to prove. So we assume $m \geq 2$. By the condition Frontier, $E^-(\boldsymbol{\rho}) = E^-(\rho,\sigma_1)$, and hence all components of $\theta^-(\rho,\sigma_1)$ are non-negative. Let s_1 be the vertex of $\sigma_1 = \langle s_0, s_1, s_2\rangle$ which is not a vertex of σ_2, and let $\{P_j\}$ be the sequence of elliptic generators associated with σ_1. Then (P_0,P_1,P_2) is a 2-simplex of $\mathcal{L}(\boldsymbol{\nu})$ and hence $F_{\boldsymbol{\rho}}(P_0,P_1,P_2) = \cap_{j=0}^2 Ih(\rho(P_j))$ is a vertex of $\partial Eh(\boldsymbol{\rho})$. Hence $\cap_{j=0}^2 Ih(\rho(P_j))$ is a singleton. Thus by Lemma 4.3.6, we have either $\theta^+(\rho,\sigma_1;s_1) < 0 < \theta^-(\rho,\sigma_1;s_1)$ or $\theta^-(\rho,\sigma_1;s_1) < 0 < \theta^+(\rho,\sigma_1;s_1)$. Since $\theta^-(\rho,\sigma_1;s_1) \geq 0$ as observed in the above, we must have $\theta^+(\rho,\sigma_1;s_1) < 0 < \theta^-(\rho,\sigma_1;s_1)$. Hence $\mathcal{L}(\rho,\sigma_1)$ is convex to the below at $c(\rho(P_1))$ by Corollary 4.2.15. Hence the edge $\overline{c(\rho(P_0))c(\rho(P_2))}$ of $\mathcal{L}(\rho,\sigma_2)$ lies above $\mathcal{L}(\rho,\sigma_1)$ by Lemma 3.1.7. Hence $\mathcal{L}(\rho,\sigma_2)$ lies above $\mathcal{L}(\rho,\sigma_1)$. Since the open 2-simplices of $\mathcal{L}(\rho,\boldsymbol{\nu})$ are disjoint, $\mathcal{L}(\rho,\sigma_3)$ lies above $\mathcal{L}(\rho,\sigma_2)$. By repeating this argument, we see that the last condition is satisfied. Hence $\mathcal{L}(\rho,\boldsymbol{\nu})$ is simple.

Proof (Proof of Theorem 6.1.12). Let $\boldsymbol{\rho} = (\rho,\boldsymbol{\nu})$ be a good labeled representation, and set $\Gamma = \rho(\pi_1(T))$. Then Γ is quasifuchsian and Ford(Γ) is equivalent to Spine$(\boldsymbol{\nu})$ by Theorem 6.1.8. By the construction of Trg$(\boldsymbol{\nu})$ in Sect. 1.4, we can see that Trg$(\boldsymbol{\nu})$ is a "topological dual" to Ford(Γ). (See Fig. 1.1. By collapsing each vertical line contained in the side of the cube into a point in the figure, we obtain a topological ideal tetrahedron dual to the vertex of Spine(σ_1,σ_2).) On the other hand, $\Delta_{\mathbb{E}}(\Gamma)$ is the geometric dual to Ford(Γ) (cf. Sect. 1.1), which in tern implies that $\Delta_{\mathbb{E}}(\Gamma)$ is a topological dual to Trg$(\boldsymbol{\nu})$. Thus $\Delta_{\mathbb{E}}(\Gamma)$ should be combinatorially equivalent to Spine$(\boldsymbol{\nu})$. This is the idea of the proof. To make this idea explicit, we give more precise descriptions of $\mathrm{Trg}(\boldsymbol{\nu})$ and $\Delta_{\mathbb{E}}(\Gamma)$. For simplicity we assume $\boldsymbol{\nu}$ is thick. (The thin case can be easily treated.) Set $\Sigma(\boldsymbol{\nu}) = (\sigma_1, \sigma_2, \cdots, \sigma_n)$. Then we have the following one-to-one correspondences for $\mathrm{Trg}(\boldsymbol{\nu})$.

- The 3-cell of $\mathrm{Trg}(\boldsymbol{\nu})$ corresponding to (σ_i, σ_{i+1}) is the image of any of the topological ideal tetrahedra in $\widetilde{\mathrm{Trg}}(\sigma_i, \sigma_{i+1})$.
- The 2-cell of $\mathrm{Trg}(\boldsymbol{\nu})$ corresponding to (s_0, s_1) (resp (s_1, s_2)) is the image of the face $\langle (0,0), (p_0, q_0), (p_1, q_1)\rangle$ (resp. $\langle (0,0), (p_1, q_1), (p_2, q_2)\rangle$) of $\widetilde{\mathrm{trg}}(\sigma_i)$, where $\sigma_i = \langle s_0, s_1, s_2\rangle = \langle q_0/p_0, q_1/p_1, q_2/p_2\rangle$.
- The 1-cell of $\mathrm{Trg}(\boldsymbol{\nu})$ corresponding to a slope $s \in \sigma_i^{(0)} \subset \Sigma(\boldsymbol{\nu})^{(0)}$ is the image of any of the edges of $\widetilde{\mathrm{trg}}(\sigma_i)$ of slope s.

For $\Delta_{\mathbb{E}}(\Gamma)$, we have the following one-to-one correspondences.

- The 3-cell of $\Delta_{\mathbb{E}}(\Gamma)$ corresponding to (σ_i, σ_{i+1}) is the image of the ideal tetrahedron $\langle \infty, \rho(P)(\infty), \rho(Q)(\infty), \rho(R)(\infty)\rangle$, where (P, Q, R) is an elliptic generator triple such that $\sigma_i = \langle s(P), s(Q), s(R)\rangle$ and $\sigma_{i+1} = \langle s(P), s(R), s(Q^R)\rangle$.
- The 2-cell of $\Delta_{\mathbb{E}}(\Gamma)$ corresponding to (s_0, s_1) (resp (s_1, s_2)) is the image of the ideal triangle $\langle \infty, \rho(P)(\infty), \rho(Q)(\infty)\rangle$ (resp. $\langle \infty, \rho(Q)(\infty), \rho(R)(\infty)\rangle$), where (P, Q, R) is an elliptic generator triple such that $\sigma_i = \langle s(P), s(Q), s(R)\rangle$.
- The 1-cell of $\mathrm{Trg}(\boldsymbol{\nu})$ corresponding to a slope $s \in \sigma_i^{(0)} \subset \Sigma(\boldsymbol{\nu})^{(0)}$ is the image of the ideal edge $\langle \infty, \rho(P)(\infty)\rangle$, where P is an elliptic generator of slope s.

Thus we have a natural correspondence between cells of $\mathrm{Trg}(\boldsymbol{\nu})$ and $\Delta_{\mathbb{E}}(\Gamma)$. Moreover we can easily check that it respects the incidence relation. Hence $\mathrm{Trg}(\boldsymbol{\nu})$ and $\Delta_{\mathbb{E}}(\Gamma)$ are combinatorially equivalent.

6.7 Characterization of $\Sigma(\boldsymbol{\nu})$ for good labeled representations

In this section, we prove Propositions 6.7.1 and 6.7.2 below, by refining the proof of Proposition 6.6.1. These are useful in the actual computation of the Ford domains, because they enable us to reconstruct the chain $\Sigma(\boldsymbol{\nu})$ of a good (or quasifuchsian) labeled representation $(\rho, \boldsymbol{\nu})$ from a single triangle in the chain. To be precise, if we know that a triangle σ belongs to the chain $\Sigma(\boldsymbol{\nu})$ associated with a quasifuchsian labeled representation $(\rho, \boldsymbol{\nu})$, then we can reconstruct the chain $\Sigma(\boldsymbol{\nu})$ only from the representation ρ and the triangle σ. This is because these propositions enable us to find out the slope of σ which is not a slope of the successor (resp. predecessor) of σ in $\Sigma(\boldsymbol{\nu})$ and hence the successor (resp. predecessor) itself. This holds the key to the successful computer program OPTi [78] (cf. [79]) developed by the third author and is the starting point of the numerous computer experiments made by the last two authors (see [50] [79] [82]).

Proposition 6.7.1. *Let $\boldsymbol{\rho} = (\rho, \boldsymbol{\nu})$ be a thick good labeled representation, and let $(\sigma_1, \sigma_2, \cdots, \sigma_m)$ be the chain $\Sigma(\boldsymbol{\nu})$. Then for each triangle σ_i in the chain, $\mathcal{L}(\rho, \sigma_i)$ is simple and ρ satisfies the triangle inequality at σ_i. In particular $\theta^{\epsilon}(\rho, \sigma_i)$ is defined.*

Suppose $1 \le i < m$ and let s_1 be the vertex of $\sigma := \sigma_i = \langle s_0, s_1, s_2 \rangle$ which is not a vertex of the successor σ_{i+1} of σ_i. Then the following hold.

1. *$\theta^+(\rho, \sigma; s_1) < 0 < \theta^-(\rho, \sigma; s_1)$.*
2. *$\cap_{j=0}^{2} Ih(\rho(P_j))$ is a singleton which is not contained in $Dh(\rho(P_{-1}))$ nor $Dh(\rho(P_3))$, where $\{P_j\}$ is the sequence of elliptic generators associated with σ.*

Moreover, s_1 is characterized by these properties among the vertices of σ, Namely, the following hold for the remaining vertices s_0 and s_2 of σ.

1. *Either (i) $\theta^+(\rho, \sigma; s_0) \ge 0$ or (ii) $\theta^+(\rho, \sigma; s_0) < 0$ and $\cap_{j=-1}^{1} Ih(\rho(P_j))$ is a singleton which is contained in* int *$Dh(\rho(P_2))$.*
2. *Either (i) $\theta^+(\rho, \sigma; s_2) \ge 0$ or (ii) $\theta^+(\rho, \sigma; s_2) < 0$ and $\cap_{j=1}^{3} Ih(\rho(P_j))$ is a singleton which is contained in* int *$Dh(\rho(P_0))$*

Proof. By Proposition 6.6.1, $\mathcal{L}(\rho, \sigma)$ is simple. The condition Duality implies that $F_{\boldsymbol{\rho}}(P_j, P_{j+1})$ is a 1-dimensional convex polyhedron and hence $Ih(\rho(P_j)) \cap Ih(\rho(P_{j+1}))$ is a geodesic for every integer j. Thus ρ satisfies the triangle inequality at σ by Lemma 4.2.1.

Now suppose that $1 \le i < m$ and s_1 is the vertex of $\sigma = \sigma_i$ which is not a vertex of σ_{i+1}. Then (P_0, P_1, P_2) is a 2-simplex of $\mathcal{L}(\boldsymbol{\nu})$ and hence $F_{\boldsymbol{\rho}}(P_0, P_1, P_2) = \cap_{j=0}^{2} Ih(\rho(P_j))$ is a vertex of $\partial Eh(\boldsymbol{\rho})$. Hence $\cap_{j=0}^{2} Ih(\rho(P_j))$ is a singleton. Thus by Lemma 4.3.6, we have either $\theta^+(\rho, \sigma; s_1) < 0 < \theta^-(\rho, \sigma; s_1)$ or $\theta^-(\rho, \sigma; s_1) < 0 < \theta^+(\rho, \sigma; s_1)$. On the other hand, since $\mathcal{L}(\boldsymbol{\rho})$ is simple by Proposition 6.6.1 (and Theorem 6.1.8), $\mathcal{L}(\rho, \sigma_{i+1})$ lies above $\mathcal{L}(\rho, \sigma) = \mathcal{L}(\rho, \sigma_i)$ (cf. Definition 3.2.2(3)). Since $\overline{c(\rho(P_0))c(\rho(P_2))}$ is an edge of $\mathcal{L}(\rho, \sigma_{i+1})$, this implies that $\mathcal{L}(\rho, \sigma)$ is convex to the below at $c(\rho(P_1))$ by Lemma 3.1.7. Hence we see $\theta^+(\rho, \sigma; s_1) < \theta^-(\rho, \sigma; s_1)$ by Corollary 4.2.15, and therefore $\theta^+(\rho, \sigma; s_1) < 0 < \theta^-(\rho, \sigma; s_1)$ by the preceding observation.

Next, we show that the singleton $v(0,1,2) := \cap_{j=0}^{2} Ih(\rho(P_j))$ is not contained in $Dh(\rho(P_{-1}))$ nor $Dh(\rho(P_3))$. Suppose to the contrary that $v(0,1,2)$ is contained in, say $Dh(\rho(P_3))$. Since $v(0,1,2)$ is contained in $\partial Eh(\boldsymbol{\rho})$ by the condition Duality, it must be contained $Ih(\rho(P_3))$ and hence $\cap_{j=0}^{3} Ih(\rho(P_j)) \ne \emptyset$. Thus $\rho(KP_0)$ is elliptic by Lemma 5.5.1. This is a contradiction (cf. Lemma 2.5.4(1)), because ρ is quasifuchsian by Theorem 6.1.8. We come to a similar contradiction if we assume that $v(0,1,2)$ is contained in $Dh(\rho(P_{-1}))$. Hence $\cap_{j=0}^{2} Ih(\rho(P_j))$ is not contained in $Dh(\rho(P_{-1}))$ nor $Dh(\rho(P_3))$.

Next we prove the assertions for the remaining slopes s_0 and s_2. Since the arguments are parallel, we study only the slope s_2. Suppose that $\theta^+(\rho, \sigma; s_2) < 0$. Then we have $\theta^-(\rho, \sigma; s_2) > 0$ by Lemma 4.3.5, because the condition Duality implies that $F_{\boldsymbol{\rho}}(P_2)$ is a 2-dimensional convex polyhedron and hence $Ih(\rho(P_2)) \cap Eh(\rho(P_1)) \cap Eh(\rho(P_2))$ has dimension 2. Thus $v(1,2,3) :=$

$\cap_{j=1}^{3} Ih(\rho(P_j))$ is a singleton by Lemma 4.3.6, and we can see as in the previous argument appealing to Lemma 2.5.4(1) that it is different from $v(0,1,2)$. On the other hand, since $\theta^+(\rho,\sigma;s_2) < 0$, we see by Lemma 4.3.1 that $v^+(1,2) := v^+(\rho;P_1,P_2)$ lies in $\operatorname{int} D(\rho(P_3))$. Hence the interior of the half geodesic $[v(1,2,3),v^+(1,2)]$ is contained in $\operatorname{int} Dh(\rho(P_3))$. Since $v(0,1,2) = F_{\boldsymbol{\rho}}(P_0,P_1,P_2)$ lies in $\partial Eh(\boldsymbol{\rho})$, it cannot be contained in the interior of the half geodesic. Thus the points $v^-(1,2) := v^-(\rho;P_1,P_2)$, $v(0,1,2)$, $v(1,2,3)$, and $v^+(1,2)$ lie in the complete geodesic $\overline{Ih}(\rho(P_1)) \cap \overline{Ih}(\rho(P_2))$ in this order. In particular, $v(1,2,3)$ is contained in the interior of the half geodesic $[v(0,1,2),v^+(1,2)]$, which in tern is contained in $\operatorname{int} Dh(\rho(P_0))$, because the condition $\theta^+(\rho,\sigma;s_1) < 0$ implies that $v^+(1,2)$ is contained in $\operatorname{int} D(\rho(P_0))$ by Lemma 4.3.1. Hence $v(1,2,3)$ is contained in $\operatorname{int} Dh(\rho(P_0))$. This completes the proof of Proposition 6.7.1.

By parallel arguments, we obtain the following twin to the above proposition.

Proposition 6.7.2. *Let $\boldsymbol{\rho} = (\rho,\boldsymbol{\nu})$ be a thick good labeled representation, and let $(\sigma_1,\sigma_2,\cdots,\sigma_m)$ be the chain $\Sigma(\boldsymbol{\nu})$. Suppose $1 < i \le m$ and let s_1 be the vertex of $\sigma := \sigma_i = \langle s_0,s_1,s_2\rangle$ which is not a vertex of the predecessor σ_{j-1} of σ_i. Then the following hold.*

1. *$\theta^-(\rho,\sigma;s_1) < 0 < \theta^+(\rho,\sigma;s_1)$.*
2. *$Ih(\rho(P_0)) \cap Ih(\rho(P_1)) \cap Ih(\rho(P_2))$ is a singleton which is not contained in $Dh(\rho(P_{-1}))$ nor $Dh(\rho(P_3))$, where $\{P_j\}$ is the sequence of elliptic generators associated with σ.*

Moreover, s_1 is characterized by these properties among the vertices of σ. Namely, the following hold for the remaining vertices s_0 and s_2 of σ.

1. *Either (i) $\theta^-(\rho,\sigma;s_0) \ge 0$ or (ii) $\theta^-(\rho,\sigma;s_0) < 0$ and $Ih(\rho(P_{-1})) \cap Ih(\rho(P_0)) \cap Ih(\rho(P_1))$ is a singleton which is contained in $\operatorname{int} Dh(\rho(P_2))$.*
2. *Either (i) $\theta^-(\rho,\sigma;s_2) \ge 0$ or (ii) $\theta^-(\rho,\sigma;s_2) < 0$ and $Ih(\rho(P_1)) \cap Ih(\rho(P_2)) \cap Ih(\rho(P_3))$ is a singleton which is contained in $\operatorname{int} Dh(\rho(P_0))$*

For thin good labeled representations, we have the following proposition.

Proposition 6.7.3. *For a type-preserving representation ρ, the following conditions are equivalent under Notation 2.1.14 (Adjacent triangles).*

1. *There is a thin label $\boldsymbol{\nu} = (\nu^-,\nu^+)$ with $\nu^\pm \in \operatorname{int}\tau$ such that $(\rho,\boldsymbol{\nu})$ is a good labeled representation.*
2. *The side parameter $\theta^\epsilon(\rho,\sigma)$ is defined and $\theta^\epsilon(\rho,\sigma) = (+,0,+)$ for each $\epsilon \in \{-,+\}$.*
3. *The side parameter $\theta^\epsilon(\rho,\sigma')$ is defined and $\theta^\epsilon(\rho,\sigma') = (+,+,0)$ for each $\epsilon \in \{-,+\}$.*

Moreover if the above mutually equivalent conditions are satisfied, then $\theta^\epsilon(\rho,\sigma)$ *and* $\theta^\epsilon(\rho,\sigma')$ *determine the same point,* ν^ϵ*, of* $\operatorname{int}\tau$ *for each* $\epsilon \in \{-,+\}$*, and* $\boldsymbol{\nu} = (\nu^-,\nu^+)$ *is the unique label such that* $(\rho,\boldsymbol{\nu})$ *is a good labeled representation. Furthermore* (ρ,τ) *is an isosceles representation.*

Proof. (1) $\rightarrow$ (2). Suppose $(\rho,\boldsymbol{\nu})$ is a good labeled representation for some thin label $\boldsymbol{\nu} = (\nu^-,\nu^+)$ with $\nu^\pm \in \operatorname{int}\tau$. Then $\theta^\epsilon(\rho,\sigma)$ gives the barycentric coordinate of ν^ϵ. Since $\nu^\epsilon \in \operatorname{int}\tau$, this implies that $\theta^\epsilon(\rho,\sigma) = (+,0,+)$.

(2) $\rightarrow$ (1). Suppose the condition (2) is satisfied. Then (ρ,τ) is an isosceles representation by Proposition 5.2.3. Since $\theta^\epsilon(\rho,\sigma) = (+,0,+)$ for each ϵ, Proposition 5.2.8 implies that ρ is quasifuchsian and its Ford domain is equal to $Eh(\tau)$. Hence we see that $(\rho,\boldsymbol{\nu})$ is a good labeled representation, where $\boldsymbol{\nu} = (\nu^-,\nu^+)$ is the thin label with $\nu^\epsilon = \theta^\epsilon(\rho,\sigma) \in \operatorname{int}\tau$.

Similarly we see the equivalence among the conditions (1) and (3). The additional assertions follow from the above proof and Proposition 6.2.5 (Uniqueness).

7

Openness

This chapter is devoted to the proof of Proposition 6.2.1 (Openness). The main ingredient of the proof is the study of the behavior of *hidden isometric hemispheres*, namely those isometric hemispheres which intersect the Ford domain but do not support 2-dimensional faces of the Ford domain (Definition 7.1.1).

In Sect. 7.1, we give a recipe to list all hidden isometric hemispheres in the general setting (Lemma 7.1.3). By using the result, we show that the Ford domain of a good labeled representation $(\rho, \boldsymbol{\nu})$ has a hidden isometric hemisphere if and only if either one of the components of $\boldsymbol{\nu}$ is *non-generic* (Definition 7.1.4) or $\boldsymbol{\nu}$ is thin (Lemma 7.1.6).

In Sects. 7.2 and 7.3, respectively, we prove Proposition 6.2.1 (Openness) around thick and thin good labeled representation.

7.1 Hidden isometric hemispheres

Throughout this section, except the last part, Γ denotes a Kleinian group such that Γ_∞ contains parabolic transformations.

Definition 7.1.1. *We call an isometric hemisphere* $\overline{Ih}(A)$ $(A \in \Gamma - \Gamma_\infty)$ *a* hidden isometric hemisphere *for the Ford domain* $\overline{Ph}(\Gamma)$, *if* $\overline{Ph}(\Gamma) \cap \overline{Ih}(A) \neq \emptyset$ *while* $Ih(A)$ *does not support a* 2*-dimensional face of the convex polyhedron* $Ph(\Gamma)$.

We begin by explaining the following key observation, which enables us to determine all hidden isometric hemispheres for the Ford domain. This is a reformulation of [72, Lemma 5.45], and is essentially equal to Lemma 6.6.2.

Lemma 7.1.2. *Let* x *be a point in* $\overline{Ph}(\Gamma)$, *and let* A *be an element of* $\Gamma - \Gamma_\infty$ *such that* $x \in \overline{Ih}(A)$. *Then* $A(x)$ *is also contained in* $\overline{Ph}(\Gamma)$.

Proof. We begin by presenting a characterization of $\overline{Ph}(\Gamma)$ which is suitable for our purpose. For each $A \in PSL(2,\mathbb{C})$ and $x \in \overline{\mathbb{H}}^3$, the derivative $dA_x : T_x\overline{\mathbb{H}}^3 \to T_{A(x)}\overline{\mathbb{H}}^3$ is equal to a positive constant times an orthogonal transformation with respect to the Euclidean metric of $\overline{\mathbb{H}}^3$. We denote the positive constant by $|A'(x)|$. Then we have the following characterization of $\overline{Ph}(\Gamma)$.

$$\overline{Ph}(\Gamma) = \{x \in \overline{\mathbb{H}}^3 \,|\, |A'(x)| \leq 1 \text{ for every } A \in \Gamma\}.$$

Now let x be a point in $\overline{Ph}(\Gamma)$ and let A be an element of Γ such that $x \in \overline{Ih}(A)$. Then for every $B \in \Gamma$,

$$\begin{aligned} |B'(A(x))| &= \frac{|(BA)'(x)|}{|A'(x)|} \\ &= |(BA)'(x)| && \text{because } x \in \overline{Ih}(A) \\ &\leq 1 && \text{because } x \in \overline{Ph}(\Gamma). \end{aligned}$$

Hence $A(x) \in \overline{Ph}(\Gamma)$ by the preceding characterization of $\overline{Ph}(\Gamma)$. This completes the proof.

By using this lemma, we can determine all isometric hemispheres containing a point of $\overline{Ph}(\Gamma)$. To describe the result, we use a slight generalization of the notation in Sect. 6.6. For each point p in $\overline{Ph}(\Gamma)$, set

$$[p] := \{x \in \overline{Ph}(\Gamma) \,|\, x = A(p) \text{ for some } A \in \Gamma\}.$$

Then we have the following companion to Lemma 6.6.2.

Lemma 7.1.3. *Suppose that Γ is geometrically finite. Then for each point $p \in \overline{Ph}(\Gamma)$, the following hold.*

1. *$[p]/\Gamma_\infty$ is a finite set.*
2. *Let $\{p = p_1, p_2, \cdots, p_n\} \subset [p]$ be a representative for $[p]/\Gamma_\infty$, and let $A_i \in \Gamma$ be an element such that $A_i(p) = p_i$ and $A_1 = 1$. Then the set of the isometric hemispheres of elements of $\Gamma - \Gamma_\infty$ containing the point p is equal to*

$$\left\{\overline{Ih}(A) \,\middle|\, A \in \left(\bigsqcup_{i=1}^{n} A_i\Gamma_p\right) - \{1\}\right\}.$$

where Γ_p is the stabilizer of p with respect to the action of Γ. In particular, the number of the isometric hemispheres containing p is infinite if and only if p is a parabolic fixed point. If p is not a parabolic fixed point, then the number is equal to $n|\Gamma_p| - 1$.

Proof. (1) follows from Lemma A.1.10.

(2) For each $p \in \overline{Ph}(\Gamma)$ and $A \in \Gamma - \Gamma_\infty$, we have the following equivalence.

$$\begin{aligned}
&p \in \overline{Ih}(A),\\
&\Leftrightarrow A(p) \in \overline{Ph}(\Gamma),\\
&\Leftrightarrow A(p) \in [p],\\
&\Leftrightarrow A(p) = BA_i(p) \text{ for some } i \in \{1, \cdots, n\} \text{ and } B \in \Gamma_\infty\\
&\Leftrightarrow A = BA_iC \text{ for some } i \in \{1, \cdots, n\},\ B \in \Gamma_\infty \text{ and } C \in \Gamma_p,\\
&\Leftrightarrow Ih(A) = Ih(A_iC) \text{ for some some } i \in \{1, \cdots, n\} \text{ and } C \in \Gamma_p.
\end{aligned}$$

Here the first equivalence follows from Lemma 7.1.2, and the last equivalence follows from Lemma 4.1.1(3). Hence we obtain the first assertion. Moreover, Lemma 4.1.1(3) also implies that the isometric hemispheres $Ih(A_iC)$ ($i \in \{1, \cdots, n\}$ and $C \in \Gamma_p$) are mutually distinct. Hence there are infinitely many isometric hemispheres which contain p if and only if Γ_p is infinite. Since Γ is geometrically finite and $p \in P(\Gamma)$, this is equivalent to the condition that p is a parabolic fixed point by Lemma A.1.4. This completes the proof.

By using the above lemma, we can completely determine the hidden isometric hemispheres for good labeled representations. Roughly speaking, those hidden isometric hemispheres described in Lemma 4.5.4(4) and Corollary 5.2.5 are the only hidden isometric hemispheres for good labeled representations. To present the explicit statement, we introduce the following definition and notation.

Definition 7.1.4. *Let $\boldsymbol{\nu} = (\nu^-, \nu^+) \in \mathbb{H}^2 \times \mathbb{H}^2$ be a thick label.*

(1) The component ν^ϵ of the label $\boldsymbol{\nu}$ is said to be non-generic *or* generic *according as it is contained in the 1-skeleton of the Farey triangulation or not, namely, $\dim \delta^\epsilon(\boldsymbol{\nu}) = 1$ or 2. Here $\delta^\epsilon(\boldsymbol{\nu})$ is the edge or the triangle of $\mathcal{D}$ whose (relative) interior contains ν^ϵ (Definition 1.3.4(2)).*

(2) Let $\epsilon \in \{-, +\}$. If ν^ϵ is non-generic, then $\sigma^{\epsilon,*}(\boldsymbol{\nu})$ denotes the triangle of $\mathcal{D}$ such that $\sigma^\epsilon(\boldsymbol{\nu}) \cap \sigma^{\epsilon,*}(\boldsymbol{\nu}) = \delta^\epsilon(\boldsymbol{\nu})$. If ν^ϵ is generic, $\sigma^{\epsilon,*}(\boldsymbol{\nu})$ is undefined.

Notation 7.1.5 (Non-generic label). Let $\boldsymbol{\nu} = (\nu^-, \nu^+) \in \mathbb{H}^2 \times \mathbb{H}^2$ be a thick label such that ν^ϵ is non-generic. Then the symbols s_j and s'_j denote the elements of $\hat{\mathbb{Q}}$ such that $\sigma^\epsilon(\boldsymbol{\nu}) = \langle s_0, s_1, s_2\rangle$ and $\sigma^{\epsilon,*}(\boldsymbol{\nu}) = \langle s'_0, s'_1, s'_2\rangle$ such that $s'_0 = s_0$ and $s'_1 = s_2$. As in Notation 2.1.14 (Adjacent triangles), $\{P_j\}$ and $\{P'_j\}$, respectively, denote the sequences of elliptic generators associated with $\sigma^\epsilon(\boldsymbol{\nu})$ and $\sigma^{\epsilon,*}(\boldsymbol{\nu})$ such that $P'_0 = P_0$, $P'_1 = P_2$ and $P'_2 = P_1^{P_2}$. (See the upper right figure in Fig. 7.3).

Then we have the following classification of the hidden isometric hemispheres for good labeled representations (cf. [72, Corollary 6.11]).

Lemma 7.1.6 (Hidden isometric hemisphere). *Let $\boldsymbol{\rho} = (\rho, \boldsymbol{\nu})$ be a good labeled representation. Then the following is the complete list of the hidden isometric hemispheres for $\overline{Ph}(\rho)$ modulo $\langle \rho(K)\rangle$.*

1. *If ν is thick and ν^ϵ is non-generic, then $\overline{Ih}(\rho(P_2'))$ is a hidden isometric hemisphere for $\overline{Ph}(\rho)$ under Notation 7.1.5 (Non-generic label). In fact $\overline{Ih}(\rho(P_2')) \cap \overline{Ph}(\rho) = v^\epsilon(\rho; P_2, P_3)$ (see Fig. 7.1(a)).*
2. *If ν is thin and $\nu^\pm \in \tau$ under Notation 2.1.14 (Adjacent triangles). Then $\overline{Ih}(\rho(P_1))$ and $Ih(\rho(P_2'))$ are hidden isometric hemispheres for $\overline{Ph}(\rho)$. In fact $\overline{Ih}(\rho(P_1)) \cap \overline{Ph}(\rho) = \overline{\mathrm{Axis}}(\rho(P_1))$ and $\overline{Ih}(\rho(P_2')) \cap \overline{Ph}(\rho) = \overline{\mathrm{Axis}}(\rho(P_2'))$ (see Fig. 7.1(b)).*

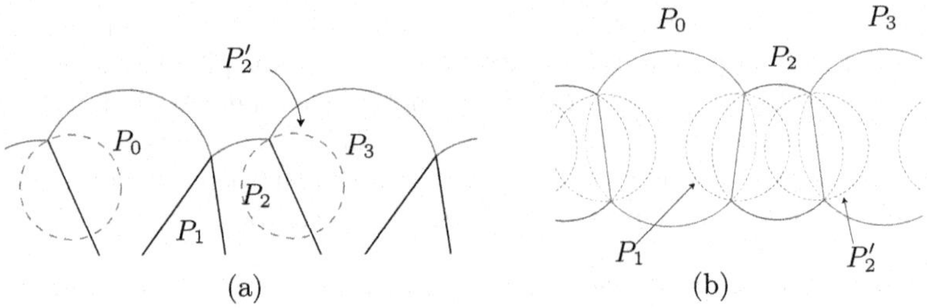

Fig. 7.1. Hidden isometric hemispheres

Proof. Let p be a point in $\overline{Ph}(\rho)$, and let $Hidden(p)$ (resp. $Visible(p)$) be the number of the hidden isometric hemispheres for $\overline{Ph}(\rho)$ containing p (resp. the number of the isometric hemispheres containing p which support a 2-dimensional face of $Ph(\rho)$). Then by Lemma 7.1.3, we have the following formula.

$$Hidden(p) = |[p]/\Gamma_\infty| \cdot |\Gamma_p| - Visible(p) - 1.$$

Then we obtain the following table (cf. Proof of Lemmas 6.6.3 and 6.6.4), which implies the desired result. In the row starting from $F_\rho(P)$ in the table, for example, the first line is for the case when p is generic, i.e., $p \notin \mathrm{Axis}(\rho(P))$, and the second line is for the case when $p \in \mathrm{Axis}(\rho(P))$. Similar conventions are employed for the rows with two lines.

The case when ν is thick				
The face whose interior contains p	$\lvert\Gamma_p\rvert$	$\lvert[p]/\Gamma_\infty\rvert$	$visible(p)$	$hidden(p)$
$F_\rho(P,Q,R)$	2	2	3	0
$F_\rho(P,Q)$	1	3	2	0
$F_\rho(P)$	1	2	1	0
	2	1	1	0
non-degenerate outer edge $f^\epsilon_\rho(P)$	1	2	1	0
	2	1	1	0
outer vertex $f^\epsilon_\rho(P_j, P_{j+1})$ (ν^ϵ: generic)	1	3	2	0
degenerate outer edge $f^\epsilon_\rho(P_1)$	2	2	3	0
outer vertex $f^\epsilon_\rho(P_2, P_3)$ (ν^ϵ: non-generic)	2	2	2	1

The case when $\boldsymbol{\nu}$ is thin				
The face whose interior contains p	$\lvert\Gamma_p\rvert$	$\lvert[p]/\Gamma_\infty\rvert$	$visible(p)$	$hidden(p)$
$F_{\boldsymbol{\rho}}(P_0, P_2)$	2	2	2	1
$F_{\boldsymbol{\rho}}(P_2, P_3)$	2	2	2	1
$F_{\boldsymbol{\rho}}(P_j)$ $j \not\equiv 1 \pmod 3$	1	2	1	0
	2	1	1	0
$f^\epsilon_{\boldsymbol{\rho}}(P_j)$ $j \not\equiv 1 \pmod 3$	1	2	1	0
	2	1	1	0
$f^\epsilon_{\boldsymbol{\rho}}(P_0, P_2)$	2	2	2	1
$f^\epsilon_{\boldsymbol{\rho}}(P_2, P_3)$	2	2	2	1

7.2 Proof of Proposition 6.2.1 (Openness) - Thick Case -

In this section, we prove Proposition 6.2.1 (Openness) in the case when $\boldsymbol{\rho}_0 = (\rho_0, \boldsymbol{\nu}_0)$ is thick. The essence of the proof is the study the behavior of the hidden isometric hemispheres under small deformation. To describe these isometric hemispheres, we introduce the "augmentation" $\mathcal{L}^*(\boldsymbol{\nu})$ of the simplicial complex $\mathcal{L}(\boldsymbol{\nu}) = \mathcal{L}(\Sigma(\boldsymbol{\nu}))$ defined in Definitions 3.2.3 and 3.3.2 (Definition 7.2.1). Its vertex set $\mathcal{L}^*(\boldsymbol{\nu})^{(0)}$ consists of those elliptic generators which are 'involved' in the Ford domain $\overline{Ph}(\rho_0)$, and the proof of Proposition 6.2.1 (Openness) consists of detailed analysis of the behavior of the isometric hemispheres of $\rho(P)$ with $P \in \mathcal{L}^*(\boldsymbol{\nu})^{(0)}$ for nearby representations ρ.

Here is a rough sketch of the idea of the proof. By Lemmas 4.6.1 and 4.6.2, we can construct a continuous map $\boldsymbol{\nu}$ from a neighborhood U_0 of ρ_0 to $\mathbb{H}^2 \times \mathbb{H}^2$, such that $\sigma^\epsilon(\boldsymbol{\nu}(\rho))$ is an ϵ-terminal triangle of ρ for every $\rho \in U_0$. By using Lemma 7.2.3 (Disjointness), which is obtained as a corollary to Lemma 7.1.6 (Hidden isometric hemisphere), we can find a smaller neighborhood U_2 such that $(\rho, \boldsymbol{\nu}(\rho))$ satisfies the conditions NonZero and Frontier for every $\rho \in U_2$ (Lemma 7.2.5). We note that $\mathcal{L}(\boldsymbol{\nu}(\rho))$ is a subcomplex of the augmentation $\mathcal{L}^*(\boldsymbol{\nu})$ every $\rho \in U_2$. Finally, by using Lemma 7.2.3 (Disjointness) again, we find a smaller neighborhood U_3 such that $(\rho, \boldsymbol{\nu}(\rho))$ also satisfies the condition Duality (Lemma 7.2.7). The idea of this last step is as follows. For each $\rho \in U_2$, $\mathcal{L}(\boldsymbol{\nu}(\rho))$ is a subcomplex of $\mathcal{L}^*(\boldsymbol{\nu}_0)$.

- If ξ is a simplex of $\mathcal{L}(\boldsymbol{\nu}_0)$ then $F_{\boldsymbol{\rho}_0}(\xi)$ is a transversal intersection of hyperplanes and and half spaces. Thus $F_{\boldsymbol{\rho}}(\xi)$ with $\boldsymbol{\rho} = (\rho, \boldsymbol{\nu}(\rho))$ continues to be a convex polyhedron of the same dimension. On the other hand, Lemma 7.2.3 (Disjointness) guarantees that those isometric hemispheres which are not involved in the definition of $F_{\boldsymbol{\rho}_0}(\xi)$, except the hidden isometric hemispheres, are disjoint from $F_{\boldsymbol{\rho}}(\xi)$. Moreover the behavior of the hidden isometric hemispheres are controlled by the side parameter. These imply that $F_{\boldsymbol{\rho}}(\xi)$ is contained in $Eh(\boldsymbol{\rho})$, after choosing a smaller neighborhood U_3.
- If ξ is not contained in $\mathcal{L}(\boldsymbol{\nu}_0)$, then hidden isometric hemispheres are involved in the definition of $F_{\boldsymbol{\rho}}(\xi)$. However, the behavior of the hidden

isometric hemispheres are controlled by the side parameter. This fact enables us to show that $F_{\rho}(\xi)$ is a convex polyhedron contained in $Eh(\rho)$ of the correct dimension, after choosing a smaller neighborhood U_3.

Before starting the formal proof of Proposition 6.2.1 (Openness), we prepare a few notations.

Definition 7.2.1. *Let $\boldsymbol{\nu} = (\nu^-, \nu^+) \in \mathbb{H}^2 \times \mathbb{H}^2$ be a thick label.*

(1) $\partial^{\epsilon}\mathcal{L}(\boldsymbol{\nu})$ denotes the 1-dimensional subcomplex of $\mathcal{L}(\boldsymbol{\nu})$ spanned by the vertices $\langle P \rangle$ with $s(P) \in \sigma^{\epsilon}(\boldsymbol{\nu})$.

(2) $\Sigma^*(\boldsymbol{\nu})$ denotes the chain of triangles obtained from $\Sigma(\boldsymbol{\nu})$ by adding $\sigma^{\epsilon,*}(\boldsymbol{\nu})$ ($\epsilon \in \{-,+\}$) whenever it is defined (Definition 7.1.4).

(3) $\mathcal{L}^*(\boldsymbol{\nu})$ denotes the abstract simplicial complex $\mathcal{L}(\Sigma^*(\boldsymbol{\nu}))$ (see Definition 3.1.1).

(4) $\partial^{\epsilon}_{aug}\mathcal{L}^*(\boldsymbol{\nu})$ denotes the subcomplex $\mathcal{L}(\sigma^{\epsilon}(\boldsymbol{\nu}))$ or $\mathcal{L}((\sigma^{\epsilon}(\boldsymbol{\nu}), \sigma^{\epsilon,*}(\boldsymbol{\nu})))$ of $\mathcal{L}^*(\boldsymbol{\nu})$ according as ν^{ϵ} is generic or non-generic.

(5) $\mathcal{L}(\delta^{\epsilon}(\boldsymbol{\nu}))$ denotes the subcomplex of $\partial^{\epsilon}_{aug}\mathcal{L}^*(\boldsymbol{\nu})$ spanned by the vertices P such that $s(P) \in \delta^{\epsilon}(\boldsymbol{\nu})$.

We note that these definitions are also motivated by the results in Sections 4.5 and 4.6. In particular, $\partial^{\epsilon}_{aug}\mathcal{L}^*(\boldsymbol{\nu})$ corresponds to the complex $\mathcal{L}(\rho, \{\sigma, \sigma'\}) \subset \mathbb{C}$ in these sections. If ν^{ϵ} is generic, then $\delta^{\epsilon}(\boldsymbol{\nu}) = \sigma^{\epsilon}(\boldsymbol{\nu})$ and $\partial^{\epsilon}_{aug}\mathcal{L}^*(\boldsymbol{\nu}) = \mathcal{L}(\sigma^{\epsilon}(\boldsymbol{\nu})) = \mathcal{L}(\delta^{\epsilon}(\boldsymbol{\nu}))$. If ν^{ϵ} is non-generic, then these three complexes are related as illustrated in Fig. 7.2.

In the proof of Proposition 6.2.1 (Openness) we need to study the behavior of (the closure of) the isometric hemispheres near the ideal boundary of $\overline{Eh}(\rho)$ systematically. To this end we introduce the following definition.

Definition 7.2.2. *Let $\boldsymbol{\rho} = (\rho, \boldsymbol{\nu})$ be a labeled representation, such that $\phi_{\rho}^{-1}(0) \cap \mathcal{L}^*(\boldsymbol{\nu})^{(0)} = \emptyset$. Then, for $\xi \in \mathcal{L}^*(\boldsymbol{\nu})^{(\leq 2)}$ and $\epsilon \in \{-,+\}$, $f^{\epsilon}_{\boldsymbol{\rho}}(\xi)$ denotes the (possibly empty) compact subset of $\mathbb{C}$ defined by*

$$f^{\epsilon}_{\boldsymbol{\rho}}(\xi) = \left(\bigcap \{I(\rho(P)) \,|\, P \in \xi^{(0)}\}\right) \cap E^{\epsilon}(\boldsymbol{\rho}).$$

The above definition is related to Definition 4.2.5 and Notation 4.3.7 as follows. If $\{P_j\}$ is the sequence of elliptic generators associated with $\sigma^{\epsilon}(\boldsymbol{\nu})$, then:

$$f^{\epsilon}_{\boldsymbol{\rho}}(P_j) = e^{\epsilon}(\rho, \sigma^{\epsilon}(\boldsymbol{\nu}); P_j),$$
$$f^{\epsilon}_{\boldsymbol{\rho}}(P_j, P_{j+1}) = v^{\epsilon}(\rho; P_j, P_{j+1}).$$

The following corollary to Lemma 7.1.6 (Hidden isometric hemisphere) plays an important role in the proof of Proposition 6.2.1 (Openness).

Lemma 7.2.3 (Disjointness). *Let $\boldsymbol{\rho} = (\rho, \boldsymbol{\nu})$ be a thick good labeled representation. Then the following hold.*

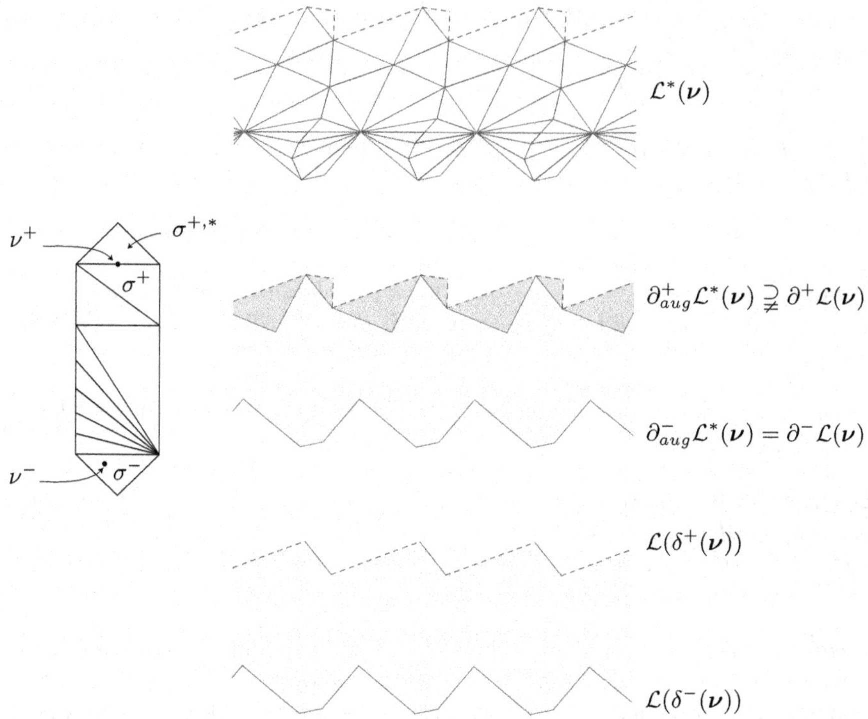

Fig. 7.2. $\mathcal{L}^*(\boldsymbol{\nu})$, $\partial^{\pm}_{aug}\mathcal{L}^*(\boldsymbol{\nu})$ and $\mathcal{L}(\delta^{\pm}(\boldsymbol{\nu}))$ in the case when ν^+ is non-generic and ν^- is generic.

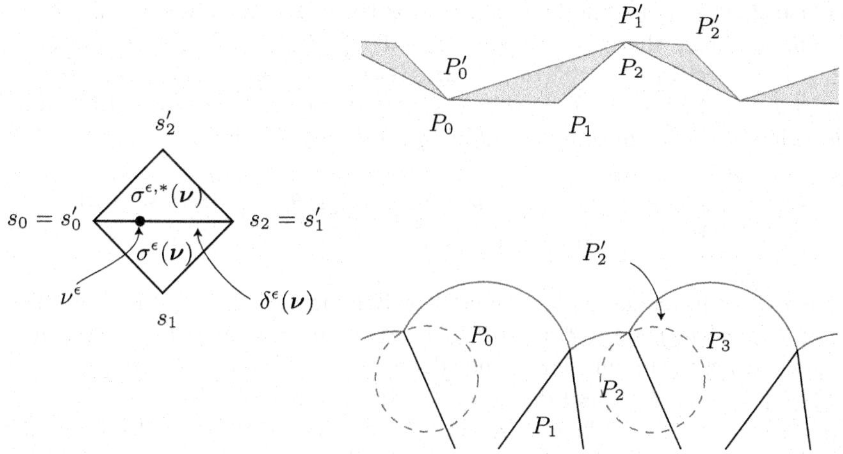

Fig. 7.3. $\partial^{\epsilon}_{aug}\mathcal{L}^*(\boldsymbol{\nu})$ and $\overline{Eh}(\rho, \boldsymbol{\nu})$

(1) If $\xi \in \mathcal{L}(\boldsymbol{\nu})^{(\le 2)}$ *and* $X \in \mathcal{L}^*(\boldsymbol{\nu})^{(0)} - \mathrm{st}_0(\xi, \mathcal{L}^*(\boldsymbol{\nu}))^{(0)}$*, then* $\overline{Dh}(\rho(X))$ *is disjoint from* $\overline{F}_{\boldsymbol{\rho}}(\xi)$*.*

(2) If $\xi \in \partial^{\epsilon}_{aug}\mathcal{L}^*(\boldsymbol{\nu})^{(\le 2)}$ *and* $X \in \mathcal{L}^*(\boldsymbol{\nu})^{(0)} - \mathrm{st}_0(\xi, \partial^{\epsilon}_{aug}\mathcal{L}^*(\boldsymbol{\nu}))^{(0)}$*, then* $D(\rho(X))$ *is disjoint from* $f^{\epsilon}_{\boldsymbol{\rho}}(\xi)$*.*

(3) If $X \in \mathcal{L}^*(\boldsymbol{\nu})^{(0)} - (\partial^{\epsilon}_{aug}\mathcal{L}^*(\boldsymbol{\nu}))^{(0)}$*, then* $D(\rho(X))$ *is disjoint from* $\mathrm{fr}\, E^{\epsilon}(\boldsymbol{\rho})$*.*

We now start the formal proof of Proposition 6.2.1 (Openness).

Lemma 7.2.4. *Let* $\boldsymbol{\rho}_0 = (\rho_0, \boldsymbol{\nu}_0)$ *be a thick good labeled representation. Then there is a neighborhood* U_1 *of* ρ_0 *in* $\mathcal{X}$ *such that* $\phi_{\rho}^{-1}(0) \cap \Sigma^*(\boldsymbol{\nu}_0)^{(0)} = \emptyset$ *for every* $\rho \in U_1$*.*

Proof. Recall that ρ_0 is quasifuchsian by Theorem 6.1.8. So we may set $U_1 = \mathcal{QF}$ (cf. Lemma 2.5.4).

Lemma 7.2.5. *Let* $\boldsymbol{\rho}_0$ *and* U_1 *be as in Lemma 7.2.4. Then there is a neighborhood* U_2 *of* ρ_0 *in* U_1 *and a continuous map* $U_2 \ni \rho \mapsto \boldsymbol{\nu}(\rho) \in \mathbb{H}^2 \times \mathbb{H}^2$*, such that* $\boldsymbol{\nu}(\rho_0) = \boldsymbol{\nu}_0$ *and that* $(\rho, \boldsymbol{\nu}(\rho))$ *satisfies the conditions NonZero and Frontier for every* $\rho \in U_2$*.*

Proof. We first choose a neighborhood U_1^{ϵ} of ρ_0 in $\mathcal{X}$ and construct a continuous map $U_1^{\epsilon} \ni \rho \mapsto \nu^{\epsilon}(\rho) \in \mathbb{H}^2$, for each ϵ, as follows:

Case 1. ν_0^{ϵ} is generic, i.e., $\nu_0^{\epsilon} \in \mathrm{int}\, \sigma^{\epsilon}(\boldsymbol{\nu}_0)$. Then by Lemma 4.6.1, there is a neighborhood U_1^{ϵ} of ρ_0 in U_1, such that $\sigma^{\epsilon}(\boldsymbol{\nu}_0)$ is an ϵ-terminal triangle of ρ for every $\rho \in U_1^{\epsilon}$. We define the continuous map $U_1^{\epsilon} \ni \rho \mapsto \nu^{\epsilon}(\rho) \in \mathbb{H}^2$ by $\nu^{\epsilon}(\rho) := \theta^{\epsilon}(\rho, \sigma^{\epsilon}(\boldsymbol{\nu}_0)) \in \sigma^{\epsilon}(\boldsymbol{\nu}_0) \cap \mathbb{H}^2$.

Case 2. ν_0^{ϵ} is non-generic, i.e., $\delta^{\epsilon}(\boldsymbol{\nu}_0)$ is the edge $\sigma^{\epsilon}(\boldsymbol{\nu}_0) \cap \sigma^{\epsilon,*}(\boldsymbol{\nu}_0)$. Then by Lemma 4.6.2, there is a neighborhood U_1^{ϵ} of ρ_0 in U_1, such that, for each $\rho \in U_1^{\epsilon}$, $\sigma^{\epsilon}(\boldsymbol{\nu}_0)$ or $\sigma^{\epsilon,*}(\boldsymbol{\nu}_0)$ is an ϵ-terminal triangle of ρ according as $\theta^{\epsilon}(\rho, \sigma^{\epsilon}(\boldsymbol{\nu}_0); s_1) \ge 0$ or $\theta^{\epsilon}(\rho, \sigma^{\epsilon}(\boldsymbol{\nu}_0); s_1) < 0$, under Notation 7.1.5 (Non-generic label). We define the map $U_2^{\epsilon} \ni \rho \mapsto \nu^{\epsilon}(\rho) \in \mathbb{H}^2$ by

$$\nu^{\epsilon}(\rho) := \begin{cases} \theta^{\epsilon}(\rho, \sigma^{\epsilon}(\boldsymbol{\nu}_0)) \in \sigma^{\epsilon}(\boldsymbol{\nu}_0) \cap \mathbb{H}^2 & \text{if } \theta^{\epsilon}(\rho, \sigma^{\epsilon}(\boldsymbol{\nu}_0); s_1) \ge 0, \\ \theta^{\epsilon}(\rho, \sigma^{\epsilon,*}(\boldsymbol{\nu}_0)) \in \sigma^{\epsilon,*}(\boldsymbol{\nu}_0) \cap \mathbb{H}^2 & \text{if } \theta^{\epsilon}(\rho, \sigma^{\epsilon}(\boldsymbol{\nu}_0); s_1) \le 0. \end{cases}$$

This map is well-defined and continuous by Lemma 4.5.3(2), because it implies that if $\theta^{\epsilon}(\rho, \sigma^{\epsilon}(\boldsymbol{\nu}_0); s_1) = 0$ then $\theta^{\epsilon}(\rho, \sigma^{\epsilon}(\boldsymbol{\nu}_0))$ and $\theta^{\epsilon}(\rho, \sigma^{\epsilon,*}(\boldsymbol{\nu}_0))$ determine the same point of the interior of the edge $\delta^{\epsilon}(\boldsymbol{\nu}_0) = \sigma^{\epsilon}(\boldsymbol{\nu}_0) \cap \sigma^{\epsilon,*}(\boldsymbol{\nu}_0)$.

Set $U_1' = U_1^- \cap U_1^+$, and define the continuous map $\boldsymbol{\nu} : U_1' \to \mathbb{H}^2 \times \mathbb{H}^2$ by $\boldsymbol{\nu}(\rho) = (\nu^-(\rho), \nu^+(\rho))$. Then $\sigma^{\epsilon}(\boldsymbol{\nu}(\rho))$ is an ϵ-terminal triangle of ρ for every $\rho \in U_1'$ and $\epsilon \in \{-,+\}$. Thus the pair $(\rho, \boldsymbol{\nu}(\rho))$ satisfies the first condition of the condition ϵ-Frontier (Definition 6.1.4). It is obvious that $(\rho, \boldsymbol{\nu}(\rho))$ also satisfies the second condition. To show that $(\rho, \boldsymbol{\nu}(\rho))$ satisfies the last condition, we need the following lemma.

Claim 7.2.6. There is a neighborhood U_2 of ρ_0 in U_1' such that

$$D(\rho(X)) \cap E^\epsilon(\rho, \sigma^\epsilon(\boldsymbol{\nu}(\rho))) = \emptyset$$

for every $\rho \in U_2$ and $X \in \mathcal{L}(\boldsymbol{\nu}(\rho))^{(0)} - \mathcal{L}(\sigma^\epsilon(\boldsymbol{\nu}(\rho)))^{(0)}$.

Proof. Pick an element $X \in \mathcal{L}^*(\boldsymbol{\nu}_0)^{(0)}$. We show that there is a neighborhood $U_2(X)$ of ρ_0 in U_1', such that if $\rho \in U_2(X)$ and $X \in \mathcal{L}(\boldsymbol{\nu}(\rho))^{(0)} - \mathcal{L}(\sigma^\epsilon(\boldsymbol{\nu}(\rho)))^{(0)}$ then $D(\rho(X)) \cap E^\epsilon(\rho, \sigma^\epsilon(\boldsymbol{\nu}(\rho))) = \emptyset$.

Suppose $X \in \partial^\epsilon_{aug}\mathcal{L}^*(\boldsymbol{\nu}_0)^{(0)}$ for some ϵ. Then we may set $U_2(X) = U_1'$. To see this, pick an element $\rho \in U_1'$ such that $X \in \mathcal{L}(\boldsymbol{\nu}(\rho))^{(0)} - \mathcal{L}(\sigma^\epsilon(\boldsymbol{\nu}(\rho)))^{(0)}$. Then ν_0^ϵ is non-generic, $s(X) = s_1$ and $\nu^\epsilon(\rho) \in \operatorname{int} \sigma^{\epsilon,*}(\boldsymbol{\nu}_0)$ under Notation 7.1.5 (Non-generic label). So $\theta^\epsilon(\rho, \sigma^\epsilon(\boldsymbol{\nu}_0); s_1) < 0$ by the definition of $\boldsymbol{\nu}(\rho)$. Thus $D(\rho(X))$ is disjoint from $E^\epsilon(\rho, \sigma^\epsilon(\boldsymbol{\nu}(\rho)))$ by Lemma 4.6.2(2).

Suppose $X \notin \partial^\epsilon_{aug}\mathcal{L}^*(\boldsymbol{\nu}_0)^{(0)}$ for each $\epsilon \in \{-,+\}$. Then $D(\rho_0(X))$ is disjoint from $E^\epsilon(\rho_0, \sigma^\epsilon(\boldsymbol{\nu}(\rho)))$ by Lemma 7.2.3(3) (Disjointness). Hence we can find a desired neighborhood $U_2(X)$.

We may assume $U_2(X)$ depends only on the slope $s(X)$ and hence the intersection, U_2, of all $U_2(X)$ ($X \in \mathcal{L}^*(\boldsymbol{\nu}_0)^{(0)}$) is a neighborhood of ρ_0. Since $\mathcal{L}(\boldsymbol{\nu}(\rho))$ is a subcomplex of $\mathcal{L}^*(\boldsymbol{\nu}_0)$, we obtain the claim.

The above claim implies that $E^\epsilon(\boldsymbol{\rho}) = E^\epsilon(\rho, \sigma^\epsilon(\boldsymbol{\nu}(\rho)))$ for every $\rho \in U_2$, where $\boldsymbol{\rho} = (\rho, \boldsymbol{\nu}(\rho))$. Hence $(\rho, \boldsymbol{\nu}(\rho))$ satisfies the condition Frontier. This completes the proof of Lemma 7.2.5.

The following lemma enables us to treat the condition Duality (Definition 6.1.3).

Lemma 7.2.7. *Let $\boldsymbol{\rho}_0 = (\rho_0, \boldsymbol{\nu}_0)$, U_2 and $\boldsymbol{\nu} : U_2 \to \mathbb{H}^2 \times \mathbb{H}^2$ be as in Lemma 7.2.5. Then there is a neighborhood U_3 of ρ_0 in U_2, such that the following condition is satisfied for every $\rho \in U_3$: $F_{\boldsymbol{\rho}}(\xi)$ is a convex polyhedron of dimension $2 - \dim \xi$ contained in $Eh(\boldsymbol{\rho})$ for every $\xi \in \mathcal{L}(\boldsymbol{\nu}(\rho))^{(\leq 2)}$, where $\boldsymbol{\rho}$ denotes the labeled representation $(\rho, \boldsymbol{\nu}(\rho))$.*

Proof. We show that for every $\xi \in \mathcal{L}^*(\boldsymbol{\nu}_0)^{(\leq 2)}$ there is a neighborhood $U_3(\xi)$ of ρ_0 in U_2, such that if $\rho \in U_3(\xi)$ and $\xi \in \mathcal{L}(\boldsymbol{\nu}(\rho))^{(\leq 2)}$ then $F_{\boldsymbol{\rho}}(\xi)$ is a convex polyhedron of dimension $2 - \dim \xi$ contained in $Eh(\boldsymbol{\rho})$.

Case 1. $\xi = (P, Q, R) \in \mathcal{L}(\boldsymbol{\nu}_0)^{(2)}$. Then $F_{\boldsymbol{\rho}_0}(\xi)$ is the transversal intersection of the three isometric hemispheres $Ih(\rho_0(P))$, $Ih(\rho_0(Q))$, and $Ih(\rho_0(R))$. Hence we can find a neighborhood $U_3(\xi)$ of ρ_0 in U_2, such that $F_{\boldsymbol{\rho}}(\xi) = Ih(\rho(P)) \cap Ih(\rho(Q)) \cap Ih(\rho(R))$ is a 0-dimensional convex polyhedron for every $\rho \in U_3(\xi)$. On the other hand, by Lemma 7.2.3(1) (Disjointness), $F_{\boldsymbol{\rho}_0}(\xi)$ is disjoint from $Dh(\rho_0(X))$ for every $X \in \mathcal{L}^*(\boldsymbol{\nu}_0)^{(0)} - \xi^{(0)} = \mathcal{L}^*(\boldsymbol{\nu}_0)^{(0)} - \mathrm{st}_0(\xi, \mathcal{L}^*(\boldsymbol{\nu}))^{(0)}$. Hence we can choose $U_3(\xi)$, so that if $\rho \in U_3(\xi)$ and $\xi \in \mathcal{L}(\boldsymbol{\nu}(\rho))^{(2)}$, then $F_{\boldsymbol{\rho}}(\xi)$ is disjoint from $Dh(\rho(X))$ for every $X \in \mathcal{L}^*(\boldsymbol{\nu}_0)^{(0)} - \mathrm{st}_0(\xi, \mathcal{L}^*(\boldsymbol{\nu}_0))^{(0)}$. Since $\mathcal{L}(\boldsymbol{\nu}(\rho))^{(0)} - \mathrm{st}_0(\xi, \mathcal{L}(\boldsymbol{\nu}(\rho)))^{(0)} \subset$

$\mathcal{L}^*(\boldsymbol{\nu}_0)^{(0)} - \mathrm{st}_0(\xi, \mathcal{L}^*(\boldsymbol{\nu}_0))^{(0)}$, we can conclude that $F_{\boldsymbol{\rho}}(\xi) \subset Eh(\boldsymbol{\rho})$ whenever $\rho \in U_3(\xi)$ and $\xi \in \mathcal{L}(\boldsymbol{\nu}(\rho))^{(2)}$.

Case 2. $\xi \in \mathcal{L}^*(\boldsymbol{\nu}_0)^{(2)} - \mathcal{L}(\boldsymbol{\nu}_0)^{(2)}$. Then we may assume that ν_0^ϵ is non-generic for some ϵ and $\xi = (P_0, P_1, P_2)$ or (P_1', P_2', P_3') under Notation 7.1.5 (Non-generic label). For simplicity we assume $\xi = (P_0, P_1, P_2)$. Pick an element $\rho \in U_2$ such that $\xi \in \mathcal{L}(\boldsymbol{\nu}(\rho))^{(2)}$. Then $\nu^\epsilon(\rho) \in \mathrm{int}\, \sigma^{\epsilon,*}(\boldsymbol{\nu}_0)$. Hence, by the definition of $\nu^\epsilon(\rho)$, we have $\theta^\epsilon(\rho, \sigma^\epsilon(\boldsymbol{\nu}_0); s_1) < 0$ and therefore $F_{\boldsymbol{\rho}}(\xi) = \cap_{j=0}^2 Ih(\rho(P_j))$ is a 0-dimensional convex polyhedron by Lemma 4.6.2(2). By Lemma 7.2.3(2) (Disjointness), $\overline{Dh}(\rho_0(X))$ is disjoint from the singleton $f^\epsilon_{\rho_0}(\xi) \subset \cap_{j=0}^2 I(\rho_0(P_j))$ for every $X \in \mathcal{L}^*(\boldsymbol{\nu}_0)^{(0)} - \{P_0, P_1, P_2\}$. Hence we can find a neighborhood $U_3(\xi)$ of ρ_0 in U_2, such that $F_{\boldsymbol{\rho}}(\xi) \subset Eh(\boldsymbol{\rho})$ whenever $\rho \in U_3(\xi)$ and $\xi \in \mathcal{L}(\boldsymbol{\nu}(\rho))^{(2)}$.

Case 3. $\xi = (P, Q) \in \mathcal{L}(\boldsymbol{\nu}_0)^{(1)}$. Since $F_{\boldsymbol{\rho}_0}(\xi)$ is a 1-dimensional convex polyhedron obtained as the transversal intersection

$$F_{\boldsymbol{\rho}}(\xi) = (Ih(\rho_0(P)) \cap Ih(\rho_0(Q))) \cap (\cap\{Eh(\rho_0(X)) \mid X \in \mathrm{lk}(\xi, \mathcal{L}(\boldsymbol{\nu}(\rho)))\}),$$

there is a neighborhood $U_3(\xi)$ of ρ_0 in U_2 such that $F_{\boldsymbol{\rho}}(\xi)$ continues to be a 1-dimensional convex polyhedron for every $\rho \in U_3(\xi)$. By Lemma 7.2.3(1) (Disjointness), $\overline{Dh}(\rho_0(X))$ is disjoint from $\overline{F}_{\boldsymbol{\rho}_0}(\xi)$ for every $X \in \mathcal{L}^*(\boldsymbol{\nu}_0)^{(0)} - \mathrm{st}_0(\xi, \mathcal{L}^*(\boldsymbol{\nu}_0))^{(0)}$. Thus we can choose $U_3(\xi)$, so that every $\rho \in U_3(\xi)$ has the same property. Since

$$\mathcal{L}(\boldsymbol{\nu}(\rho))^{(0)} - \mathrm{st}_0(\xi, \mathcal{L}(\boldsymbol{\nu}(\rho)))^{(0)} \subset \mathcal{L}^*(\boldsymbol{\nu}_0)^{(0)} - \mathrm{st}_0(\xi, \mathcal{L}^*(\boldsymbol{\nu}_0))^{(0)},$$

we can conclude that $F_{\boldsymbol{\rho}}(\xi) \subset Eh(\boldsymbol{\rho})$ for every $\rho \in U_3(\xi)$.

Case 4. $\xi = (P, Q) \in \mathcal{L}^*(\boldsymbol{\nu}_0)^{(1)} - \mathcal{L}(\boldsymbol{\nu}_0)^{(1)}$. As in Case 2, we may assume ν_0^ϵ is non-generic for some ϵ and $\xi = (P_j', P_{j+1}')$ for some $j \in \{0, 1, 2\}$ under Notation 7.1.5 (Non-generic label).

First, we study the case when $\xi = (P_0', P_1') = (P_0, P_2)$. Note that $\hat{\xi} := (P_0, P_1, P_2)$ belongs to $\mathcal{L}^*(\boldsymbol{\nu}_0)^{(2)}$. By the argument in Case 2, there is a neighborhood $U_3(\hat{\xi})$ of ρ_0 in U_2, such that if $\rho \in U_3(\hat{\xi})$ and $\xi \in \mathcal{L}(\boldsymbol{\nu}(\rho))^{(1)}$ (and hence $\hat{\xi} \in \mathcal{L}(\boldsymbol{\nu}(\rho))^{(2)}$) then $F_{\boldsymbol{\rho}}(\hat{\xi})$ is a 0-dimensional convex polyhedron. We show that we may set $U_3(\xi) = U_3(\hat{\xi})$. To this end, pick an element $\rho \in U_3(\hat{\xi})$ such that $\xi \in \mathcal{L}(\boldsymbol{\nu}(\rho))^{(1)}$. Then $\hat{\xi} \in \mathcal{L}(\boldsymbol{\nu}(\rho))^{(2)}$ and hence $v^\epsilon(\rho; P_0', P_1') \in E^\epsilon(\boldsymbol{\rho})$ by Lemma 7.2.5. Hence the convex hull of $v^\epsilon(\rho; P_0', P_1')$ and $F_{\boldsymbol{\rho}}(\hat{\xi})$ in $\overline{\mathbb{H}}^3$ is contained in $\overline{Eh}(\boldsymbol{\rho})$ and has dimension 1. Since $\mathrm{lk}(\xi, \mathcal{L}(\boldsymbol{\nu}(\rho))) = \{P_1\}$, the convex hull is equal to

$$\overline{F}_{\boldsymbol{\rho}}(\xi) = (\overline{Ih}(\rho(P_0')) \cap \overline{Ih}(\rho(P_1'))) \cap \overline{Eh}(\rho(P_1)).$$

Hence $U_3(\xi) = U_3(\hat{\xi})$ has the desired property for $\xi = (P_0', P_1')$.

Next, we show that the same neighborhood also satisfies the desired condition for (P_1', P_2') and (P_2', P_3'). Since $\rho(P_j')(F_{\boldsymbol{\rho}}(P_j', P_{j+1}')) = F_{\boldsymbol{\rho}}(P_{j-1}', P_j')$ by Lemma 4.1.3 (Chain rule), we see $\dim F_{\boldsymbol{\rho}}(P_2', P_3') = \dim F_{\boldsymbol{\rho}}(P_1', P_2') =$

$\dim F_{\boldsymbol{\rho}}(P_0', P_1') = 1$ for every ρ in the neighborhood such that $\boldsymbol{\nu}(\rho) \in \operatorname{int} \sigma^{\epsilon,*}(\boldsymbol{\nu}_0)$. Moreover, $\overline{F}_{\boldsymbol{\rho}}(P_1', P_2')$ (resp. $\overline{F}_{\boldsymbol{\rho}}(P_2', P_3')$) is the convex hull of $v^{\epsilon}(\rho; P_1', P_2')$ (resp. $v^{\epsilon}(\rho; P_2', P_3')$) and $F_{\boldsymbol{\rho}}(P_1', P_2', P_3')$ in $\overline{\mathbb{H}}^3$. Hence we can conclude that both $\overline{F}_{\boldsymbol{\rho}}(P_1', P_2')$ and $\overline{F}_{\boldsymbol{\rho}}(P_2', P_3')$ are 1-dimensional convex polyhedra in $\overline{Eh}(\boldsymbol{\rho})$ as in the above.

Case 5. $\xi = (P) \in \mathcal{L}(\boldsymbol{\nu}_0)^{(0)}$. As in Case 3, we can find a neighborhood $U_3(\xi)$ of ρ_0 in U_2 such that $\dim F_{\boldsymbol{\rho}}(\xi) = 2$ for every $\rho \in U_3(\xi)$. By using Lemma 7.2.3 (Disjointness)(1), we can choose $U_3(\xi)$, so that $F_{\boldsymbol{\rho}}(\xi) \subset Eh(\boldsymbol{\rho})$ for every $\rho \in U_3(\xi)$.

Case 6. $\xi = (P) \in \mathcal{L}^*(\boldsymbol{\nu}_0)^{(0)} - \mathcal{L}(\boldsymbol{\nu}_0)^{(0)}$. As in Case 2, we may assume that ν_0^{ϵ} is non-generic for some ϵ and that $\xi = (P_2')$ under Notation 7.1.5 (Non-generic label). Let $\hat{\xi} = (P_1', P_2', P_3')$. We show that we may set $U_3(\xi) = U_3(\hat{\xi})$. To this end, pick an element $\rho \in U_3(\hat{\xi})$ such that $\xi \in \mathcal{L}(\boldsymbol{\nu}(\rho))^{(0)}$. Then $\hat{\xi} \in \mathcal{L}(\boldsymbol{\nu}(\rho))^{(2)}$ and hence $F_{\boldsymbol{\rho}}(\hat{\xi})$ is a 0-dimensional convex polyhedron in $Eh(\boldsymbol{\rho})$ by the definition of $U_3(\hat{\xi})$. Moreover $e^{\epsilon}(\rho, \sigma^{\epsilon,*}(\boldsymbol{\nu}_0); P_2') \subset E^{\epsilon}(\boldsymbol{\rho})$ by Lemma 7.2.5. Hence the convex hull of $e^{\epsilon}(\rho, \sigma^{\epsilon,*}(\boldsymbol{\nu}_0); P_2')$ and $F_{\boldsymbol{\rho}}(\hat{\xi})$ in $\overline{\mathbb{H}}^3$ is contained in $\overline{Eh}(\boldsymbol{\rho})$ and has dimension 2. Since $\operatorname{lk}(\xi, \mathcal{L}(\boldsymbol{\nu}(\rho)))^{(0)} = \{P_1', P_3'\}$, the convex hull is equal to

$$\overline{F}_{\boldsymbol{\rho}}(\xi) = \overline{Ih}(\rho(P_2')) \cap \left(\overline{Eh}(\rho(P_1')) \cap \overline{Eh}(\rho(P_3'))\right).$$

Hence $U_3(\xi) = U_3(\hat{\xi})$ has the desired property.

We have thus constructed the desired neighborhood $U_3(\xi)$ for every $\xi \in \mathcal{L}^*(\boldsymbol{\nu}_0)^{(\leq 2)}$. We may assume $U_3(\xi)$ depends only on the image of ξ in $\mathcal{L}(\boldsymbol{\nu}_0)/\langle K \rangle$. Hence the intersection $U_3 = \cap\{U_3(\xi) \mid \xi \in \mathcal{L}^*(\boldsymbol{\nu}_0)^{(2)}\}$ is a neighborhood of ρ_0 in $\mathcal{X}$. Since $\mathcal{L}(\boldsymbol{\nu}(\rho))$ is a subcomplex of $\mathcal{L}^*(\boldsymbol{\nu}_0)$ for every $\rho \in U_3$, U_3 has the desired property. This completes the proof of Lemma 7.2.7.

Proof (Proof of Proposition 6.2.1 (Openness)). Let U be the neighborhood U_3 of ρ_0 in Lemma 7.2.7 and $\boldsymbol{\nu} : U \to \mathbb{H}^2 \times \mathbb{H}^2$ be the restriction of the continuous map in Lemma 7.2.5. Then by Lemmas 7.2.4, 7.2.5 and 7.2.7, $(\rho, \boldsymbol{\nu}(\rho))$ satisfies the conditions NonZero, Frontier and Duality for every $\rho \in U$. Hence $(\rho, \boldsymbol{\nu}(\rho))$ is good.

7.3 Proof of Proposition 6.2.1 (Openness) - Thin case -

In this section, we prove Proposition 6.2.1 (Openness) in the case when $\boldsymbol{\rho}_0 = (\rho_0, \boldsymbol{\nu}_0)$ is thin. The difference between the thick and thin cases is that there is interaction between the two components of the side parameter. Because of this interaction, various patterns appear after a small perturbation of ρ_0, as illustrated in Fig. 7.4. Though various patterns appear, they are controlled by the side parameter.

However, the idea of the proof is essentially the same as that for the thick case. By using Lemma 4.6.7, we construct a continuous map $\boldsymbol{\nu}$ from

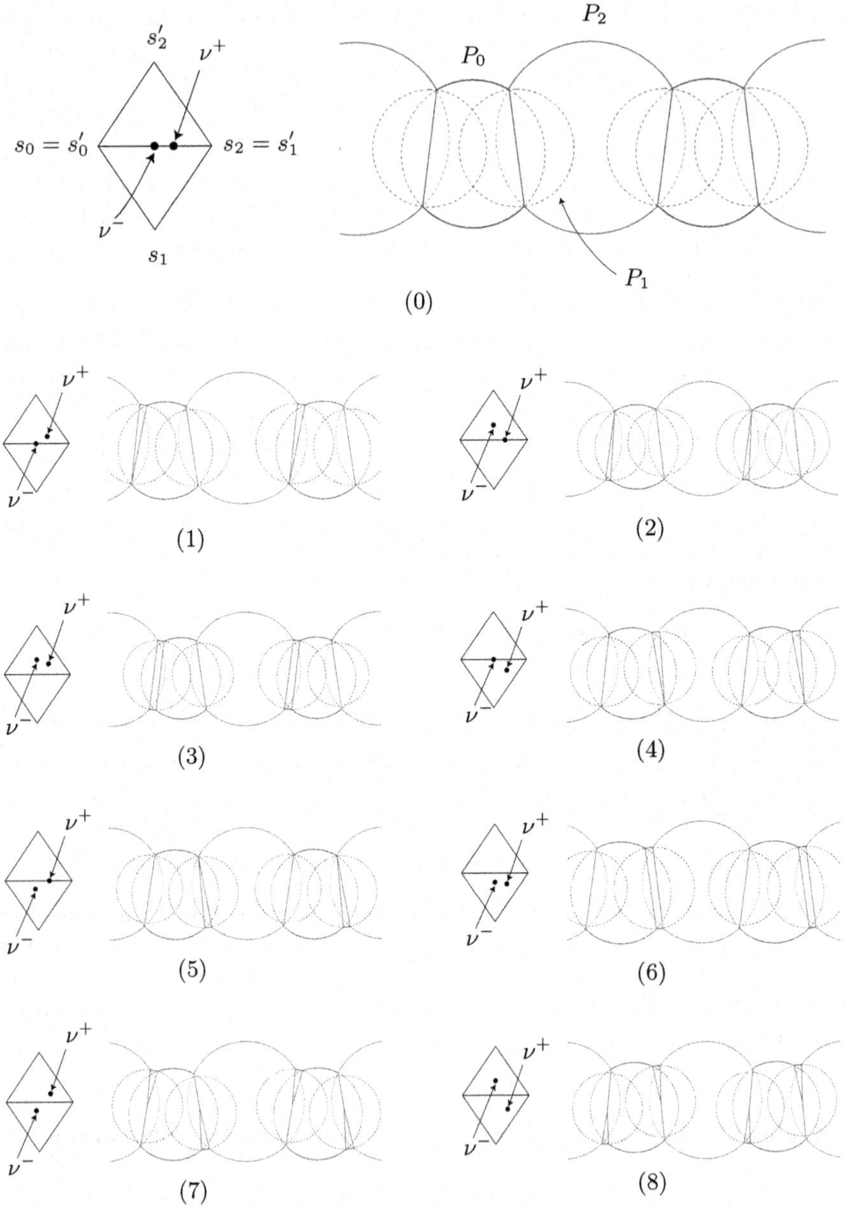

Fig. 7.4. Ford domains of representation in a neighborhood of an isosceles representation

a neighborhood U_0 of ρ_0 to $\mathbb{H}^2 \times \mathbb{H}^2$, such that $\sigma^\epsilon(\boldsymbol{\nu}(\rho))$ is an ϵ-terminal triangle of ρ for every $\rho \in U_0$. By using Lemma 7.3.1 (Disjointness), an analogy of Lemma 7.2.3 (Disjointness), we can find a smaller neighborhood U_2 such that $(\rho, \boldsymbol{\nu}(\rho))$ satisfies the conditions NonZero and Frontier for every $\rho \in U_2$ (Lemma 7.3.3). Finally, by using Lemma 7.3.1 (Disjointness) again, we find a smaller neighborhood U_3 such that $(\rho, \boldsymbol{\nu}(\rho))$ also satisfies the condition Duality (Lemma 7.3.5).

We now give a formal proof for the thin case. Throughout the proof, we employ Notation 2.1.14 (Adjacent triangles), and assume that $\nu_0^\pm \in \tau = \sigma \cap \sigma'$, where $\boldsymbol{\nu}_0 = (\nu_0^-, \nu_0^+)$. We adapt Definition 7.1.4 for the thin case, and define $\Sigma^*(\boldsymbol{\nu}_0)$ to be the chain (σ, σ') and $\mathcal{L}^*(\boldsymbol{\nu}_0) := \mathcal{L}(\Sigma^*(\boldsymbol{\nu}_0))$. Then the following corollary to Lemma 7.1.6 (Hidden isometric hemisphere) plays the role of Lemma 7.2.3 (Disjointness).

Lemma 7.3.1 (Disjointness). *Let $\boldsymbol{\rho}_0 = (\rho_0, \boldsymbol{\nu}_0)$ be a thin good labeled representation as above. Then (ρ_0, τ) is an isosceles representation and the following holds.*

1. *If $X \in \mathcal{L}^*(\boldsymbol{\nu}_0)^{(0)} - \{P_0, P_1, P_2\}$, then $\overline{Dh}(\rho(X))$ is disjoint from $\overline{F}_{\boldsymbol{\rho}}(P_0, P_2)$.*
2. *If $X \in \mathcal{L}^*(\boldsymbol{\nu}_0)^{(0)} - \{P_1', P_2', P_3'\}$, then $\overline{Dh}(\rho(X))$ is disjoint from $\overline{F}_{\boldsymbol{\rho}}(P_1', P_3')$ $= \overline{F}_{\boldsymbol{\rho}}(P_2, P_3)$.*

The proof of Lemma 7.2.4 works in the thin case, and we have the following lemma.

Lemma 7.3.2. *Let $\boldsymbol{\rho}_0 = (\rho_0, \boldsymbol{\nu}_0)$ be a thin good labeled representation as above. Then there is a neighborhood U_1 of ρ_0 in $\mathcal{X}$ such that $\phi_\rho^{-1}(0) \cap \Sigma^*(\boldsymbol{\nu}_0)^{(0)} = \emptyset$ for every $\rho \in U_1$.*

We also have the following analogy of Lemma 7.2.5.

Lemma 7.3.3. *Let $\boldsymbol{\rho}_0$ and U_1 be as in Lemma 7.3.2. Then there is a neighborhood U_2 of ρ_0 in U_1 and a continuous map $U_2 \ni \rho \mapsto \boldsymbol{\nu}(\rho) \in \mathbb{H}^2 \times \mathbb{H}^2$, such that $\boldsymbol{\nu}(\rho_0) = \boldsymbol{\nu}_0$ and $\boldsymbol{\rho} = (\rho, \boldsymbol{\nu}(\rho))$ satisfies the conditions NonZero and Frontier for every $\rho \in U_2$. Moreover, if $\boldsymbol{\nu}(\rho)$ is thin, then $\boldsymbol{\rho}$ is good.*

Proof. As in the proof of Lemma 7.2.5, by using Lemma 4.6.7 instead of Lemma 4.6.2, we can find a neighborhood U_2 and a continuous map $\boldsymbol{\nu} : U_2 \to \mathbb{H}^2 \times \mathbb{H}^2$ such that $\sigma^\epsilon(\boldsymbol{\nu}(\rho))$ is an ϵ-terminal triangle of $\boldsymbol{\rho} = (\rho, \boldsymbol{\nu}(\rho))$ for every $\rho \in U_1'$. If $\boldsymbol{\nu}(\rho)$ is thin. Then (ρ, τ) is an isosceles representation by Proposition 5.2.3. Thus, since we have chosen U_2 so that it is contained in $\mathcal{QF}$, it follows that $\boldsymbol{\rho} = (\rho, \boldsymbol{\nu}(\rho))$ is good by Proposition 5.2.8(2). Thus we may assume $\boldsymbol{\nu}(\rho)$ is thick.

Claim 7.3.4. If $\rho \in U_2$ and and $\boldsymbol{\nu}(\rho)$ is thick, then

$$D(\rho(X)) \cap E^\epsilon(\rho, \sigma^\epsilon(\boldsymbol{\nu}(\rho))) = \emptyset$$

for every $X \in \mathcal{L}(\boldsymbol{\nu}(\rho))^{(0)} - \mathcal{L}(\sigma^\epsilon(\boldsymbol{\nu}(\rho)))^{(0)}$. (See Fig. 7.4(7) and (8).)

Proof (Proof of Claim 7.3.4). Suppose $\boldsymbol{\nu}(\rho)$ is thick and $X \in \mathcal{L}(\boldsymbol{\nu}(\rho))^{(0)} - \mathcal{L}(\sigma^{\epsilon}(\boldsymbol{\nu}(\rho)))^{(0)}$. Then the components, $\nu^{\epsilon}(\rho)$ and $\nu^{-\epsilon}(\rho)$, of $\boldsymbol{\nu}(\rho)$ are contained in the interior of distinct triangles. (Otherwise, $\mathcal{L}(\boldsymbol{\nu}(\rho)) = \mathcal{L}(\sigma^{\epsilon}(\boldsymbol{\nu}(\rho)))$ and hence such X does not exist.) So we may assume $\nu^{\epsilon}(\rho) \in \operatorname{int}\sigma$, $\nu^{-\epsilon}(\rho) \in \operatorname{int}\sigma'$ and $X = P_2'$. Then $\theta^{\epsilon}(\rho,\sigma) = (+,+,+)$ and hence $\theta^{\epsilon}(\rho,\sigma';s_2') < 0$ by Lemma 4.5.3(1). Hence $D(\rho(P_2'))$ is disjoint from $E^{\epsilon}(\rho,\sigma) = E^{\epsilon}(\rho,\sigma^{\epsilon}(\boldsymbol{\nu}(\rho)))$ by Lemma 4.6.7(1).

By using the above claim, we see as in the proof of Lemma 7.2.5 that $(\rho,\boldsymbol{\nu}(\rho))$ satisfies the condition Frontier (Definition 6.1.4) for every $\rho \in U_2$ whenever $\boldsymbol{\nu}(\rho)$ is thick, where $\boldsymbol{\rho} = (\rho,\boldsymbol{\nu}(\rho))$. Thus we have proved that $\boldsymbol{\rho} = (\rho,\boldsymbol{\nu}(\rho))$ satisfies the condition Frontier for every $\rho \in U_2$. This completes the proof of Lemma 7.3.3.

Now the proof of Proposition 6.2.1 for the thin case is completed by the following analogy of Lemma 7.2.7.

Lemma 7.3.5. *Let $\boldsymbol{\rho}_0 = (\rho_0,\boldsymbol{\nu}_0)$, U_2 and $\boldsymbol{\nu} : U_2 \to \mathbb{H}^2 \times \mathbb{H}^2$ be as in Lemma 7.3.3. Then there is a neighborhood U_3 of ρ_0 in U_2, such that the following condition is satisfied for every $\rho \in U_3$: $F_{\boldsymbol{\rho}}(\xi)$ is a convex polyhedron of dimension $2-\dim\xi$ contained in $Eh(\boldsymbol{\rho})$ for every $\xi \in \mathcal{L}(\boldsymbol{\nu}(\rho))^{(\le 2)}$, where $\boldsymbol{\rho}$ denotes the labeled representation $(\rho,\boldsymbol{\nu}(\rho))$.*

Proof. As in the proof of Lemma 7.2.7, we show that for every $\xi \in \mathcal{L}^*(\boldsymbol{\nu}_0)^{(\le 2)}$ there is a neighborhood $U_3(\xi)$ of ρ_0 in U_2, such that if $\rho \in U_3(\xi)$ and $\xi \in \mathcal{L}(\boldsymbol{\nu}(\rho))^{(\le 2)}$, then $F_{\boldsymbol{\rho}}(\xi)$ is a convex polyhedron of dimension $2-\dim\xi$ contained in $Eh(\boldsymbol{\rho})$.

Case 1. $\xi \in \mathcal{L}^*(\boldsymbol{\nu}_0)^{(2)}$. Then we may assume $\xi = (P_0,P_1,P_2)$ or (P_1',P_2',P_3'). For simplicity, we assume $\xi = (P_0,P_1,P_2)$. Pick an element $\rho \in U_2$ such that $\xi \in \mathcal{L}(\boldsymbol{\nu}(\rho))^{(2)}$. Then by the definition of $\mathcal{L}(\boldsymbol{\nu}(\rho))$, we have $\Sigma(\boldsymbol{\nu}(\rho)) = \{\sigma,\sigma'\}$, $\theta^{\epsilon}(\rho,\sigma) = (+,+,+)$ and $\theta^{-\epsilon}(\rho,\sigma') = (+,+,+)$ for some $\epsilon \in \{-,+\}$. Since $\theta^{-\epsilon}(\rho,\sigma';s_2') > 0$, we have $\theta^{-\epsilon}(\rho,\sigma;s_1) < 0$ by Lemma 4.5.3(1). Since $\theta^{\epsilon}(\rho,\sigma;s_1) > 0$, this implies that $F_{\boldsymbol{\rho}}(\xi) = \cap_{j=0}^{2} Ih(\rho(P_j))$ is a 0-dimensional convex polyhedron by Lemma 4.3.6. By Lemma 7.3.1 (Disjointness) (1), $Ih(\rho_0(X))$ is disjoint from the 1-dimensional convex polyhedron $F_{\boldsymbol{\rho}_0}(P_0,P_2) = Ih(\rho_0(P_0)) \cap Ih(\rho_0(P_2))$ for every $X \in \mathcal{L}(\boldsymbol{\nu}(\rho))^{(0)} - \{P_0,P_1,P_2\} = \mathcal{L}^*(\boldsymbol{\nu}_0)^{(0)} - \{P_0,P_1,P_2\}$. Hence we can find a neighborhood $U_3(\xi)$ of ρ_0 in U_2, such that $F_{\boldsymbol{\rho}}(\xi) \subset Eh(\boldsymbol{\rho})$ whenever $\rho \in U_3(\xi)$ and $\xi \in \mathcal{L}(\boldsymbol{\nu}(\rho))^{(\le 2)}$.

Case 2. $\xi \in \mathcal{L}(\boldsymbol{\nu}_0)^{(1)}$. Then we may assume $\xi = (P_0,P_2)$ or (P_2,P_3). For simplicity we assume $\xi = (P_0,P_2) = (P_0',P_1')$. Pick an element $\rho \in U_2$ such that $\xi \in \mathcal{L}(\boldsymbol{\nu}(\rho))^{(1)}$. By Lemma 7.3.3, we may assume $\boldsymbol{\nu}(\rho)$ is thick. Then $\Sigma(\boldsymbol{\nu}(\rho)) = \{\sigma,\sigma'\}$ or $\{\sigma'\}$.

Suppose $\Sigma(\boldsymbol{\nu}(\rho)) = \{\sigma,\sigma'\}$. Then by the argument in Case 1, $\cap_{j=0}^{2} Ih(\rho(P_j))$ is a 0-dimensional convex polyhedron. Since $\operatorname{lk}(\xi,\mathcal{L}(\boldsymbol{\nu}(\rho))) = \{P_1\}$, $F_{\boldsymbol{\rho}}(\xi) = Ih(\rho(P_0)) \cap Ih(\rho(P_2)) \cap Eh(\rho(P_1))$ is a 1-dimensional convex polyhedron. By using Lemma 7.3.1 (Disjointness)(1) as in Case 1, we can find a neighborhood $U_3'(\xi)$ of ρ_0 in U_2, such that $F_\rho(\xi) \subset Eh(\boldsymbol{\rho})$ whenever $\rho \in U_3'(\xi)$ and $\Sigma(\boldsymbol{\nu}(\rho)) = \{\sigma, \sigma'\}$.

Suppose $\Sigma(\boldsymbol{\nu}(\rho)) = \{\sigma'\}$. Since $Ih(\rho_0(P_0)) \cap Ih(\rho_0(P_2))$ is a 1-dimensional subspace of $\mathbb{H}^3$, there is a neighborhood $U_2(\xi)$ of ρ_0 in U_2 such that $F_{\boldsymbol{\rho}}(\xi) = Ih(\rho(P_0)) \cap Ih(\rho(P_2))$ continues to be a 1-dimensional subspace of $\mathbb{H}^3$ for every $\rho \in U_2(\xi)$ such that $\Sigma(\boldsymbol{\nu}(\rho)) = \{\sigma'\}$. (Here we use the fact that $\mathrm{lk}(\xi, \mathcal{L}(\boldsymbol{\nu}(\rho))) = \emptyset$.) By using Lemma 7.3.1 (Disjointness)(1) as in Case 1, we can find a neighborhood $U_3''(\xi)$ of ρ_0 in U_2, such that $F_{\boldsymbol{\rho}}(\xi) \subset Eh(\boldsymbol{\rho})$ whenever $\rho \in U_3''(\xi)$ and $\Sigma(\boldsymbol{\nu}(\rho)) = \{\sigma'\}$.

Thus $U_3(\xi) := U_3'(\xi) \cap U_3''(\xi)$ satisfies the desired property.

Case 3. $\xi \in \mathcal{L}^*(\boldsymbol{\nu}_0)^{(1)} - \mathcal{L}(\boldsymbol{\nu}_0)^{(1)}$. Then we may assume $\xi = (P_0, P_1)$, (P_1, P_2), (P_1', P_2') or (P_2', P_3'). For simplicity we assume $\xi = (P_0, P_1)$. If $\xi \in \mathcal{L}(\boldsymbol{\nu}(\rho))^{(1)}$, then we have $\Sigma(\boldsymbol{\nu}(\rho)) = \{\sigma, \sigma'\}$ or $\{\sigma\}$; thus $F_{\boldsymbol{\rho}}(\xi) = Ih(\rho(P_0)) \cap Ih(\rho(P_1)) \cap Eh(\rho(P_2))$ or $Ih(\rho(P_0)) \cap Ih(\rho(P_1))$ according as the former or latter holds. By using the argument in Case 1 and Lemma 7.3.1 (Disjointness)(1) as in Case 2, we can find the desired neighborhood $U_3(\xi)$ of ρ_0.

Case 4. $\xi \in \mathcal{L}(\boldsymbol{\nu}_0)^{(0)}$, i.e., $\xi = (P_j)$ for some j such that $j \not\equiv 1 \pmod 3$. For simplicity we assume $P = P_0$. Pick an element $\rho \in U_2$. By Lemma 7.3.3, we may assume $\boldsymbol{\nu}(\rho)$ is thick. Then $\mathrm{lk}(\xi, \mathcal{L}(\boldsymbol{\nu}(\rho)))^{(0)} = \{P_{-1}, P_1\}$, $\{P_{-1}', P_1'\}$ or $\{P_{-1}, P_{-1}', P_1, P_1'\}$. By using the argument in Case 1 and Lemma 7.3.1 (Disjointness) as in Case 2, we can find the desired neighborhood $U_3'(\xi)$ of ρ_0.

Case 5. $\xi \in \mathcal{L}^*(\boldsymbol{\nu}_0)^{(0)} - \mathcal{L}(\boldsymbol{\nu}_0)^{(0)}$. Then $\xi = (P_j)$ ($j \equiv 1 \pmod 3$) or $\xi = (P_j')$ ($j \equiv 2 \pmod 3$). For simplicity we assume $\xi = (P_1)$. Pick an element $\rho \in U_2$ such that $\xi \in \mathcal{L}(\boldsymbol{\nu}(\rho))^{(2)}$. Then $\mathrm{lk}(\xi, \mathcal{L}(\boldsymbol{\nu}(\rho)))^{(0)} = \{P_0, P_2\}$. By using the argument in Case 3, we can find the desired neighborhood $U_3'(\xi)$ of ρ_0.

We have thus constructed the desired neighborhood $U_3(\xi)$ for every $\xi \in \mathcal{L}^*(\boldsymbol{\nu}_0)^{(\le 2)}$. As in the proof of Lemma 7.2.7, the intersection $U_3 = \cap\{U_3(\xi) \,|\, \xi \in \mathcal{L}^*(\boldsymbol{\nu}_0)^{(2)}\}$ is a neighborhood of ρ_0 satisfying the desired property. This completes the proof of Lemma 7.3.5.

8

Closedness

In this chapter, we study what happens at the limit of a sequence $\{\boldsymbol{\rho}_n\} = \{(\rho_n, \boldsymbol{\nu}_n)\}$ of good labeled representations.

In Sect. 8.1, we prove Proposition 6.2.3 (SameStratum) which guarantees that if $\lim \rho_n \in \mathcal{QF}$ exists then some subsequence $\{\boldsymbol{\rho}_n\}$ satisfies the condition SameStratum (Definition 6.2.2) and therefore we can talk about the "behavior of a face (or an edge or a vertex) of $Ph(\rho_n)$ as $n \to \infty$". The proof is based on the fact that the convergence $\rho_n \to \rho_\infty$ is strong and Lemma 8.1.1 due to Jorgensen, which is a prototype of Minsky's pivot theorem [58]. Proposition 6.2.3 (SameStratum) enables us to use Proposition 6.2.4 (Closedness), in the proof of the closedness of the image of the projection $\boldsymbol{\mu}_1 : \mathcal{J}[\mathcal{QF}] \to \mathcal{QF}$ in $\mathcal{QF}$.

In Sect. 8.2, we prove Proposition 6.2.7 (Convergence), which guarantees that, for a sequence of good labeled representations $\{(\rho_n, \boldsymbol{\nu}_n)\}$, if $\boldsymbol{\nu}_\infty := \lim \boldsymbol{\nu}_n \in \mathbb{H}^2 \times \mathbb{H}^2$ exists (and if it satisfies the condition SameStratum), then it has a subsequence such that the corresponding subsequence of $\{\rho_n\}$ converges in $\overline{\mathcal{QF}}$. To this end, we study the behavior of the complex probability $(a_{0,n}, a_{1,n}, a_{2,n})$ of ρ_n at a triangle σ_0 in the common chain Σ_0. The Shimizu-Leutbecher lemma (Lemma 2.5.4(2-ii)) implies that the sequence is bounded and hence some subsequence converges to a triple $(a_{0,\infty}, a_{1,\infty}, a_{2,\infty})$ in $\mathbb{C}^3$. Our task is to show that no component of the triple is 0. This is done by contradiction, by showing that if some component is equal to 0, then the limit label $\boldsymbol{\nu}_\infty$ is equal to (s, s) for some vertex s of σ_0, and hence $\boldsymbol{\nu}_\infty$ cannot be contained in $\mathbb{H}^2 \times \mathbb{H}^2$. This is proved by studying the shape of the Ford domains under the wrong assumption. Proposition 6.2.7 (Convergence) is used in Chap. 9, together with Propositions 6.2.2 (SameStratum) and 6.2.4 (Closedness) to prove the bijectivity of $\boldsymbol{\mu}_2 : \mathcal{J}[\mathcal{QF}] \to \mathbb{H}^2 \times \mathbb{H}^2$.

The rest of this chapter (Sects. 8.3–8.12) is devoted to the proof of Proposition 6.2.4 (Closedness), which guarantees that if $\{\boldsymbol{\rho}_n\}$ converges to a labeled representation $\boldsymbol{\rho}_\infty = (\rho_\infty, \boldsymbol{\nu}_\infty) \in \overline{\mathcal{QF}} \times (\mathbb{H}^2 \times \mathbb{H}^2)$ and if $\{\boldsymbol{\rho}_n\}$ satisfies the condition SameStratum, then the limit $\boldsymbol{\rho}_\infty$ is a good labeled representation and hence belongs to $\mathcal{J}[\mathcal{QF}]$. The main task in the proof is to show that no unexpected degeneration of a face of $Ph(\rho_n)$ happens as $n \to \infty$. This is the most involved part of this paper, and a route map is given in Sect. 8.3. A reason why it is so involved is that we have to list all possible degenerations, before showing degenerations do not happen. However, as is found in Jorgensen's original argument [40], the idea to prohibit degenerations consists of the following three observations.

1. If a 'generic' edge of the Ford domain shrinks to a point at the limit, then ρ_∞ has an 'accidental' elliptic or parabolic transformation. Since $\rho_\infty \in \overline{\mathcal{QF}}$, an accidental elliptic transformation does not exist. The existence of an accidental parabolic transformation is also prohibited by Proposition 8.7.1, which shows that those elements of $\rho_\infty(\pi_1(T))$ which support faces of the Ford domain cannot be parabolic. (If we know that $\rho_\infty \in \mathcal{QF}$ in advance, then it is obvious that an accidental parabolic transformation does not exist. Thus we do not need Proposition 8.7.1 in the proof of surjectivity of $\boldsymbol{\mu}_1 : \mathcal{J}[\mathcal{QF}] \to \mathcal{QF}$ in $\mathcal{QF}$. But in the proof of the bijectivity of $\boldsymbol{\mu}_2 : \mathcal{J}[\mathcal{QF}] \to \mathbb{H}^2 \times \mathbb{H}^2$, we do need Proposition 8.7.1.)
2. If a vertex of the Ford domain $Ph(\rho_n)$ drops onto the complex plane which forms an isolated point of the Ford region $P(\rho_\infty) \subset \mathbb{C}$ of ρ_∞, then we have a contradiction to the chain rule of isometric circles (Fig. 8.9).
3. By virtue of the chain rule and by a topological argument, no unexpected degeneration occur on the boundary of the Ford region $P(\rho_\infty)$ in the complex plane (Sect. 8.12). (The proof of the corresponding assertion in [72, Lemma 6.14] seems not to be correct.)

Another reason for the complication in this step (and the previous step) lies in the treatment of the 'thin' case, i.e., the case when both components of $\boldsymbol{\nu}_\infty = \lim \boldsymbol{\nu}_n$ belong to the interior of a single edge τ of the Farey triangulation. In this case, it turns out that the pair (ρ_∞, τ) forms an 'isosceles representation', which is extensively studied in Sect. 5.2. In a neighborhood of isosceles representations, various phenomena occur. However, we can see by a direct calculation that these are essentially controlled by side parameters (Proposition 5.2.13) and we treat the special case by using this proposition.

8.1 Proof of Proposition 6.2.3 (SameStratum)

The following lemma is due to Jorgensen [40, Lemma 4.3]. Bowditch's Theorem 1 in [17] and Minsky's pivot theorem in [58] may be regarded as a refined variation of this lemma.

Lemma 8.1.1. *Let $(\rho, \boldsymbol{\nu})$ be an element of $\mathcal{J}[\mathcal{QF}]$, s a pivot of $\Sigma(\boldsymbol{\nu})$, and P an elliptic generator with $s(P) = s$. Then $r(\rho(P)) > 1/(4 + 2\sqrt{5})$.*

Proof. For completeness, we give a proof following Jorgensen (cf. [40, Lemma 4.3]). From the assumption that s is a pivot of $\Sigma(\boldsymbol{\nu})$, there exists an elliptic generator triple (P_0, P_1, P_2) such that $s(P_0) = s$ and that both (P_0, P_1, P_2) and (P_1, P_2, P_3) are contained in $\mathcal{L}(\boldsymbol{\nu})^{(2)}$ (cf. Lemma 3.2.4(2)). For simplicity, we denote $\phi(s(P_j))$ by x_j $(j = 0, 1, 2)$. In what follows we prove that $|x_0| < 4 + 2\sqrt{5}$. This completes the proof because $r(\rho(P)) = r(\rho(P_0)) = 1/|x_0|$.

Since (P_0, P_1) (resp. (P_1, P_3)) is contained in $\mathcal{L}(\boldsymbol{\nu})^{(1)}$, $Ih(\rho(P_0)) \cap Ih(\rho(P_1))$ (resp. $Ih(\rho(P_1)) \cap Ih(\rho(P_3))$) is nonempty (cf. Definition 6.1.3). Thus we have

$$\begin{aligned} 2(r(\rho(P_0)) + r(\rho(P_1))) &= (r(\rho(P_0)) + r(\rho(P_1))) + (r(\rho(P_1)) + r(\rho(P_3))) \\ &> |c(\rho(P_1)) - c(\rho(P_0))| + |c(\rho(P_3)) - c(\rho(P_1))| \\ &\geq |c(\rho(P_3)) - c(\rho(P_0))| = 1. \end{aligned}$$

Hence $|x_0|^{-1} + |x_1|^{-1} > 1/2$. Since both (P_0, P_2) and (P_2, P_3) are contained in $\mathcal{L}(\boldsymbol{\nu})$, we also have $|x_0|^{-1} + |x_2|^{-1} > 1/2$. Thus, for each $j = 1, 2$, it follows that $(|x_0| - 2)|x_j| < 2|x_0|$.

In what follows we may suppose that $|x_0| > 2$, otherwise $|x_0| \leq 2 < 4+2\sqrt{5}$ which is the desired inequality. Then $|x_j| < 2|x_0|/(|x_0| - 2)$ for each $j = 1, 2$. From the Markoff equation, we have $-x_0^2 = x_1^2 + x_2^2 - x_0 x_1 x_2$. Thus

$$\begin{aligned} |x_0|^2 &= |x_1^2 + x_2^2 - x_0 x_1 x_2| \leq |x_1|^2 + |x_2|^2 + |x_0||x_1||x_2| \\ &< \left(\frac{2|x_0|}{|x_0| - 2}\right)^2 + \left(\frac{2|x_0|}{|x_0| - 2}\right)^2 + |x_0| \left(\frac{2|x_0|}{|x_0| - 2}\right)^2 \\ &= 4|x_0|^2 \frac{|x_0| + 2}{(|x_0| - 2)^2}. \end{aligned}$$

Therefore $|x_0|^2 - 8|x_0| - 4 < 0$, and hence $|x_0| < 4 + 2\sqrt{5}$.

Notation 8.1.2. Let $\{\boldsymbol{\rho}_n\} = \{(\rho_n, \boldsymbol{\nu}_n)\}$ be a sequence in $\mathcal{J}[\mathcal{QF}]$. Then we will use the following notation under Notation 5.5.5.

(1) ϕ_n denotes the upward Markoff map inducing ρ_n.
(2) $(a_{0,n}^{(k)}, a_{1,n}^{(k)}, a_{2,n}^{(k)})$ denotes the complex probability of ρ_n at $\sigma^{(k)}$.
(3) When the sequence $\{\rho_n\}$ converges to $\rho_\infty \in \overline{\mathcal{QF}}$, ϕ_∞ denotes the limit of the sequence $\{\phi_n\}$, and $(a_{0,\infty}^{(k)}, a_{1,\infty}^{(k)}, a_{2,\infty}^{(k)})$ denotes the complex probability of ρ_∞ at $\sigma^{(k)}$.

We begin the proof of Proposition 6.2.3 by showing the following lemma.

Lemma 8.1.3. *Let $\{(\rho_n, \boldsymbol{\nu}_n)\}$ be a sequence in $\mathcal{J}[\mathcal{QF}]$, such that $\{\rho_n\}$ converges to $\rho_\infty \in \mathcal{QF}$. Let X be an element of $\pi_1(\mathcal{O})$ such that $Ih(\rho_\infty(X))$ supports a face of the Ford domain $Ph(\rho_\infty)$. Then $Ih(\rho_n(X))$ supports a face of the Ford domain $Ph(\rho_n)$ for all sufficiently large n. Moreover, there is an elliptic generator, P, such that $Ih(\rho_\infty(X)) = Ih(\rho_\infty(P))$ and $Ih(\rho_n(X)) = Ih(\rho_n(P))$ for every n.*

Proof. Suppose the first assertion does not hold, that is, there is a subsequence of $\{\rho_n\}$, which we denote by the same symbol, such that each $Ih(\rho_n(X))$ does not support a face of $Ph(\rho_n)$. Let x_∞ be a point in the interior of the face of $Ph(\rho_\infty)$ supported by $Ih(\rho_\infty(X))$. Then we can find a sequence $\{x_n\}$ in $\mathbb{H}^3$ converging to x_∞, such that each x_n is contained in $Ih(\rho_n(X))$. Since $(\rho_n, \boldsymbol{\nu}_n) \in \mathcal{J}[\mathcal{QF}]$, there is an elliptic generator $P^{(n)}$, such that $x_n \in Dh(\rho_n(P^{(n)}))$ and that $Ih(\rho_n(P^{(n)}))$ supports a face of $Ph(\rho_n)$. Then $Ih(\rho_n(P^{(n)})) \neq Ih(\rho_n(X))$, because $Ih(\rho_n(X))$ does not support a face of $Ph(\rho_n)$. Since $\{x_n\}$ converges to $x_\infty \in \mathbb{H}^3$ and since the radii $r(\rho_n(P^{(n)}))$ are bounded above by 1 (see Lemma 2.5.2(1)), we may assume, by taking a subsequence, that $\{c(\rho_n(P^{(n)}))\}$ converges to some point in the complex plane and that $\{r(\rho_n(P^{(n)}))\}$ converges to some positive number. Hence, by taking a subsequence again, we may assume that the axis of the elliptic transformation $\rho_n(P^{(n)})$ converges to some geodesic as $n \to \infty$. This means that the sequence $\{\rho_n(P^{(n)})\}$ in $PSL(2,\mathbb{C})$ converges to an elliptic transformation, say Y', of order 2. Since $\{\rho_n\}$ converges strongly to ρ_∞ (see for example [55, Proposition 7.39]), we may assume, after taking a subsequence, that $P^{(n)}$ is equal to a fixed element, Y, and $Y' = \rho_\infty(Y)$. Note that

$$x_\infty = \lim_{n\to\infty} x_n \in \lim_{n\to\infty} Dh(\rho_n(P^{(n)})) = Dh(\rho_\infty(Y)).$$

This implies that $Ih(\rho_\infty(X)) = Ih(\rho_\infty(Y))$, because x_∞ is contained in the interior of the face of $Ph(\rho_\infty)$ supported by $Ih(\rho_\infty(X))$. In particular, we have $\lim_{n\to\infty} |c(\rho_n(P^{(n)})) - c(\rho_n(X))| = 0$. On the other hand, since $\lim_{n\to\infty} r(\rho_n(P^{(n)})) = r(\rho_\infty(X))$, $r(\rho_n(P^{(n)}))$ is bounded below by a positive number. Hence, by Corollary 2.5.3, we have $Ih(\rho_n(P^{(n)})) = Ih(\rho_n(X))$ for sufficiently large n. This is a contradiction. Hence, the first assertion of the lemma holds, that is, $Ih(\rho_n(X))$ supports a face of the Ford domain $Ph(\rho_n)$ for all sufficiently large n.

Since $(\rho_n, \boldsymbol{\nu}_n) \in \mathcal{J}[\mathcal{QF}]$, there is an elliptic generator $P^{(n)}$ such that $Ih(\rho_n(X)) = Ih(\rho_n(P^{(n)}))$. Since the image of two different elliptic generators by a quasifuchsian representation cannot share the same isometric hemisphere (see Lemma 2.5.4 (2-3)), the above elliptic generator $P^{(n)}$ is uniquely determined by X and does not depend on n. Hence we also obtain the second assertion.

Corollary 8.1.4. *Under the assumption of Lemma 8.1.3, any face of the Ford domain $Ph(\rho_\infty)$ is supported by the isometric hemisphere $Ih(\rho_\infty(P))$ for some elliptic generator P.*

Lemma 8.1.5. *Under the assumption of Lemma 8.1.3, there is an edge τ of $\mathcal{D}$ such that $\Sigma(\boldsymbol{\nu}_n)$ contains τ for all sufficiently large n.*

Proof. Let V be the set of slopes of elliptic generators P such that $Ih(\rho_\infty(P))$ supports a face of the Ford domain $Ph(\rho_\infty)$. Then V is non-empty by Corollary 8.1.4. Since $\rho_\infty \in \mathcal{QF}$, we see by Lemma 5.3.10 and Corollary 8.1.4 that

V contains at least two elements, say s_1 and s_2. By Lemma 8.1.3, $\Sigma(\boldsymbol{\nu}_n)$ contains s_1 and s_2 for sufficiently large n. If s_1 and s_2 span an edge of $\mathcal{D}$, then let τ be the edge. Otherwise, let τ be an edge of a triangle of σ whose interior intersects the geodesic in $\mathbb{H}^2$ connecting s_1 and s_2. Since $(\rho_n, \boldsymbol{\nu}_n) \in \mathcal{J}[\mathcal{QF}]$, τ is an edge of $\Sigma(\boldsymbol{\nu}_n)$.

Proof (Proof of Proposition 6.2.3). We first show that there is a subsequence of $\{(\rho_n, \boldsymbol{\nu}_n)\}$, which we denote by the same symbol, such that $\Sigma(\boldsymbol{\nu}_n)$ does not depend on n. Suppose this does not hold. Then infinitely many mutually different chains appear in any subsequence of $\{\Sigma(\boldsymbol{\nu}_n)\}$. By using this fact, we construct an ascending sequence $\{\Sigma_k\}$ of finite chains and a descending series of subsequences $\{\Sigma_n^{(k)}\}$ of $\{\Sigma(\boldsymbol{\nu}_n)\}$ (n is the suffix for the members of each subsequence and k is the suffix for the subsequences), inductively as follows.

Step 1. Let τ be the edge of $\mathcal{D}$ satisfying the conclusion of Lemma 8.1.5. By the assumption, $\Sigma(\boldsymbol{\nu}_n)$ must contain τ for all sufficiently large n. So, at least one of the two triangles containing τ is contained in infinitely many members of the sequence $\{\Sigma(\boldsymbol{\nu}_n)\}$. Let $\sigma_1^{(1)}$ be such a triangle. Put $\Sigma_1 = \{\sigma_1^{(1)}\}$, and let $\{\Sigma_n^{(1)}\}$ be the subsequence of $\{\Sigma(\boldsymbol{\nu}_n)\}$ consisting of those $\Sigma(\boldsymbol{\nu}_n)$ containing the triangle $\sigma_1^{(1)}$.

Step $k+1$. Suppose we have constructed the chain $\Sigma_k = (\sigma_1^{(k)}, \sigma_2^{(k)}, \cdots, \sigma_k^{(k)})$ and the subsequence $\{\Sigma_n^{(k)}\}$ of $\{\Sigma(\boldsymbol{\nu}_n)\}$, such that $\Sigma_k \subset \Sigma_n^{(k)}$ for any n. Since infinitely many mutually different chains appear in the subsequence $\{\Sigma_n^{(k)}\}$, at least one of the four triangles adjacent to the union $\cup_{i=1}^k \sigma_i^{(k)}$ and not contained in Σ_k is contained in infinitely many members of the subsequence $\{\Sigma_n^{(k)}\}$. If a triangle σ adjacent to $\sigma_k^{(k)}$ satisfies this condition, then define

$$(\sigma_1^{(k+1)}, \cdots, \sigma_k^{(k+1)}, \sigma_{k+1}^{(k+1)}) := (\sigma_1^{(k)}, \cdots, \sigma_k^{(k)}, \sigma).$$

If not, then let σ be a triangle adjacent to $\sigma_1^{(k)}$ satisfying the condition, and define

$$(\sigma_1^{(k+1)}, \sigma_2^{(k+1)} \cdots, \sigma_{k+1}^{(k+1)}) := (\sigma, \sigma_1^{(k)}, \cdots, \sigma_k^{(k)}).$$

Set $\Sigma_{k+1} = (\sigma_1^{(k+1)}, \sigma_2^{(k+1)} \cdots, \sigma_{k+1}^{(k+1)})$, and let $\{\Sigma_n^{(k+1)}\}$ be the subsequence of $\{\Sigma_n^{(k)}\}$ consisting of those $\Sigma_n^{(k)}$ containing the chain Σ_{k+1}.

Now, let Σ_∞ be the infinite chain obtained as the union of the ascending chains $\{\Sigma_k\}$, and let $\{\Sigma_n^{(\infty)}\}$ be the subsequence of $\{\Sigma(\boldsymbol{\nu}_n)\}$, defined by $\Sigma_n^{(\infty)} := \Sigma_n^{(n)}$. Consider the subsequence of $\{(\rho_n, \boldsymbol{\nu}_n)\}$ corresponding to the subsequence $\{\Sigma_n^{(\infty)}\}$, and denote it by $\{(\rho_n, \boldsymbol{\nu}_n)\}$ so that $\Sigma_n^{(\infty)} = \Sigma(\boldsymbol{\nu}_n)$. Then, from the construction of the chains and the fact that $(\rho_n, \boldsymbol{\nu}_n) \in \mathcal{J}$, for each k and n with $k \leq n$, Σ_k is a subchain of $\Sigma(\boldsymbol{\nu}_n)$. In particular, if an elliptic generator P satisfies $s(P) \in \Sigma_k$, then $Ih(\rho_n(P))$ supports a face of $Ph(\rho_n)$ for any n with $n \geq k$. These properties together with the geometrical finiteness of ρ_∞ implies the following claim.

Claim 8.1.6. (1) The number of pivots of Σ_∞ is finite.

(2) For each pivot of Σ_∞, the number of triangles in Σ_∞ containing the pivot is finite.

Proof. (1) Suppose to the contrary that Σ_∞ contains infinitely many pivots s_k $(k \in \mathbb{N})$. Then, for each $k \in \mathbb{N}$, we can find an elliptic generator $P^{(k)}$ with $s(P^{(k)}) = s_k$, such that $c(\rho_\infty(P^{(k)}))$ is contained in the strip $\{z \in \mathbb{C} \,|\, 0 \le \Re z \le 1\}$. By Lemma 2.5.4(2.3), $c(\rho_\infty(P^{(k)}))$ $(k \in \mathbb{N})$ are distinct from one another. Since s_k is a pivot of $\Sigma(\boldsymbol{\nu}_n)$ (Definition 3.2.1(4)) for all sufficiently large n, we see $r(\rho_\infty(P^{(k)})) \ge 1/(4+2\sqrt{5})$ by Lemma 8.1.1. Thus, for any $k \ne k'$, we have $|c(\rho_\infty(P^{(k)})) - c(\rho_\infty(P^{(k')}))| \ge 1/(4+2\sqrt{5})^2$ by Lemma 2.5.2. Put $R = 1/\{3(4+2\sqrt{5})^2\}$, and let $D^{(k)}$ be the disk in $\mathbb{C}$ with center $c(\rho_\infty(P^{(k)}))$ and radius R. Then, by the above observation, $D^{(k)}$ $(k \in \mathbb{N})$ are mutually disjoint. On the other hand, since ρ_∞ is geometrically finite, there are two lines $l^\pm = \{z \in \mathbb{C} \,|\, \Im z = t^\pm\}$ $(t^- < t^+)$ such that the limit set $\Lambda(\mathrm{Im}\,\rho_\infty)$ of $\mathrm{Im}\,\rho_\infty$ is contained in the region bounded by l^- and l^+. Note that each $c(\rho_\infty(P^{(k)}))$ is contained in $\Lambda(\mathrm{Im}\,\rho_\infty)$, for it is the fixed point of the parabolic element $\rho_\infty(P^{(k)}KP^{(k)})$. Thus each $D^{(k)}$ is contained in the rectangular region $\{z \in \mathbb{C} \,|\, -R \le \Re z \le 1+R,\ t^- - R \le \Im z \le t^+ + R\}$. This is a contradiction.

(2) Suppose to the contrary that there exists an elliptic generator P_0 such that $s(P_0)$ is contained in infinitely many triangles in Σ_∞. Choose elliptic generators P_1 and P_2 so that the triplet (P_0, P_1, P_2) is an elliptic generator triple and that the triangle $\langle s(P_0), s(P_1), s(P_2)\rangle$ is contained in Σ_∞. In what follows, we follow Notation 8.1.2. Then we may suppose that the triangle $\sigma^{(k)}$ is contained in Σ_∞ for all positive integer k. (Actually we must also consider the case when $\sigma^{(k)}$ is contained in Σ_∞ for all negative integer k. However a parallel argument works for the case.) From the construction of Σ_∞, every $\sigma^{(k)}$ $(k \in \mathbb{N})$ is contained in $\Sigma(\boldsymbol{\nu}_n)$ for all sufficiently large n. Thus, for any $k \in \mathbb{N}$, ρ_n satisfies the triangle inequality at $\sigma^{(k)}$ for all sufficiently large n by Proposition 6.7.1. Hence $I(\rho_\infty(P_0)) \cap I(\rho_\infty(P_1^{(k)})) \ne \emptyset$ for any $k \in \mathbb{N}$ (cf. Lemma 4.2.1). Therefore $|c(\rho_\infty(P_0)) - c(\rho_\infty(P_1^{(k)}))| \le 1+1 = 2$ by Lemma 2.5.2(1). We can see that $\lim_{k\to\infty} r(\rho_\infty(P_1^{(k)})) = 0$ as follows. Suppose to the contrary that $r(\rho_\infty(P_1^{(k_l)})) \ge r_0$ for some subsequence $\{k_l\} \subset \{k\}$ and for some $r_0 > 0$. Then, by Lemma 2.5.4(2-3), the closed disks D_l with center $c(\rho_\infty(P_1^{(k_l)}))$ and with radius $1/(3r_0^2)$ are disjoint from one another. On the other hand, they are contained in a compact region in $\mathbb{C}$ because $|c(\rho_\infty(P_0)) - c(\rho_\infty(P_1^{(k)}))| \le 2$. This is a contradiction.

Since $P_2^{(k)} = P_1^{(k+1)}$ by definition, we also have $\lim_{k\to\infty} r(\rho_\infty(P_2^{(k)})) = 0$.

Since $\sigma^{(k)}$ is contained in Σ_∞, the edge $\langle P_1^{(k)}, P_2^{(k)}\rangle$ is contained in $\Sigma(\boldsymbol{\nu}_n)$ for all sufficiently large n. Thus $I(\rho_\infty(P_1^{(k)})) \cap I(\rho_\infty(P_2^{(k)})) \ne \emptyset$. Since both $r(\rho_\infty(P_1^{(k)}))$ and $r(\rho_\infty(P_2^{(k)}))$ tend to 0 as $k \to \infty$, this implies that $a_{0,\infty}^{(k)} = c(\rho_\infty(P_2^{(k)}))) - c(\rho_\infty(P_1^{(k)})))$ tends to 0 as $k \to \infty$. Since

$\left|\sqrt{|a_{1,\infty}^{(k)}|} - \sqrt{|a_{2,\infty}^{(k)}|}\right| \le \sqrt{|a_{0,\infty}^{(k)}|}$ by Lemma 4.2.1, we see $\lim \left||a_{1,\infty}^{(k)}| - |a_{2,\infty}^{(k)}|\right| = 0$. On the other hand, we have $\sqrt{|a_{1,\infty}^{(k)} a_{2,\infty}^{(k)}|} = r(\rho_\infty(P_0)) \le 1$ by Lemma 2.5.4. Thus both $|a_{1,\infty}^{(k)}|$ and $|a_{2,\infty}^{(k)}|$ are bounded, and therefore by taking a subsequence, $a_{1,\infty}^{(k)}$ and $a_{2,\infty}^{(k)}$ converge to some complex numbers, $a_{1,\infty}^{(\infty)}$ and $a_{2,\infty}^{(\infty)}$, respectively. Since $|a_{1,\infty}^{(\infty)}| - |a_{2,\infty}^{(\infty)}| = \lim(|a_{1,\infty}^{(k)}| - |a_{2,\infty}^{(k)}|) = 0$ and since $a_{1,\infty}^{(\infty)} + a_{2,\infty}^{(\infty)} = 1 - \lim_{k\to\infty} a_{0,\infty}^{(k)} = 1$, we have $a_{1,\infty}^{(\infty)} = \frac{1}{2} + it$ and $a_{2,\infty}^{(\infty)} = \frac{1}{2} - it$ for some $t \in \mathbb{R}$. Thus the Markoff map ϕ_∞ satisfies

$$\phi_\infty(s(P_0)) = \pm 1/\sqrt{a_{1,\infty}^{(k)} a_{2,\infty}^{(k)}} = \pm 1/\sqrt{a_{1,\infty}^{(\infty)} a_{2,\infty}^{(\infty)}} = \pm 1/\sqrt{(1/4)^2 + t^2} \in [-2, 2].$$

Hence $\rho_\infty(KP_0)$ is either parabolic or elliptic. This contradicts the assumption that $\rho_\infty \in \mathcal{QF}$.

The above claim implies Σ_∞ is finite, a contradiction. Hence there is a subsequence of $\{(\rho_n, \boldsymbol{\nu}_n)\}$, which we denote by the same symbol, such that $\Sigma(\boldsymbol{\nu}_n)$ does not depend on n. Hence, we may assume, by taking a subsequence again, that $\{(\rho_n, \boldsymbol{\nu}_n)\}$ satisfies the condition SameStratum. Thus we have proved the second assertion of Proposition 6.2.3.

To show the first assertion, we note that we may assume, by taking a further subsequence, that $\{\boldsymbol{\nu}_n\}$ converges to a label $\boldsymbol{\nu}_\infty$ in $(\mathbb{H}^2 \cup \hat{\mathbb{Q}}) \times (\mathbb{H}^2 \cup \hat{\mathbb{Q}}) = \mathcal{D} \times \mathcal{D}$. We show that the limit $\boldsymbol{\nu}_\infty$ belongs to $\mathbb{H}^2 \times \mathbb{H}^2$. Suppose to the contrary that $\boldsymbol{\nu}_\infty \notin \mathbb{H}^2 \times \mathbb{H}^2$. Then $\nu_\infty^\epsilon = s$ for some $\epsilon = \pm$ and for some vertex s of σ^ϵ. This contradicts the following Lemma 8.1.7. Thus we have completed the proof of Proposition 6.2.3.

Lemma 8.1.7. *Let $\{\boldsymbol{\rho}_n\} = \{(\rho_n, \boldsymbol{\nu}_n)\}$ be a sequence in $\mathcal{J}[\mathcal{QF}]$ such that $\{\boldsymbol{\rho}_n\}$ converges to $\boldsymbol{\rho}_\infty = (\rho_\infty, \boldsymbol{\nu}_\infty) \in \mathcal{QF} \times (\overline{\mathbb{H}}^2 \times \overline{\mathbb{H}}^2)$. Suppose that the sequence $\{\boldsymbol{\rho}_n\}$ satisfies the condition SameStratum. Then $\nu_\infty^\epsilon \notin \hat{\mathbb{Q}}$ for each $\epsilon \in \{-, +\}$.*

Proof. Suppose to the contrary that $\nu_\infty^\epsilon \in \hat{\mathbb{Q}}$. Then there is a sequence of elliptic generators $\{P_j\}$ associated with the ϵ-terminal triangle, σ^ϵ, of the common chain $\Sigma(\boldsymbol{\nu}_n)$ such that $s(P_0) = \nu_\infty^\epsilon$. By the assumption that the sequence $\{\nu_n^\epsilon\}$ converges to $\nu_\infty^\epsilon = s(P_0)$, the sequence of side parameters $\{\theta^\epsilon(\rho_n, \sigma)\}$ converges to the triplet $(\pi/2, 0, 0)$ (cf. Definition 4.2.9). Since the radius $r(\rho_n(P_j))$ is bounded above by 1 (Lemma 2.5.2(1)), the Euclidean length of $e^\epsilon(\rho_n, \sigma^\epsilon; P_j)$ is bounded above by $2\theta^\epsilon(\rho_n, \sigma^\epsilon; s(P_j))$. Thus the union $e^\epsilon(\rho_n, \sigma^\epsilon; P_1) \cup e^\epsilon(\rho_n, \sigma^\epsilon; P_2)$ converges to a point in $\mathbb{C}$ as $n \to \infty$ with respect to the Hausdorff topology. The limit point is a common fixed point of the involutions $\rho_n(P_1)$ and $\rho_n(P_2)$, because $e^\epsilon(\rho_n, \sigma^\epsilon; P_j)$ contains a fixed point of $\rho_n(P_j)$. Hence $\rho_\infty(KP_0) = \rho_\infty(P_2P_1)$ is either parabolic or the identity. This contradicts the assumption that $\rho_\infty \in \mathcal{QF}$.

8.2 Proof of Proposition 6.2.7 (Convergence)

Proof (Proof of Proposition 6.2.7). Let $\{\boldsymbol{\rho}_n\} = \{(\rho_n, \boldsymbol{\nu}_n)\}$ be a sequence of good labeled representations satisfying the following conditions:

1. $\{\boldsymbol{\nu}_n\}$ converges to $\boldsymbol{\nu}_\infty \in \mathbb{H}^2 \times \mathbb{H}^2$.
2. $\{\boldsymbol{\rho}_n\}$ satisfies the condition SameStratum.

First, suppose that $\boldsymbol{\nu}_\infty$ is thin. Then, by Proposition 5.2.11, there is a unique element $\rho_\infty \in \mathcal{QF}$ which realizes $\boldsymbol{\nu}_\infty$. By Proposition 5.2.13, we can see that the sequence $\{\rho_n\}$ converges to ρ_∞. This completes the proof of Proposition 6.2.7 in the case when $\boldsymbol{\nu}_\infty$ is thin.

Next, suppose that $\boldsymbol{\nu}_\infty$, and hence all $\boldsymbol{\nu}_n$, is thick. Pick an arbitrary triangle σ_0 in the chain $\Sigma(\boldsymbol{\nu}_n)$. Let $\{P_j\}$ be the sequence of elliptic generators associated with σ_0, and let $(a_{0,n}, a_{1,n}, a_{2,n})$ be the complex probability of ρ_n at σ_0. Since σ_0 is contained in $\Sigma(\boldsymbol{\nu}_n)$, the 1-dimensional convex polyhedron $F_{\rho_n}(P_j, P_{j+1})$ is contained in $Ih(\rho_n(P_j)) \cap Ih(\rho_n(P_{j+1}))$ for every $j \in \mathbb{Z}$, and hence ρ_n satisfies the triangle inequality at σ_0 by Lemma 4.2.1. This implies that any two successive isometric hemispheres $Ih(\rho_n(P_{j+1}))$ and $Ih(\rho_n(P_{j+2}))$ have a non-trivial intersection (cf. Lemma 4.2.1). Hence $|a_{j,n}| = |c(\rho_n(P_{j+2})) - c(\rho_n(P_{j+1}))| < 2$ by Lemma 2.5.4(2-2). So, we may assume, by taking a subsequence if necessary, that the sequence $\{(a_{0,n}, a_{1,n}, a_{2,n})\}$ of complex probabilities converges to some triplet $(a_{0,\infty}, a_{1,\infty}, a_{2,\infty})$ in $\mathbb{C}^3$ which also satisfies $a_{0,\infty} + a_{1,\infty} + a_{2,\infty} = 1$.

In what follows, we will see that the triplet $(a_{0,\infty}, a_{1,\infty}, a_{2,\infty})$ is in fact contained in $(\mathbb{C} - \{0\})^3$. Suppose one of the components, say $a_{0,\infty}$, is equal to 0. Then the following claim holds.

Claim 8.2.1. Every triangle of $\Sigma(\boldsymbol{\nu}_n)$ contains the slope $s(P_0)$ as a vertex. Moreover, we have $\lim_{n\to\infty} \theta^\epsilon(\rho_n, \sigma^\epsilon) = (\pi/2, 0, 0)$ for each $\epsilon \in \{-, +\}$.

Proof. Suppose to the contrary that $\Sigma(\boldsymbol{\nu}_n)$ contains a neighbor σ' of σ_0 with $s(P_0) \notin (\sigma')^{(0)}$. Then $\sigma' = \langle s(P_1), s(P_1P_0P_1), s(P_2)\rangle$. Let $(a'_{0,n}, a'_{1,n}, a'_{2,n})$ be the complex probability of ρ_n at σ'. Then we have:

$$a'_{0,n} = \frac{a_{0,n}a_{1,n}}{a_{1,n} + a_{2,n}} \to \frac{0}{1}, \quad a'_{1,n} = a_{1,n} + a_{2,n} \to 1, \quad a'_{2,n} = \frac{a_{0,n}a_{2,n}}{a_{1,n} + a_{2,n}} \to \frac{0}{1}.$$

Hence the triplet $(\sqrt{|a'_{0,n}|}, \sqrt{|a'_{1,n}|}, \sqrt{|a'_{2,n}|})$ does not satisfy the triangle inequality for sufficiently large $n \in \mathbb{N}$. On the other hand, ρ_n must satisfy the condition σ'-Simple (cf. Assumption 4.2.4) because $\sigma' \in \Sigma(\boldsymbol{\nu}_n)$ and $\boldsymbol{\rho}_n = (\rho_n, \boldsymbol{\nu}_n) \in \mathcal{J}[\mathcal{QF}]$. This is a contradiction by Lemma 4.2.1. Thus any neighbor of σ in $\Sigma(\boldsymbol{\nu}_n)$ contains $s(P_0)$ as a vertex, and hence we can see inductively that every triangle of $\Sigma(\boldsymbol{\nu}_n)$ contains $s(P_0)$ as a vertex.

In what follows, we will follow Notation 8.1.2. By the observation in the preceding paragraph, we have $\Sigma(\boldsymbol{\nu}_n) = \{\sigma^{(k)} \mid k^- \le k \le k^+\}$ for some $k^\pm \in \mathbb{Z}$. We also have, for each k,

$$a_{0,n}^{(k+1)} = \frac{a_{0,n}^{(k)} a_{1,n}^{(k)}}{a_{0,n}^{(k)} + a_{2,n}^{(k)}}, \quad a_{1,n}^{(k+1)} = \frac{a_{1,n}^{(k)} a_{2,n}^{(k)}}{a_{0,n}^{(k)} + a_{2,n}^{(k)}}, \quad a_{2,n}^{(k+1)} = a_{0,n}^{(k)} + a_{2,n}^{(k)},$$

$$a_{0,n}^{(k)} = \frac{a_{0,n}^{(k+1)} a_{2,n}^{(k+1)}}{a_{0,n}^{(k+1)} + a_{1,n}^{(k+1)}}, \quad a_{1,n}^{(k)} = a_{0,n}^{(k+1)} + a_{1,n}^{(k+1)}, \quad a_{2,n}^{(k)} = \frac{a_{1,n}^{(k+1)} a_{2,n}^{(k+1)}}{a_{0,n}^{(k+1)} + a_{1,n}^{(k+1)}}.$$

From the assumption, we have $(a_{0,n}^{(0)}, a_{1,n}^{(0)}, a_{2,n}^{(0)}) = (a_{0,n}, a_{1,n}, a_{2,n})$. Thus $a_{0,n}^{(0)} = a_{0,n}$ tends to 0 as $n \to \infty$ by assumption. Hence we can see inductively that every complex probability $(a_{0,n}^{(k)}, a_{1,n}^{(k)}, a_{2,n}^{(k)})$ converges to the common $(a_{0,\infty}, a_{1,\infty}, a_{2,\infty}) = (0, a_{1,\infty}, a_{2,\infty})$ as $n \to \infty$. In particular, we can see for any k that $r(\rho_n(P_1^{(k)})) = \sqrt{|a_{0,n}^{(k)} a_{2,n}^{(k)}|} \to 0$ as $n \to \infty$. Since each pair of isometric hemispheres $Ih(\rho_n(P_1^{(k)}))$ and $Ih(\rho_n(P_2^{(k)})) = Ih(\rho_n(P_1^{(k+1)}))$ $(k^- \le k \le k^+)$ have a non-trivial intersection, the union of isometric circles $\cup_{k=k^-}^{k^++1} I(\rho_n(P_1^{(k)}))$ converges to a point in $\mathbb{C}$ in the Hausdorff topology as $n \to \infty$. On the other hand, the union $\cup_{k=k^-}^{k^++1} I(\rho_n(P_1^{(k)}))$ has a non-trivial intersection with both $E^{\pm}(\boldsymbol{\rho}_n)$ for every $n \in \mathbb{N}$, because $\boldsymbol{\rho}_n$ satisfies the condition Frontier (cf. Definition 6.1.4). Hence the distance $d_E(E^-(\boldsymbol{\rho}_n), E^+(\boldsymbol{\rho}_n))$ converges to 0 as $n \to \infty$.

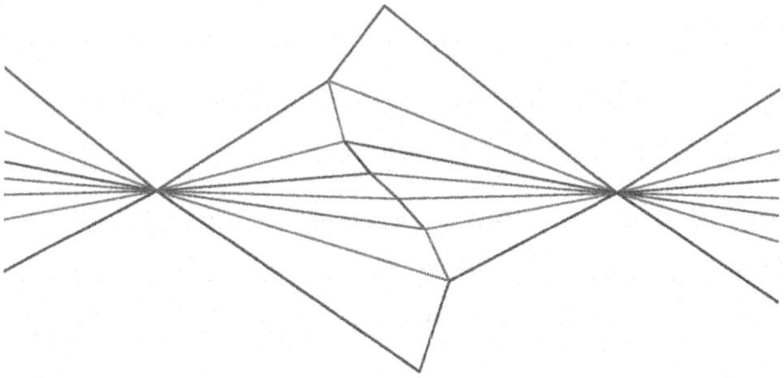

Fig. 8.1. Complex probability taken from OPTi

In what follows, we prove $a_{1,\infty} = a_{2,\infty} = 1/2$ by using the above observation. By Lemma 4.2.1, we have $\left|\sqrt{|a_{1,n}|} - \sqrt{|a_{2,n}|}\right| < \sqrt{|a_{0,n}|}$ for any $n \in \mathbb{N}$ and hence $|a_{1,\infty}| = |a_{2,\infty}|$. Since $a_{1,\infty} + a_{2,\infty} = 1 - a_{0,\infty} = 1$, it follows that $\Re(a_{1,\infty}) = \Re(a_{2,\infty}) = 1/2$ and that $\Im(a_{2,\infty}) = -\Im(a_{1,\infty})$. Thus $r(\rho_n(P_0)) = \sqrt{|a_{1,n} a_{2,n}|}$ converges to $\sqrt{1/4 + y^2}$ where $y = |\Im(a_{1,\infty})| = |\Im(a_{2,\infty})|$. So the distance between the two components of the common exterior $\cap_{k \in \mathbb{Z}} E(\rho_n(P_{3k}))$ is at least y for all sufficiently large $n \in \mathbb{N}$, because it converges to $\cap_{k \in \mathbb{Z}} E(\rho_\infty(P_{3k}))$ in the Hausdorff topology whose components are separated by the y-neighborhood of the real axis. Since

$E^{\pm}(\boldsymbol{\rho}_n)$ are contained in the different components of $\cap_{k\in\mathbb{Z}}E(\rho_n(P_{3k}))$, we see $d_E(E^-(\boldsymbol{\rho}_n), E^+(\boldsymbol{\rho}_n)) \ge y$ for sufficiently large $n \in \mathbb{N}$. This implies $y = 0$, because $d_E(E^-(\boldsymbol{\rho}_n), E^+(\boldsymbol{\rho}_n))$ converges to 0 as observed in the preceding paragraph. Hence $a_{0,\infty} = a_{2,\infty} = 1/2$.

Since $a_{0,\infty} = a_{2,\infty} = 1/2$, it follows that $r(\rho_n(P_0)) = \sqrt{|a_{0,n}a_{2,n}|} \to 1/2$ as $n \to \infty$. Thus, by using also the fact that that $r(\rho_n(P_1^{(k)}))$ converges to 0 for every k, we can see that $\lim_{n\to\infty} \theta^{\epsilon}(\rho_n, \sigma^{\epsilon}; s(P_0)) = \pi/2$ for each $\epsilon \in \{-,+\}$. Hence we have $\lim_{n\to\infty} \theta^{\epsilon}(\rho_n, \sigma^{\epsilon}) = (\pi/2, 0, 0)$ for each $\epsilon \in \{-,+\}$, by the inequality $\theta^{\epsilon}(\rho_n, \sigma^{\epsilon}; s(P_j)) \ge 0$ (cf. Definition 4.3.8(1)) and by the identity $\sum_{j=0}^{2} \theta^{\epsilon}(\rho_n, \sigma^{\epsilon}; s(P_j)) = \pi/2$ (cf. Proposition 4.2.16). This completes the proof of Claim 8.2.1.

By Claim 8.2.1, the sequence $\{\boldsymbol{\nu}_n\}$ converges to the pair $(s(P_0), s(P_0)) \in \partial\mathbb{H}^2 \times \partial\mathbb{H}^2$. This contradicts the assumption that the limit is contained in $\mathbb{H}^2 \times \mathbb{H}^2$. Hence the triplet $(a_{0,\infty}, a_{1,\infty}, a_{2,\infty})$ is contained in $(\mathbb{C} - \{0\})^3$. So the triplet $(a_{0,\infty}, a_{1,\infty}, a_{2,\infty})$ defines a representation, ρ_∞, in $\mathcal{X}$. By Proposition 2.4.2, the sequence $\{\rho_n\}$ converges to ρ_∞. Since the sequence $\{\boldsymbol{\nu}_n\}$ converges in $\mathbb{H}^2 \times \mathbb{H}^2$ by assumption, it follows that the sequence $\{\boldsymbol{\rho}_n\}$ converges in $\overline{\mathcal{QF}} \times (\mathbb{H}^2 \times \mathbb{H}^2)$. This completes the proof of Proposition 6.2.7.

8.3 Route map of the proof of Proposition 6.2.4 (Closedness)

In this section, we give a route map of the proof of Proposition 6.2.4. We first recall the assumption of the proposition.

Assumption 8.3.1. We assume that $\{\boldsymbol{\rho}_n\} = \{(\rho_n, \boldsymbol{\nu}_n)\}$ is a sequence in $\mathcal{J}[\mathcal{QF}]$ satisfying the following conditions:

(1) $\{\boldsymbol{\rho}_n\}$ converges to a labeled representation $\boldsymbol{\rho}_\infty = (\rho_\infty, \boldsymbol{\nu}_\infty) \in \overline{\mathcal{QF}} \times (\mathbb{H}^2 \times \mathbb{H}^2)$.
(2) $\{\boldsymbol{\rho}_n\}$ satisfies the condition SameStratum, i.e., every chain $\Sigma(\boldsymbol{\nu}_n)$ is equal to a fixed chain Σ_0, and hence every $\mathcal{L}(\boldsymbol{\nu}_n)$ is equal to the fixed elliptic generator complex $\mathcal{L}_0 = \mathcal{L}(\Sigma_0)$.

For each sign $\epsilon \in \{-,+\}$, $\sigma^{\epsilon} = \langle s_0^{\epsilon}, s_1^{\epsilon}, s_2^{\epsilon}\rangle$ denotes the ϵ-terminal triangle of Σ_0 and $\{P_j^{\epsilon}\}$ denotes the sequence of elliptic generators associated with σ^{ϵ}.

We note that Proposition 6.2.4 in the case where $\boldsymbol{\nu}_\infty$ is thin can be easily proved.

Proposition 8.3.2 (ThinGoodConvergence implies GoodLimit). *Under Assumption 8.3.1, assume that $\boldsymbol{\nu}_\infty \in \mathbb{H}^2 \times \mathbb{H}^2$ is thin. Then the limit $\boldsymbol{\rho}_\infty$ is a good labeled representation and hence belongs to $\mathcal{J}[\mathcal{QF}]$.*

Proof. Since $\boldsymbol{\nu}_\infty$ is thin, we see by Proposition 5.2.11 that there is a unique representation $\rho' \in \mathcal{QF}$ such that $(\rho', \boldsymbol{\nu}_\infty)$ is a good labeled representation. Let τ be the edge of $\mathcal{D}$ such that $\nu_\infty^\pm \in \operatorname{int}\tau$. Since $\lim \boldsymbol{\nu}_n = \boldsymbol{\nu}_\infty$, we may assume $\boldsymbol{\nu}_n$ lies in the neighborhood U of $\operatorname{int}\tau \times \operatorname{int}\tau$ constructed in Proposition 5.2.13. Since ρ_n (resp. ρ') is induced by an algebraic root for $\boldsymbol{\nu}_n$ (resp. $\boldsymbol{\nu}_\infty$), Proposition 5.2.13 guarantees that ρ_n converges to ρ'. Thus we obtain $\rho' = \rho_\infty$, and hence $\boldsymbol{\rho}_\infty = (\rho_\infty, \boldsymbol{\nu}_\infty) = (\rho', \boldsymbol{\nu}_\infty)$ is a good labeled representation.

Thus Proposition 6.2.4 is reduced to the following proposition.

Proposition 8.3.3 (ThickGoodConvergence implies GoodLimit). *Under Assumption 8.3.1, assume that $\boldsymbol{\nu}_\infty \in \mathbb{H}^2 \times \mathbb{H}^2$ is thick. Then the limit $\boldsymbol{\rho}_\infty$ is a good labeled representation and hence belongs to $\mathcal{J}[\mathcal{QF}]$.*

In the remainder of this section, we give an outline for the proof of Proposition 8.3.3.

Reduction of Proposition 8.3.3. We introduce a condition which we call *HausdorffConvergence* in Definition 8.4.1, and reduce Proposition 8.3.3 to the following three Propositions 8.3.4, 8.3.5 and 8.3.6.

Proposition 8.3.4 (ThickGoodConvergence implies Hausdorff-Convergence). *Under Assumption 8.3.1, $\{\boldsymbol{\rho}_n\}$ contains a subsequence which satisfies the condition HausdorffConvergence.*

Proposition 8.3.5 (ThickGoodConvergence implies Duality). *Under Assumption 8.3.1, assume that $\boldsymbol{\nu}_\infty$ is thick and that $\{\boldsymbol{\rho}_n\}$ satisfies the condition HausdorffConvergence. Then $\boldsymbol{\rho}_\infty$ satisfies the condition Duality.*

Proposition 8.3.6 (ThickGoodConvergence implies Frontier). *Under Assumption 8.3.1, assume that $\boldsymbol{\nu}_\infty$ is thick and that $\{\boldsymbol{\rho}_n\}$ satisfies the condition HausdorffConvergence. Then $\boldsymbol{\rho}_\infty$ satisfies the condition Frontier.*

Assuming the above three propositions, the proof of Proposition 8.3.3 is completed as follows:

Proof (Proof of Proposition 8.3.3). Under Assumption 8.3.1, assume that $\boldsymbol{\nu}_\infty$ is thick. Since $\rho_\infty \in \overline{\mathcal{QF}}$, $\boldsymbol{\rho}_\infty$ satisfies the condition NonZero (cf. Lemma 2.5.4). By Proposition 8.3.4, we may assume after taking a subsequence that $\{\boldsymbol{\rho}_n\}$ also satisfies the condition HausdorffConvergence. Thus, by Propositions 8.3.5 and 8.3.6, $\boldsymbol{\rho}_\infty$ satisfies the conditions Duality and Frontier. Hence $\boldsymbol{\rho}_\infty$ is good (cf. Definitions 6.1.7).

The proof of Proposition 8.3.4 is given in Sect. 8.4. An outline of the proof of Proposition 8.3.5 is given at the end of Sect. 8.4, where it is reduced to Propositions 8.4.4 and 8.4.5: Proposition 8.4.4 is proved in Sect. 8.6, and Proposition 8.4.5 is proved in Sects. 8.8–8.11. Finally, the proof of Proposition 8.3.6 is given in Sect. 8.12.

8.4 Reduction of Proposition 8.3.5 - The condition HausdorffConvergence -

We refer the definitions of Hausdorff metric and Chabauty topology, to [15, Sect. E.1] and [16, Sect. 9.11].

Definition 8.4.1 (HausdorffConvergence). *Suppose* $\{\boldsymbol{\rho}_n\} = \{(\rho_n, \boldsymbol{\nu}_n)\}$ *satisfies the condition SameStratum. Then we say that* $\{\boldsymbol{\rho}_n\}$ *satisfies the condition* HausdorffConvergence *if the following conditions are satisfied with respect to the Hausdorff metric on the space of compact sets of* $\overline{\mathbb{H}}^3$, *with respect to the Euclidean metric of* $\overline{\mathbb{H}}^3$.

1. *For each simplex* ξ *of* $\mathcal{L}_0$ *(Definition 6.2.2), the sequence* $\{\overline{F}_{\rho_n}(\xi)\}$ *converges. We denote the limit by* $\overline{F}_\infty(\xi)$.
2. *For each simplex* ξ *of* $\partial^\epsilon \mathcal{L}_0$ *(cf. Definition 7.2.1), the sequence* $\{f^\epsilon_{\rho_n}(\xi)\}$ *converges. We denote the limit by* $f^\epsilon_\infty(\xi)$.

Remark 8.4.2. As a matter of fact, this is a condition for a sequence in $\mathrm{Hom}_{\mathrm{tp}}(\pi_1(\mathcal{O}), PSL(2,\mathbb{C}))$ and not for a sequence in $\mathcal{X}$. Thus the precise meaning of the above definition is that $\{\boldsymbol{\rho}_n\}$ contains a subsequence which lifts to a sequence in $\mathrm{Hom}_{\mathrm{tp}}(\pi_1(\mathcal{O}), PSL(2,\mathbb{C}))$ satisfying the required conditions.

To prove Proposition 8.3.4, we make the following simple observation.

Lemma 8.4.3. *Under Assumption 8.3.1, the following hold for any elliptic generator* P.

1. $\overline{Ih}(\rho_\infty(P))$ *is well-defined, and the sequence* $\{\overline{Ih}(\rho_n(P))\}$ *converges to* $\overline{Ih}(\rho_\infty(P))$ *with respect to the Hausdorff metric. In particular, there is a compact subset, say* Z, *of* $\overline{\mathbb{H}}^3$ *which contains* $\overline{Ih}(\rho_\infty(P))$ *and* $\overline{Ih}(\rho_n(P))$ *for all* $n \in \mathbb{N}$.
2. $\overline{Eh}(\rho_\infty(P))$ *is well-defined, and the sequence* $\{\overline{Eh}(\rho_n(P))\}$ *converges to* $\overline{Eh}(\rho_\infty(P))$ *with respect to the Chabauty topology on the closed sets of* $\overline{\mathbb{H}}^3$. *Moreover, for any compact set* Z, $\{\overline{Eh}(\rho_n(P)) \cap Z\}$ *converges to* $\overline{Eh}(\rho_\infty(P)) \cap Z$ *with respect to the Hausdorff metric.*

Proof. Recall that we employ Convention 2.2.7(3) and hence ρ_n converges to ρ_∞ as elements of $\mathrm{Hom}_{\mathrm{tp}}(\pi_1(\mathcal{O}), PSL(2,\mathbb{C}))$ (not only as elements of $\mathcal{X}$). By Lemma 2.5.4, $\phi_n(s(P))$ $(n \in \mathbb{N})$ and $\phi_\infty(s(P))$ are non-zero. Hence $I(\rho_n(P))$ and $I(\rho_\infty(P))$ are well-defined by Lemma 2.4.4(1.1). Thus $\{c(\rho_n(P))\}$ and $\{r(\rho_n(P))\}$ converge to $c(\rho_\infty(P))$ and $r(\rho_\infty(P))$, respectively. This implies the desired results.

Proof (Proof of Proposition 8.3.4). Let ξ be a simplex of $\mathcal{L}_0$. Pick a vertex, P, of ξ. By Lemma 8.4.3, $\overline{Ih}(\rho_\infty(P))$ is well-defined, $\{\overline{Ih}(\rho_n(P))\}$ converges to $\overline{Ih}(\rho_\infty(P))$, and there is a compact subset, say Z, of $\overline{\mathbb{H}}^3$ which contains $\overline{Ih}(\rho_\infty(P))$ and $\overline{Ih}(\rho_n(P))$ for all $n \in \mathbb{N}$. Since $\overline{F}_{\rho_n}(\xi) \subset \overline{Ih}(\rho_n(P))$, $\overline{F}_{\rho_n}(\xi)$

is contained in Z for all sufficiently large $n \in \mathbb{N}$. Hence $\{\overline{F}_{\rho_n}(\xi)\}$ contains a convergent subsequence. Similarly, we see that, for each simplex ξ of $\partial^{\epsilon}\mathcal{L}_0$, $\{f^{\epsilon}_{\rho_n}(\xi)\}$ contains a convergent subsequence. Since $\mathcal{L}_0$ and $\partial^{\pm}\mathcal{L}_0$ contain only finitely many simplices modulo the action of K, we obtain the desired result.

Proposition 8.3.5 is reduced to the following Propositions 8.4.4 and 8.4.5. Proposition 8.4.4 is proved in Sect. 8.6. Propositions 8.4.5 is divided into two cases according as the length of the chain Σ_0 is equal to 1 or ≥ 2; the length 1 case is treated in Sect. 8.8 and the length ≥ 2 case is treated in Sects. 8.9–8.11.

Proposition 8.4.4 **($\overline{F}_\infty(\xi) \subset \partial\overline{Eh}(\rho_\infty, \mathcal{L}_0)$).** *Under Assumption 8.3.1, assume that $\{\boldsymbol{\rho}_n\}$ satisfies the condition HausdorffConvergence. Then the following hold.*

(1) For any simplex ξ of $\mathcal{L}_0$, $\overline{F}_\infty(\xi)$ is contained in $\partial\overline{Eh}(\rho_\infty, \mathcal{L}_0)$.
(2) For any simplex ξ of $\partial^{\epsilon}\mathcal{L}_0$, $f^{\epsilon}_\infty(\xi)$ is contained in $\mathrm{fr}\, E(\rho_\infty, \mathcal{L}_0)$.

Proposition 8.4.5. *Under Assumption 8.3.1, assume that $\boldsymbol{\nu}_\infty$ is thick and that $\{\boldsymbol{\rho}_n\}$ satisfies the condition HausdorffConvergence. Then the following hold:*

(1) For any $\xi \in \mathcal{L}(\boldsymbol{\nu}_\infty)^{(i)}$, $\overline{F}_{\boldsymbol{\rho}_\infty}(\xi)$ is equal to $\overline{F}_\infty(\xi)$, and it is a convex polyhedron of dimension $2-i$.

(2) For any vertex $(P) \in \mathcal{L}_0^{(0)} - \mathcal{L}(\boldsymbol{\nu}_\infty)^{(0)}$, $\overline{Ih}(\rho_\infty(P))$ is invisible in $\overline{Eh}(\boldsymbol{\rho}_\infty)$. Namely,

$$\overline{Eh}(\rho_\infty(P)) \supset \overline{Eh}(\boldsymbol{\rho}_\infty).$$

Assuming Propositions 8.4.4 and 8.4.5, we can prove Proposition 8.3.5 as follows.

Proof (Proof of Proposition 8.3.5 assuming Propositions 8.4.4 and 8.4.5). Since $\rho_\infty \in \overline{\mathcal{QF}}$ by the assumption, $\boldsymbol{\rho}_\infty$ satisfies the condition NonZero (cf. Lemma 2.5.4). Thus we have:

$$\begin{aligned}
\overline{Eh}(\rho_\infty, \mathcal{L}_0) &= \cap\{\overline{Eh}(\rho_\infty(P)) \mid P \in \mathcal{L}_0^{(0)}\} \\
&= \left(\cap\{\overline{Eh}(\rho_\infty(P)) \mid P \in \mathcal{L}(\boldsymbol{\nu}_\infty)^{(0)}\}\right) \\
&\qquad \cap \left(\cap\{\overline{Eh}(\rho_\infty(P)) \mid P \in \mathcal{L}_0^{(0)} - \mathcal{L}(\boldsymbol{\nu}_\infty)^{(0)}\}\right) \\
&= \overline{Eh}(\boldsymbol{\rho}_\infty) \cap \left(\cap\{\overline{Eh}(\rho_\infty(P)) \mid P \in \mathcal{L}_0^{(0)} - \mathcal{L}(\boldsymbol{\nu}_\infty)^{(0)}\}\right) \\
&= \overline{Eh}(\boldsymbol{\rho}_\infty).
\end{aligned}$$

Here, the last equality follows from Proposition 8.4.5(2).

Let ξ be any element of $\mathcal{L}(\boldsymbol{\nu}_\infty)^{(i)}$. Then, by Proposition 8.4.5, $\overline{F}_{\boldsymbol{\rho}_\infty}(\xi)$ is equal to $\overline{F}_\infty(\xi)$ and it is a $(2-i)$-dimensional convex polyhedron in $\overline{\mathbb{H}}^3$. By Proposition 8.4.4, $\overline{F}_\infty(\xi)$ is contained in $\partial\overline{Eh}(\rho_\infty, \mathcal{L}_0)$, which is equal to

$\partial \overline{Eh}(\rho_\infty)$ by the preceding observation. Thus ρ_∞ satisfies the first condition in Definition 6.1.3 (Duality).

Next we prove that ρ_∞ satisfies the second condition in Definition 6.1.3 (Duality). For mutually distinct vertices P and P' of $\mathcal{L}(\nu_\infty)$, suppose to the contrary that $F_{\rho_\infty}(P) = F_{\rho_\infty}(P')$. Then the two isometric hemispheres $Ih(\rho_\infty(P))$ and $Ih(\rho_\infty(P'))$ coincide, because $\dim F_{\rho_\infty}(P) = \dim F_{\rho_\infty}(P') = 2$. This contradicts Lemma 2.5.4(3). Thus ρ_∞ also satisfies the second condition in Definition 6.1.3 (Duality). This completes the proof of Proposition 8.3.5.

8.5 Classification of simplices of $\mathcal{L}(\nu)$

In this section, we divide the edges of the abstract simplicial complex $\mathcal{L}(\nu)$ into three classes, generic edges and ϵ-extreme edges ($\epsilon \in \{-,+\}$), and give certain characterizations of the three classes. This result is used in the following sections.

Definition 8.5.1 (Classification of the edges of $\mathcal{L}(\nu)$). *Let $\nu = (\nu^-, \nu^+)$ be a thick label, and let $\xi = (P_j, P_{j+1})$ be a 1-cell of $\mathcal{L}(\nu)$, where $\{P_j\}$ is the sequence of elliptic generators associated with a triangle σ in the chain $\Sigma(\nu)$.*

(1) (ϵ-extreme edge) We say that ξ is an ϵ-extreme edge, *if it is an edge of $\partial^\epsilon \mathcal{L}(\nu)$. namely, σ is the ϵ-terminal triangle of $\Sigma(\nu)$.*

(2) (Generic edge) We say that ξ is a generic edge, *if ξ is not an ϵ-extreme edge for any $\epsilon = \pm$, namely $\sigma \in \Sigma(\nu) - \{\sigma^-(\nu), \sigma^+(\nu)\}$.*

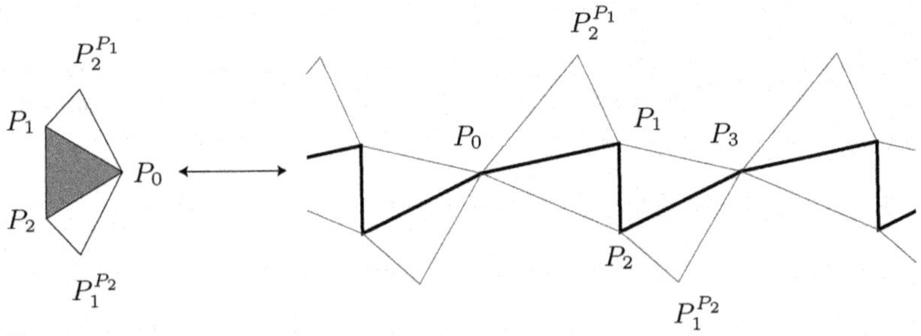

Fig. 8.2. Generic edges: the thick broken line consists of generic edges

Lemma 8.5.2. *Let ξ be a generic edge of $\mathcal{L}(\nu)$. Then there is a triangle σ of $\Sigma(\nu) - \{\sigma^-(\nu), \sigma^+(\nu)\}$ satisfying the following conditions (see Fig. 8.2).*

1. *$\xi = (P_j, P_{j+1})$ for some j, where $\{P_j\}$ is the sequence of elliptic generators associated with σ.*
2. *The elliptic generator triples (P_0, P_1, P_2), (P_1, P_2, P_3), $(P_0, P_2^{P_1}, P_1)$, and $(P_2, P_1^{P_2}, P_3)$ determine 2-simplices of $\mathcal{L}(\nu)$.*

3. Set $\sigma' = \langle s(P_0), s(P_2^{P_1}), s(P_1)\rangle$ *and* $\sigma'' = \langle s(P_2), s(P_1^{P_2}), s(P_3)\rangle$. *If* σ' *lies on the* ϵ*-side of* σ, *then* σ'' *lies on the* $(-\epsilon)$*-side of* σ *and the following hold.*

a) (P_1, P_2, P_3) *and* $(P_0, P_2^{P_1}, P_1)$ *lie on the* ϵ*-side of* $\mathcal{L}(\sigma)$.
b) (P_0, P_1, P_2) *and* $(P_2, P_1^{P_2}, P_3)$ *lie on the* $(-\epsilon)$*-side of* $\mathcal{L}(\sigma)$.

Proof. Let ξ be a generic edge of $\mathcal{L}(\boldsymbol{\nu})$. Then, by definition, there is a triangle σ of $\Sigma(\boldsymbol{\nu}) - \{\sigma^-, \sigma^+\}$ satisfying the first condition, namely, $\xi = (P_j, P_{j+1})$ for some j where $\{P_j\}$ is the sequence of elliptic generators associated with σ. Since σ is different from $\sigma^\pm(\boldsymbol{\nu})$, we may assume, after a shift of indices, that $\langle s(P_0), s(P_1)\rangle$ (resp. $\langle s(P_0), s(P_2)\rangle$) is an edge of a triangle, say σ' (resp. σ''), in $\Sigma(\boldsymbol{\nu})$ different from σ. We see that the sequence of elliptic generators associated with σ' (resp. σ'') is as follows (cf. Proposition 2.1.6(1)):

$$\cdots, P_{-2}, P_0, P_{-1}^{P_0} = P_2^{P_1}, P_1, P_3, P_2^{P_3}, \cdots$$
$$(\text{resp. } \cdots, P_0, P_2, P_1^{P_2}, P_3, P_5, P_4^{P_5}, \cdots).$$

Hence we see that (P_0, P_1, P_2), (P_1, P_2, P_3), $(P_0, P_2^{P_1}, P_1)$, and $(P_2, P_1^{P_2}, P_3)$ are 2-simplices of $\mathcal{L}(\boldsymbol{\nu})$. Thus the second condition is satisfied. The last assertion follows from the definition (Definition 3.2.5).

We can easily observe the following lemma.

Lemma 8.5.3. *Let* $\boldsymbol{\rho} = (\rho, \boldsymbol{\nu})$ *be a good labeled representation, and let* ξ *be an edge of* $\mathcal{L}(\boldsymbol{\nu})$. *Then the following hold.*

1. *If* ξ *is generic, then* $\overline{F}_{\boldsymbol{\rho}}(\xi)$ *is a geodesic segment in* $\mathbb{H}^3$.
2. *If* ξ *is* ϵ*-extreme for an unique sign* ϵ, *then* $\overline{F}_{\boldsymbol{\rho}}(\xi)$ *is a half geodesic.*
3. *If* ξ *is* ϵ*-extreme for for each sign* ϵ, *then* $\overline{F}_{\boldsymbol{\rho}}(\xi)$ *is a complete geodesic.*

8.6 Proof of Proposition 8.4.4 ($\overline{F}_\infty(\xi) \subset \partial\overline{Eh}(\rho_\infty, \mathcal{L}_0)$)

Lemma 8.6.1. *Under Assumption 8.3.1, assume that* $\{\boldsymbol{\rho}_n\}$ *satisfies the condition HausdorffConvergence. Then the following hold.*

(1) $f_\infty^\epsilon(P_j^\epsilon)$ *is equal to the (possibly degenerate) subarc of* $I(\rho_\infty(P_j^\epsilon)) \cap (E(\rho_\infty(P_{j-1}^\epsilon)) \cap E(\rho_\infty(P_{j+1}^\epsilon)))$ *which contains* $\lim \mathrm{Fix}_{\sigma^\epsilon}^\epsilon(\rho_n(P_j^\epsilon))$ *and which is bounded by* $f_\infty^\epsilon(P_{j-1}^\epsilon, P_j^\epsilon) \in I(\rho_\infty(P_{j-1}^\epsilon)) \cap I(\rho_\infty(P_j^\epsilon))$ *and* $f_\infty^\epsilon(P_j^\epsilon, P_{j+1}^\epsilon) \in I(\rho_\infty(P_j^\epsilon)) \cap I(\rho_\infty(P_{j+1}^\epsilon))$. *Moreover, the angle of* $f_\infty^\epsilon(P_j^\epsilon)$ *is equal to the double of the* $[j]$*-th barycentric coordinate of the point* $\nu_\infty^\epsilon = \theta^\epsilon(\rho, \sigma^\epsilon)$ *in the* ϵ*-terminal triangle* σ^ϵ *of* Σ_0.

(2) Let ξ *be an element of* $\mathcal{L}_0^{(i)}$, *where* $i = 0$ *or* 1. *Then* $\overline{F}_\infty(\xi)$ *is equal to the convex polyhedron in* $\overline{\mathbb{H}}^3$ *spanned by the limits of the vertices of* $\overline{F}_{\boldsymbol{\rho}_n}(\xi)$ *and the limits of the ideal boundaries of* $\overline{F}_{\boldsymbol{\rho}_n}(\xi)$. *To be precise,* $\overline{F}_\infty(\xi)$ *is the convex polyhedron in* $\overline{\mathbb{H}}^3$ *spanned by the union,* V, *of the following subsets:*

1. $\overline{F}_\infty(\eta)$ *where* η *is a 2-simplex of* $\mathcal{L}_0$ *containing* ξ.
2. $f^\epsilon_\infty(P^\epsilon_{j-1}, P^\epsilon_j)$ *where* $(P^\epsilon_{j-1}, P^\epsilon_j)$ *is an* ϵ*-extreme edge of* $\mathcal{L}_0$ *containing* ξ.
3. $f^\epsilon_\infty(P^\epsilon_j)$ *where* (P^ϵ_j) *is an* ϵ*-extreme vertex of* $\mathcal{L}_0$ *containing* ξ *(i.e.,* $\xi = (P^\epsilon_j)$*).*

Proof. (1) This follows from the fact that $f^\epsilon_{\rho_n}(P^\epsilon_j)$ is the component of $I(\rho_n(P^\epsilon_j)) \cap (E(\rho_n(P^\epsilon_{j-1})) \cap E(\rho_n(P^\epsilon_{j+1})))$ containing $\mathrm{Fix}^\epsilon_{\sigma^\epsilon}(\rho_n(P^\epsilon_j))$, and has the angle π times the $[j]$-th component of ν^ϵ_n.

(2) We prove only the assertion for vertices of $\mathcal{L}_0$, because the assertion for edges can be proved similarly. Let $\xi = (P)$ be a vertex of $\mathcal{L}_0$. Let V_n be the union of the vertices and ideal boundaries of $\overline{F}_{\rho_n}(\xi)$, that is, V_n is the union of the following sets:

1. $\overline{F}_{\rho_n}(\eta)$ where η is a 2-simplex of $\mathcal{L}_0$ containing ξ.
2. $f^\epsilon_{\rho_n}(P^\epsilon_{j-1}, P^\epsilon_j)$ where $(P^\epsilon_{j-1}, P^\epsilon_j)$ is an ϵ-extreme edge of $\mathcal{L}_0$ containing ξ.
3. $f^\epsilon_{\rho_n}(P^\epsilon_j)$ where (P^ϵ_j) is an ϵ-extreme vertex of $\mathcal{L}_0$ containing ξ.

Then $\overline{F}_{\rho_n}(\xi)$ is the closed convex hull of V_n in $\overline{\mathbb{H}}^3$. Since $V_n \subset \overline{F}_{\rho_n}(\xi)$ and $V = \lim V_n$, V is contained in $\overline{F}_\infty(\xi) = \lim \overline{F}_{\rho_n}(\xi)$. On the other hand, since $\overline{F}_{\rho_n}(\xi)$ is convex, its limit, $\overline{F}_\infty(\xi)$, is also convex. Hence the closed convex hull of V is contained in $\overline{F}_\infty(\xi)$. To see the converse, we show that the complement (in $\overline{\mathbb{H}}^3$) of the closed convex hull of V is contained in the complement of $\overline{F}_\infty(\xi)$. Let x be a point in the complement of the closed convex hull of V. Then there is a closed half space H of $\overline{\mathbb{H}}^3$ such that $V \subset H$ and $x \notin H$. We can find a closed half space H' such that $H \subset \mathrm{int}\, H'$ and $x \notin H'$. Since $\lim V_n = V \subset H \subset \mathrm{int}\, H'$, V_n is contained in $\mathrm{int}\, H'$ for all sufficiently large n. Since $\overline{F}_{\rho_n}(\xi)$ is the convex hull of V_n, this implies $\overline{F}_{\rho_n}(\xi) \subset \mathrm{int}\, H'$ for all sufficiently large n. Since $x \notin H'$, this implies that x is not contained in $\overline{F}_\infty(\xi) = \lim \overline{F}_{\rho_n}(\xi)$, i.e., x is contained in the complement of $\overline{F}_\infty(\xi)$. Thus we have proved that $\overline{F}_\infty(\xi)$ is equal to the closed convex hull of V. Since V is a subset of $\overline{Ih}(\rho_\infty(P))$ which consists of finitely many points in $Ih(\rho_\infty(P))$ and one or two (possibly degenerate) circular arcs in $I(\rho_\infty(P))$, we see that $\overline{F}_\infty(\xi)$ is a convex polyhedron (cf. Definition 3.4.1).

Proof (Proof of Proposition 8.4.4). Let ξ be a k-simplex of $\mathcal{L}_0$ and let $\{P_j \,|\, 0 \le j \le k\}$ be the vertices of ξ. Then $\overline{F}_\infty(\xi) \subset \cap_{j=0}^k \overline{Ih}(\rho_\infty(P_j))$. Since $\overline{Ih}(\rho_\infty(P_j)) \cap \mathrm{int}\, \overline{Eh}(\rho_\infty, \mathcal{L}_0) = \emptyset$, we have $\overline{F}_\infty(\xi) \cap \mathrm{int}\, \overline{Eh}(\rho_\infty, \mathcal{L}_0) = \emptyset$. So, we have only to show that $\overline{F}_\infty(\xi) \subset \overline{Eh}(\rho_\infty, \mathcal{L}_0)$.

Suppose to the contrary that there is a point, x_∞, of $\overline{F}_\infty(\xi)$ which is not contained in $\overline{Eh}(\rho_\infty, \mathcal{L}_0)$. Then $x_\infty \in \mathrm{int}\, \overline{Dh}(\rho_\infty(X))$ for some vertex (X) of $\mathcal{L}_0$. Let $\epsilon > 0$ be the Euclidean distance $d_E(x_\infty, \overline{Ih}(\rho_\infty(X)))$. Since $x_\infty \in \overline{F}_\infty(\xi) = \lim \overline{F}_{\rho_n}(\xi)$, there is a sequence $\{x_n\}$ in $\overline{\mathbb{H}}^3$ converging to x_∞ such that $x_n \in \overline{F}_{\rho_n}(\xi)$. Then we have $d_E(x_n, x_\infty) < \epsilon/2$ for all sufficiently large n. On the other hand, we see $d_E(\overline{Ih}(\rho_n(X)), \overline{Ih}(\rho_\infty(X))) < \epsilon/2$ for all sufficiently large n (cf. Lemma 8.4.3). Hence $x_n \in \mathrm{int}\, \overline{Dh}(\rho_n(X))$ for all sufficiently large

n. This contradicts the fact that $x_n \in \overline{F}_{\boldsymbol{\rho}_n}(\xi) \subset \partial\overline{Eh}(\boldsymbol{\rho}_n) = \partial\overline{Eh}(\rho_n, \mathcal{L}_0) \subset \overline{Eh}(\rho_n(X))$. Hence we have $\overline{F}_\infty(\xi) \subset \overline{Eh}(\rho_\infty, \mathcal{L}_0)$. This completes the proof of the first assertion of Proposition 8.4.4. The second assertion is proved by a similar argument.

Lemma 8.6.2. *Under Assumption 8.3.1, assume that $\{\boldsymbol{\rho}_n\}$ satisfies the condition HausdorffConvergence. Then, for any simplex ξ of $\mathcal{L}(\boldsymbol{\nu}_\infty)$, the following holds:*

$$\overline{F}_\infty(\xi) \subset \overline{F}_{(\rho_\infty, \mathcal{L}_0)}(\xi) \subset \overline{F}_{\boldsymbol{\rho}_\infty}(\xi).$$

Proof. Since $\mathcal{L}_0 \supset \mathcal{L}(\boldsymbol{\nu}_\infty)$, we have $lk(\xi, \mathcal{L}_0) \supset lk(\xi, \mathcal{L}(\boldsymbol{\nu}_\infty))$ and therefore $\overline{Eh}(\rho_\infty, lk(\xi, \mathcal{L}_0)) \subset \overline{Eh}(\rho_\infty, lk(\xi, \mathcal{L}(\boldsymbol{\nu}_\infty)))$. Hence we obtain the second inclusion as follows:

$$\begin{aligned}
\overline{F}_{(\rho_\infty, \mathcal{L}_0)}(\xi) &= \left(\bigcap\{\overline{Ih}(\rho_\infty(P)) \mid P \in \xi^{(0)}\}\right) \cap \overline{Eh}(\rho_\infty, lk(\xi, \mathcal{L}_0)) \\
&\subset \left(\bigcap\{\overline{Ih}(\rho_\infty(P)) \mid P \in \xi^{(0)}\}\right) \cap \overline{Eh}(\rho_\infty, lk(\xi, \mathcal{L}(\boldsymbol{\nu}_\infty))) \\
&= \overline{F}_{\boldsymbol{\rho}_\infty}(\xi).
\end{aligned}$$

Next we prove the first inclusion. Since $\mathrm{st}_0(\xi, \mathcal{L}_0)^{(0)}$ is finite, Lemma 8.4.3 shows the existence of a compact subset Z of $\overline{\mathbb{H}}^3$ which contains $\overline{Ih}(\rho_\infty(P))$ and $\overline{Ih}(\rho_n(P))$ for every $P \in \mathrm{st}_0(\xi, \mathcal{L}_0)^{(0)}$ and every $n \in \mathbb{N}$. Hence we see by lemma 8.4.3

$$\begin{aligned}
\overline{F}_\infty(\xi) &= \overline{F}_\infty(\xi) \cap Z \\
&= \lim(\overline{F}_{\boldsymbol{\rho}_n}(\xi) \cap Z) \\
&= \lim\left(\left(\bigcap\{\overline{Ih}(\rho_n(P)) \mid P \in \xi^{(0)}\}\right)\right. \\
&\qquad \left.\cap \left(\bigcap\{\overline{Eh}(\rho_n(X)) \cap Z \mid X \in \mathrm{lk}(\xi, \mathcal{L}_0)^{(0)}\}\right)\right) \\
&\subset \left(\bigcap\{\lim \overline{Ih}(\rho_n(P)) \mid P \in \xi^{(0)}\}\right) \\
&\qquad \cap \left(\bigcap\{\lim(\overline{Eh}(\rho_n(X)) \cap Z) \mid X \in \mathrm{lk}(\xi, \mathcal{L}_0)^{(0)}\}\right) \\
&= \left(\bigcap\{\overline{Ih}(\rho_\infty(P)) \mid P \in \xi^{(0)}\}\right) \\
&\qquad \cap \left(\bigcap\{\overline{Eh}(\rho_\infty(X)) \cap Z \mid X \in \mathrm{lk}(\xi, \mathcal{L}_0)^{(0)}\}\right) \\
&= \overline{F}_{(\rho_\infty, \mathcal{L}_0)}(\xi) \cap Z \\
&= \overline{F}_{(\rho_\infty, \mathcal{L}_0)}(\xi).
\end{aligned}$$

8.7 Accidental parabolic transformation

Proposition 8.7.1. *Under Assumption 8.3.1, $\rho_\infty(KP)$ is not parabolic for any elliptic generator satisfying $s(P) \in \Sigma_0^{(0)}$.*

Proof. By Proposition 8.3.2, we may suppose that $\boldsymbol{\nu}_\infty$ is thick. Suppose to the contrary that $\rho_\infty(KP_0)$ is parabolic for an elliptic generator triple (P_0, P_1, P_2) such that $\langle s(P_0), s(P_1), s(P_2)\rangle$ is a triangle in Σ_0. In what follows, we will use Notations 5.5.5 (InfiniteFan) and 8.1.2. Then, by Lemma 5.5.6, either (i) $\arg a_{1,\infty}^{(k)} = -\arg a_{2,\infty}^{(k)} \in (0,\pi)$ for every $k \in \mathbb{Z}$, or (ii) $\arg a_{1,\infty}^{(k)} = -\arg a_{2,\infty}^{(k)} \in (-\pi, 0)$ for every $k \in \mathbb{Z}$. For simplicity, we suppose that $\arg a_{1,\infty}^{(k)} = -\arg a_{2,\infty}^{(k)} \in (0,\pi)$ for every $k \in \mathbb{Z}$. (A similar argument works for the remaining case.)

Claim 8.7.2. The following hold for every integer k.

1. If $\sigma^{(k)} \in \Sigma_0$, then $\arg a_{1,\infty}^{(k)} = -\arg a_{2,\infty}^{(k)} \in (0, \pi/2]$.
2. If $\sigma^{(k)} \in \Sigma_0$ and $\sigma^{(k)}$ is not the $(+)$-terminal triangle of Σ_0, then its successor is equal to $\sigma^{(k-1)}$ or $\sigma^{(k+1)}$.

Proof. Suppose $\sigma^{(k)} \in \Sigma_0$. Suppose to the contrary that $\arg a_{1,\infty}^{(k)} = -\arg a_{2,\infty}^{(k)} \in (\pi/2, \pi)$, Then we can see that $\mathcal{L}(\rho_n, \sigma^{(k)})$ is not simple for sufficiently large n. This contradicts the assumption that $\sigma^{(k)} \in \Sigma_0 = \Sigma(\boldsymbol{\nu}_n)$ and $(\rho_n, \boldsymbol{\nu}_n)$ is good. Hence we obtain the first assertion.

To see the second assertion, note that the first assertion implies that $\mathcal{L}(\rho_n, \sigma^{(k)})$ is convex to the above at $c(\rho_n(P_0^{(k)}))$ for all sufficiently large n. Thus $\theta^+(\rho_n, \sigma^{(k)}; s(P_0^{(k)})) > \theta^-(\rho_n, \sigma^{(k)}; s(P_0^{(k)}))$ by Corollary 4.2.15. Therefore, by Proposition 6.7.1, if $\sigma^{(k)}$ is not the $(+)$-terminal triangle of Σ_0, its successor in $\Sigma_0 = \Sigma(\boldsymbol{\nu}_n)$ must contain $s(P_0^{(k)})$ (cf. Definition 3.2.1(1)). Hence it is equal to either $\sigma^{(k-1)}$ or $\sigma^{(k+1)}$.

Claim 8.7.3. ν_∞^+ is equal to $s(P_0)$.

Proof. By Claim 8.7.2, the $(+)$-terminal triangle of Σ_0 is equal to $\sigma^{(k)}$ for some $k \in \mathbb{Z}$. Reset $\sigma := \sigma^{(k)}$, $s_j = s_j^{(k)}$, $a_{j,\infty} = a_{j,\infty}^{(k)}$ $(j = 0, 1, 2)$ and $P_j = P_j^{(k)}$ $(k \in \mathbb{Z})$. Since $\rho_\infty(P_2P_1) = \rho_\infty(KP_0)$ is parabolic by the assumption, $\rho_\infty(P_1)$ and $\rho_\infty(P_2)$ share the unique fixed point $f := c(\rho_\infty(P_0)) + 1/2$, and it is the parabolic fixed point of $\rho_\infty(KP_0)$ (cf. Lemma 5.5.3). We show that $f = \lim \mathrm{Fix}_\sigma^+(\rho_n(P_1)) = \lim \mathrm{Fix}_\sigma^+(\rho_n(P_2))$. For simplicity we prove only $f = \lim \mathrm{Fix}_\sigma^+(\rho_n(P_1))$. Note that there is a sequence $\{\epsilon_n\}$ in $\{-,+\}$ such that $f = \lim \mathrm{Fix}_\sigma^{\epsilon_n}(\rho_n(P_1))$. Our task is to prove $\epsilon_n = +$ for all sufficiently large n. Recall that $\arg a_{1,\infty} = -\arg a_{2,\infty} \in (0, \pi/2]$ by Claim 8.7.2. This implies

$$\Im(f) = \Im(c(\rho_\infty(P_0))) > \Im(c(\rho_\infty(P_1))).$$

Hence

$$\arg\left(\frac{c(\rho_\infty(P_0)) - c(\rho_\infty(P_1))}{f - c(\rho_\infty(P_1))}\right) \in (0, \pi).$$

Thus

$$\arg\left(\frac{c(\rho_n(P_0)) - c(\rho_n(P_1))}{\mathrm{Fix}_\sigma^{\epsilon_n}(\rho_n(P_1)) - c(\rho_n(P_1))}\right) \in (0, \pi)$$

for all sufficiently large n. This implies $\epsilon_n = +$ for all sufficiently large n, and therefore $f = \lim \mathrm{Fix}_\sigma^+(\rho_n(P_1))$. Since $f = I(\rho_\infty(P_1)) \cap I(\rho_\infty(P_2))$, this in turn implies $f = \lim v^+(\rho_n; P_0, P_1)$. Thus $\lim \mathrm{Fix}_\sigma^+(\rho_n(P_1)) = v^+(\rho_n; P_0, P_1)$, and hence $\lim \theta^\epsilon(\rho_n, \sigma; s(P_1)) = 0$. Similarly, by using $f = \lim \mathrm{Fix}_\sigma^+(\rho_n(P_2))$, we see $\lim \theta^\epsilon(\rho_n, \sigma; s(P_2)) = 0$. Hence, ν_∞^+ is equal to the point $s(P_0)$ in σ represented by $(\pi/2, 0, 0) = \lim \theta^+(\rho_n, \sigma)$.

Claim 8.7.3 contradicts Assumption 8.3.1(1). This completes the proof of Proposition 8.7.1.

Proposition 8.7.4. *Under Assumption 8.3.1, assume that $\boldsymbol{\nu}_\infty$ is thick and that $\{\boldsymbol{\rho}_n\}$ satisfies the condition HausdorffConvergence. Then (ρ_∞, τ) is not an isosceles representation for any edge τ of Σ_0.*

Proof. Suppose to the contrary that there is an edge τ of Σ_0 such that (ρ_∞, τ) is an isosceles representation. Set $\sigma = \langle s_0, s_1, s_2\rangle = \langle s(P_0), s(P_1), s(P_2)\rangle$ and $\tau = \langle s_0, s_2\rangle$. Since $\rho_\infty \in \overline{\mathcal{QF}}$, none of $\rho_\infty(KP_0)$ and $\rho_\infty(KP_2)$ is elliptic (cf. Lemma 2.5.4). Thus, by Proposition 5.2.8(1), they are either purely-hyperbolic or parabolic. Hence, by Proposition 8.7.1, both $\rho_\infty(KP_0)$ and $\rho_\infty(KP_2)$ must be purely-hyperbolic. Then $\rho_\infty \in \mathcal{QF}$ and $(\rho_\infty, \boldsymbol{\mu}) \in \mathcal{J}[\mathcal{QF}]$ for some thin label $\boldsymbol{\mu} \in \mathrm{int}\,\tau \times \mathrm{int}\,\tau$ by Proposition 5.2.8(2).

By Proposition 6.2.1, there are a neighborhood U of ρ_∞ in $\mathcal{QF}$ and a continuous map $U \ni \rho \mapsto \boldsymbol{\nu}(\rho) \in \mathbb{H}^2 \times \mathbb{H}^2$ with $\boldsymbol{\nu}(\rho_\infty) = \boldsymbol{\mu}$ such that $(\rho, \boldsymbol{\nu}(\rho))$ is good for any $\rho \in U$. We may assume, by taking a subsequence, that $\rho_n \in U$ for all $n \in \mathbb{N}$. Since both $(\rho_n, \boldsymbol{\nu}_n)$ and $(\rho_n, \boldsymbol{\nu}(\rho_n))$ are good, we have $\boldsymbol{\nu}_n = \boldsymbol{\nu}(\rho_n)$ by Corollary 6.2.5. Hence $\boldsymbol{\nu}_\infty = \lim \boldsymbol{\nu}_n = \lim \boldsymbol{\nu}(\rho_n) = \boldsymbol{\nu}(\rho_\infty) = \boldsymbol{\mu}$. This contradicts the assumption that $\boldsymbol{\nu}_\infty$ is thick.

8.8 Proof of Proposition 8.4.5 - length 1 case -

In this section, we prove Proposition 8.4.5 in the case where the length of Σ_0 is 1. We begin by proving the following general fact.

Proof (Proof of Proposition 8.4.5 - the length 1 *case -).* Under the assumption of Proposition 8.4.5, assume further that the length of the common chain Σ_0 is equal to 1. Then Σ_0 consists of precisely one triangle, say $\sigma = \langle s_0, s_1, s_2\rangle$, $\mathcal{L}_0$ contains no 2-simplices, and $\mathcal{L}(\boldsymbol{\nu}_\infty)$ is equal to $\mathcal{L}_0$. Thus the assertion (2) obviously holds.

To prove the assertion (1), let $\{P_j\}$ be the sequence of elliptic generators associated with σ. Then $\mathcal{L}_0^{(0)} = \{P_j \mid j \in \mathbb{Z}\}$ and $\mathcal{L}_0^{(1)} = \{(P_j, P_{j+1}) \mid j \in \mathbb{Z}\}$.

First, we prove the assertion (1) for an edge $\xi = (P_j, P_{j+1})$ of $\mathcal{L}_0 = \mathcal{L}(\boldsymbol{\nu}_\infty)$. We may assume $j = 0$ without loss of generality. We show that

$\overline{F}_\infty(\xi)$ is a 1-dimensional convex polyhedron in $\overline{\mathbb{H}}^3$ and that $\overline{F}_{\rho_\infty}(\xi)$ is equal to $\overline{F}_\infty(\xi)$. To this end, we have only to show that $\overline{Ih}(\rho_\infty(P_0)) \cap \overline{Ih}(\rho_\infty(P_1))$ is a complete geodesic. In fact, if this is proved, then $I(\rho_\infty(P_0)) \cap I(\rho_\infty(P_1)) = \{f_\infty^-(P_0,P_1), f_\infty^+(P_0,P_1)\}$. This implies that $\overline{F}_\infty(\xi)$ is equal to the 1-dimensional convex polyhedron $\overline{Ih}(\rho_\infty(P_0)) \cap \overline{Ih}(\rho_\infty(P_1))$, because $\overline{F}_\infty(\xi) = [f_\infty^-(P_0,P_1), f_\infty^+(P_0,P_1)]$ by Lemma 8.6.1(2). Since $\overline{F}_{\rho_\infty}(\xi)$ is also equal to $\overline{Ih}(\rho_\infty(P_0)) \cap \overline{Ih}(\rho_\infty(P_1))$ by definition, we have $\overline{F}_{\rho_\infty}(\xi) = \overline{F}_\infty(\xi)$.

In what follows we prove by contradiction that $\overline{Ih}(\rho_\infty(P_0)) \cap \overline{Ih}(\rho_\infty(P_1))$ is a complete geodesic. So, suppose this is not the case. Then either $I(\rho_\infty(P_0)) = I(\rho_\infty(P_1))$ or $I(\rho_\infty(P_0))$ and $I(\rho_\infty(P_1))$ are tangent at a point, because $I(\rho_n(P_0)) \cap I(\rho_n(P_1)) \neq \emptyset$ for every $n \in \mathbb{N}$ and hence $I(\rho_\infty(P_0)) \cap I(\rho_\infty(P_1)) \neq \emptyset$. By Lemma 2.5.4(3), we have $I(\rho_\infty(P_0)) \neq I(\rho_\infty(P_1))$. Therefore $I(\rho_\infty(P_0))$ and $I(\rho_\infty(P_1))$ are tangent at a point, and hence we have

(i) $I(\rho_\infty(P_0)) \subset D(\rho_\infty(P_1))$,
(ii) $I(\rho_\infty(P_1)) \subset D(\rho_\infty(P_0))$, or
(iii) $\mathrm{int}(D(\rho_\infty(P_0))) \cap \mathrm{int}(D(\rho_\infty(P_1))) = \emptyset$.

If the condition (i) holds, then $I(\rho_\infty(P_0)) \cap E(\rho_\infty(P_1))$ consists of precisely one point. Thus one of the two fixed points of $\rho_\infty(P_0)$ in $\hat{\mathbb{C}}$ lies in $\mathrm{int}(D(\rho_\infty(P_1)))$. Hence $\mathrm{Fix}_\sigma^\epsilon(\rho_n(P_0)) \in \mathrm{int}(D(\rho_n(P_1)))$ for some $\epsilon \in \{-,+\}$ and for sufficiently large $n \in \mathbb{N}$. On the other hand, since $\Sigma(\boldsymbol{\nu}_n) = \Sigma_0 = \{\sigma\}$ and since $\boldsymbol{\rho}_n = (\rho_n, \boldsymbol{\nu}_n)$ is good, σ is the ϵ-terminal triangle of ρ_n and hence $\mathrm{Fix}_\sigma^\epsilon(\rho_n(P_0)) \in E(\rho_n(P_1))$. This is a contradiction. We have a similar contradiction if the condition (ii) holds. If the condition (iii) holds, then we see that $\lim(\theta^-(\rho_n,\sigma;s_j) + \theta^+(\rho_n,\sigma;s_j)) = \pi$ for $j = 0,1$ (see Fig. 8.3). On the

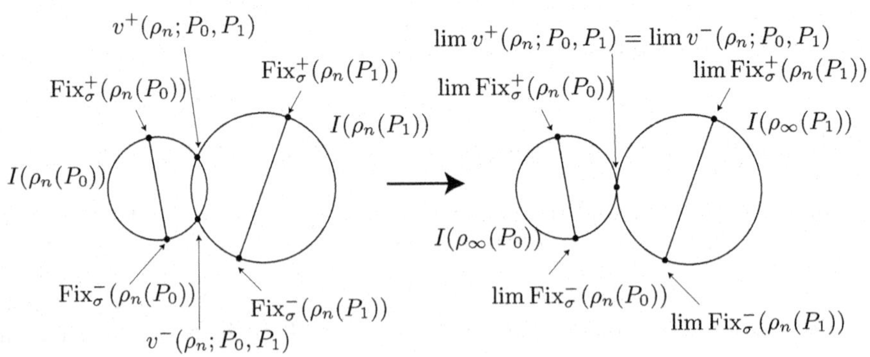

Fig. 8.3. Case (iii): $\mathrm{int}(D(\rho_\infty(P_0))) \cap \mathrm{int}(D(\rho_\infty(P_1))) = \emptyset$.

other hand, since σ is the ϵ-terminal triangle of ρ_n for each $\epsilon \in \{-,+\}$, we have $0 \leq \theta^\epsilon(\rho_n,\sigma;s_j) \leq \pi/2$ for $j = 0,1,2$ by Proposition 4.2.16. Hence $\lim \theta^\epsilon(\rho_n,\sigma;s_j) = \pi/2$ for $j = 0,1$. Therefore $\lim\left(\sum_{j=0}^2 \theta^\epsilon(\rho_n,\sigma;s_j)\right) \geq \pi >$

$\pi/2$. This contradicts Proposition 4.2.16. Hence $\overline{Ih}(\rho_\infty(P_0)) \cap \overline{Ih}(\rho_\infty(P_1))$ is a complete geodesic. Thus we have proved that the assertion (1) holds for any edge of $\mathcal{L}_0$.

Next, we prove the assertion (1) for a vertex $\xi = (P_j)$ of $\mathcal{L}_0 = \mathcal{L}(\boldsymbol{\nu}_\infty)$. We may suppose $j = 1$ without loss of generality. By Lemma 8.6.1, $\overline{F}_\infty(P_1)$ is the convex hull of $f_\infty^-(P_1) \cup f_\infty^+(P_1)$. Since $f_\infty^\epsilon(P_1)$ contains $f_\infty^\epsilon(P_0, P_1)$ and $f_\infty^\epsilon(P_1, P_2)$ for each ϵ, $\overline{F}_\infty(P_1)$ contains the 1-dimensional convex polyhedra $\overline{F}_\infty(P_0, P_1)$ and $\overline{F}_\infty(P_1, P_2)$. Hence $\dim \overline{F}_\infty(P_1) \neq 2$ if and only if $\overline{F}_\infty(P_0, P_1) = \overline{Ih}(\rho_\infty(P_0)) \cap \overline{Ih}(\rho_\infty(P_1))$ is equal to $\overline{F}_\infty(P_1, P_2) = \overline{Ih}(\rho_\infty(P_1)) \cap \overline{Ih}(\rho_\infty(P_2))$. Suppose to the contrary that this happens. Then $\overline{Ih}(\rho_\infty(P_0)) \cap \overline{Ih}(\rho_\infty(P_1)) \cap \overline{Ih}(\rho_\infty(P_2))$ is a complete geodesic, and $f_\infty^\epsilon(P_1)$ is equal to the singleton $f_\infty^\epsilon(P_0, P_1) = f_\infty^\epsilon(P_1, P_2)$, which in turn is equal to the singleton $\lim \mathrm{Fix}_\sigma^\epsilon(\rho_n(P_1))$. Thus $\overline{Ih}(\rho_\infty(P_0)) \cap \overline{Ih}(\rho_\infty(P_1)) \cap \overline{Ih}(\rho_\infty(P_2))$ is equal to the geodesic joining the two fixed points of $\rho_\infty(P_1)$. Hence we have $\overline{\mathrm{Axis}}(\rho_\infty(P_1)) = \overline{Ih}(\rho_\infty(P_0)) \cap \overline{Ih}(\rho_\infty(P_2))$. Thus (ρ_∞, τ) with $\tau = \langle s(P_0), s(P_2)\rangle$ is an isosceles representation by Proposition 5.2.3. This contradicts Proposition 8.7.4. Hence $\dim \overline{F}_\infty(P_1) = 2$. In order to prove $\overline{F}_{\boldsymbol{\rho}_\infty}(P_1) = \overline{F}_\infty(P_1)$, we note that $\overline{F}_\infty(P_1)$ is the subspace of $\overline{Ih}(\rho_\infty(P_1))$ bounded by the two geodesics $\overline{Ih}(\rho_\infty(P_0)) \cap \overline{Ih}(\rho_\infty(P_1))$ and $\overline{Ih}(\rho_\infty(P_1)) \cap \overline{Ih}(\rho_\infty(P_2))$. By Lemma 8.6.2, $\overline{F}_\infty(P_1) \subset \overline{F}_{\boldsymbol{\rho}_\infty}(P_1)$, and by definition $\overline{F}_{\boldsymbol{\rho}_\infty}(P_1) = \overline{Ih}(\rho_\infty(P_1)) \cap \overline{Eh}(\rho_\infty(P_0)) \cap \overline{Eh}(\rho_\infty(P_2))$. Thus $\overline{F}_{\boldsymbol{\rho}_\infty}(P_1)$ is also the subspace of $\overline{Ih}(\rho_\infty(P_1))$ bounded by the two geodesics $\overline{Ih}(\rho_\infty(P_0)) \cap \overline{Ih}(\rho_\infty(P_1))$ and $\overline{Ih}(\rho_\infty(P_1)) \cap \overline{Ih}(\rho_\infty(P_2))$. Hence we have $\overline{F}_{\boldsymbol{\rho}_\infty}(P_1) = \overline{F}_\infty(P_1)$. Thus we have proved that the assertion (1) holds for any vertex of $\mathcal{L}_0$. This completes the proof of Proposition 8.4.5 in the case when the length of Σ_0 is 1.

8.9 Proof of Proposition 8.4.5 - length ≥ 2 case - (Step 1)

Sections 8.9–8.11 are devoted to the proof of Proposition 8.4.5 in the case where the length of Σ_0 is ≥ 2. Thus we presume the following assumption throughout these sections.

Assumption 8.9.1. We assume that $\{\boldsymbol{\rho}_n\} = \{(\rho_n, \boldsymbol{\nu}_n)\}$ is a sequence in $\mathcal{J}[\mathcal{QF}]$ satisfying the following conditions:

(1) $\{\boldsymbol{\rho}_n\}$ converges to a labeled representation $\boldsymbol{\rho}_\infty = (\rho_\infty, \boldsymbol{\nu}_\infty) \in \overline{\mathcal{QF}} \times (\mathbb{H}^2 \times \mathbb{H}^2)$.
(2) $\{\boldsymbol{\rho}_n\}$ satisfies the condition SameStratum, i.e., every chain $\Sigma(\boldsymbol{\nu}_n)$ is equal to a fixed chain Σ_0, and hence every $\mathcal{L}(\boldsymbol{\nu}_n)$ is equal to the fixed elliptic generator complex $\mathcal{L}_0 = \mathcal{L}(\Sigma_0)$.
(3) $\{\boldsymbol{\rho}_n\} = \{(\rho_n, \boldsymbol{\nu}_n)\}$ satisfies the condition HausdorffConvergence.
(4) $\boldsymbol{\nu}_\infty$ is thick.
(5) The length of Σ_0 is ≥ 2, or equivalently, $\mathcal{L}_0$ contains a 2-simplex.

The purpose of this section is to study the behavior of the limit of three isometric hemispheres whose intersection constitute a vertex of $\overline{Eh}(\boldsymbol{\rho}_n)$. Thus we also presume the following assumption throughout this section (except in Lemma 8.9.9)

Assumption 8.9.2. Under Assumption 8.9.1, $\xi = (P_0, P_1, P_2)$ denotes a 2-simplex of $\mathcal{L}_0$, where $\{P_j\}$ is the sequence of elliptic generators associated with a triangle $\sigma = \langle s_0, s_1, s_2 \rangle$ of $\mathcal{D}$. As in Notation 2.1.14 (Adjacent triangles), $\sigma' = \langle s'_0, s'_1, s'_2 \rangle = \langle s_0, s_2, s'_2 \rangle$ denotes the triangle of $\mathcal{D}$ sharing the edge $\langle s_0, s_2 \rangle = \langle s'_0, s'_1 \rangle$ with σ. We note that both σ and σ' belong to the chain Σ_0 by Lemma 3.2.4(2). The symbol ϵ_0 denotes the sign such that σ lies in the ϵ_0-side of σ'.

We first prove the following consequence of Lemma 4.8.7, which plays a key role in the proof of Lemma 8.9.8.

Lemma 8.9.3 (NoBadFold). *Under Assumptions 8.9.1 and 8.9.2, $\mathcal{L}(\rho_\infty, \sigma)$ is not folded at $s(P_1)$. Namely, the centers $c(\rho_\infty(P_j))$ $(j = 0, 1, 2)$ cannot sit on the line containing $\pi(\mathrm{Axis}(\rho_\infty(P_1)))$.*

Proof. From the assumption, either (σ, σ') or (σ', σ) forms a subchain, Σ, of Σ_0. Since $\phi_\infty^{-1}(0) \cap \sigma^{(0)} = \emptyset$, the complex probability $(a_{0,\infty}, a_{1,\infty}, a_{2,\infty}) \in (\mathbb{C} - \{0\})^3$ of ρ_∞ at σ is defined. Suppose to the contrary that $c(\rho_\infty(P_j))$ $(j = 0, 1, 2)$ sit on the line containing $\pi(\mathrm{Axis}(\rho_\infty(P_1)))$. Then we have $a_{2,\infty} = \lambda a_{0,\infty}$ for some negative real number λ. By Proposition 5.4.3, ρ_∞ is not pure-imaginary, and therefore $\mathcal{L}(\rho_\infty, \sigma)$ is not doubly folded by Lemma 5.4.2. Hence, if $\lambda \neq -1$, then $\mathcal{L}(\rho_n, \Sigma)$ is not simple for all sufficiently large n by Lemma 4.8.7. So we have $\lambda = -1$ and hence $I(\rho_\infty(P_0)) = I(\rho_\infty(P_2))$. This contradicts Lemma 2.5.4(2-iii).

To state the main result of this section, we need the following lemma.

Lemma 8.9.4. *Under Assumptions 8.9.1 and 8.9.2, precisely one of the following three situations occurs,*

(1) $\xi \in \mathcal{L}(\boldsymbol{\nu}_\infty)^{(2)}$.
(2) $\nu_\infty^{\epsilon_0} \in \mathrm{int}\langle s_0, s_2 \rangle$, $s_1 \notin \Sigma(\boldsymbol{\nu}_\infty)^{(0)}$ and $\xi \notin \mathcal{L}(\boldsymbol{\nu}_\infty)^{(2)}$. In this case σ and σ', respectively, are the ϵ_0-terminal triangles of Σ_0 and $\Sigma(\boldsymbol{\nu}_\infty)$ (see Fig. 8.4(a)).
(3) $\nu_\infty^{-\epsilon_0} \in \mathrm{int}\langle s_0, s_2 \rangle$, $s_1 \in \Sigma(\boldsymbol{\nu}_\infty)^{(0)}$ and $\xi \notin \mathcal{L}(\boldsymbol{\nu}_\infty)^{(2)}$. In this case σ' and σ, respectively, are the $(-\epsilon_0)$-terminal triangles of Σ_0 and $\Sigma(\boldsymbol{\nu}_\infty)$ (see Fig. 8.4(b)).

Proof. Since $\xi = (P_0, P_1, P_2)$ is a 2-simplex of $\mathcal{L}_0$, both σ and σ' belong to the chain Σ_0 by Lemma 3.2.4(2). Suppose that the first condition does not hold, i.e., $\xi \notin \mathcal{L}(\boldsymbol{\nu}_\infty)^{(2)}$. Then σ or σ' does not belong to $\Sigma(\boldsymbol{\nu}_\infty)$. If σ does not belong to $\Sigma(\boldsymbol{\nu}_\infty)$, then the geodesic segment $[\nu_\infty^-, \nu_\infty^+]$ is disjoint from

$\mathrm{int}\,\sigma$, whereas each $[\nu_n^-, \nu_n^+]$ intersects both $\mathrm{int}\,\sigma$ and $\mathrm{int}\,\sigma'$. Thus $\nu_n^{\epsilon_0} \in \sigma$ and $\nu_\infty^{\epsilon_0} \in \sigma \cap \sigma' = \langle s_0, s_2 \rangle$, because σ lies in the ϵ_0-side of σ'. In particular, $s_1 \notin \Sigma(\boldsymbol{\nu}_\infty)^{(0)}$. Since $\boldsymbol{\nu}_\infty$ is thick by assumption, the triangle σ' must be contained in $\Sigma(\boldsymbol{\nu}_\infty)$. Hence, by using the fact that $\nu_\infty^{\epsilon_0} \in \mathbb{H}^2$, we see that the second condition in the lemma holds. If σ' does not belong to $\Sigma(\boldsymbol{\nu}_\infty)$, then by a similar argument we see that the third condition in the lemma holds.

Under Assumptions 8.9.1 and 8.9.2, the three isometric hemispheres $\overline{Ih}(\rho_n(P_j))$ $(j = 0, 1, 2)$ intersect transversely as illustrated in Fig. 8.5, because $\boldsymbol{\rho}_n = (\rho_n, \boldsymbol{\nu}_n)$ is good. Namely, the following hold.

1. $\overline{Ih}(\rho_n(P_0)) \cap \overline{Ih}(\rho_n(P_1)) \cap \overline{Ih}(\rho_n(P_2))$ is equal to the singleton $\overline{F}_{\boldsymbol{\rho}_n}(\xi)$ in $\mathbb{H}^3$.
2. For each $k \in \{0,1,2\}$ set $L_n(k) := \overline{Ih}(\rho_n(P_i)) \cap \overline{Ih}(\rho_n(P_j)) \cap \overline{Eh}(\rho_n(P_k))$ where $\{i,j,k\} = \{0,1,2\}$ and $i < j$. Then $L_n(k)$ is equal to the half geodesic $[\overline{F}_{\boldsymbol{\rho}_n}(\xi), v^{\epsilon(k)}(\rho_n; P_i, P_j)]$ contained in the complete geodesic $\overline{Ih}(\rho_n(P_i)) \cap \overline{Ih}(\rho_n(P_j))$, where $\epsilon(k) = \epsilon_0$ or $-\epsilon_0$ according as $k \in \{0, 2\}$ or $k = 1$.

Proposition 8.9.5 describes what happens to the configuration of the three isometric hemispheres at the limit. Roughly speaking, we show that the limit configuration is combinatorially equivalent to the original one or that illustrated in Figs. 8.6 or 8.7 according to the three situations in Lemma 8.9.4. Throughout Sects. 8.9 and 8.10 we denote $\lim v^\epsilon(\rho_n; P, Q)$ (resp. $\lim \mathrm{Fix}_\sigma^\epsilon(\rho_n(P))$) by $v_\infty^\epsilon(P,Q)$ (resp. $\mathrm{Fix}_\sigma^\epsilon(\rho_\infty(P))$).

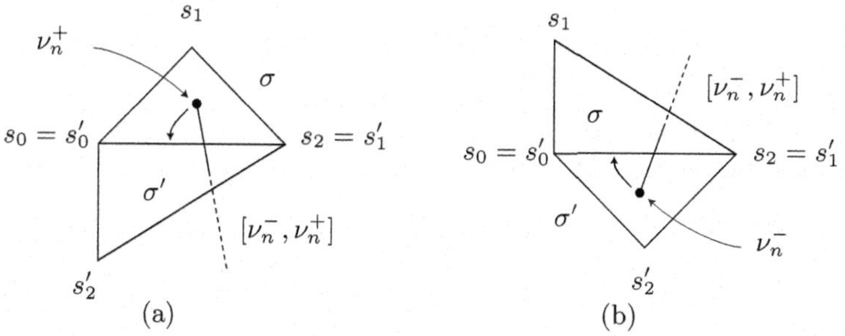

Fig. 8.4. (a) $\nu_\infty^{\epsilon_0} \in \mathrm{int}\langle s_0, s_2\rangle$, (b) $\nu_\infty^{-\epsilon_0} \in \mathrm{int}\langle s_0, s_2\rangle$; the figures are for $\epsilon_0 = +$.

Proposition 8.9.5. *Under Assumptions 8.9.1 and 8.9.2, set* $L_\infty(k) := \overline{Ih}(\rho_\infty(P_i)) \cap \overline{Ih}(\rho_\infty(P_j)) \cap \overline{Eh}(\rho_\infty(P_k))$ *for each* $k \in \{0,1,2\}$*, where* $\{i,j,k\} = \{0,1,2\}$ *and* $i < j$. *Let* $\epsilon(k)$ *be the sign defined in the above, i.e.,* $\epsilon(k) = \epsilon_0$ *or* $-\epsilon_0$ *according as* $k \in \{0,2\}$ *or* $k = 1$

(1) If $\xi \in \mathcal{L}(\boldsymbol{\nu}_\infty)^{(2)}$, *then the combinatorial structure of the configuration of the three isometric hemispheres does not change at the limit. Namely, the following hold (cf. Fig. 8.5).*

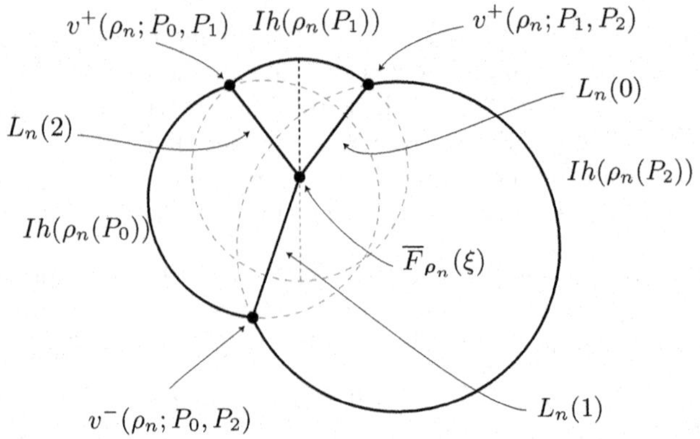

Fig. 8.5. The combinatorial structure in the case when $\xi = (P_0, P_1, P_2) \in \mathcal{L}_0$; the figure is for $\epsilon_0 = +$.

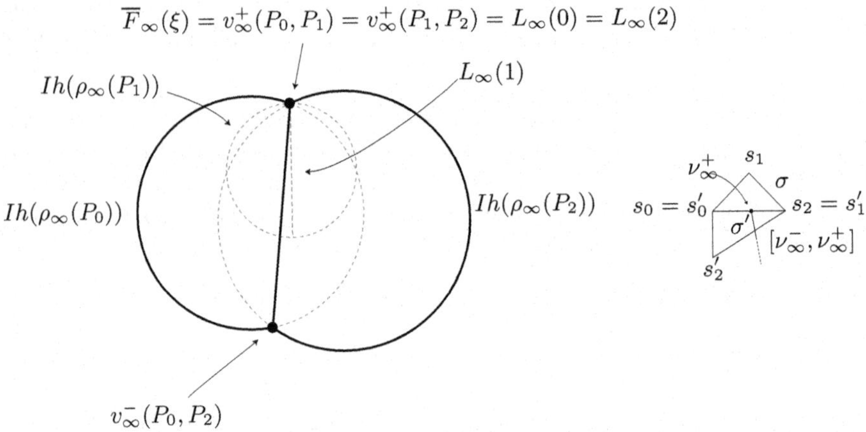

Fig. 8.6. The degeneration of the combinatorial structure in the case when $\nu_\infty^{\epsilon_0} \in \mathrm{int}\langle s_0, s_2\rangle$; the figure is for $\epsilon_0 = +$.

(i) $\overline{Ih}(\rho_\infty(P_0)) \cap \overline{Ih}(\rho_\infty(P_1)) \cap \overline{Ih}(\rho_\infty(P_2))$ *is equal to the singleton* $\overline{F}_\infty(\xi)$ *in* $\mathbb{H}^3$.

(ii) For each $k \in \{0, 1, 2\}$, $L_\infty(k)$ *is equal to the half geodesic* $[\overline{F}_\infty(\xi), v_\infty^{\epsilon(k)}(P_i, P_j)]$ *contained in the complete geodesic* $\overline{Ih}(\rho_\infty(P_i)) \cap \overline{Ih}(\rho_\infty(P_j))$.

(2) If $\nu_\infty^{\epsilon_0} \in \mathrm{int}\langle s_0, s_2\rangle$, *i.e., if the second condition in Lemma 8.9.4 holds, then as* $n \to \infty$, $\overline{F}_{\rho_n}(\xi)$ *drops onto the point* $\mathrm{Fix}_\sigma^{\epsilon_0}(\rho_\infty(P_1))$ *in* $\mathbb{C}$, $\overline{Ih}(\rho_n(P_1))$ *becomes invisible, and* $L_n(0)$ *and* $L_n(2)$ *shrink to the point in* $\mathbb{C}$, *whereas* $L_n(1)$ *turns into a complete geodesic. To be precise, the following hold (see Fig. 8.6).*

(i) $\overline{Ih}(\rho_\infty(P_0)) \cap \overline{Ih}(\rho_\infty(P_1)) \cap \overline{Ih}(\rho_\infty(P_2))$ *is equal to the singleton* $\overline{F}_\infty(\xi)$ *in* $\mathbb{C}$, *and*

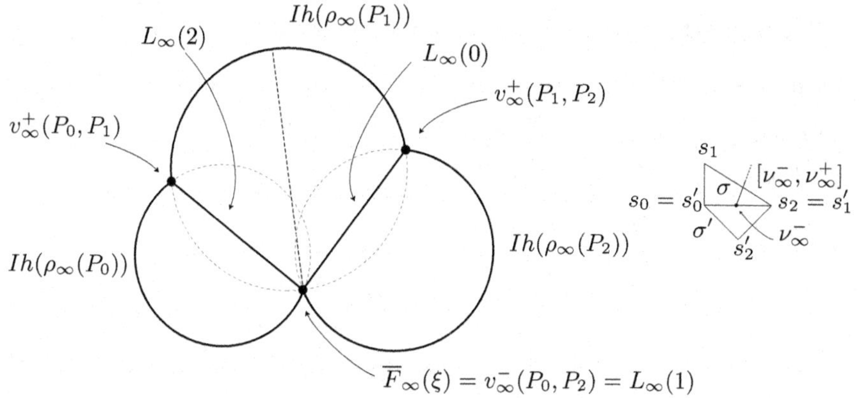

Fig. 8.7. The degeneration of the combinatorial structure in the case when $\nu_\infty^{-\epsilon_0}\in \mathrm{int}\langle s_0,s_2\rangle$; the figure is for $\epsilon_0=+$.

$$\overline{F}_\infty(\xi)=\mathrm{Fix}_\sigma^{\epsilon_0}(\rho_\infty(P_1))=v_\infty^{\epsilon_0}(P_0,P_1)=v_\infty^{\epsilon_0}(P_1,P_2)=v_\infty^{\epsilon_0}(P_0,P_2).$$

(*ii*) $\overline{Ih}(\rho_\infty(P_1))$ *is covered by the remaining two isometric hemispheres. To be precise:*
  (*a*) $\overline{Dh}(\rho_\infty(P_1))\cap\overline{Eh}(\rho_\infty(P_0))\cap\overline{Eh}(\rho_\infty(P_2))$ *is equal to the singleton* $\overline{F}_\infty(\xi)$ *in* $\mathbb{C}$.
  (*b*) $\overline{Eh}(\rho_\infty(P_1))\supset\overline{Eh}(\rho_\infty(P_0))\cap\overline{Eh}(\rho_\infty(P_2))$.

(*iii*) $L_\infty(k)=[\overline{F}_\infty(\xi),v_\infty^{\epsilon(k)}(P_i,P_j)]$. *Thus* $L_\infty(0)$ *and* $L_\infty(2)$ *are equal to the singleton* $\overline{F}_\infty(\xi)$ *in* $\mathbb{C}$, *whereas* $L_\infty(1)$ *is equal to the complete geodesic* $\overline{Ih}(\rho_\infty(P_0))\cap\overline{Ih}(\rho_\infty(P_2))$.

(3) If $\nu_\infty^{-\epsilon_0}\in\mathrm{int}\langle s_0,s_2\rangle$, *i.e., if the third condition in Lemma 8.9.4 holds, then, as* $n\to\infty$, $\overline{F}_{\rho_n}(\xi)$ *drops onto the point* $\mathrm{Fix}_\sigma^{-\epsilon_0}(\rho_\infty(P_1))$ *in* $\mathbb{C}$, *and* $L_n(1)$ *shrinks to the point in* $\mathbb{C}$, *whereas* $L_n(0)$ *and* $L_n(2)$ *turn into complete geodesics. To be precise, we have the following (see Fig. 8.7).*

(*i*) $\overline{Ih}(\rho_\infty(P_0))\cap\overline{Ih}(\rho_\infty(P_1))\cap\overline{Ih}(\rho_\infty(P_2))$ *is equal to the singleton* $\overline{F}_\infty(\xi)$ *in* $\mathbb{C}$, *and*

$$\overline{F}_\infty(\xi)=\mathrm{Fix}_\sigma^{-\epsilon_0}(\rho_\infty(P_1))=v_\infty^{-\epsilon_0}(P_0,P_1)=v_\infty^{-\epsilon_0}(P_1,P_2)=v_\infty^{-\epsilon_0}(P_0,P_2).$$

(*ii*) $L_\infty(k)=[\overline{F}_\infty(\xi),v_\infty^{\epsilon(k)}(P_i,P_j)]$. *Thus* $L_\infty(1)$ *is equal to the singleton* $\overline{F}_\infty(\xi)$ *in* $\mathbb{C}$, *whereas* $L_\infty(2)$ *is equal to the complete geodesic* $\overline{Ih}(\rho_\infty(P_0))\cap\overline{Ih}(\rho_\infty(P_1))$ *and* $L_\infty(0)$ *is equal to the complete geodesic* $\overline{Ih}(\rho_\infty(P_1))\cap\overline{Ih}(\rho_\infty(P_2))$.

As an immediate corollary, we obtain the following.

Corollary 8.9.6. *Under Assumptions 8.9.1 and 8.9.2, we have*

$$L_\infty(k)=\lim L_n(k)=[\overline{F}_\infty(\xi),v_\infty^{\epsilon(k)}(P_i,P_j)].$$

for each $k\in\{0,1,2\}$.

We begin the proof of this proposition by proving a few lemmas.

Lemma 8.9.7. *Under Assumptions 8.9.1 and 8.9.2, the following hold.*
(1) $\overline{F}_\infty(\xi) \subset \overline{Ih}(\rho_\infty(P_0)) \cap \overline{Ih}(\rho_\infty(P_1)) \cap \overline{Ih}(\rho_\infty(P_2))$.
(2) $[\overline{F}_\infty(\xi), v_\infty^{\epsilon(k)}(P_i, P_j)] \subset L_\infty(k)$.

Proof. (1) This is proved as follows.

$$\begin{aligned}\overline{F}_\infty(\xi) &= \lim \overline{F}_{\rho_n}(\xi)\\ &= \lim \left(\overline{Ih}(\rho_n(P_0)) \cap \overline{Ih}(\rho_n(P_1)) \cap \overline{Ih}(\rho_n(P_2))\right)\\ &\subset \overline{Ih}(\rho_\infty(P_0)) \cap \overline{Ih}(\rho_\infty(P_1)) \cap \overline{Ih}(\rho_\infty(P_2)).\end{aligned}$$

(2) Since $\overline{F}_\infty(\xi) = \lim \overline{F}_{\rho_n}(\xi)$ and $v_\infty^{\epsilon(k)}(P_i, P_j) = \lim v^{\epsilon(k)}(\rho_n; P_i, P_j)$, we have:

$$\begin{aligned}[\overline{F}_\infty(\xi), v_\infty^{\epsilon(k)}(P_i, P_j)] &= \lim[\overline{F}_{\rho_n}(\xi), v^{\epsilon(k)}(\rho_n; P_i, P_j)]\\ &= \lim \left(\overline{Ih}(\rho_n(P_i)) \cap \overline{Ih}(\rho_n(P_j)) \cap \overline{Eh}(\rho_n(P_k))\right)\\ &\subset \overline{Ih}(\rho_\infty(P_i)) \cap \overline{Ih}(\rho_\infty(P_j)) \cap \overline{Eh}(\rho_\infty(P_k))\\ &= L_\infty(k).\end{aligned}$$

Lemma 8.9.8. *Under Assumptions 8.9.1 and 8.9.2, the following hold.*
(1) $\overline{Ih}(\rho_\infty(P_0)) \cap \overline{Ih}(\rho_\infty(P_1))$ *and* $\overline{Ih}(\rho_\infty(P_1)) \cap \overline{Ih}(\rho_\infty(P_2))$ *are complete geodesics.*
(2) $\overline{Ih}(\rho_\infty(P_0)) \cap \overline{Ih}(\rho_\infty(P_1)) \neq \overline{Ih}(\rho_\infty(P_1)) \cap \overline{Ih}(\rho_\infty(P_2))$.
(3) $\overline{Ih}(\rho_\infty(P_0)) \cap \overline{Ih}(\rho_\infty(P_1)) \cap \overline{Ih}(\rho_\infty(P_2))$ *is equal to the singleton* $\overline{F}_\infty(\xi)$.

Proof. (1) By Lemma 8.9.7(1), we have

$$\overline{F}_\infty(\xi) \subset \overline{Ih}(\rho_\infty(P_0)) \cap \overline{Ih}(\rho_\infty(P_1)) \cap \overline{Ih}(\rho_\infty(P_2)) \subset \overline{Ih}(\rho_\infty(P_0)) \cap \overline{Ih}(\rho_\infty(P_1)).$$

Since $\overline{Ih}(\rho_\infty(P_0)) \neq \overline{Ih}(\rho_\infty(P_1))$ (cf. Lemma 2.5.4(2-iii)), the above inclusion implies that $\overline{Ih}(\rho_\infty(P_0)) \cap \overline{Ih}(\rho_\infty(P_1))$ is either the singleton $\overline{F}_\infty(\xi)$ or a complete geodesic. Suppose to the contrary that $\overline{Ih}(\rho_\infty(P_0)) \cap \overline{Ih}(\rho_\infty(P_1))$ is the singleton $\overline{F}_\infty(\xi)$. Then, by Lemma 4.1.3(2) (Chain rule), $\overline{Ih}(\rho_\infty(P_1)) \cap \overline{Ih}(\rho_\infty(P_2)) = \rho_\infty(P_1)(\overline{Ih}(\rho_\infty(P_0)) \cap \overline{Ih}(\rho_\infty(P_1)))$ is also a singleton. Since $\overline{F}_\infty(\xi)$ is contained in $\overline{\mathrm{Axis}}(\rho_\infty(P_1))$, this implies that the centers $c(\rho_\infty(P_j))$ $(j = 0, 1, 2)$ lie in the line in $\mathbb{C}$ containing $\mathrm{proj}(\overline{\mathrm{Axis}}(\rho_\infty(P_1)))$. This contradicts Lemma 8.9.3 (NoBadFold). Hence $\overline{Ih}(\rho_\infty(P_0)) \cap \overline{Ih}(\rho_\infty(P_1))$ is a geodesic. By Lemma 4.1.3(2), $\overline{Ih}(\rho_\infty(P_1)) \cap \overline{Ih}(\rho_\infty(P_2))$ is also a geodesic.

(2) Suppose to the contrary that $\overline{Ih}(\rho_\infty(P_0)) \cap \overline{Ih}(\rho_\infty(P_1)) = \overline{Ih}(\rho_\infty(P_1)) \cap \overline{Ih}(\rho_\infty(P_2))$. Then, by Lemma 5.2.6, either (ρ_∞, τ) is an isosceles representation where $\tau = \langle s_0, s_2 \rangle$ or $\mathcal{L}(\rho_\infty, \sigma)$ is folded at $c(\rho_\infty(P_1))$. By Lemma 8.9.3 (NoBadFold), the latter cannot happen. So, (ρ_∞, τ) is an isosceles representation. This contradicts Proposition 8.7.4.

(3) By (1) and (2), $\overline{Ih}(\rho_\infty(P_0)) \cap \overline{Ih}(\rho_\infty(P_1))$ and $\overline{Ih}(\rho_\infty(P_1)) \cap \overline{Ih}(\rho_\infty(P_2))$ are different geodesics, and hence $\overline{Ih}(\rho_\infty(P_0)) \cap \overline{Ih}(\rho_\infty(P_1)) \cap \overline{Ih}(\rho_\infty(P_2))$ is either empty or a singleton. Hence we obtain the desired result by Lemma 8.9.7(1).

Lemma 8.9.9. *Under Assumption 8.9.1, let η be a generic edge of $\mathcal{L}_0$ (Definition 8.5.1). Then η is also an edge of $\mathcal{L}(\boldsymbol{\nu}_\infty)$, and $\overline{F}_\infty(\eta)$ is a 1-dimensional convex polyhedron in $\overline{\mathbb{H}}^3$.*

Proof. Since η is a generic edge of $\mathcal{L}_0$, we see by Lemma 8.5.2 that there is a triangle σ of $\Sigma_0 - \{\sigma^-, \sigma^+\}$ such that (i) $\eta = (P_j, P_{j+1})$ for some j and (ii) both (P_0, P_1, P_2) and (P_1, P_2, P_3) belong to $\mathcal{L}_0^{(2)}$. Here $\{P_j\}$ is the sequence of elliptic generators associated with $\sigma = \langle s_0, s_1, s_2 \rangle$. Since σ belongs to $\Sigma_0 - \{\sigma^-, \sigma^+\}$, it also belongs to $\Sigma(\boldsymbol{\nu}_\infty)$. Hence η is also an edge of $\mathcal{L}(\boldsymbol{\nu}_\infty)$. Suppose that $\overline{F}_\infty(\eta)$ is not a 1-dimensional convex polyhedron. Then it must be a singleton, because it is the convex hull of two (possibly identical) points by Lemma 8.6.1(2). By Lemma 6.4.2(2), this implies that $\overline{F}_\infty(P_j, P_{j+1})$ is equal to a singleton for every j. Hence, we have $\overline{F}_\infty(P_0, P_1, P_2) = \overline{F}_\infty(P_1, P_2, P_3)$, because both points are contained in the singleton $\overline{F}_\infty(P_1, P_2)$. On the other hand, $\overline{F}_\infty(P_0, P_1, P_2)$ and $\overline{F}_\infty(P_1, P_2, P_3)$, respectively, are contained in $\overline{\mathrm{Axis}}(\rho_\infty(P_1))$ and $\overline{\mathrm{Axis}}(\rho_\infty(P_2))$ by Lemma 6.4.2(3). Thus $\overline{\mathrm{Axis}}(\rho_\infty(P_1))$ and $\overline{\mathrm{Axis}}(\rho_\infty(P_2))$ share the point $\overline{F}_\infty(P_0, P_1, P_2) = \overline{F}_\infty(P_1, P_2, P_3)$. Hence $\rho_\infty(KP_0) = \rho_\infty(P_2 P_1)$ is elliptic, parabolic or the identity. This contradicts Lemma 2.5.4(2) and Proposition 8.7.1.

Lemma 8.9.10. *Under Assumptions 8.9.1 and 8.9.2, $\overline{F}_\infty(\xi)$ belongs to $\mathbb{C}$ if and only if $\xi \notin \mathcal{L}(\boldsymbol{\nu}_\infty)^{(2)}$. Moreover, in this case, precisely one of the following holds.*

(1) $\overline{F}_\infty(\xi) = \mathrm{Fix}_\sigma^{\epsilon_0}(\rho_\infty(P_1))$ and it is equal to $v_\infty^{\epsilon_0}(P_0, P_1) = v_\infty^{\epsilon_0}(P_1, P_2) = v_\infty^{\epsilon_0}(P_0, P_2)$. Moreover $\nu_\infty^{\epsilon_0} \in \mathrm{int}\langle s_0, s_2 \rangle$, $s_1 \notin \Sigma(\boldsymbol{\nu}_\infty)^{(0)}$, and the condition in Lemma 8.9.4(2) holds (see Fig. 8.6).

(2) $\overline{F}_\infty(\xi) = \mathrm{Fix}_\sigma^{-\epsilon_0}(\rho_\infty(P_1))$ and it is equal to $v_\infty^{-\epsilon_0}(P_0, P_1) = v_\infty^{-\epsilon_0}(P_1, P_2) = v_\infty^{-\epsilon_0}(P_0, P_2)$. Moreover $\nu_\infty^{-\epsilon_0} \in \mathrm{int}\langle s_0, s_2 \rangle$, $s_1 \in \Sigma(\boldsymbol{\nu}_\infty)^{(0)}$, and the condition in Lemma 8.9.4(3) holds (see Fig. 8.7).

Proof. Note that $\mathcal{L}(\rho_n, \sigma)$ lies above or below $\mathcal{L}(\rho_n, \sigma')$ according as $\epsilon_0 = +$ or $-$, because σ lies on the ϵ_0-side of σ' and because $\boldsymbol{\rho}_n$ satisfies the condition Simple (cf. Definition 3.2.2). Since the three isometric hemispheres $Ih(\rho_n(P_j))$ ($j = 0, 1, 2$) intersect at the singleton $F_{\rho_n}(\xi)$, the following hold by Lemma 4.3.6 and Corollary 4.2.15.

1. $\theta^{\epsilon_0}(\rho_n, \sigma; s_1) > 0$ and $\theta^{-\epsilon_0}(\rho_n, \sigma; s_1) < 0$.
2. $\mathrm{Fix}_\sigma^{\epsilon_0}(\rho_n(P_1)) \notin D(\rho_n(P_0)) \cup D(\rho_n(P_2))$ and $\mathrm{Fix}_\sigma^{-\epsilon_0}(\rho_n(P_1)) \in \mathrm{int}(D(\rho_n(P_0)) \cap D(\rho_n(P_2)))$.

Note that the three points $v^{\epsilon_0}(\rho_n; P_0, P_1)$, $v^{\epsilon_0}(\rho_n; P_1, P_2)$, and $v^{-\epsilon_0}(\rho_n; P_0, P_2)$ lie on the boundary of $\cup_{j=0}^{2} D(\rho_n(P_j))$ (cf. Fig. 8.5).

First, suppose that $\overline{F}_\infty(\xi)$ belongs to $\mathbb{C}$. Then $\overline{F}_\infty(\xi)$ is equal to $\mathrm{Fix}_\sigma^{\epsilon_0}(\rho_\infty(P_1))$ or $\mathrm{Fix}_\sigma^{-\epsilon_0}(\rho_\infty(P_1))$, because $\overline{F}_\infty(\xi) \in \overline{\mathrm{Axis}}(\rho_\infty(P_1)) \cap \mathbb{C} = \{\mathrm{Fix}_\sigma^{\pm}(\rho_\infty(P_1))\}$.

Case 1. Suppose that $\overline{F}_\infty(\xi) = \mathrm{Fix}_\sigma^{\epsilon_0}(\rho_\infty(P_1))$.

Claim 8.9.11. The 2-dimensional convex polyhedron $\overline{Ih}(\rho_n(P_1)) \cap \overline{Eh}(\rho_n(P_0)) \cap \overline{Eh}(\rho_n(P_2))$ shrinks to the singleton $\overline{F}_\infty(\xi) = \mathrm{Fix}_\sigma^{\epsilon_0}(\rho_\infty(P_1))$ as $n \to \infty$ (cf. Fig. 8.6). Namely,

$$\lim \left(\overline{Ih}(\rho_n(P_1)) \cap \overline{Eh}(\rho_n(P_0)) \cap \overline{Eh}(\rho_n(P_2))\right) = \overline{F}_\infty(\xi).$$

In particular, the following hold.

(1) The geodesic segments $[F_{\rho_n}(\xi), v^{\epsilon_0}(\rho_n; P_0, P_1)]$ and $[F_{\rho_n}(\xi), v^{\epsilon_0}(\rho_n; P_1, P_2)]$ shrink to the point $\overline{F}_\infty(\xi)$ as $n \to \infty$. In particular $\overline{F}_\infty(P_0, P_1)$ and $\overline{F}_\infty(P_1, P_2)$ are singletons.
(2) The circular arc $e^{\epsilon_0}(\rho_n, \sigma; P_1)$ shrinks to the point $\overline{F}_\infty(\xi)$ as $n \to \infty$.

Proof. Let ℓ_n and ℓ'_n, respectively, be the perpendicular to $\mathrm{proj}(\overline{\mathrm{Axis}}(\rho_n(P_1)))$ at $\mathrm{Fix}_\sigma^{\epsilon_0}(\rho_n(P_1))$ and at $\mathrm{proj}(\overline{F}_{\rho_n}(\xi))$ (see Fig. 8.8). Then $\mathrm{proj}(\overline{Ih}(\rho_n(P_1)) \cap$

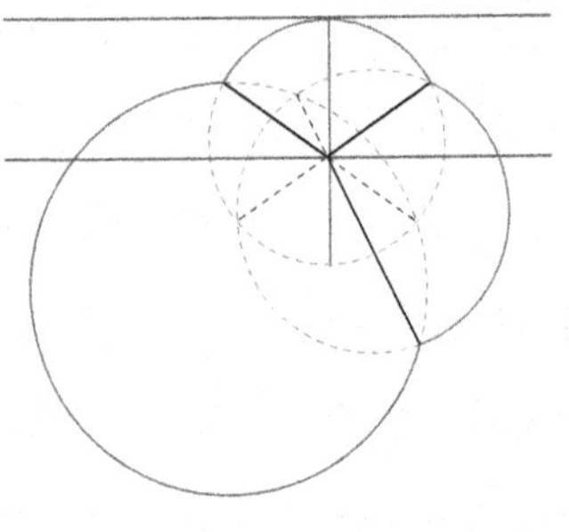

Fig. 8.8. Figure of three isometric circles and two lines

$\overline{Eh}(\rho_n(P_0)) \cap \overline{Eh}(\rho_n(P_2)))$ is contained in the region in $D(\rho_n(P_1))$ bounded by ℓ_n and ℓ'_n. Since $\lim \overline{F}_{\rho_n}(\xi) = \overline{F}_\infty(\xi) = \mathrm{Fix}_\sigma^{\epsilon_0}(\rho_\infty(P_1)) = \lim \mathrm{Fix}_\sigma^{\epsilon_0}(\rho_n(P_1))$, both ℓ_n and ℓ'_n converge to the perpendicular, ℓ_∞, to $\mathrm{proj}(\overline{\mathrm{Axis}}(\rho_\infty(P_1)))$ at $\overline{F}_\infty(\xi)$. Hence $\mathrm{proj}(\overline{Ih}(\rho_n(P_1)) \cap \overline{Eh}(\rho_n(P_0)) \cap \overline{Eh}(\rho_n(P_2)))$ converges to $\ell_\infty \cap D(\rho_\infty(P_1)) = \overline{F}_\infty(\xi)$. This implies that $\overline{Ih}(\rho_n(P_1)) \cap \overline{Eh}(\rho_n(P_0)) \cap \overline{Eh}(\rho_n(P_2))$ converges to $\overline{F}_\infty(\xi)$. The remaining assertions follow from the fact that $[F_{\rho_n}(\xi), v^{\epsilon_0}(\rho_n; P_0, P_1)]$, $[F_{\rho_n}(\xi), v^{\epsilon_0}(\rho_n; P_1, P_2)]$ and $e^{\epsilon_0}(\rho_n, \sigma; P_1)$ are contained in the 2-dimensional convex polyhedron $\overline{Ih}(\rho_n(P_1)) \cap \overline{Eh}(\rho_n(P_0)) \cap \overline{Eh}(\rho_n(P_2))$.

By Claim 8.9.11(1) and Lemma 8.9.9, (P_0, P_1) is not a generic edge of $\mathcal{L}_0$. Hence the interior of the half geodesic $[F_{\rho_n}(\xi), v^{\epsilon_0}(\rho_n; P_0, P_1)]$ does not contain a vertex of $Eh(\boldsymbol{\rho}_n)$ by Lemma 8.5.3. Thus $v^{\epsilon_0}(\rho_n; P_0, P_1) \in \mathrm{fr}\, E^{\epsilon_0}(\boldsymbol{\rho}_n)$. Similarly, we also see that (P_1, P_2) is not a generic edge of $\mathcal{L}_0$ and $v^{\epsilon_0}(\rho_n; P_1, P_2) \in \mathrm{fr}\, E^{\epsilon_0}(\boldsymbol{\rho}_n)$. Hence the ϵ_0-terminal triangle of Σ_0 is equal to σ, and therefore $\nu_n^{\epsilon_0} \in \sigma$. On the other hand, we have $\lim \theta^{\epsilon_0}(\rho_n, \sigma; s_1) = 0$ by Claim 8.9.11(2). Hence $\nu_\infty^{\epsilon_0} \in \mathrm{int}\langle s_0, s_2\rangle$ and the condition in Lemma 8.9.4(2) holds. We also see that $\overline{F}_\infty(\xi) = \mathrm{Fix}_\sigma^{\epsilon_0}(\rho_\infty(P_1)) = v_\infty^{\epsilon_0}(P_0, P_1) = v_\infty^{\epsilon_0}(P_1, P_2)$. Thus we have shown that if $\overline{F}_\infty(\xi) = \mathrm{Fix}_\sigma^{\epsilon_0}(\rho_\infty(P_1))$ then the condition in Lemma 8.9.10(1) holds.

Case 2. Suppose that $\overline{F}_\infty(\xi) = \mathrm{Fix}_\sigma^{-\epsilon_0}(\rho_\infty(P_1))$. First, we prove that $\overline{F}_\infty(\xi) = v_\infty^{-\epsilon_0}(P_0, P_2)$. Suppose this is not the case. Then we have $\overline{F}_\infty(\xi) = v_\infty^{\epsilon_0}(P_0, P_2)$ and that $v_\infty^{-\epsilon_0}(P_0, P_2) \neq v_\infty^{\epsilon_0}(P_0, P_2)$ (see Fig. 8.9), because

$$\begin{aligned}\overline{F}_\infty(\xi) &\in \overline{Ih}(\rho_\infty(P_0)) \cap \overline{Ih}(\rho_\infty(P_2)) \cap \mathbb{C}\\ &= I(\rho_\infty(P_0)) \cap I(\rho_\infty(P_2)) = \{v_\infty^{\pm}(P_0, P_2)\}.\end{aligned}$$

Let $\theta(i,j)$ be the dihedral angle of $E(\rho_\infty(P_i)) \cap E(\rho_\infty(P_j))$. Then $\theta(0,2) > 0$, because $v_\infty^{-\epsilon_0}(P_0, P_2) \neq v_\infty^{\epsilon_0}(P_0, P_2)$. Moreover we can see that $\theta(0,1) + \theta(1,2) + \theta(0,2) = \pi$. Hence we have $\theta(0,1) + \theta(1,2) < \pi$. This contra-

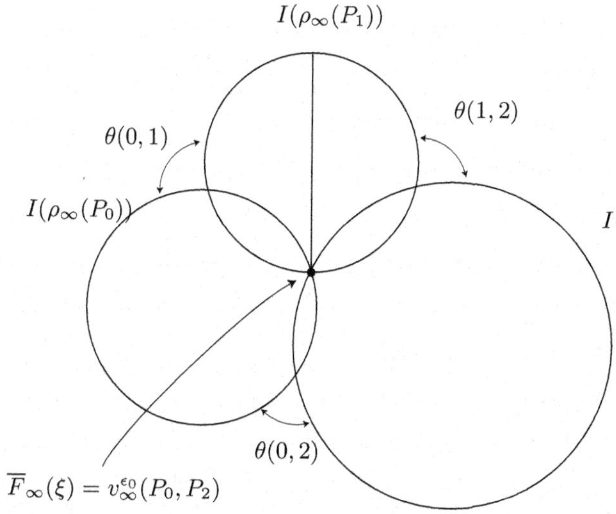

Fig. 8.9. $\theta(0,1)+\theta(1,2)+\theta(0,2) = \pi$ and hence $\theta(0,1)+\theta(1,2) < \pi$. This contradicts Lemma 4.1.3(3) (Chain rule).

dicts Lemma 4.1.3(3) (Chain rule). Hence $\overline{F}_\infty(\xi) = v_\infty^{-\epsilon_0}(P_0, P_2)$. This implies that the geodesic segment $[F_{\rho_n}(\xi), v^{-\epsilon_0}(\rho_n; P_0, P_2)]$ shrinks to the point $\overline{F}_\infty(\xi) \in \mathbb{C}$ as $n \to \infty$ (see Fig. 8.7). We can see by using Lemmas 8.5.3 and 8.9.9 that the interior of the geodesic segment $[F_{\rho_n}(\xi), v^{-\epsilon_0}(\rho_n; P_0, P_2)]$

does not contain a vertex of $Eh(\boldsymbol{\rho}_n)$. So, $v^{-\epsilon_0}(\rho_n; P_0, P_2) \in \mathrm{fr}\, E^{-\epsilon_0}(\boldsymbol{\rho}_n)$, and hence s_0 and s_2 are vertices of the $(-\epsilon_0)$-terminal triangle of Σ_0. On the other hand, since $\theta^{-\epsilon_0}(\rho_n, \sigma; s_1) < 0$, σ is not the $(-\epsilon_0)$-terminal triangle of Σ_0. Hence the $(-\epsilon_0)$-terminal triangle of Σ_0 is equal to σ' and $\nu_n^{-\epsilon_0}$ is contained in σ'. Notice that $\rho_n(P_1') = \rho_n(P_2)$ acts on $\overline{Ih}(\rho_n(P_1')) = \overline{Ih}(\rho_n(P_2))$ as a Euclidean isometry. Since $\rho_n(P_1')([F_{\boldsymbol{\rho}_n}(\xi), v^{-\epsilon_0}(\rho_n; P_0', P_1')]) = \rho_n(P_1')(\overline{F}_{\boldsymbol{\rho}_n}(P_0', P_1')) = \overline{F}_{\boldsymbol{\rho}_n}(P_1', P_2') = [F_{\boldsymbol{\rho}_n}(P_1', P_2', P_3'), v^{-\epsilon_0}(\rho_n; P_1', P_2')]$ by Lemma 6.4.2(2), and since $[F_{\boldsymbol{\rho}_n}(\xi), v^{-\epsilon_0}(\rho_n; P_0, P_2)]$ shrinks to a point, we see that $[F_{\boldsymbol{\rho}_n}(P_1', P_2', P_3'), v^{-\epsilon_0}(\rho_n; P_1', P_2')]$ also shrinks to a point. Similarly, we also see that $[F_{\boldsymbol{\rho}_n}(P_1', P_2', P_3'), v^{-\epsilon_0}(\rho_n; P_2', P_3')]$ shrinks to a point. Thus we have $\overline{F}_\infty(P_1', P_2', P_3') = v_\infty^{-\epsilon_0}(P_1', P_2') = v_\infty^{-\epsilon_0}(P_2', P_3')$. On the other hand, since $\overline{F}_{\boldsymbol{\rho}_n}(P_1', P_2', P_3') \in \overline{\mathrm{Axis}(\rho_n(P_2'))}$ for every n, the point is equal to $\mathrm{Fix}^-_{\sigma'}(\rho_\infty(P_2'))$ or $\mathrm{Fix}^+_{\sigma'}(\rho_\infty(P_2'))$. Let ℓ_n be the perpendicular at $c(\rho_n(P_2'))$ to $\mathrm{proj}(\mathrm{Axis}(\rho_n(P_2')))$. Since $0 \le \theta^{-\epsilon_0}(\rho_n, \sigma'; P_2') \le \pi/2$, $e^{-\epsilon_0}(\rho_n, \sigma'; P_2')$ is contained in the closure of the component of $I(\rho_n(P_2')) - \ell_n$ containing $\mathrm{Fix}^{-\epsilon_0}_{\sigma'}(\rho_n(P_2'))$. Thus $v_\infty^{-\epsilon_0}(P_1', P_2') = v_\infty^{-\epsilon_0}(P_2', P_3')$ must be equal to $\mathrm{Fix}^{-\epsilon_0}_{\sigma'}(\rho_\infty(P_2'))$. Therefore $\lim \theta^{-\epsilon_0}(\rho_n, \sigma'; P_2') = 0$, and hence $\nu_\infty^{-\epsilon_0} \in \mathrm{int}\langle s_0, s_2\rangle$. We also see that the condition in Lemma 8.9.4(3) holds. Hence we have shown that if $\overline{F}_\infty(\xi) = \mathrm{Fix}^{-\epsilon_0}_\sigma(\rho_\infty(P_1))$ then the condition in Lemma 8.9.10(2) holds.

We have shown that if $\overline{F}_\infty(\xi)$ belongs to $\mathbb{C}$ then precisely one of the conditions (1) and (2) holds and hence $\xi \notin \mathcal{L}(\boldsymbol{\nu}_\infty)^{(2)}$.

In what follows we prove that the condition $\xi \notin \mathcal{L}(\boldsymbol{\nu}_\infty)^{(2)}$ implies the condition $\overline{F}_\infty(\xi) \in \mathbb{C}$. So, suppose that $\xi \notin \mathcal{L}(\boldsymbol{\nu}_\infty)^{(2)}$. Then precisely one of the following holds by Lemma 8.9.4.

1. $\nu_n^{\epsilon_0} \in \sigma \cap \mathbb{H}^2$ and $\nu_\infty^{\epsilon_0} \in \mathrm{int}\langle s_0, s_2\rangle$.
2. $\nu_n^{-\epsilon_0} \in \sigma' \cap \mathbb{H}^2$ and $\nu_\infty^{-\epsilon_0} \in \mathrm{int}\langle s_0, s_2\rangle$.

In the first case, we have $\lim \theta^{\epsilon_0}(\rho_n, \sigma; s(P_1)) = 0$. Thus $e^{\epsilon_0}(\rho_n, \sigma; P_1)$ shrinks to the point $\mathrm{Fix}^{\epsilon_0}_\sigma(\rho_\infty(P_1))$. This implies that $\mathrm{Fix}^{\epsilon_0}_\sigma(\rho_\infty(P_1)) \in I(\rho_\infty(P_0)) \cap I(\rho_\infty(P_1)) \cap I(\rho_\infty(P_2)) \subset \overline{Ih}(\rho_\infty(P_0)) \cap \overline{Ih}(\rho_\infty(P_1)) \cap \overline{Ih}(\rho_\infty(P_2))$. Thus $\mathrm{Fix}^{\epsilon_0}_\sigma(\rho_\infty(P_1)) = \overline{F}_\infty(\xi)$ by Lemma 8.9.8(3). Hence $\overline{F}_\infty(\xi) \in \mathbb{C}$. In the second case, we can see by a similar argument that $\overline{F}_\infty(P_1', P_2', P_3')$ is contained in $\mathbb{C}$. Since $\bigcap_{i=0}^2 \overline{Ih}(\rho_\infty(P_j))$ is the singleton $\overline{F}_\infty(\xi)$ by Lemma 8.9.8(3), $\bigcap_{i=1}^3 \overline{Ih}(\rho_\infty(P_j'))$ is equal to the singleton $\rho_\infty(P_1')(\overline{F}_\infty(\xi))$ by Lemma 4.1.3(4) (Chain rule). Therefore $\bigcap_{i=1}^3 \overline{Ih}(\rho_\infty(P_j'))$ must be equal to $\overline{F}_\infty(P_1', P_2', P_3')$ in $\mathbb{C}$, and hence $\overline{F}_\infty(\xi) = \rho_\infty(P_1')(\overline{F}_\infty(P_1', P_2', P_3')) \in \mathbb{C}$. This completes the proof Lemma 8.9.10.

Proof (Proof of Proposition 8.9.5(1)). Suppose that ξ is contained in $\mathcal{L}(\boldsymbol{\nu}_\infty)$.

(1-i) By Lemma 8.9.8(3), $\overline{Ih}(\rho_\infty(P_0)) \cap \overline{Ih}(\rho_\infty(P_1)) \cap \overline{Ih}(\rho_\infty(P_2))$ is equal to the singleton $\overline{F}_\infty(\xi)$. By Lemma 8.9.10, this lies in $\mathbb{H}^3$. Hence we obtain the conclusion.

(1-ii) By Lemma 8.9.8(1), $\overline{Ih}(\rho_\infty(P_0)) \cap \overline{Ih}(\rho_\infty(P_1))$ is a complete geodesic. Moreover, by the above (1-i), it intersects the hyperplane $\overline{Ih}(\rho_\infty(P_2)) =$

$\partial \overline{Eh}(\rho_\infty(P_2))$ at the single point $\overline{F}_\infty(\xi)$ in $\mathbb{H}^3$. Thus the intersection $L_\infty(2) = \overline{Ih}(\rho_\infty(P_0)) \cap \overline{Ih}(\rho_\infty(P_1)) \cap \overline{Eh}(\rho_\infty(P_2))$ is a half geodesic in $\overline{Ih}(\rho_\infty(P_0)) \cap \overline{Ih}(\rho_\infty(P_1))$ which has $\overline{F}_\infty(\xi)$ as the endpoint in $\mathbb{H}^3$. On the other hand, $L_\infty(2)$ contains the half geodesic $[\overline{F}_\infty(\xi), v_\infty^{\epsilon(2)}(P_0, P_1)]$ by Lemma 8.9.7(2). Hence we have $L_\infty(2) = [\overline{F}_\infty(\xi), v_\infty^{\epsilon(2)}(P_0, P_1)]$. By a parallel argument, we can prove the assertion for $L_\infty(0)$.

We show the assertion for $L_\infty(1)$. Since $\overline{F}_\infty(\xi)$ is a subset of $\overline{Ih}(\rho_\infty(P_0)) \cap \overline{Ih}(\rho_\infty(P_2))$ and since $\overline{F}_\infty(\xi)$ is contained in $\mathbb{H}^3$, we see that $\overline{Ih}(\rho_\infty(P_0)) \cap \overline{Ih}(\rho_\infty(P_2))$ contains a point in $\mathbb{H}^3$. Since these two isometric hemispheres are not identical by Lemma 2.5.4(2), this implies that $\overline{Ih}(\rho_\infty(P_0)) \cap \overline{Ih}(\rho_\infty(P_2))$ is a complete geodesic. So, we can obtain the assertion for $L_\infty(1)$ by the above argument for $L_\infty(2)$.

Proof (Proof of Proposition 8.9.5(2)). Suppose $\nu_\infty^{\epsilon_0} \in \mathrm{int}\langle s_0, s_2 \rangle$. Then $\xi \notin \mathcal{L}(\boldsymbol{\nu}_\infty)^{(2)}$ and hence we see that the conclusion of Lemma 8.9.10(1) holds. Hence we obtain the assertion (2-i).

We show the assertions (2-ii) and (2-iii). Since $\nu_\infty^{\epsilon_0} \in \mathrm{int}\langle s_0, s_2 \rangle$, we have $\lim \theta^{\epsilon_0}(\rho_n, \sigma; s_1) = 0$. On the other hand, we see $\lim \theta^{-\epsilon_0}(\rho_n, \sigma; s_1) \leq 0$ by Lemma 4.3.6 (cf. the first paragraph in the proof of Lemma 8.9.10). Hence, by Lemma 4.3.5, we have $\overline{Ih}(\rho_\infty(P_1)) \subset \overline{Dh}(\rho_\infty(P_0)) \cup \overline{Dh}(\rho_\infty(P_2))$. Thus $\overline{Ih}(\rho_\infty(P_1)) \cap \overline{Eh}(\rho_\infty(P_0)) \cap \overline{Eh}(\rho_\infty(P_2))$ is equal to $\overline{Ih}(\rho_\infty(P_0)) \cap \overline{Ih}(\rho_\infty(P_1)) \cap \overline{Ih}(\rho_\infty(P_2))$, which in turn is equal to $\overline{F}_\infty(\xi)$ by Lemma 8.9.8(3). Hence we see that $\overline{Dh}(\rho_\infty(P_1)) \cap \overline{Eh}(\rho_\infty(P_0)) \cap \overline{Eh}(\rho_\infty(P_2))$ is equal to the singleton $\overline{F}_\infty(\xi)$ in $\mathbb{C}$. This implies that $\overline{Ih}(\rho_\infty(P_0)) \cap \overline{Ih}(\rho_\infty(P_2))$ lies in $\overline{Eh}(\rho_\infty(P_1))$, and hence

$$\overline{Eh}(\rho_\infty(P_1)) \supset \overline{Eh}(\rho_\infty(P_0)) \cap \overline{Eh}(\rho_\infty(P_2)).$$

Thus we have proved the assertion (2-ii). Moreover, the above argument shows

$$L_\infty(1) = \overline{Ih}(\rho_\infty(P_0)) \cap \overline{Ih}(\rho_\infty(P_2)) \cap \overline{Eh}(\rho_\infty(P_1)) = \overline{Ih}(\rho_\infty(P_0)) \cap \overline{Ih}(\rho_\infty(P_2)).$$

Since $\overline{F}_\infty(\xi) = v_\infty^{\epsilon_0}(P_0, P_2) = v_\infty^{-\epsilon(1)}(P_0, P_2)$ by (2-i), this implies

$$L_\infty(1) = \overline{Ih}(\rho_\infty(P_0)) \cap \overline{Ih}(\rho_\infty(P_2)) = [\overline{F}_\infty(\xi), v_\infty^{\epsilon(1)}(P_0, P_2)].$$

Thus we have obtained the assertion (2-iii) for $L_\infty(1)$. We note that the above argument also implies

$$\begin{aligned} L_\infty(2) &= \overline{Ih}(\rho_\infty(P_0)) \cap \overline{Ih}(\rho_\infty(P_1)) \cap \overline{Eh}(\rho_\infty(P_2)) \\ &\subset \overline{Eh}(\rho_\infty(P_0)) \cap \overline{Ih}(\rho_\infty(P_1)) \cap \overline{Eh}(\rho_\infty(P_2)) = \overline{F}_\infty(\xi). \end{aligned}$$

Hence $L_\infty(2) = \overline{F}_\infty(\xi)$. Since $\overline{F}_\infty(\xi) = v_\infty^{\epsilon_0}(P_0, P_1) = v_\infty^{\epsilon(2)}(P_0, P_1)$ by (2-i), we also have $L_\infty(2) = [\overline{F}_\infty(\xi), v_\infty^{\epsilon(2)}(P_0, P_1)]$. This proves the assertion (2-iii) for $L_\infty(2)$. By a parallel argument, we can also prove the assertion (2-iii) for $L_\infty(0)$.

Proof (Proof of Proposition 8.9.5(3)). Suppose $\nu_\infty^{-\epsilon_0} \in \mathrm{int}\langle s_0, s_2\rangle$. Then $\xi \notin \mathcal{L}(\boldsymbol{\nu}_\infty)^{(2)}$ and hence we see that the conclusion of Lemma 8.9.10(2) holds. Hence we obtain the assertion (3-i).

We show the assertion (3-ii) for $L_\infty(1)$. If $\overline{Ih}(\rho_\infty(P_0)) \cap \overline{Ih}(\rho_\infty(P_2))$ is a singleton, then the assertion obviously holds. So, we assume that $\overline{Ih}(\rho_\infty(P_0)) \cap \overline{Ih}(\rho_\infty(P_2))$ is a complete geodesic. Since $\overline{Ih}(\rho_\infty(P_0)) \cap \overline{Ih}(\rho_\infty(P_1)) \cap \overline{Ih}(\rho_\infty(P_2))$ is equal to the singleton $\overline{F}_\infty(\xi) = \mathrm{Fix}_\sigma^{-\epsilon_0}(\rho_\infty(P_1))$ in $\mathbb{C}$, we see that $L_\infty(1) = \overline{Ih}(\rho_\infty(P_0)) \cap \overline{Ih}(\rho_\infty(P_2)) \cap \overline{Eh}(\rho_\infty(P_1))$ is either the entire geodesic or the singleton $\overline{F}_\infty(\xi)$. Since $v^{\epsilon_0}(\rho_n; P_0, P_2)$ is contained in $D(\rho_n(P_1))$ for every $n \in \mathbb{N}$, the endpoint $v_\infty^{\epsilon_0}(P_0, P_2)$ of $\overline{Ih}(\rho_\infty(P_0)) \cap \overline{Ih}(\rho_\infty(P_2))$ is contained in $D(\rho_\infty(P_1))$. Since $v_\infty^{-\epsilon_0}(P_0, P_2) = \overline{F}_\infty(\xi)$ is the unique point in $\overline{Ih}(\rho_\infty(P_0)) \cap \overline{Ih}(\rho_\infty(P_1)) \cap \overline{Ih}(\rho_\infty(P_2))$ and since it is distinct from $v_\infty^{\epsilon_0}(P_0, P_2)$ by the assumption, $v_\infty^{\epsilon_0}(P_0, P_2)$ is not contained in $\overline{Eh}(\rho_\infty(P_1))$. Therefore $L_\infty(1)$ is equal to the singleton $\overline{F}_\infty(\xi)$. Since $\overline{F}_\infty(\xi) = v_\infty^{-\epsilon_0}(P_0, P_2) = v_\infty^{\epsilon(1)}(P_0, P_2)$ by (3-i), we also have $L_\infty(1) = [\overline{F}_\infty(\xi), v_\infty^{\epsilon(1)}(P_0, P_2)]$. Thus we have obtained the assertion (3-ii) for $L_\infty(1)$.

We show the assertion (3-ii) for $L_\infty(2)$. By Lemma 8.9.8(1), $\overline{Ih}(\rho_\infty(P_0)) \cap \overline{Ih}(\rho_\infty(P_1))$ is a complete geodesic. This complete geodesic is equal to $[\overline{F}_\infty(\xi), v_\infty^{\epsilon(2)}(P_0, P_1)]$, because $\overline{F}_\infty(\xi) = v_\infty^{-\epsilon_0}(P_0, P_1) = v_\infty^{-\epsilon(2)}(P_0, P_1)$ by (3-i). Since $L_\infty(2)$ contains $[\overline{F}_\infty(\xi), v_\infty^{\epsilon(2)}(P_0, P_1)]$ by Lemma 8.9.7(2), we see

$$L_\infty(2) = [\overline{F}_\infty(\xi), v_\infty^{\epsilon(2)}(P_0, P_1)] = \overline{Ih}(\rho_\infty(P_0)) \cap \overline{Ih}(\rho_\infty(P_1)).$$

This proves the assertion (3-ii) for $L_\infty(2)$. We obtain the assertion for $L_\infty(0)$ by a parallel argument. This completes the proof of Proposition 8.9.5(3).

Proposition 8.9.5 is used in the following forms in Sects. 8.10 and 8.11.

Corollary 8.9.12. *Under Assumptions 8.9.1 and 8.9.2, the following hold.*

(1) $\overline{Ih}(\rho_\infty(P_0)) \cap \overline{Ih}(\rho_\infty(P_1)) \cap \overline{Ih}(\rho_\infty(P_2))$ *is equal to the singleton* $\overline{F}_\infty(\xi)$ *in* $\overline{\mathbb{H}}^3$.

(2) Suppose $\nu_\infty^\epsilon \in \mathrm{int}\langle s_0, s_2\rangle$ *for some* ϵ. *Then* $\xi \notin \mathcal{L}(\boldsymbol{\nu}_\infty)^{(2)}$ *and* $\overline{F}_\infty(\xi) = \mathrm{Fix}_\sigma^\epsilon(\rho_\infty(P_1))$. *Moreover, if* $s_1 \notin \Sigma(\boldsymbol{\nu}_\infty)^{(0)}$, *or equivalently* $(P_1) \notin \mathcal{L}(\boldsymbol{\nu}_\infty)^{(0)}$, *then* $\overline{Ih}(\rho_\infty(P_1))$ *is covered by the remaining two isometric hemispheres. To be precise, the following hold (see Fig. 8.5).*

- *(i)* $\overline{Dh}(\rho_\infty(P_1)) \cap \overline{Eh}(\rho_\infty(P_0)) \cap \overline{Eh}(\rho_\infty(P_2))$ *is equal to the singleton* $\overline{F}_\infty(\xi)$ *in* $\mathbb{C}$.
- *(ii)* $\overline{Eh}(\rho_\infty(P_1)) \supset \overline{Eh}(\rho_\infty(P_0)) \cap \overline{Eh}(\rho_\infty(P_2))$.

Proof. We have only to prove (2). Suppose $\nu_\infty^\epsilon \in \mathrm{int}\langle s_0, s_2\rangle$ for some ϵ. Then $\xi \notin \mathcal{L}(\boldsymbol{\nu}_\infty)^{(2)}$. If $\epsilon = \epsilon_0$, then $s_1 \notin \Sigma(\boldsymbol{\nu}_\infty)^{(0)}$ by Lemma 8.9.4(2), and we obtain the desired result by Proposition 8.9.5(2). If $\epsilon = -\epsilon_0$, then $s_1 \in \Sigma(\boldsymbol{\nu}_\infty)^{(0)}$ by Lemma 8.9.4(3), and we see by Proposition 8.9.5(3-i) that

$$\overline{F}_\infty(\xi) = \mathrm{Fix}_\sigma^{-\epsilon_0}(\rho_\infty(P_1)) = \mathrm{Fix}_\sigma^\epsilon(\rho_\infty(P_1)).$$

This completes the proof.

Corollary 8.9.13. *Under Assumption 8.9.1, let ξ be a triangle of $\mathcal{L}_0$, and let $\{P,Q,R\}$ be the vertex set of ξ, i.e., $\xi = ((P,Q,R))$ (cf. Definition 3.2.6). Then we have the following.*

(1) $\overline{Ih}(\rho_\infty(P)) \cap \overline{Ih}(\rho_\infty(Q)) \cap \overline{Ih}(\rho_\infty(R))$ is equal to the singleton $\overline{F}_\infty(\xi)$.

(2) Set $L_\infty(R) := \overline{Ih}(\rho_\infty(P)) \cap \overline{Ih}(\rho_\infty(Q)) \cap \overline{Eh}(\rho_\infty(R))$. Then $L_\infty(R)$ is equal to the (possibly degenerate) geodesic segment $[\overline{F}_\infty(\xi), v_\infty^{\epsilon(R)}(P,Q)]$, where $\epsilon(R)$ is the sign such that $((P,Q))$ lies on the $\epsilon(R)$-side of ξ (cf. Definition 3.2.5). Moreover, we have the following:

(i) Suppose $\xi \in \mathcal{L}(\boldsymbol{\nu}_\infty)^{(2)}$. Then $\overline{F}_\infty(\xi) \in \mathbb{H}^3$ and $L_\infty(R)$ is a half geodesic.
(ii) Suppose $\xi \notin \mathcal{L}(\boldsymbol{\nu}_\infty)^{(2)}$ and $((P,Q)) \notin \mathcal{L}(\boldsymbol{\nu}_\infty)^{(1)}$. Then $\overline{F}_\infty(\xi) = v_\infty^{\epsilon(R)}(P,Q)$ and $L_\infty(R)$ is the singleton $\overline{F}_\infty(\xi) = v_\infty^{\epsilon(R)}(P,Q)$ in $\mathbb{C}$ (see Figs. 8.5 and 8.6).
(iii) Suppose $\xi \notin \mathcal{L}(\boldsymbol{\nu}_\infty)^{(2)}$ and $((P,Q)) \in \mathcal{L}(\boldsymbol{\nu}_\infty)^{(1)}$. Then $\overline{F}_\infty(\xi) = v_\infty^{-\epsilon(R)}(P,Q)$ and $L_\infty(R)$ is the complete geodesic $\overline{Ih}(\rho_\infty(P)) \cap \overline{Ih}(\rho_\infty(Q))$ (see Figs. 8.5 and 8.6).

(3) If $((P,Q)) \in \mathcal{L}(\boldsymbol{\nu}_\infty)^{(1)}$, then $\overline{Ih}(\rho_\infty(P)) \cap \overline{Ih}(\rho_\infty(Q))$ is a complete geodesic.

Proof. We may assume ξ is equal to the 2-simplex ξ in Assumption 8.9.2, namely $\{P,Q,R\} = \{P_0,P_1,P_2\}$. Then (P_0,P_1) and (P_1,P_2) lie on the ϵ_0-side of ξ, whereas (P_0,P_2) lies on the $(-\epsilon_0)$-side of (P_0,P_2). This means that the sign $\epsilon(P_k)$ introduced in the corollary is equal to the sign $\epsilon(k)$ in Proposition 8.9.5. Since $L_\infty(k) = [\overline{F}_\infty(\xi), v_\infty^{\epsilon(k)}(P_i,P_j)]$ for every $k \in \{0,1,2\}$ by Corollary 8.9.6, we have $L_\infty(R) = [\overline{F}_\infty(\xi), v_\infty^{\epsilon(R)}(P,Q)]$. The remaining assertions follows from Proposition 8.9.5 and the fact that $\epsilon(P_k) = \epsilon(k)$.

8.10 Proof of Proposition 8.4.5 - length ≥ 2 case - (Step 2)

In this section, we prove the following Propositions 8.10.1 and 8.10.2, which show that Proposition 8.4.5(1), in the case where the length of Σ_0 is ≥ 2, is valid for the edges of $\mathcal{L}_0$.

Proposition 8.10.1. *Under Assumption 8.9.1, let ξ be a generic edge of $\mathcal{L}_0$. Then ξ is an edge of $\mathcal{L}(\boldsymbol{\nu}_\infty)$, $\overline{F}_{\rho_\infty}(\xi)$ is equal to $\overline{F}_\infty(\xi)$, and it is a 1-dimensional convex polyhedron in $\overline{\mathbb{H}}^3$.*

Proposition 8.10.2. *Under Assumption 8.9.1, let ξ be an ϵ-extreme edge of $\mathcal{L}_0$ for some sign $\epsilon \in \{-,+\}$. Then the following hold.*

(1) If ξ is an edge of $\mathcal{L}(\boldsymbol{\nu}_\infty)$, then $\overline{F}_{\rho_\infty}(\xi)$ is equal to $\overline{F}_\infty(\xi)$, and it is a 1-dimensional convex polyhedron in $\overline{\mathbb{H}}^3$.
(2) If ξ is not contained in $\mathcal{L}(\boldsymbol{\nu}_\infty)$, then $\overline{F}_\infty(\xi)$ is a singleton in $\mathbb{C}$.

Proof (Proof of Proposition 8.10.1). Under Assumption 8.9.1, let ξ be a generic edge of $\mathcal{L}_0$. Then $\xi = (P_j, P_{j+1})$ where $\{P_j\}$ is the sequence of elliptic generators associated with a triangle $\sigma = \langle s_0, s_1, s_2\rangle$ in $\Sigma_0 - \{\sigma^-, \sigma^+\}$, where σ^ϵ ($\epsilon \in \{-,+\}$) is the ϵ-terminal triangle of Σ_0. Let σ'_- and σ'_+, respectively, be the predecessor and the successor of σ in Σ_0. We assume that $\sigma'_- \cap \sigma = \langle s_0, s_2\rangle$ and $\sigma'_+ \cap \sigma = \langle s_0, s_1\rangle$. (To be precise, we also need to consider the case where $\sigma'_- \cap \sigma = \langle s_0, s_2\rangle$ and $\sigma'_+ \cap \sigma = \langle s_0, s_1\rangle$, because of Convention 2.1.12. But this case is equivalent to the above case after changing the roles of K and K^{-1}.) Then we have the following by Lemma 8.5.2.

1. (P_1, P_2, P_3) and $(P_0, P_2^{P_1}, P_1)$ are 2-simplices of $\mathcal{L}_0$ which lie on the $(+)$-side of $\mathcal{L}(\sigma)$, and we have $\sigma'_+ = \langle s_0, s_2^+, s_1\rangle$ where $s_2^+ = s(P_2^{P_1})$.
2. (P_0, P_1, P_2) and $(P_2, P_1^{P_2}, P_3)$ are 2-simplices of $\mathcal{L}_0$ which lie on the $(-)$-side of $\mathcal{L}(\sigma)$, and we have $\sigma'_- = \langle s_0, s_2, s_1^-\rangle$ where $s_1^- = s(P_1^{P_2})$.

By Lemma 8.9.9, ξ is an edge of $\mathcal{L}(\boldsymbol{\nu}_\infty)$, and $\overline{F}_\infty(\xi)$ is a 1-dimensional convex polyhedron in $\overline{\mathbb{H}}^3$. In what follows we prove that $\overline{F}_{\boldsymbol{\rho}_\infty}(\xi)$ is equal to $\overline{F}_\infty(\xi)$. To this end we show that $\overline{F}_{\boldsymbol{\rho}_\infty}(\xi) = \overline{F}_{(\rho_\infty, \mathcal{L}_0)}(\xi)$ and $\overline{F}_{(\rho_\infty, \mathcal{L}_0)}(\xi) = \overline{F}_\infty(\xi)$.

We first show $\overline{F}_{\boldsymbol{\rho}_\infty}(\xi) = \overline{F}_{(\rho_\infty, \mathcal{L}_0)}(\xi)$. In order to use Corollary 8.9.13, we set $P = P_j$ and $Q = P_{j+1}$. Let $\hat{\xi}^\epsilon$ ($\epsilon \in \{-,+\}$) be the 2-simplex of $\mathcal{L}_0$ containing ξ and lying on the ϵ-side of ξ. Since $(P,Q) \in \mathcal{L}(\boldsymbol{\nu}_n)^{(1)}$ and $\boldsymbol{\rho}_n$ is good, we see that $L_n := \overline{Ih}(\rho_n(P)) \cap \overline{Ih}(\rho_n(Q))$ is a complete geodesic and that the four points $v^-(\rho_n; P, Q)$, $\overline{F}_{\boldsymbol{\rho}_n}(\hat{\xi}^-)$, $\overline{F}_{\boldsymbol{\rho}_n}(\hat{\xi}^+)$ and $v^+(\rho_n; P, Q)$ lie in the geodesic L_n in this order. Since $\xi = (P,Q) \in \mathcal{L}(\boldsymbol{\nu}_\infty)^{(1)}$, we see by Corollary 8.9.13(3) that $L_\infty := \overline{Ih}(\rho_\infty(P)) \cap \overline{Ih}(\rho_\infty(Q))$ is a complete geodesic. By continuity we can also see that the four points $v_\infty^-(P,Q)$, $\overline{F}_\infty(\hat{\xi}^-)$, $\overline{F}_\infty(\hat{\xi}^+)$ and $v_\infty^+(P,Q)$ sit in the geodesic L_∞ in this order, though some of them may coincide. Note that $\overline{F}_\infty(\hat{\xi}^-)$ and $\overline{F}_\infty(\hat{\xi}^+)$ are different because they are the endpoints of a 1-dimensional polyhedron $\overline{F}_\infty(\xi)$ (cf. Lemma 8.9.9). Let R^ϵ be the elliptic generator such that $(\hat{\xi}^\epsilon)^{(0)} = \{P, Q, R^\epsilon\}$. By definition $\overline{F}_{(\rho_\infty, \mathcal{L}_0)}(\xi)$ is equal to $L_\infty \cap \overline{Eh}(\rho_\infty(R^-)) \cap \overline{Eh}(\rho_\infty(R^+))$, and hence it is equal to the intersection of the two subsets $L_\infty \cap \overline{Eh}(\rho_\infty(R^-))$ and $L_\infty \cap \overline{Eh}(\rho_\infty(R^+))$ of $\mathcal{L}_0$. Since (P,Q) lies on the $(-\epsilon)$-side of $\hat{\xi}^\epsilon$, we see by Corollary 8.9.13(2)

$$L_\infty \cap \overline{Eh}(\rho_\infty(R^\epsilon)) = [\overline{F}_\infty(\hat{\xi}^\epsilon), v_\infty^{-\epsilon}(P,Q)] \subset L_\infty$$

for each $\epsilon \in \{-,+\}$. By considering the order of the four points $v_\infty^-(P,Q)$, $\overline{F}_\infty(\hat{\xi}^-)$, $\overline{F}_\infty(\hat{\xi}^+)$ and $v_\infty^+(P,Q)$ in L_∞, we can conclude that $\overline{F}_{(\rho_\infty, \mathcal{L}_0)}(\xi)$ is equal to $\overline{F}_\infty(\xi) = [\overline{F}_\infty(\hat{\xi}^-), \overline{F}_\infty(\hat{\xi}^+)]$.

Next we show $\overline{F}_{\boldsymbol{\rho}_\infty}(\xi) = \overline{F}_{(\rho_\infty, \mathcal{L}_0)}(\xi)$. We divide the proof into four cases.

Case 1. $\sigma'_+, \sigma'_- \in \Sigma(\boldsymbol{\nu}_\infty)$. Then $\mathrm{lk}(\xi, \mathcal{L}(\boldsymbol{\nu}_\infty)) = \mathrm{lk}(\xi, \mathcal{L}_0) = \{R^-, R^+\}$. Thus $\overline{F}_{(\rho_\infty, \mathcal{L}_0)}(\xi)$ is by definition equal to $\overline{F}_{\boldsymbol{\rho}_\infty}(\xi)$.

Case 2. $\sigma'_+ \in \Sigma(\boldsymbol{\nu}_\infty)$ and $\sigma'_- \notin \Sigma(\boldsymbol{\nu}_\infty)$. Then $\hat{\xi}^- \notin \mathcal{L}(\boldsymbol{\nu}_\infty)^{(2)}$ and $lk(\xi, \mathcal{L}(\boldsymbol{\nu}_\infty)) = \{R^+\}$. Since $(P,Q) \in \mathcal{L}(\boldsymbol{\nu}_\infty)^{(1)}$, we see by Corollary 8.9.13

(2-iii) that

$$\overline{Ih}(\rho_\infty(P)) \cap \overline{Ih}(\rho_\infty(Q)) \cap \overline{Eh}(\rho_\infty(R^-)) = \overline{Ih}(\rho_\infty(P)) \cap \overline{Ih}(\rho_\infty(Q)).$$

Hence

$$\begin{aligned}\overline{F}_{\rho_\infty}(\xi) &= \overline{Ih}(\rho_\infty(P)) \cap \overline{Ih}(\rho_\infty(Q)) \cap \overline{Eh}(\rho_\infty(R^+)) \\ &= \overline{Ih}(\rho_\infty(P)) \cap \overline{Ih}(\rho_\infty(Q)) \cap \overline{Eh}(\rho_\infty(R^-)) \cap \overline{Eh}(\rho_\infty(R^+)) \\ &= \overline{F}_{(\rho_\infty, \mathcal{L}_0)}(\xi).\end{aligned}$$

Case 3. $\sigma'_- \in \Sigma(\boldsymbol{\nu}_\infty)$ and $\sigma'_+ \notin \Sigma(\boldsymbol{\nu}_\infty)$. The proof for this case is parallel to that for Case 2.

Case 4. $\sigma'_+ \notin \Sigma(\boldsymbol{\nu}_\infty)$ and $\sigma'_- \notin \Sigma(\boldsymbol{\nu}_\infty)$. Then $\hat{\xi}^\pm \notin \mathcal{L}(\boldsymbol{\nu}_\infty)^{(2)}$ and $\mathrm{lk}(\xi, \mathcal{L}(\boldsymbol{\nu}_\infty)) = \emptyset$. Since $(P,Q) \in \mathcal{L}(\boldsymbol{\nu}_\infty)^{(1)}$, we see by Corollary 8.9.13(2-iii) that

$$\overline{Ih}(\rho_\infty(P)) \cap \overline{Ih}(\rho_\infty(Q)) \cap \overline{Eh}(\rho_\infty(R^\pm)) = \overline{Ih}(\rho_\infty(P)) \cap \overline{Ih}(\rho_\infty(Q)).$$

Hence

$$\begin{aligned}\overline{F}_{\rho_\infty}(\xi) &= \overline{Ih}(\rho_\infty(P)) \cap \overline{Ih}(\rho_\infty(Q)) \\ &= \overline{Ih}(\rho_\infty(P)) \cap \overline{Ih}(\rho_\infty(Q)) \cap \overline{Eh}(\rho_\infty(R^-)) \cap \overline{Eh}(\rho_\infty(R^+)) \\ &= \overline{F}_{(\rho_\infty, \mathcal{L}_0)}(\xi).\end{aligned}$$

Proof (Proof of Proposition 8.10.2). Let $\xi = ((P,Q))$ be an ϵ-extreme edge of $\mathcal{L}_0$. Then there is a unique 2-simplex $\hat{\xi} = ((P,Q,R))$ containing ξ. Moreover ξ lies on the ϵ-side of $\hat{\xi}$.

Case 1. $\xi \in \mathcal{L}(\boldsymbol{\nu}_\infty)^{(1)}$ and $\hat{\xi} \in \mathcal{L}(\boldsymbol{\nu}_\infty)^{(2)}$. Then $\mathrm{lk}(\xi, \mathcal{L}(\boldsymbol{\nu}_\infty)) = \{R\}$. Thus by using Corollary 8.9.13(2-i) and Lemma 8.6.1(2), we see

$$\begin{aligned}\overline{F}_{\rho_\infty}(\xi) &= \overline{Ih}(\rho_\infty(P)) \cap \overline{Ih}(\rho_\infty(Q)) \cap \overline{Eh}(\rho_\infty(R)) \\ &= [\overline{F}_\infty(\hat{\xi}), v^\epsilon_\infty(P,Q)] \\ &= \overline{F}_\infty(\xi).\end{aligned}$$

Case 2. $\xi \in \mathcal{L}(\boldsymbol{\nu}_\infty)^{(1)}$ and $\hat{\xi} \notin \mathcal{L}(\boldsymbol{\nu}_\infty)^{(2)}$. Then $\mathrm{lk}(\xi, \mathcal{L}(\boldsymbol{\nu}_\infty)) = \emptyset$. Thus by using Corollary 8.9.13(2-iii) and Lemma 8.6.1(2), we see

$$\begin{aligned}\overline{F}_{\rho_\infty}(\xi) &= \overline{Ih}(\rho_\infty(P)) \cap \overline{Ih}(\rho_\infty(Q)) \\ &= [\overline{F}_\infty(\hat{\xi}), v^\epsilon_\infty(P,Q)] \\ &= \overline{F}_\infty(\xi).\end{aligned}$$

Case 3. $\xi \notin \mathcal{L}(\boldsymbol{\nu}_\infty)^{(1)}$. Then we have $\overline{F}_{\rho_\infty}(\xi)$ is a singleton in $\mathbb{C}$ by Corollary 8.9.13(2-ii).

8.11 Proof of Proposition 8.4.5 - length ≥ 2 case - (Step 3)

In this section, we prove the following Proposition 8.11.1, which shows that Proposition 8.4.5, in the case where the length of Σ_0 is ≥ 2, is valid for the vertices of $\mathcal{L}_0$.

Proposition 8.11.1. *Under Assumption 8.9.1, the following hold for each* $P \in \mathcal{L}_0^{(0)}$.

(1) If $P \in \mathcal{L}(\boldsymbol{\nu}_\infty)^{(0)}$*, then* $\overline{F}_{\rho_\infty}(P)$ *is equal to* $\overline{F}_\infty(P)$ *and it is a convex polyhedron in* $\overline{\mathbb{H}}^3$ *of dimension* 2.

(2) If $P \notin \mathcal{L}(\boldsymbol{\nu}_\infty)^{(0)}$*, then* $\overline{Ih}(\rho_\infty(P))$ *is invisible in* $\overline{Eh}(\boldsymbol{\rho}_\infty)$*. Namely,* $\overline{Eh}(\rho_\infty(P)) \supset \overline{Eh}(\boldsymbol{\rho}_\infty)$.

Proof. (1) Let P be an element of $\mathcal{L}(\boldsymbol{\nu}_\infty)^{(0)}$. First, we prove that $\overline{F}_\infty(P)$ is a 2-dimensional convex polyhedron in $\overline{\mathbb{H}}^3$. Since $\overline{F}_\infty(P)$ is a convex polyhedron in $\overline{\mathbb{H}}^3$ of dimension at most 2 by Lemma 8.6.1, it suffices to prove that $\dim \overline{F}_\infty(P) \geq 2$.

Case 1. Suppose that there is a 2-simplex $\xi = (P_0, P_1, P_2)$ of $\mathcal{L}(\boldsymbol{\nu}_\infty)$ such that $P = P_1$. Then (P_0, P_1) and (P_1, P_2) are edges of $\mathcal{L}(\boldsymbol{\nu}_\infty)$, and therefore $\overline{F}_\infty(P_0, P_1)$ and $\overline{F}_\infty(P_1, P_2)$ are 1-dimensional convex polyhedra by Propositions 8.10.1 and 8.10.2. Since $\overline{F}_\infty(P) = \overline{F}_\infty(P_1)$ contains $\overline{F}_\infty(P_0, P_1)$ and $\overline{F}_\infty(P_1, P_2)$, we see that $\dim \overline{F}_\infty(P) < 2$ only when $\overline{F}_\infty(P_0, P_1)$ and $\overline{F}_\infty(P_1, P_2)$ are contained in a 1-dimensional subspace in $\overline{\mathbb{H}}^3$. If this happens, then $\overline{Ih}(\rho_\infty(P_0)) \cap \overline{Ih}(\rho_\infty(P_1))$ and $\overline{Ih}(\rho_\infty(P_1)) \cap \overline{Ih}(\rho_\infty(P_2))$ are identical geodesics. This contradicts Corollary 8.9.12(1) (cf. Lemma 8.9.8(2)). Hence we have $\dim \overline{F}_\infty(P) = 2$.

Case 2. Suppose that no 2-simplex (P_0, P_1, P_2) of $\mathcal{L}(\boldsymbol{\nu}_\infty)$ satisfies $P = P_1$.

Claim 8.11.2. Every triangle of $\Sigma(\boldsymbol{\nu}_\infty)$ contains $s(P)$ as a vertex.

Proof. Suppose that $\Sigma(\boldsymbol{\nu}_\infty)$ contains a triangle, say σ'', which does not contain $s(P)$ as a vertex. Then, by considering a path in $\Sigma(\boldsymbol{\nu}_\infty)$ joining $s(P)$ to σ'', we can find a pair of triangles $\sigma = \langle s_0, s_1, s_2 \rangle$ and $\sigma' = \langle s_0', s_1', s_2' \rangle$ sharing the edge $\langle s_0, s_2 \rangle = \langle s_0', s_1' \rangle$ such that $s_1 = s(P) \neq s_2'$ (see Fig. 8.10). Then the elliptic generator triple (P_0, P_1, P_2), with $P = P_1$, associated with σ determines a 2-simplex of $\mathcal{L}(\boldsymbol{\nu}_\infty)$ by Lemma 3.2.4(2), a contradiction.

It should also be noted that some triangle of Σ_0 may not contain $s(P)$ as a vertex. Since $\boldsymbol{\nu}_\infty$ is thick, $\Sigma(\boldsymbol{\nu}_\infty)$ contains at least one triangle, $\sigma = \langle s_0, s_1, s_2 \rangle$ with $s_1 = s(P)$.

Claim 8.11.3. $\overline{F}_\infty(P)$ contains $\mathrm{Fix}_\sigma^\epsilon(\rho_\infty(P))$ for each $\epsilon \in \{-, +\}$, and hence $\overline{F}_\infty(P) \supset \overline{\mathrm{Axis}}(\rho_\infty(P))$.

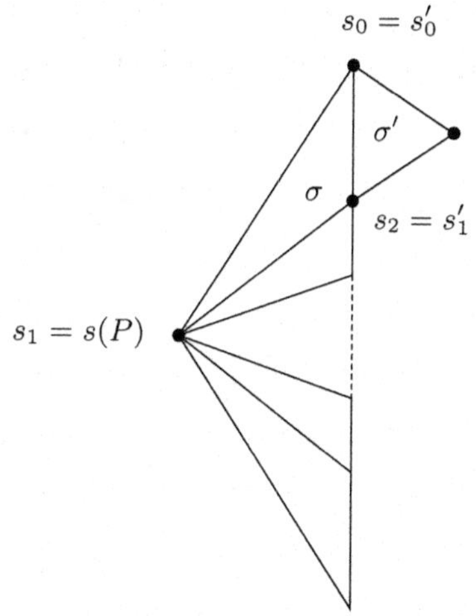

Fig. 8.10. Figure of $\Sigma(\boldsymbol{\nu}_\infty)$ for Claim 8.11.2

Proof. To show the claim, we have only to show that $\overline{F}_\infty(P)$ contains both $\mathrm{Fix}_\sigma^\pm(\rho_\infty(P))$, because $\overline{F}_\infty(P)$ is convex. Let ϵ be an arbitrary sign. First, suppose $s(P)$ is a vertex of the ϵ-terminal triangle σ^ϵ of Σ_0. Then $\mathrm{Fix}_\sigma^\epsilon(\rho_n(P)) = \mathrm{Fix}_{\sigma^\epsilon}^\epsilon(\rho_n(P)) \in \overline{F}_{\boldsymbol{\rho}_n}(P)$ for every $n \in \mathbb{N}$, and therefore

$$\mathrm{Fix}_\sigma^\epsilon(\rho_\infty(P)) = \lim \mathrm{Fix}_\sigma^\epsilon(\rho_n(P)) \in \lim \overline{F}_{\boldsymbol{\rho}_n}(P) = \overline{F}_\infty(P).$$

Next, suppose $s(P)$ is not a vertex of σ^ϵ. Let $\sigma' = \langle s_0', s_1', s_2' \rangle$ be the ϵ-terminal triangle of $\Sigma(\boldsymbol{\nu}_\infty)$, where $s_1' = s(P)$ (see Claim 8.11.2). Then $\sigma^\epsilon \cap \sigma'$ is equal to the edge $\langle s_0', s_2' \rangle$ (see Fig. 8.11). Let $\{P_j'\}$ be the sequence of elliptic generators associated with the triangle σ' such that $P = P_1'$. Then $\xi := (P_0', P_1', P_2')$ is a 2-simplex of $\mathcal{L}_0$. By the assumption, ξ is not a 2-simplex of $\mathcal{L}(\boldsymbol{\nu}_\infty)$, and therefore $\nu_\infty^\epsilon \in \mathrm{int}\langle s_0', s_2' \rangle$. Thus we see by Corollary 8.9.12(2) that $\overline{F}_\infty(\xi) = \mathrm{Fix}_{\sigma'}^\epsilon(\rho_\infty(P_1')) = \mathrm{Fix}_\sigma^\epsilon(\rho_\infty(P))$. Hence, $\mathrm{Fix}_\sigma^\epsilon(\rho_\infty(P)) = \overline{F}_\infty(\xi) \subset \overline{F}_\infty(P)$. Thus we have shown that $\overline{F}_\infty(P)$ contains both $\mathrm{Fix}_\sigma^\pm(\rho_\infty(P))$. This completes the proof of the claim.

Suppose to the contrary that $\dim \overline{F}_\infty(P) < 2$. Then the above claim implies $\overline{F}_\infty(P) = \overline{\mathrm{Axis}}(\rho_\infty(P))$ and hence $\overline{Ih}(\rho_\infty(P_0)) \cap \overline{Ih}(\rho_\infty(P_1))) \cap \overline{Ih}(\rho_\infty(P_2))) = \overline{\mathrm{Axis}}(\rho_\infty(P_1))$. This contradicts Corollary 8.9.12(1). Hence we have $\dim \overline{F}_\infty(P) = 2$.

In what follows we prove the assertion that $\overline{F}_\infty(P)$ is equal to $\overline{F}_{\boldsymbol{\rho}_\infty}(P)$. By Lemma 8.6.2, $\overline{F}_\infty(P)$ is contained in $\overline{F}_{\boldsymbol{\rho}_\infty}(P)$. Hence it suffices to prove that $\overline{F}_{\boldsymbol{\rho}_\infty}(P)$ is contained in $\overline{F}_\infty(P)$. Suppose this does not hold. Then there is a point, x_∞, contained in the interior of the 2-dimensional convex polyhedron $\overline{F}_{\boldsymbol{\rho}_\infty}(P)$ which is not contained in $\overline{F}_\infty(P) = \lim \overline{F}_{\boldsymbol{\rho}_n}(P)$. Take a sequence $\{x_n\}$ consisting of points $x_n \in \overline{Ih}(\rho_n(P))$ and converging to x_∞.

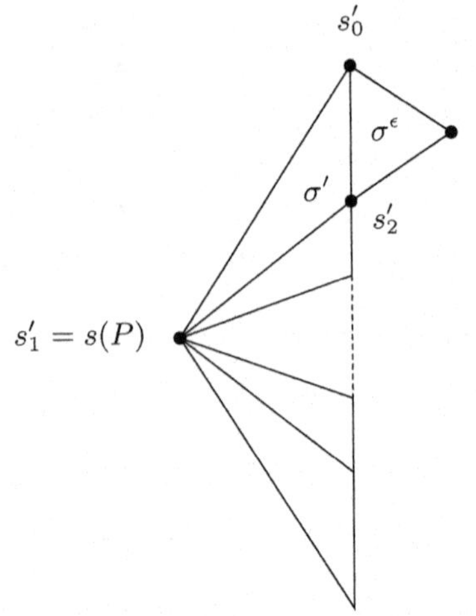

Fig. 8.11. Figure of Σ_0 for Claim 8.11.3

Since $x_\infty \notin \overline{F}_\infty(P) = \lim \overline{F}_{\rho_n}(P)$, we may assume $x_n \notin \overline{F}_{\rho_n}(P)$ for every $n \in \mathbb{N}$, after taking a subsequence. Thus for each $n \in \mathbb{N}$, there is some $X_n \in \mathrm{lk}(P, \mathcal{L}_0)^{(0)}$ such that $\mathrm{int}(\overline{Dh}(\rho_n(X_n)))$ contains x_n. Since $\mathrm{lk}(P, \mathcal{L}_0)^{(0)}$ is finite, we may suppose, by further taking a subsequence, that every x_n is contained in $\mathrm{int}(\overline{Dh}(\rho_n(X)))$ for some $X \in \mathrm{lk}(P, \mathcal{L}_0)^{(0)}$. Note that $x_\infty = \lim x_n$ is contained in $\overline{Dh}(\rho_\infty(X)) = \lim \overline{Dh}(\rho_n(X))$.

First, suppose that $X \in \mathcal{L}(\boldsymbol{\nu}_\infty)^{(0)}$. Then $X \in \mathrm{lk}(P, \mathcal{L}(\boldsymbol{\nu}_\infty))^{(0)}$. Since x_∞ is contained in $\overline{F}_{\boldsymbol{\rho}_\infty}(P)$, it is contained in $\overline{Eh}(\rho_\infty(X))$. Hence x_∞ is contained in $\overline{Ih}(\rho_\infty(X)) = \overline{Dh}(\rho_\infty(X)) \cap \overline{Eh}(\rho_\infty(X))$. Thus x_∞ is contained in $\overline{Ih}(\rho_\infty(P)) \cap \overline{Ih}(\rho_\infty(X))$, which is a geodesic by Corollary 8.9.13(2-iii), because (P, X) is an edge of $\mathcal{L}(\boldsymbol{\nu}_\infty)$. Thus x_∞ cannot be contained in the interior of the 2-dimensional convex polyhedron $\overline{F}_{\boldsymbol{\rho}_\infty}(P)$, a contradiction.

Next, suppose that $X \notin \mathcal{L}(\boldsymbol{\nu}_\infty)^{(0)}$. Then $s(X)$ is a vertex of the ϵ-terminal triangle σ^ϵ of Σ_0 for some $\epsilon \in \{-,+\}$, and any other triangle of Σ_0 does not contain $s(X)$. Set $\sigma^\epsilon = \langle s_0, s_1, s_2 \rangle$ where $s_1 = s(X)$, and let $\{P_j\}$ be the sequence of elliptic generators associated with σ^ϵ such that $P_1 = X$. Then, since the length of Σ_0 is ≥ 2 by the assumption, (P_0, P_1, P_2) is a 2-simplex of $\mathcal{L}_0$. Moreover, (P_0, P_2) is an edge of $\mathcal{L}(\boldsymbol{\nu}_\infty)$. Since X is contained in $\mathrm{lk}(P, \mathcal{L}_0)$, P is equal to either P_0 or P_2. We may assume without loss of generality that $P = P_0$. Then, since $P_2 \in \mathrm{lk}(P_0, \mathcal{L}(\boldsymbol{\nu}_\infty))^{(0)}$, we have

$$\begin{aligned} x_\infty &\in \overline{F}_{\boldsymbol{\rho}_\infty}(P) \cap \overline{Dh}(\rho_\infty(X)) \\ &\subset \left(\overline{Ih}(\rho_\infty(P_0)) \cap \overline{Eh}(\rho_\infty(P_2))\right) \cap \overline{Dh}(\rho_\infty(P_1)) \\ &\subset \overline{Dh}(\rho_\infty(P_1)) \cap \overline{Eh}(\rho_\infty(P_0)) \cap \overline{Eh}(\rho_\infty(P_2)). \end{aligned}$$

Since $\xi := (P_0, P_1, P_2)$ is a 2-simplex of $\mathcal{L}_0$, and since $s_1 = s(X) \notin \Sigma(\boldsymbol{\nu}_\infty)^{(0)}$, Corollary 8.9.12(2-i) implies that the last term in the above formula is equal to the singleton $\overline{F}_\infty(\xi)$ in $\mathbb{C}$. This contradicts the assumption that x_∞ is contained in the interior of $\overline{F}_{\rho_\infty}(P)$. This completes the proof Proposition 8.11.1(1).

(2) Let P be an element of $\mathcal{L}_0^{(0)} - \mathcal{L}(\boldsymbol{\nu}_\infty)^{(0)}$. Then $s(P)$ is a vertex of the ϵ-terminal triangle σ^ϵ of Σ_0 for some $\epsilon \in \{-,+\}$, and $s(P)$ does not belong to $\Sigma(\boldsymbol{\nu}_\infty)^{(0)}$. Set $\sigma^\epsilon = \langle s_0, s_1, s_2\rangle$ where $s_1 = s(P)$, and let $\{P_j\}$ be the sequence of elliptic generators associated with σ^ϵ such that $P_1 = P$. Since the length of Σ_0 is ≥ 2 by the assumption, we see that $\xi := (P_0, P_1, P_2)$ is a 2-simplex of $\mathcal{L}_0$. Since $s_1 = s(P) \notin \Sigma(\boldsymbol{\nu}_\infty)^{(0)}$, we have $\nu_\infty^\epsilon \in \mathrm{int}\langle s_0, s_2\rangle$. Hence we see by Corollary 8.9.12(2-iii) that

$$\begin{aligned}\overline{Eh}(\rho_\infty(P)) &= \overline{Eh}(\rho_\infty(P_1))\\ &\supset \overline{Eh}(\rho_\infty(P_0)) \cap \overline{Eh}(\rho_\infty(P_2))\\ &\supset \cap\{\overline{Eh}(\rho_\infty(X)) \,|\, X \in \mathcal{L}(\boldsymbol{\nu}_\infty)^{(0)}\}\\ &= \overline{Eh}(\boldsymbol{\rho}_\infty).\end{aligned}$$

This completes the proof of Proposition 8.11.1(2).

8.12 Proof of Proposition 8.3.6

In this section, we give a proof of Proposition 8.3.6. Throughout this section, we presume the following assumption.

Assumption 8.12.1. Under Assumption 8.3.1, assume that $\boldsymbol{\nu}_\infty$ is thick and that $\{\boldsymbol{\rho}_n\}$ satisfies the condition HausdorffConvergence. Moreover, if $\nu_\infty^\epsilon \in \partial\sigma^\epsilon(\boldsymbol{\nu}_\infty)$, then $\nu_\infty^\epsilon \in \langle s_0^\epsilon, s_1^\epsilon\rangle$. In this case $\sigma^\epsilon \cap \sigma^\epsilon(\boldsymbol{\nu}_\infty) = \langle s_0^\epsilon, s_1^\epsilon\rangle$.

Lemma 8.12.2. *Under Assumption 8.12.1, $\mathcal{L}(\rho_\infty, \sigma^\epsilon(\boldsymbol{\nu}_\infty))$ is simple.*

Proof. Let $\{P_j\}$ be the sequence of elliptic generators associated with $\sigma^\epsilon(\boldsymbol{\nu}_\infty)$. Then (P_j, P_{j+1}) is an edge of $\mathcal{L}(\boldsymbol{\nu}_\infty)$ for every $j \in \mathbb{Z}$. By Proposition 8.3.5, $\boldsymbol{\rho}_\infty$ satisfies the condition Duality. Hence $F_{\rho_\infty}(P_j, P_{j+1})$ is a well-defined 1-dimensional convex polyhedron in $\partial Eh(\boldsymbol{\rho}_\infty)$. Now suppose to the contrary that $\mathcal{L}(\rho_\infty, \sigma^\epsilon(\boldsymbol{\nu}_\infty))$ is not simple. Then, by Lemma 4.8.2, it is folded at some vertex $c(\rho_\infty(P_j))$. Hence $Ih(\rho_\infty(P_{j-1})) \cap Ih(\rho_\infty(P_j)) = Ih(\rho_\infty(P_j)) \cap Ih(\rho_\infty(P_{j+1}))$ by Lemma 4.8.5, and therefore we can see that $F_{\rho_\infty}(P_{j-1}, P_j) = F_{\rho_\infty}(P_j, P_{j+1})$. This contradicts Proposition 6.3.1.

Lemma 8.12.3. *Under Assumption 8.12.1, $f_\infty^\epsilon(P_j^\epsilon) \subset \mathrm{fr}\, E(\boldsymbol{\rho}_\infty)$ for every $j \in \mathbb{Z}$. Here $\{P_j^\epsilon\}$ is the sequence of elliptic generators associated with the ϵ-terminal triangle, σ^ϵ, of the common chain Σ_0.*

Proof. By Proposition 8.4.4, we have $f_\infty^\epsilon(P_j^\epsilon) \subset \mathrm{fr}\, E(\rho_\infty, \mathcal{L}_0)$. On the other hand, we see $E(\rho_\infty, \mathcal{L}_0) = E(\boldsymbol{\rho}_\infty)$ by the proof of of Proposition 8.3.5 in Sect. 8.4. Hence we obtain the desired result.

Lemma 8.12.4. *Under Assumption 8.12.1, there exists a local homeomorphism $h\colon \mathbb{R} \to \mathbb{C}$, with the symmetry $h(t+1) = h(t)+1$ $(t \in \mathbb{R})$, which satisfies the following condition.*

1. *If $\nu^{\epsilon}_{\infty} \in \operatorname{int}\sigma^{\epsilon}(\boldsymbol{\nu}_{\infty})$, then $h([j/3,(j+1)/3]) = f^{\epsilon}_{\infty}(P^{\epsilon}_j)$ for every $j \in \mathbb{Z}$.*
2. *If $\nu^{\epsilon}_{\infty} \in \partial\sigma^{\epsilon}(\boldsymbol{\nu}_{\infty})$, then for any $k \in \mathbb{Z}$,*

$$h([k,k+1/2]) = f^{\epsilon}_{\infty}(P^{\epsilon}_{3k}),$$
$$h([k+1/2,k+1]) = f^{\epsilon}_{\infty}(P^{\epsilon}_{3k+1}).$$

Proof. Suppose first that $\nu^{\epsilon}_{\infty} \in \operatorname{int}\sigma^{\epsilon}(\boldsymbol{\nu}_{\infty})$. Then, for every $j \in \mathbb{Z}$, $f^{\epsilon}_{\infty}(P^{\epsilon}_j)$ is a non-degenerate circular arc by Lemma 8.6.1(1). We define the restriction of the map h to the interval $[j/3,(j+1)/3]$ to be the unit speed path, with respect to the Euclidean metric on $\mathbb{C}$, onto $f^{\epsilon}_{\infty}(P^{\epsilon}_j)$ connecting $h(j/3) := v^{\epsilon}(\rho_{\infty}; P^{\epsilon}_{j-1}, P^{\epsilon}_j)$ and $h((j+1)/3) := v^{\epsilon}(\rho_{\infty}; P^{\epsilon}_j, P^{\epsilon}_{j+1})$. Then h is a well-defined continuous map from $\mathbb{R}$ to $\mathbb{C}$, whose restriction to each interval $[j/3,(j+1)/3]$ is a homeomorphism onto $f^{\epsilon}_{\infty}(P^{\epsilon}_j)$. Since $I(\rho_{\infty}(P^{\epsilon}_{j-1}))$ and $I(\rho_{\infty}(P^{\epsilon}_j))$ are different circles (cf. Lemma 2.5.4(2)) and $h(j/3)$ is an isolated intersection point of these two circles, we see that h is a local homeomorphism near the point $j/3$. Hence h is a local homeomorphism. It is clear from the construction that h has the symmetry $h(t+1) = h(t)+1$ $(t \in \mathbb{R})$.

A similar argument works for the case where $\nu^{\epsilon}_{\infty} \in \partial\sigma^{\epsilon}(\boldsymbol{\nu}_{\infty})$, by noticing the fact that $f^{\epsilon}_{\infty}(P^{\epsilon}_j)$ is degenerated if and only if $j \equiv 2 \pmod 3$.

Lemma 8.12.5. *Under Assumption 8.12.1, h is a homeomorphism onto its image.*

Proof. We have only to show that h is injective. In fact, if h is injective then h induces a continuous injective map $\bar{h}$ from $\mathbb{R}/\mathbb{Z}$ to $\mathbb{C}/\mathbb{Z}$. Since $\mathbb{R}/\mathbb{Z}$ is compact, $\bar{h}$ is a homeomorphism onto its image. This implies that h is a homeomorphism onto its image.

We give a proof of the injectivity of h only for the case when $\nu^{\epsilon}_{\infty} \in \operatorname{int}\sigma^{\epsilon}(\boldsymbol{\nu}_{\infty})$. A similar argument works for the case when $\nu^{\epsilon}_{\infty} \in \partial\sigma^{\epsilon}(\boldsymbol{\nu}_{\infty})$. Suppose to the contrary that h is not injective. Then $\delta := \min\{|s-t| \mid s \neq t,\ h(s) = h(t)\}$ is a positive real number, because h is a local homeomorphism. Take $t_0 \in \mathbb{R}$ such that $h(t_0) = h(t_0+\delta)$. Then $h([t_0, t_0+\delta])$ is a circle in $\mathbb{C}$ and hence it bounds a disk, D_0, in $\mathbb{C}$.

Claim 8.12.6. The disk D_0 is contained in $E(\boldsymbol{\rho}_{\infty})$.

Proof. Note that

$$\mathbb{C} - E(\boldsymbol{\rho}_{\infty}) = \cup\{\operatorname{int} D(\rho_{\infty}(P)) \mid P \in \mathcal{L}(\boldsymbol{\nu}_{\infty})^{(0)}\} = \operatorname{proj}(\partial Eh(\boldsymbol{\rho}_{\infty})).$$

Since $\boldsymbol{\rho}_{\infty}$ satisfies the condition Duality by Proposition 8.3.5, we see that $\partial Eh(\boldsymbol{\rho}_{\infty})$ is arcwise connected by Proposition 6.3.1. Hence $\operatorname{proj}(\partial Eh(\boldsymbol{\rho}_{\infty}))$ is arcwise connected. Now suppose that the claim does not hold. Then there is a point, x_0, in D_0 such that $x_0 \in \mathrm{proj}(\partial Eh(\boldsymbol{\rho}_\infty))$. Since D_0 is compact, there is an integer k such that $x_0 + k \notin D_0$. Since $\mathrm{proj}(\partial Eh(\boldsymbol{\rho}_\infty))$ is invariant by the translation $z \mapsto z+1$, the point $x_0 + k$ is also contained in $\mathrm{proj}(\partial Eh(\boldsymbol{\rho}_\infty))$. Since this set is arcwise connected, there is a path, γ, in $\mathrm{proj}(\partial Eh(\boldsymbol{\rho}_\infty))$ connecting x_0 and x_0+k. Since $x_0 \in D_0$ and $x_0+k \notin D_0$, γ intersects $\partial D_0 \subset \cup\{f^\epsilon_\infty(P^\epsilon_j) \mid j \in \mathbb{Z}\}$. Thus some $f^\epsilon_\infty(P^\epsilon_j)$ contains a point in $\mathrm{proj}(\partial Eh(\boldsymbol{\rho}_\infty)) = \mathbb{C} - E(\boldsymbol{\rho}_\infty)$. On the other hand, Lemma 8.12.3 implies $f^\epsilon_\infty(P^\epsilon_j) \subset E(\boldsymbol{\rho}_\infty)$, a contradiction.

We may assume without loss of generality that $t_0 \in [0, 1/3)$. Let k be the integer such that $t_0 + \delta \in (k/3, (k+1)/3]$.

Then $k \neq 0$, because the restriction of h to the subinterval $[0, 1/3]$ is injective by the construction of h.

We show that $k \neq 1$. By the temporary assumption that $\nu^\epsilon_\infty \in \mathrm{int}\,\sigma^\epsilon(\boldsymbol{\nu}_\infty)$, we see that $(P^\epsilon_j, P^\epsilon_{j+1})$ is an edge of $\mathcal{L}(\boldsymbol{\nu}_\infty)$, and hence $F_{\boldsymbol{\rho}_\infty}(P^\epsilon_j, P^\epsilon_{j+1})$ is a well-defined 1-dimensional convex polyhedron for every $j \in \mathbb{Z}$, because $\boldsymbol{\rho}_\infty$ satisfies the condition Duality by Proposition 8.3.5. In particular, $I(\rho_\infty(P^\epsilon_0))$ and $I(\rho_\infty(P^\epsilon_1))$ intersect in two points, and $f^\epsilon_\infty(P^\epsilon_0)$ and $f^\epsilon_\infty(P^\epsilon_1)$, respectively, are subarcs of the arcs $I(\rho_\infty(P^\epsilon_0)) \cap E(\rho_\infty(P^\epsilon_1))$ and $E(\rho_\infty(P^\epsilon_0)) \cap I(\rho_\infty(P^\epsilon_1))$. Now suppose to the contrary that that $k = 1$. Then $f^\epsilon_\infty(P^\epsilon_0)$ and $f^\epsilon_\infty(P^\epsilon_1)$ intersect in two points, and therefore we have $f^\epsilon_\infty(P^\epsilon_0) = I(\rho_\infty(P^\epsilon_0)) \cap E(\rho_\infty(P^\epsilon_1))$ and $f^\epsilon_\infty(P^\epsilon_1) = E(\rho_\infty(P^\epsilon_0)) \cap I(\rho_\infty(P^\epsilon_1))$. Thus the disk D_0 is equal to the union $D(\rho_\infty(P^\epsilon_0)) \cup D(\rho_\infty(P^\epsilon_1))$. This contradicts Claim 8.12.6.

We show that $k \neq 2$. Suppose to the contrary that $k = 2$. Then the boundary of the disk D_0 consists of three circular arcs. Since $D_0 \subset E(\boldsymbol{\rho}_\infty)$ by Claim 8.12.6, the sum of the angles of D_0 at the three vertices is less than π. Now let θ_j be the dihedral angle of $E(\rho_\infty(P^\epsilon_j)) \cap E(\rho_\infty(P^\epsilon_{j+1}))$. Then two of the three angles of the circular triangle D_0 are equal to the angles θ_0 and θ_1, respectively. Hence we have $\theta_0 + \theta_1 < \pi$. This contradicts Lemma 4.1.3(3) (Chain rule).

We show that $k \neq 3$. Suppose to the contrary that $k = 3$. Then as in the preceding argument, the boundary of the disk D_0 consists of four circular arcs, and the sum of the angles of D_0 at the four vertices is less than 2π. Three of the four angles of the circular triangle D_0 are equal to the angles θ_0, θ_1 and θ_2, respectively. Hence we have $\theta_0 + \theta_1 + \theta_2 < 2\pi$. This contradicts Lemma 4.1.3(3).

Finally suppose that $k \geq 4$. Then $\delta > 1$, and hence $x_0 := h(t_0 + 1) \in \partial D_0 - \{h(t_0)\}$. By the symmetry, we have $h(t_0 + \delta + 1) = h(t_0 + \delta) + 1 = h(t_0) + 1 = h(t_0 + 1) = x_0$. Since ∂D_0 is a finite union of circular arcs, we can find a small disk neighborhood U of x_0 in $\mathbb{C}$, such that $U - \partial D_0$ consists of two components, U^ϵ and $U^{-\epsilon}$. We may assume $U^\epsilon \subset \mathrm{int}\, D_0$. Then $U^\epsilon \subset \mathrm{int}\, E(\boldsymbol{\rho}_\infty)$ by Claim 8.12.6. Moreover, we may also assume $U^{-\epsilon} \subset \cup_j \mathrm{int}\, D(\rho_\infty(P^\epsilon_j)) \subset \mathbb{C} - E(\boldsymbol{\rho}_\infty)$. This is obvious when x_0 lies in the interior of a circular edge of ∂D_0. The assertion for the case when x_0 is a vertex of ∂D_0 follows from the fact that two successive isometric circles $I(\rho_\infty(P^\epsilon_j))$ and

$I(\rho_\infty(P^\epsilon_{j+1}))$ intersect transversely as observed in the proof of $k \neq 1$. On the other hand, $t_0 + \delta + 1$ is an isolated point of the inverse image by h of the intersection $\partial D_0 \cap h([t_0 + \delta, \infty))$, because ∂D_0 is contained in a finite union of circular arcs $\{f^\epsilon_\infty(P^\epsilon_j)\}$ and because $\{I(\rho_\infty(P^\epsilon_j))\}$ are mutually distinct (cf. Lemma 2.5.4). Hence there is an open interval J having $t_0 + \delta + 1$ as an endpoint, such that

$$h(J) \subset U - \partial D_0 = U^\epsilon \sqcup U^{-\epsilon} \subset (\operatorname{int} E(\boldsymbol{\rho}_\infty)) \cup (\mathbb{C} - E(\boldsymbol{\rho}_\infty)) = \mathbb{C} - \operatorname{fr} E(\boldsymbol{\rho}_\infty).$$

This contradicts Lemma 8.12.3. Thus $k \geq 4$ cannot happen.

Hence we have proved that h is injective. This completes the proof of Lemma 8.12.5.

Set $\ell_\infty = h(\mathbb{R}) = \cup_j f^\epsilon_\infty(P^\epsilon_j)$. Then ℓ_∞ projects to an essential simple loop in $\mathbb{C}/\mathbb{Z}$ by Lemma 8.12.5, and hence it separates $\mathbb{C}$ into two connected components. For each sign $\eta \in \{-,+\}$, let $E^\eta(\ell_\infty)$ be the closure of the component which contains the set $\{z \,|\, \eta\Im z > L\}$ for sufficiently large L.

Lemma 8.12.7. *$D(\rho_\infty(P)) \subset E^{-\epsilon}(\ell_\infty)$ for every vertex (P) of $\mathcal{L}(\boldsymbol{\nu}_\infty)$.*

Proof. Suppose this is not the case. Then $D(\rho_\infty(P)) \cap \operatorname{int} E^\epsilon(\ell_\infty) \neq \emptyset$ and hence there is a point $x \in \operatorname{int} D(\rho_\infty(P))$ which is contained in $\operatorname{int} E^\epsilon(\ell_\infty)$. Since $x \in \operatorname{int} E^\epsilon(\ell_\infty)$, there is a path, γ, in $\mathbb{C} - \ell_\infty$ joining x to a point in a region $\{z \,|\, \epsilon\Im z > L\} \subset \operatorname{int} E^\epsilon(\ell_\infty)$. By the periodicity of ℓ_∞, the Euclidean distance between the (finite) path γ and ℓ_∞ is positive. Since $\operatorname{fr} E^\epsilon(\boldsymbol{\rho}_n)$ converges to ℓ_∞, this implies that γ is disjoint from $\operatorname{fr} E^\epsilon(\boldsymbol{\rho}_n)$ for all sufficiently large n. Moreover if we choose L sufficiently large in advance, then the endpoint of γ lies in $E^\epsilon(\boldsymbol{\rho}_n)$ for all n. Hence it follows that $x \in \operatorname{int} E^\epsilon(\boldsymbol{\rho}_n)$ for all sufficiently large n. On the other hand, since $x \in \operatorname{int} D(\rho_\infty(P))$ and $D(\rho_\infty(P)) = \lim D(\rho_n(P))$, we have $x \in \operatorname{int} D(\rho_n(P))$ for all sufficiently large n. This is a contradiction. Hence we obtain the desired result.

Lemma 8.12.8. *$E^\epsilon(\ell_\infty) = E^\epsilon(\boldsymbol{\rho}_\infty) = E^\epsilon(\rho_\infty, \sigma^\epsilon(\boldsymbol{\nu}_\infty))$ and hence $\ell_\infty = \operatorname{fr} E^\epsilon(\rho_\infty, \sigma^\epsilon(\boldsymbol{\nu}_\infty))$.*

Proof. By Lemma 8.12.7, $\cup\{D(\rho_\infty(P)) \,|\, (P) \in \mathcal{L}(\boldsymbol{\nu}_\infty)^{(0)}\}$ is contained in $E^{-\epsilon}(\ell_\infty)$. Hence $\operatorname{int} E^\epsilon(\ell_\infty)$ is disjoint from

$$\begin{aligned}\operatorname{int}(\cup\{D(\rho_\infty(P)) \,|\, (P) \in \mathcal{L}(\boldsymbol{\nu}_\infty)^{(0)}\}) &= \cup\{\operatorname{int} D(\rho_\infty(P)) \,|\, (P) \in \mathcal{L}(\boldsymbol{\nu}_\infty)^{(0)}\} \\ &= \mathbb{C} - E(\boldsymbol{\rho}_\infty).\end{aligned}$$

Thus $\operatorname{int} E^\epsilon(\ell_\infty)$ and hence $E^\epsilon(\ell_\infty)$ are contained in a connected component of $E(\boldsymbol{\rho}_\infty)$. Since both $E^\epsilon(\ell_\infty)$ and $E^\epsilon(\boldsymbol{\rho}_\infty)$ contain a region $\{z \,|\, \epsilon\Im z > L\}$ for sufficiently large L, $E^\epsilon(\ell_\infty)$ is contained in $E^\epsilon(\boldsymbol{\rho}_\infty)$. On the other hand, every point $x \in \ell_\infty$ has a neighborhood U in $\mathbb{C}$ such that $U - E^\epsilon(\ell_\infty) \subset \cup\{\operatorname{int} D(\rho_\infty(P)) \,|\, (P) \in \mathcal{L}(\boldsymbol{\nu}_\infty)^{(0)}\}$. (This is obvious if x lies in the interior of a nondegenerate circular arc $f^\epsilon_\infty(P^\epsilon_j)$. The assertion for the case when $x =$

$f_\infty^\epsilon(P_j^\epsilon, P_{j+1}^\epsilon)$ follows from the fact that $\operatorname{int} D(\rho_\infty(P_j^\epsilon)) \cap \operatorname{int} D(\rho_\infty(P_{j+1}^\epsilon)) \neq \emptyset$.) Thus $E^\epsilon(\ell_\infty)$ is open in $E^\epsilon(\boldsymbol{\rho}_\infty)$. Since $E^\epsilon(\ell_\infty)$ is also closed in $E^\epsilon(\boldsymbol{\rho}_\infty)$, it follows that $E^\epsilon(\ell_\infty)$ is equal to a component of $E^\epsilon(\boldsymbol{\rho}_\infty)$. Since $E^\epsilon(\ell_\infty)$ is contained in $E^\epsilon(\boldsymbol{\rho}_\infty)$, this implies that $E^\epsilon(\ell_\infty) = E^\epsilon(\boldsymbol{\rho}_\infty)$.

We can also prove $E^\epsilon(\ell_\infty) = E^\epsilon(\rho_\infty, \sigma^\epsilon(\boldsymbol{\nu}_\infty))$ by an argument parallel to the above, where we replace $\cup\{D(\rho_\infty(P)) \mid (P) \in \mathcal{L}(\boldsymbol{\nu}_\infty)^{(0)}\}$ with $\cup\{D(\rho_\infty(P)) \mid (P) \in \mathcal{L}(\sigma^\epsilon(\boldsymbol{\nu}_\infty))^{(0)}\}$. This completes the proof.

Lemma 8.12.9. *$\sigma^\epsilon(\boldsymbol{\nu}_\infty)$ is an ϵ-terminal triangle of ρ_∞.*

Proof. We first check that $(\rho_\infty, \sigma^\epsilon(\boldsymbol{\nu}_\infty))$ satisfies Assumption 4.2.4 (σ-Simple). Since $\rho_\infty \in \overline{\mathcal{QF}}$, the corresponding Markoff map ϕ_∞ satisfies the condition $\sigma^\epsilon(\boldsymbol{\nu}_\infty)$-NonZero. Since $\boldsymbol{\rho}_\infty = (\rho_\infty, \boldsymbol{\nu}_\infty)$ satisfies the condition Duality by Proposition 8.3.5, ϕ_∞ satisfies the triangle inequality at $\sigma^\epsilon(\boldsymbol{\nu}_\infty)$. By Lemma 8.12.2, $\mathcal{L}(\rho_\infty, \sigma^\epsilon(\boldsymbol{\nu}_\infty))$ is simple. Hence $(\rho_\infty, \sigma^\epsilon(\boldsymbol{\nu}_\infty))$ satisfies Assumption 4.2.4 (σ-Simple).

We show that $\sigma^\epsilon(\boldsymbol{\nu}_\infty)$ satisfies the first condition in Definition 4.3.8.

Case 1. $\nu_\infty^\epsilon \in \operatorname{int} \sigma^\epsilon(\boldsymbol{\nu}_\infty)$. In this case $\sigma^\epsilon(\boldsymbol{\nu}_\infty)$ is equal to the ϵ-terminal triangle σ^ϵ of the common chain Σ_0. Moreover, since $\boldsymbol{\rho}_n = (\rho_n, \boldsymbol{\nu}_n)$ with $\boldsymbol{\nu}_n = (\nu_n^-, \nu_n^+)$ is good, $\theta^\epsilon(\rho_n, \sigma^\epsilon)$ is identified with the point ν_n^ϵ (cf. Definition 4.2.17). Hence we have

$$\theta^\epsilon(\rho_\infty, \sigma^\epsilon(\boldsymbol{\nu}_\infty)) = \theta^\epsilon(\rho_\infty, \sigma^\epsilon) = \lim \theta^\epsilon(\rho_n, \sigma^\epsilon) = \lim \nu_n^\epsilon = \nu_\infty^\epsilon \in \operatorname{int} \sigma^\epsilon(\boldsymbol{\nu}_\infty).$$

Thus all components of $\theta^\epsilon(\rho_\infty, \sigma^\epsilon(\boldsymbol{\nu}_\infty))$ are positive and therefore the first condition in Definition 4.3.8 is satisfied.

Case 2. $\nu_\infty^\epsilon \in \operatorname{int}\langle s_0^\epsilon, s_1^\epsilon\rangle \subset \partial\sigma^\epsilon(\boldsymbol{\nu}_\infty)$. In this case $\sigma^\epsilon(\boldsymbol{\nu}_\infty) \cap \sigma^\epsilon = \langle s_0^\epsilon, s_1^\epsilon\rangle$. Since

$$\theta^\epsilon(\rho_\infty, \sigma^\epsilon) = \lim \theta^\epsilon(\rho_n, \sigma^\epsilon) = \lim \nu_n^\epsilon = \nu_\infty^\epsilon \in \operatorname{int}\langle s_0^\epsilon, s_1^\epsilon\rangle,$$

we have $\theta^\epsilon(\rho_\infty, \sigma^\epsilon; s_2^\epsilon) = 0$. Hence $\theta^\epsilon(\rho_\infty, \sigma^\epsilon(\boldsymbol{\nu}_\infty))$ and $\theta^\epsilon(\rho_\infty, \sigma^\epsilon)$ determine the same point in $\operatorname{int}\langle s_0^\epsilon, s_1^\epsilon\rangle$ by Lemma 4.5.3. Thus the first condition in Definition 4.3.8 is satisfied.

We do not need to check the second condition in Definition 4.3.8, because it is a consequence of the fourth condition (Remark 4.3.9(1)).

We show that $\sigma^\epsilon(\boldsymbol{\nu}_\infty)$ satisfies the third condition in Definition 4.3.8. By Lemma 8.12.8, $\operatorname{fr} E^\epsilon(\rho_\infty, \sigma^\epsilon(\boldsymbol{\nu}_\infty)) = \ell_\infty = h(\mathbb{R}) = \cup_j f_\infty^\epsilon(P_j^\epsilon)$. Moreover this is homeomorphic to $\mathbb{R}$ by Lemma 8.12.5. Thus the third condition follows from the following fact, which in tern is a consequence of Lemma 8.6.1.

1. Suppose $\nu_\infty^\epsilon \in \operatorname{int} \sigma^\epsilon(\boldsymbol{\nu}_\infty)$. Then $e^\epsilon(\rho_\infty, \sigma^\epsilon; P_j^\epsilon) = f_\infty^\epsilon(P_j^\epsilon)$.
2. Suppose $\nu_\infty^\epsilon \in \operatorname{int}\langle s_0^\epsilon, s_1^\epsilon\rangle \subset \partial\sigma^\epsilon(\boldsymbol{\nu}_\infty)$, and let $\{P_j\}$ be the sequence of elliptic generators associated with $\sigma^\epsilon(\boldsymbol{\nu}_\infty)$. Then $e^\epsilon(\rho_\infty, \sigma^\epsilon; P_j)$ is equal to $f_\infty^\epsilon(P_0^\epsilon)$, $f_\infty^\epsilon(P_0^\epsilon, P_1^\epsilon)$ or $f_\infty^\epsilon(P_1^\epsilon)$ according as $j = 0$, 1 or 2.

Finally we show that the fourth condition in Definition 4.3.8 is satisfied. Let $\{P_j\}$ be the sequence of elliptic generators associated with $\sigma^\epsilon(\boldsymbol{\nu}_\infty)$. Since $\boldsymbol{\rho}_\infty$ satisfies the condition Duality by Proposition 8.3.5, each $F_{\boldsymbol{\rho}}(P_j, P_{j+1})$ is a 1-dimensional convex polyhedron contained in $Eh(\boldsymbol{\rho}_\infty)$. In fact it is proved that $\overline{F}_{\boldsymbol{\rho}_\infty}(P_j, P_{j+1})$ is contained in $\overline{Eh}(\boldsymbol{\rho}_\infty)$. Since $\overline{F}_{\boldsymbol{\rho}_\infty}(P_j, P_{j+1})$ is either a complete or a half geodesic which has $v^\epsilon(\rho_\infty; P_j, P_{j+1})$ as an endpoint, we see that the fourth condition is satisfied.

Proof (Proof of Proposition 8.3.6). We check that the conditions in Definition 6.1.4(1) are satisfied for each $\epsilon \in \{-, +\}$. By Lemma 8.12.9, $\boldsymbol{\rho}_\infty$ satisfies the first condition. In the proof of Lemma 8.12.9, we have already observed that the second condition is also satisfied. The last condition is a consequence of Lemma 8.12.8. This completes the proof of Proposition 8.3.6.

9

Algebraic roots and geometric roots

The purpose of this chapter is to prove Proposition 6.2.6 (Unique realization), which implies the bijectivity of the map $\boldsymbol{\mu}_2 : \mathcal{J}[\mathcal{QF}] \to \mathbb{H}^2 \times \mathbb{H}^2$. To this end, we first make a careful study of the algebraic curves in the algebraic surface $\Phi \cong \{(x, y, z) \in \mathbb{C}^3 \mid x^2 + y^2 + z^2 = xyz\}$ determined by the equations in Definition 4.2.19, and find the irreducible components which contain the *geometric roots* (Definition 9.1.2) for a given label $\boldsymbol{\nu} = (\nu^-, \nu^+) \in \mathbb{H}^2 \times \mathbb{H}^2$ (Lemmas 9.1.8 and 9.1.12). We also observe that the number of the algebraic roots for $\boldsymbol{\nu}$ is finite (Proposition 9.1.13). Thus the problem is how to single out the geometric roots among the algebraic roots. Our answer is to appeal to the idea of the *geometric continuity.* By using the idea, we show that all geometric roots for a given label $\boldsymbol{\nu}$ are obtained by continuous deformation of the unique geometric root for a special label corresponding to a fuchsian group. This implies the desired result that each label has the unique geometric root. To realize this idea, we introduce the concept of the "geometric degree" $d_G(\boldsymbol{\nu})$ of a label $\boldsymbol{\nu}$, and then show that $d_G(\boldsymbol{\nu}) = 1$ for every $\boldsymbol{\nu}$ by using the argument of geometric continuity (Proposition 9.2.3).

9.1 Algebraic roots

In this chapter, we make a slight change of notation and denote by Φ the space of all Markoff maps, including the trivial one. Definition 4.2.19 is rephrased as follows.

Definition 9.1.1. *(1) Let* $\sigma = \langle s_0, s_1, s_2 \rangle$ *be a triangle of* $\mathcal{D}$ *and* $\nu = (\theta_0, \theta_1, \theta_2)$ *be a point in* $\sigma \cap \mathbb{H}^2$.

1. $\zeta^{\epsilon}_{\nu,\sigma} \colon \Phi \to \mathbb{C}$ *denotes the map defined by:*

$$\zeta^{\epsilon}_{\nu,\sigma}(\phi) = \phi(s_0) + \alpha^{\epsilon}\phi(s_1) + \beta^{\epsilon}\phi(s_2),$$

where $\alpha^{\epsilon} = \epsilon i \exp(\epsilon i \theta_2)$ *and* $\beta^{\epsilon} = -\epsilon i \exp(-\epsilon i \theta_1)$.

2. $\Phi^{\epsilon}_{\nu,\sigma}$ *denotes the subset of* Φ *defined by*

$$\Phi^{\epsilon}_{\nu,\sigma} = \{\phi \in \Phi \,|\, \zeta^{\epsilon}_{\nu,\sigma}(\phi) = 0\}.$$

(2) Let σ^{ϵ} *be a triangle of* $\mathcal{D}$ *and* ν^{ϵ} *be a point in* $\sigma^{\epsilon} \cap \mathbb{H}^2$ *for each* $\epsilon \in \{-,+\}$. *Then a Markoff map* ϕ *is called an* algebraic root for $((\nu^-,\sigma^-), (\nu^+,\sigma^+))$ *if* $\phi \in \Phi^-_{\nu^-,\sigma^-} \cap \Phi^+_{\nu^+,\sigma^+}$.

(3) For a given $\boldsymbol{\nu} = (\nu^-,\nu^+) \in \mathbb{H}^2 \times \mathbb{H}^2$, *a Markoff map* ϕ *is called an* algebraic root for $\boldsymbol{\nu}$ *if it is an algebraic root for* $((\nu^-,\sigma^-(\boldsymbol{\nu})), (\nu^+,\sigma^+(\boldsymbol{\nu})))$, *where* $\sigma^{\epsilon}(\boldsymbol{\nu})$ *is the* ϵ*-terminal triangle of* $\Sigma(\boldsymbol{\nu})$ *(Definition 3.3.3).*

Definition 9.1.2. *Let* $\boldsymbol{\nu} = (\nu^-,\nu^+)$ *be an element of* $\mathbb{H}^2 \times \mathbb{H}^2$. *Then a Markoff map* ϕ *is called a* geometric root for $\boldsymbol{\nu}$ *if* $(\rho, \boldsymbol{\nu})$ *is a good labeled representation and* ϕ *is upward at every triangle of* $\Sigma(\boldsymbol{\nu})$, *where* ρ *is the type-preserving representation induced by* ϕ. *If* $\boldsymbol{\nu}$ *is thin, then we require that* ϕ *is upward at each of the two triangles of* $\mathcal{D}$ *which contain the edge that contains* $\nu^{\pm}$.

Lemma 9.1.3. *(1) Let* σ^{ϵ} *be a triangle of* $\mathcal{D}$ *and* ν^{ϵ} *be a point in* $\sigma^{\epsilon} \cap \mathbb{H}^2$ *for each* $\epsilon \in \{-,+\}$. *Then every geometric root for* $\boldsymbol{\nu} = (\nu^-,\nu^+)$ *is an algebraic root for* $((\nu^-,\sigma^-), (\nu^+,\sigma^+))$.

(2) Let $(\rho, \boldsymbol{\nu})$ *with* $\boldsymbol{\nu} = (\nu^-,\nu^+)$ *be a good labeled representation. Then there is a unique Markoff map* ϕ *inducing* ρ *which is a geometric root for* $\boldsymbol{\nu}$.

Proof. (1) Let ϕ be a geometric root for $\boldsymbol{\nu} = (\nu^-,\nu^+)$ and ρ the type-preserving representation induced by ϕ. Then $(\rho, \boldsymbol{\nu})$ is a good labeled representation and hence it satisfies the condition Frontier (Definition 6.1.4). Thus ν^{ϵ} is equal to the point $\theta^{\epsilon}(\rho, \sigma^{\epsilon}) \in \sigma^{\epsilon}$. (Here we use Lemmas 4.5.3, 4.5.4 and 4.5.6 when ν^{ϵ} is non-generic and σ^{ϵ} does not belong to $\Sigma(\boldsymbol{\nu})$.) Moreover ϕ is upward at σ^{ϵ}. Hence ϕ is an algebraic root for $((\nu^-,\sigma^-), (\nu^+,\sigma^+))$ by Lemma 4.2.18.

(2) Let $(\rho, \boldsymbol{\nu})$ be a good labeled representation. Then $(\rho, \boldsymbol{\nu})$ is quasifuchsian by Theorem 6.1.8 and $\mathcal{L}(\rho, \boldsymbol{\nu})$ is simple by Proposition 6.6.1. Suppose first that $\boldsymbol{\nu}$ is thick. Let ϕ be the unique Markoff map inducing ρ which is upward at $\sigma^-(\boldsymbol{\nu})$ (Lemma 3.1.4). Then ϕ is upward at every triangle of $\Sigma(\boldsymbol{\nu})$ by Lemma 4.5.2. Thus ϕ is a geometric root for $\boldsymbol{\nu}$, and hence it is the unique geometric root for $\boldsymbol{\nu}$ inducing the given representation ρ. Next, suppose that $\boldsymbol{\nu}$ is thin. Let τ be the edge containing $\nu^{\pm}$ and σ and σ' be the triangles containing τ. Then we see that $\mathcal{L}(\rho, \sigma)$ and $\mathcal{L}(\rho, \sigma')$ are simple, and there is a unique Markoff map ϕ inducing ρ which is upward at σ and σ' by Lemma 4.5.6. Then ϕ is a geometric root for $\boldsymbol{\nu}$, and hence it is the unique geometric root for $\boldsymbol{\nu}$ inducing the given representation ρ.

If $\sigma = \langle s_0, s_1, s_2 \rangle$ is a triangle of $\mathcal{D}$, then the correspondence

$$\Phi \ni \phi \mapsto (\phi(s_0), \phi(s_1), \phi(s_2)) \in \mathbb{C}^3$$

gives an identification of Φ with the affine algebraic variety defined by $x^2 + y^2 + z^2 = xyz$. Repeated application of Proposition 2.3.4(2) shows that this identification does not depend on the choice of σ modulo post-composition by a biregular automorphism of $\mathbb{C}^3$. It also follows that $\zeta^{\epsilon}_{\nu,\sigma} : \Phi \to \mathbb{C}$ is a polynomial function. Thus we may identify the set $\Phi^{\epsilon}_{\nu,\sigma}$ in Definition 9.1.1 with the following subvariety of Φ.

$$\begin{aligned}\Phi^{\epsilon}_{\nu,\sigma} &= \{(x,y,z) \in \mathbb{C}^3 \mid x^2+y^2+z^2 = xyz,\ x+\alpha^{\epsilon}y+\beta^{\epsilon}z = 0\} \\ &= \{(x,y,z) \in \mathbb{C}^3 \mid F^{\epsilon}_{\nu,\sigma}(y,z) = 0,\ x = -(\alpha^{\epsilon}y+\beta^{\epsilon}z)\} \\ &\cong \{(y,z) \in \mathbb{C}^2 \mid F^{\epsilon}_{\nu,\sigma}(y,z) = 0\},\end{aligned}$$

where $\alpha^{\epsilon} = \epsilon i \exp(\epsilon i\theta_2)$ and $\beta^{\epsilon} = -\epsilon i \exp(-\epsilon i\theta_1)$ and

$$F^{\epsilon}_{\nu,\sigma}(y,z) = y^2+z^2+(\alpha^{\epsilon}y+\beta^{\epsilon}z)^2+yz(\alpha^{\epsilon}y+\beta^{\epsilon}z).$$

The prime factorization of the defining polynomial $F^{\epsilon}_{\nu,\sigma}$ of $\Phi^{\epsilon}_{\nu,\sigma}$ is given by the following lemma.

Lemma 9.1.4. *The polynomial $F^{\epsilon}_{\nu,\sigma}(y,z)$ is irreducible if and only if $\nu \in \operatorname{int}\sigma$. If $\nu \in \partial\sigma$, then the prime factorization of $F^{\epsilon}_{\nu,\sigma}(x,y)$ is given by the following formula.*

1. *If $\nu \in \operatorname{int}\langle s_0, s_1\rangle$, then*

$$F^{\epsilon}_{\nu,\sigma}(y,z) = z\{(\beta^{\epsilon}y+(\beta^{\epsilon})^2+1)z+\epsilon iy(y+2\beta^{\epsilon})\}.$$

2. *If $\nu \in \operatorname{int}\langle s_1, s_2\rangle$, then*

$$F^{\epsilon}_{\nu,\sigma}(y,z) = \{z-\epsilon iy\}\{(\beta^{\epsilon}y+(\beta^{\epsilon})^2+1)z+\epsilon i(1-(\beta^{\epsilon})^2)y\}.$$

3. *If $\nu \in \operatorname{int}\langle s_2, s_0\rangle$, then*

$$F^{\epsilon}_{\nu,\sigma}(y,z) = y\{(\alpha^{\epsilon}z+(\alpha^{\epsilon})^2+1)y-\epsilon iz(z+2\alpha^{\epsilon})\}.$$

Before proving this lemma, we make the following simple observation.

Sublemma 9.1.5. *(1) If $\nu \in \operatorname{int}\sigma$, then none of α^{ϵ}, $\alpha^{\epsilon}\bar{\beta^{\epsilon}}$ and β^{ϵ} is equal to $\pm i$.*

(2) If $\nu \in \partial\sigma$, then we have the following equivalences.

1. $\nu \in \operatorname{int}\langle s_0, s_1\rangle \iff \alpha^{\epsilon} = \epsilon i \iff \alpha^{\epsilon} = \pm i$.
2. $\nu \in \operatorname{int}\langle s_1, s_2\rangle \iff \alpha^{\epsilon} = -\epsilon i\beta^{\epsilon} \iff \alpha^{\epsilon} = \pm i\beta^{\epsilon}$.
3. $\nu \in \operatorname{int}\langle s_2, s_0\rangle \iff \beta^{\epsilon} = -\epsilon i \iff \beta^{\epsilon} = \pm i$.

Proof. Since (1) is a consequence of (2), we prove (2). The first and the last assertions are direct consequence of the defining identities $\alpha^{\epsilon} = \epsilon i \exp(\epsilon i\theta_2)$ and $\beta^{\epsilon} = -\epsilon i \exp(-\epsilon i\theta_1)$ and the facts that $\theta_j \in [0, \pi/2]$ for each j. To see the second assertion, note that $\nu \in \operatorname{int}\langle s_1, s_2\rangle$ if and only if $\theta_1+\theta_2 = \pi/2$. Since $\alpha^{\epsilon}\bar{\beta^{\epsilon}} = -\exp(\epsilon i(\theta_1+\theta_2))$, this is equivalent to the condition $\alpha^{\epsilon}\bar{\beta^{\epsilon}} = -\epsilon i$. Thus we obtain the second assertion.

Proof (Proof of Lemma 9.1.4). Throughout the proof, we abbreviate $F^{\epsilon}_{\nu,\sigma}$, α^{ϵ} and β^{ϵ}, respectively, as F, α and β. Note that

$$\begin{aligned} F(y,z) &= y^2 + z^2 + (\alpha y + \beta z)^2 + yz(\alpha y + \beta z) \\ &= (1 + \beta^2 + \beta y)z^2 + (2\alpha\beta y + \alpha y^2)z + (1+\alpha^2)y^2 \\ &= (1 + \alpha^2 + \alpha z)y^2 + (2\alpha\beta z + \beta z^2)y + (1+\beta^2)z^2. \end{aligned}$$

Suppose that F has a nontrivial factorization $F = F_1F_2$. Since $\deg_z F = 2$, the pair $(\deg_z F_1, \deg_z F_2)$ is equal to $(1,1)$ or $(2,0)$ up to permutation of indices.

Case 1. $(\deg_z F_1, \deg_z F_2) = (1,1)$. Note that the discriminant of $F(y,z)$ as a polynomial in the variable z is equal to

$$\alpha^2y^2(2\beta+y)^2 - 4y^2(1+\alpha^2)(1+\beta^2+\beta y) = y^2\left\{\alpha^2y^2 - 4\beta y - 4(\alpha^2+\beta^2+1)\right\}.$$

By the assumption, it must be equal to the square of a polynomial in $\mathbb{C}[y]$. Hence, the discriminant of its second factor vanishes, i.e.,

$$(2\beta)^2 + 4\alpha^2(\alpha^2+\beta^2+1) = 4(\alpha^2+\beta^2)(\alpha^2+1) = 0.$$

Hence $\alpha = \pm i\beta$ or $\alpha = \pm i$. By Sublemma 9.1.5, $\alpha = \pm i\beta$ if and only if $\nu \in \operatorname{int}\langle s_1, s_2\rangle$. Moreover if this holds, then $\alpha = -\epsilon i\beta$ and

$$F(y,z) = \{z - \epsilon i y\}\{(\beta y + \beta^2 + 1)z + \epsilon i(1-\beta^2)y\}.$$

By Sublemma 9.1.5, the condition $\alpha = \pm i$ is equivalent to the condition $\nu \in \operatorname{int}\langle s_0, s_1\rangle$. Moreover, if this condition is satisfied, then $\alpha = \epsilon i$ and

$$F(y,z) = z\{(\beta y + \beta^2 + 1)z + \epsilon i y(y + 2\beta)\}.$$

Case 2. $(\deg_z F_1, \deg_z F_2) = (2,0)$. Then $(\deg_y F_1, \deg_y F_2)$ is equal to $(1,1)$ or $(0,2)$, because F_2 is not a constant. Suppose that $(\deg_y F_1, \deg_y F_2) = (1,1)$. Then we see, by an argument similar to the above, that ν lies in either $\langle s_1, s_2\rangle$ or $\langle s_2, s_0\rangle$. Conversely, if this condition is satisfied, then $F(x,y)$ has the non-trivial factorization explained in the lemma. Suppose $(\deg_y F_1, \deg_y F_2) = (0,2)$. Then F_1 and F_2, respectively, are elements of $\mathbb{C}[z]$ and $\mathbb{C}[y]$ of degree 2. Thus F_2 must divide the three polynomials,

$$1 + \beta^2 + \beta y, \quad \alpha y(2\beta + y), \quad (1+\alpha^2)y^2.$$

This is a contradiction, because the degree in y of the first one is 1. Hence, we see that $(\deg_z F_1, \deg_z F_2) = (2,0)$ cannot occur. Thus we have proved that F is irreducible if and only if $\nu \in \operatorname{int}\sigma$.

To complete the proof, we need to show that the second factor in each of the factorizations of F for the case when $\nu \in \partial\sigma$ is actually irreducible. We shall prove this only for the first case. Namely, we show that

$$G(y,z) := (\beta y + \beta^2 + 1)z + \epsilon i y(y + 2\beta)$$

is irreducible if $\nu \in \mathrm{int}\langle s_0, s_1\rangle$. Suppose this has a nontrivial factorization $G = G_1 G_2$. Since $\deg_z G = 1$, we may assume that $\deg_z G_1 = 0$. Then G_1 is an element of $\mathbb{C}[z]$ and hence it must be a common factor of the polynomials $\beta y + \beta^2 + 1$ and $y(y + 2\beta)$. Since $\nu \in \mathrm{int}\langle s_0, s_1\rangle$, we have $\theta_1 \neq 0$ and hence $\beta^2 + 1 \neq 0$ (cf. Sublemma 9.1.5). So, we see that the two polynomials $\beta y + \beta^2 + 1$ and $y + 2\beta$ coincide up to multiplication of a non-zero complex number. This implies that $\beta^2 = 1$ and hence $\theta_1 = \pi/2$, or equivalently, $\nu = s_1$, a contradiction. The remaining cases can be treated similarly.

As a corollary to Lemmas 9.1.4, we obtain the following.

Corollary 9.1.6. *Let $\sigma = \langle s_0, s_1, s_2\rangle$ be a triangle of $\mathcal{D}$ and $\nu = (\theta_0, \theta_1, \theta_2)$ be a point in $\sigma \cap \mathbb{H}^2$.*

(1) If $\nu \in \mathrm{int}\,\sigma$, then $\Phi^\epsilon_{\nu,\sigma}$ is irreducible.

(2) If $\nu \in \partial\sigma$, then $\Phi^\epsilon_{\nu,\sigma}$ has precisely two irreducible components, $\check{\Phi}^\epsilon_{\nu,\sigma}$ and $\Psi^\epsilon_{\nu,\sigma}$, which are defined as follows.

1. *If $\nu \in \mathrm{int}\langle s_0, s_1\rangle$, then*

$$\begin{aligned}\check{\Phi}^\epsilon_{\nu,\sigma} &:= \{(x,y,z) \mid (\beta^\epsilon y + (\beta^\epsilon)^2 + 1)z + \epsilon i y(y + 2\beta^\epsilon) = 0,\\ &\qquad x = -(\epsilon i y + \beta^\epsilon z)\},\\ \Psi^\epsilon_{\nu,\sigma} &:= \{(x,y,z) \mid z = 0,\ x = -(\epsilon i y + \beta^\epsilon z)\} = \{(x, \epsilon i x, 0) \mid x \in \mathbb{C}\}.\end{aligned}$$

2. *If $\nu \in \mathrm{int}\langle s_1, s_2\rangle$, then*

$$\begin{aligned}\check{\Phi}^\epsilon_{\nu,\sigma} &:= \{(x,y,z) \mid (\beta^\epsilon y + (\beta^\epsilon)^2 + 1)z + \epsilon i(1 - (\beta^\epsilon)^2)y = 0,\\ &\qquad x = -(\alpha^\epsilon y + \beta^\epsilon z)\},\\ \Psi^\epsilon_{\nu,\sigma} &:= \{(x,y,z) \mid z - \epsilon i y = 0,\ x = -(\alpha^\epsilon y + \beta^\epsilon z)\} = \{(0, y, \epsilon i y) \mid y \in \mathbb{C}\}.\end{aligned}$$

3. *If $\nu \in \mathrm{int}\langle s_2, s_0\rangle$, then*

$$\begin{aligned}\check{\Phi}^\epsilon_{\nu,\sigma} &:= \{(x,y,z) \mid (\alpha^\epsilon z + (\alpha^\epsilon)^2 + 1)y - \epsilon i z(z + 2\alpha^\epsilon) = 0,\\ &\qquad x = -(\alpha^\epsilon y - \epsilon i z)\},\\ \Psi^\epsilon_{\nu,\sigma} &:= \{(x,y,z) \mid y = 0,\ x = -(\alpha^\epsilon y + \beta^\epsilon z)\} = \{(\epsilon i z, 0, z) \mid z \in \mathbb{C}\}.\end{aligned}$$

Here $\alpha^\epsilon = \epsilon i \exp(\epsilon i \theta_2)$ and $\beta^\epsilon = -\epsilon i \exp(-\epsilon i \theta_1)$, and (x,y,z) represents the point $\phi \in \Phi$ such that $(x,y,z) = (\phi(s_0), \phi(s_1), \phi(s_2))$.

Remark 9.1.7. In the case when $\nu \in \partial\sigma$, the component $\Psi^\epsilon_{\nu,\sigma}$ of $\Phi^\epsilon_{\nu,\sigma}$ is distinguished from the component $\check{\Phi}^\epsilon_{\nu,\sigma}$ by the property that $\Psi^\epsilon_{\nu,\sigma}$ is contained in the subvariety $\{\phi \in \Phi \mid \phi(s) = 0\}$, where s is the vertex of σ which is not a vertex of the edge containing ν. In particular, no element of $\Psi^\epsilon_{\nu,\sigma}$ induces a quasifuchsian representation.

Lemma 4.5.3 suggests the following lemma, which shows that the subvariety $\check{\Phi}^{\epsilon}_{\nu,\sigma}$ of Φ in Corollary 9.1.6(2) does not depend on the choice of σ, though $\Psi^{\epsilon}_{\nu,\sigma}$ does.

Lemma 9.1.8. *Let σ and σ' be triangles of $\mathcal{D}$ sharing an edge τ, and let ν be a point in* $\operatorname{int}\tau$. *Then* $\check{\Phi}^{\epsilon}_{\nu,\sigma} = \check{\Phi}^{\epsilon}_{\nu,\sigma'} \subset \Phi$.

Proof. We may assume $\sigma = \langle s_0, s_1, s_2\rangle$, $\sigma' = \langle s_0', s_1', s_2'\rangle$, $\tau = \langle s_1, s_2\rangle = \langle s_2', s_1'\rangle$, where $s_1' = s_2$ and $s_2' = s_1$. Then the coordinates of ν in σ and σ', respectively, are of the form $(0, \theta_1, \theta_2)$ and $(0, \theta_1', \theta_2')$ where $\theta_1' = \theta_2$ and $\theta_2' = \theta_1$. We identify Φ with the affine algebraic variety defined by $x^2 + y^2 + z^2 = xyz$ through the correspondence

$$\Phi \ni \phi \mapsto (\phi(s_0), \phi(s_1), \phi(s_2)) \in \mathbb{C}^3.$$

Then $\check{\Phi}^{\epsilon}_{\nu,\sigma}$ is the subvariety of $\Phi \subset \mathbb{C}^3$ determined by the equations

$$G(y, z) = 0,\ x = -(\alpha y + \beta z),$$

where $G(y, z) = (\beta y + \beta^2 + 1)z + \epsilon i(1 - \beta^2)y$, $\alpha = \epsilon i \exp(\epsilon i\theta_2) = -\epsilon i\beta$ and $\beta = -\epsilon i \exp(-\epsilon i\theta_1)$. On the other hand, $\Phi^{\epsilon}_{\nu,\sigma'}$ is the subvariety of Φ determined by the equation $\phi(s_0') + \alpha'\phi(s_1') + \beta'(s_2') = 0$, where $\alpha' = \epsilon i \exp(\epsilon i\theta_2') = \bar{\beta}$ and $\beta' = -\epsilon i \exp(-\epsilon i\theta_1') = \epsilon i\bar{\beta}$. Since $(\phi(s_0'), \phi(s_1'), \phi(s_2')) = (yz - x, z, y)$, $\Phi^{\epsilon}_{\nu,\sigma'}$ is determined by the equations

$$(yz - x)^2 + y^2 + z^2 - (yz - x)yz = 0,\ (yz - x) + \bar{\beta}z + \epsilon i\bar{\beta}y = 0. \tag{9.1}$$

Let $H(y, z)$ be the polynomial obtained from the first polynomial by putting $x = yz + \bar{\beta}z + \epsilon i\bar{\beta}y$. Then it has the prime factorization

$$H(x, y) = \bar{\beta}^2(z + \epsilon iy)G(y, z).$$

Thus $\Phi^{\epsilon}_{\nu,\sigma'}$ has two irreducible components corresponding to the factors $z + \epsilon iy$ and $G(y, z)$ respectively. However the component corresponding to the factor $z + \epsilon iy$ is contained in the subvariety $\phi(s_0') = 0$ and hence it is equal to the component $\Psi^{\epsilon}_{\nu,\sigma'}$ by Remark 9.1.7. Hence $\check{\Phi}^{\epsilon}_{\nu,\sigma'}$ is equal to the component corresponding to the factor $G(y, z)$. The equation $G(y, z) = 0$ implies $yz = -\bar{\beta}\{(\beta^2 + 1)z + \epsilon i(1 - \beta^2)y\}$, which in tern implies that the second equation in (9.1) is equivalent to the equation $x = -(\alpha y + \beta z)$. Hence we have $\check{\Phi}^{\epsilon}_{\nu,\sigma'} = \check{\Phi}^{\epsilon}_{\nu,\sigma}$.

Remark 9.1.9. By the above proof, we see that $\Psi^{\epsilon}_{\nu,\sigma'}$ is the subvariety of $\Phi \subset \mathbb{C}^3$ determined by the equations $z + \epsilon iy = 0$ and $x = \epsilon iz^2$. On the other hand, $\Psi^{\epsilon}_{\nu,\sigma}$ is determined by the equations $z + \epsilon iy = 0$ and $x = 0$. Thus $\Psi^{\epsilon}_{\nu,\sigma}$ and $\Psi^{\epsilon}_{\nu,\sigma'}$ share the same equation $z + \epsilon iy = 0$, but they are not identical. In fact every nontrivial Markoff map $\phi \in \Psi^{\epsilon}_{\nu,\sigma}$ has the properties $\phi(s_0) = 0$ and $\phi(s_0') \neq 0$, whereas every nontrivial Markoff map $\phi' \in \Psi^{\epsilon}_{\nu,\sigma'}$ has the properties $\phi'(s_0) \neq 0$ and $\phi(s_0') = 0$,

By Lemma 9.1.8, we may introduce the following definition.

Definition 9.1.10. *For each $\nu \in \mathbb{H}^2$ and $\epsilon = \pm$, let $\check{\Phi}^\epsilon_\nu$ be the subvariety of Φ defined as follows.*

1. *If ν lies in the interior of a triangle σ of $\mathcal{D}$, then $\check{\Phi}^\epsilon_\nu := \Phi^\epsilon_{\nu,\sigma}$.*
2. *If ν lies in the interior of an edge τ of $\mathcal{D}$, then $\check{\Phi}^\epsilon_\nu := \check{\Phi}^\epsilon_{\nu,\sigma} = \check{\Phi}^\epsilon_{\nu,\sigma'}$, where σ and σ' are the triangles of $\mathcal{D}$ sharing the edge τ.*

Then we have the following lemma.

Lemma 9.1.11. *The singular set of the irreducible variety $\check{\Phi}^\epsilon_\nu$ is as follows:*

1. *If ν lies in the interior of a triangle of $\mathcal{D}$, then $\check{\Phi}^\epsilon_\nu$ is singular only at the trivial Markoff map.*
2. *If ν lies in the interior of an edge of $\mathcal{D}$, say $\nu = (\theta_0, \theta_1, 0) \in \sigma = \langle s_0, s_1, s_2 \rangle$, then the trivial Markoff map and the Markoff map ϕ such that $(\phi(s_0), \phi(s_1), \phi(s_2)) = (-2\epsilon i\beta^\epsilon, -2\beta^\epsilon, 0)$ are the only singular points of $\check{\Phi}^\epsilon_\nu$. Here $\beta^\epsilon = -\epsilon i \exp(-\epsilon i \theta_1)$.*

Proof. (1) Suppose that ν lies in the interior of a triangle σ of $\mathcal{D}$. Then $\check{\Phi}^\epsilon_\nu = \Phi^\epsilon_{\nu,\sigma}$ is defined by the polynomial

$$F(y,z) = F^\epsilon_{\nu,\sigma}(y,z) = y^2 + z^2 + (\alpha y + \beta z)^2 + yz(\alpha y + \beta z),$$

where $\alpha = \alpha^\epsilon$ and $\beta = \beta^\epsilon$ are as in Definition 9.1.1. We can see

$$F - \frac{y}{2}\frac{\partial F}{\partial y} - \frac{z}{2}\frac{\partial F}{\partial z} = -\frac{yz}{2}(\alpha y + \beta z).$$

Let (y_0, z_0) be a singular point of $\check{\Phi}^\epsilon_\nu$, that is,

$$F(y_0, z_0) = \frac{\partial F}{\partial y}(y_0, z_0) = \frac{\partial F}{\partial z}(y_0, z_0) = 0.$$

Then we have

$$-\frac{y_0 z_0}{2}(\alpha y_0 + \beta z_0) = 0.$$

So, one of $\alpha y_0 + \beta z_0$, y_0 and z_0 is equal to 0. On the other hand, since $\nu \in \operatorname{int}\sigma$, Sublemma 9.1.5 implies that none of α, β, and α/β are equal to $\pm i$. Through elementary calculation by using this fact. we obtain $(y_0, z_0) = (0,0)$. Hence, the origin is the unique singular point of $\check{\Phi}^\epsilon_\nu$.

(2) Suppose that $\nu = (\theta_0, \theta_1, 0) \in \sigma = \langle s_0, s_1, s_2 \rangle$. Then $\check{\Phi}^\epsilon_\nu$ is defined by

$$G(y,z) := (\beta y + \beta^2 + 1)z + \epsilon i y(y + 2\beta),$$

where $\beta = \beta^\epsilon = -\epsilon i \exp(-\epsilon i \theta_1)$ (see Corollary 9.1.6). Note that

$$G - z\frac{\partial G}{\partial z} = \epsilon i y(y + 2\beta).$$

Let (y_0, z_0) be a singular point of $\check{\Phi}^{\epsilon}_{\nu}$. Then we have either $y_0 = 0$ or $y_0 = -2\beta$. On the other hand, since ν lies in the interior of an edge of $\mathcal{D}$, we have $0 < \theta_1 < \pi/2$ and hence $\beta = -\epsilon i \exp(-\epsilon i \theta_1) \neq \pm 1$. By using this fact and the identity $G(y_0, z_0) = 0$, we see (y_0, z_0) is equal to $(0, 0)$ or $(-2\beta, 0)$.

Lemma 9.1.12. *Let $\boldsymbol{\rho} = (\rho, \boldsymbol{\nu})$ with $\boldsymbol{\nu} = (\nu^-, \nu^+)$ be a good labeled representation, and let ϕ be the geometric root for $\boldsymbol{\nu}$ inducing ρ. Then the following hold.*

1. *If ν^ϵ is non-generic and σ^ϵ is a triangle containing ν^ϵ, then ϕ is not contained in the algebraic variety $\Psi^{\epsilon}_{\nu,\sigma^\epsilon}$.*
2. *ϕ is a smooth point of the algebraic variety $\check{\Phi}^{\epsilon}_{\nu^\epsilon}$ for each $\epsilon = \pm$.*

Proof. Let σ^ϵ be a triangle containing σ^ϵ. Then $\phi \in \Phi^{\epsilon}_{\nu^\epsilon,\sigma^\epsilon}$ by Lemma 9.1.3. Suppose that ν^ϵ is non-generic, i.e., $\nu^\epsilon \in \partial\sigma^\epsilon$. Since $\boldsymbol{\rho}$ is good and hence ρ is quasifuchsian (Theorem 6.1.8), we see that ϕ cannot be contained in the component $\Psi^{\epsilon}_{\nu,\sigma^\epsilon}$ by Remark 9.1.7. Hence ϕ is contained in $\check{\Phi}^{\epsilon}_{\nu^\epsilon}$. Moreover Lemma 9.1.11 implies that ϕ is a smooth point of $\check{\Phi}^{\epsilon}_{\nu^\epsilon}$. Suppose next that $\nu^\epsilon \in \operatorname{int}\sigma^\epsilon$. Then $\check{\Phi}^{\epsilon}_{\nu^\epsilon} = \Phi^{\epsilon}_{\nu^\epsilon,\sigma^\epsilon}$ and hence $\phi \in \check{\Phi}^{\epsilon}_{\nu^\epsilon}$. Since ϕ is nontrivial (because $\boldsymbol{\rho}$ is good), it is a smooth point of $\check{\Phi}^{\epsilon}_{\nu^\epsilon}$ by Lemma 9.1.11.

Next, we prove the following proposition.

Proposition 9.1.13. *Let $\sigma^\epsilon = \langle s^\epsilon_0, s^\epsilon_1, s^\epsilon_2\rangle$ be a triangle of $\mathcal{D}$, and let $\nu^\epsilon = (\theta^\epsilon_0, \theta^\epsilon_1, \theta^\epsilon_2)$ be a point in $\sigma^\epsilon \cap \mathbb{H}^2$ for each $\epsilon \in \{-, +\}$. Then $\Phi^{-}_{\nu^-,\sigma^-}$ and $\Phi^{+}_{\nu^+,\sigma^+}$ do not share a common component.*

We begin the proof of the proposition, by proving the following lemma.

Lemma 9.1.14. *Under the assumption of Proposition 9.1.13, let Φ_0 be the subvariety of Φ defined by*

$$\Phi_0 = \{\phi \in \Phi \,|\, (\phi(s^+_0), \phi(s^+_1), \phi(s^+_2)) = (0, x, ix) \text{ for some } x \in \mathbb{C}\} \cong \mathbb{C}.$$

Then the restriction of $\zeta^-_{\nu^-,\sigma^-}$ to Φ_0 is a polynomial in the variable $x = \phi(s^+_1)$ of positive degree. Moreover, it is not a monomial except when

1. *$\sigma^- = \sigma^+$ or*
2. *$\sigma^- \cap \sigma^+$ is an edge of $\mathcal{D}$ and its interior contains ν^-.*

Proof. After a coordinate change, we may assume that $\langle s^+_0, s^+_1, s^+_2\rangle = \langle 1/0, 0/1, 1/1\rangle$ and that all vertices of σ^- are contained in $[0,1] \cup \{1/0\}$.

Suppose first that $\sigma^- = \sigma^+$. Then for each $\phi \in \Phi_0$,

$$\zeta^-_{\nu^-,\sigma^-}(\phi) = 0 + \alpha^- x + \beta^-(ix) = (\alpha^- + i\beta^-)x,$$

where $\alpha^- = -i\exp(-i\theta^-_2)$ and $\beta^- = i\exp(i\theta^-_1)$. Since $\alpha^- + i\beta^- \neq 0$ by Sublemma 9.1.5, the above polynomial is a monomial of degree 1.

Suppose that $\sigma^- \cap \sigma^+$ is an edge of $\mathcal{D}$, namely $\sigma^- = \langle 1/2, 1/1, 0/1 \rangle$. Then for each $\phi \in \Phi_0$, $(\phi(1/2), \phi(1/1), \phi(1/0)) = (ix^2, ix, x)$ and hence

$$\zeta^-_{\nu^-,\sigma^-}(\phi) = ix^2 + \alpha^-(ix) + \beta^- x = ix(x + (\alpha^- - i\beta^-)).$$

By Sublemma 9.1.5, $\alpha^- - i\beta^- = 0$ if and only if $\nu^- \in \langle 0/1, 1/1 \rangle$, namely the second condition in Lemma 9.1.14 holds. Hence the desired result holds.

Suppose $\sigma^- \neq \sigma^+$ and $\sigma^- \cap \sigma^+$ is not an edge. Then, by virtue of Lemma 5.3.12(1), we may further assume that all vertices of σ^- are contained in the interval $[0, 1/2]$. Then we may assume $\sigma^- = \langle (q_1+q_2)/(p_1+p_2), q_1/p_1, q_2/p_2 \rangle$ where p_1, p_2, q_1 and q_2 are integers such that $0 \leq q_j \leq p_j$ $(j = 1, 2)$, $1 \leq p_1 \leq p_2$ and $\begin{vmatrix} q_1 & q_2 \\ p_1 & p_2 \end{vmatrix} = \pm 1$. Then by Lemma 5.3.12, for every $\phi \in \Phi_0$, we have

$$\begin{aligned} \zeta^-_{\nu^-,\sigma^-}(\phi) &= V[(q_1+q_2)/(p_1+p_2)](x) + \alpha^- V[q_1/p_1](x) + \beta^- V[q_2/p_2](x) \\ &= i^{q_1+q_2}(x^{p_1+p_2} - cx^{p_1+p_2-2} + \text{(lower terms)}) \\ &\quad + i^{q_1}\alpha^-(x^{p_1} + \text{(lower terms)}) + i^{q_2}\beta^-(x^{p_2} + \text{(lower terms)}), \end{aligned}$$

where $c = c_{(q_1+q_2)/(p_1+p_2)}$ is a positive integer. Thus $\zeta^-_{\nu^-,\sigma^-}(\phi)$ is a polynomial in x of positive degree $p_1 + p_2$. In the following we show that it is not a monomial.

If $p_1 \geq 3$, then both p_1 and p_2 are smaller than $p_1 + p_2 - 2$ and hence the coefficient of $x^{p_1+p_2-2}$ is equal to $-i^{q_1+q_2}c \neq 0$. So the polynomial is not a monomial.

If $p_1 = 1$, then $q_1 = 0$ and $p_2 = p_1 + p_2 - 1$, and therefore the coefficient of $x^{p_1+p_2-1}$ is $i^{q_2}\beta^- \neq 0$. Thus the polynomial is not a monomial.

If $p_1 = 2$, then $q_1 = 1$ and $q_2/p_2 = n/(2n+1)$ for some integer $n \geq 1$. Thus $p_1 < p_1 + p_2 - 2$ and $p_2 = p_1 + p_2 - 2$, and therefore the coefficient of $x^{p_1+p_2-2} = x^{p_2}$ is equal to

$$-i^{q_1+q_2}c + i^{q_2}\beta^- = -i^{q_1+q_2}(c - i^{-q_1}\beta^-) = -i^{1+q_2}(c + i\beta^-).$$

Hence the polynomial is not a monomial if $\beta^- \neq i$. Since $c = c_{(n+1)/(2n+3)} = 1$ by Lemma 5.3.14, the coefficient of $x^{p_1+p_2-2}$ vanishes if $\beta^- = i$. So we study the coefficient of $x^{p_1+p_2-4} = x^{2n-1}$ in case $\beta^- = i$. By Lemma 5.3.14, we see

$$\begin{aligned} \zeta^-_{\nu^-,\sigma^-}(\phi) &= V[(n+1)/(2n+3)](x) + \alpha^- V[1/2](x) + \beta^- V[n/2n+1](x) \\ &= i^{n+1}(x^{2n+3} - x^{2n+1} - nx^{2n-1} + \text{(lower terms)}) \\ &\quad + i\alpha^- x^2 + i^n\beta^-(x^{2n+1} - x^{2n-1} + \text{(lower terms)}). \end{aligned}$$

If $\beta^- = i$, then the coefficient of $x^{p_1+p_2-4} = x^{2n-1}$ is equal to

$$-i^{n+1}n - i^n\beta^- = -i^{n+1}(n+1) \neq 0.$$

Hence the polynomial is not a monomial in this case, too. This completes the proof of Lemma 9.1.14.

By the fundamental theorem of algebra, we obtain the following corollary, which is used in the proof of Lemma 9.1.16.

Corollary 9.1.15. *Under the assumption of Proposition 9.1.13, assume that $\sigma^- \neq \sigma^+$ and that $\nu^- \notin \sigma^- \cap \sigma^+$. Then there is a non-trivial Markoff map $\phi \in \Phi^-_{\nu^-,\sigma^-}$ such that $(\phi(s_0^+), \phi(s_1^+), \phi(s_2^+)) = (0, x, ix)$ for some $x \in \mathbb{C} - \{0\}$.*

Lemma 9.1.16. *Under the assumption of Proposition 9.1.13, the restriction of the function $\zeta^+_{\nu^+,\sigma^+}$ to $\Phi^-_{\nu^-,\sigma^-}$ is not a constant function.*

Proof. If $\sigma^- = \sigma^+$ or $\sigma^- \cap \sigma^+$ is an edge, then $\Phi^-_{\nu^-,\sigma^-} \cap \Phi^+_{\nu^+,\sigma^+}$ contains at most two non-trivial Markoff maps counted with multiplicity by Lemmas 5.2.12 and 5.2.14. So we obtain the desired result. Hence we may assume that $\sigma^- \neq \sigma^+$ and that $\sigma^- \cap \sigma^+$ is not an edge. Since the image of the trivial Markoff map by $\zeta^+_{\nu^+,\sigma^+}$ is 0, we have only to show that $\zeta^+_{\nu^+,\sigma^+}(\phi) \neq 0$ for some $\phi \in \Phi^-_{\nu^-,\sigma^-}$. To this end we use Corollary 9.1.15, which guarantees the existence of a non-trivial Markoff map $\phi \in \Phi^-_{\nu^-,\sigma^-}$ such that $(\phi(s_0^+), \phi(s_1^+), \phi(s_2^+)) = (0, x, ix)$ for some $x \in \mathbb{C} - \{0\}$. Then

$$\zeta^+_{\nu^+,\sigma^+}(\phi) = \phi(s_0^+) + \alpha^+\phi(s_1^+) + \beta^+\phi(s_2^+) = (\alpha^+ + i\beta^+)x,$$

where $\alpha^+ = i\exp(i\theta_2^+)$ and $\beta^+ = -i\exp(-i\theta_1^+)$. If $\nu^+ \notin \mathrm{int}\langle s_1^+, s_2^+\rangle$, we have $\alpha^+ + i\beta^+ \neq 0$ by Sublemma 9.1.5. Thus $\zeta^+_{\nu^+,\sigma^+}(\phi) \neq 0$. If $\nu^+ \in \mathrm{int}\langle s_1^+, s_2^+\rangle$, then we make use of another non-trivial Markoff map ϕ' such that $(\phi'(s_0^+), \phi'(s_1^+), \phi'(s_2^+)) = (x, ix, 0)$ for some $x \in \mathbb{C} - \{0\}$, whose existence is guaranteed by Corollary 9.1.15. Then

$$\zeta^+_{\nu^+,\sigma^+}(\phi') = \phi'(s_0^+) + \alpha^+\phi'(s_1^+) + \beta^+\phi'(s_2^+) = (1 + i\alpha^+)x.$$

Since $\nu^+ \in \mathrm{int}\langle s_1^+, s_2^+\rangle$ by the assumption, we have $\nu^+ \notin \mathrm{int}\langle s_0^+, s_1^+\rangle$. Thus $1 + i\alpha^+ \neq 0$ by Sublemma 9.1.5. So $\zeta^+_{\nu^+,\sigma^+}(\phi') \neq 0$. This completes the proof of Lemma 9.1.16.

Before proceeding to the next lemma, we prepare the following sublemma.

Sublemma 9.1.17. *Let $V\colon \hat{\mathbb{Q}} \to k(x)$ be a map to the rational function field $k(x)$ over a field k with the property that $V(s_0) + V(s_2') = V(s_0)V(s_2)$ for any pair of adjacent triangles $\langle s_0, s_1, s_2\rangle$ and $\langle s_0', s_1', s_2'\rangle$ of $\mathcal{D}$ with $s_0' = s_0$ and $s_1' = s_2$. Denote the degree of an element $f \in k(x)$ by $\deg f$, i.e., if $f = f_1/f_2$ where f_1, f_2 belong to the polynomial ring $k[x]$, then $\deg f = \deg f_1 - \deg f_2$. Suppose that $\deg V(1/0) \leq 1$ and that $\deg V(0/1) = \deg V(1/1) = 1$. Then $\deg V(q/p) = p$ for any pair of coprime integers (p, q) with $0 \leq q \leq p$.*

Proof. This is proved by an inductive argument using the following facts as in the proof of Lemma 5.3.14 (cf. [44, Proposition 3.1]).

- $\deg(fg) = \deg f + \deg g$ for any $f, g \in k(x)$.

- If $\deg f > \deg g$, then $\deg(f+g) = \deg f$.

Lemma 9.1.18. *Under the assumption of Proposition 9.1.13, suppose that $\nu^- \in \partial\sigma^-$. Then the restriction of the function $\zeta^+_{\nu^+,\sigma^+}$ to $\check{\Phi}^-_{\nu^-}$ is not a constant function. In particular, $\check{\Phi}^-_{\nu^-,\sigma^-}$ does not share a common component with $\Phi^+_{\nu^+,\sigma^+}$.*

Proof. As in the proof of Lemma 9.1.16, we may assume that $\sigma^- \neq \sigma^+$ and that $\sigma^- \cap \sigma^+$ is not an edge by Lemmas 5.2.12 and 5.2.14. We may also assume $\sigma^- = \langle 1/0, 0/1, 1/1 \rangle$, and identify Φ with the affine algebraic variety defined by $x^2+y^2+z^2 = xyz$ through the correspondence

$$\Phi \ni \phi \mapsto (x,y,z) := (\phi(1/0), \phi(0/1), \phi(1/1)) \in \mathbb{C}^3.$$

Case 1. σ^- is equal to the $(-)$-terminal triangle $\sigma^-(\boldsymbol{\nu})$ of $\boldsymbol{\nu}$. Then may assume that $\nu^- \in \langle 1/0, 0/1 \rangle$ and that all vertices of σ^+ belong to the interval $[0,1]$. (To be precise, all vertices of σ^+ belong to either $[0,1/2]$ or $[1/2,1]$.) Then $\check{\Phi}^-_{\nu^-}$ is determined by the equations

$$(\beta^- y + (\beta^-)^2 + 1)z - iy(y+2\beta^-) = 0, \quad x = iy - \beta^- z,$$

where $\beta^- = i\exp(i\theta_1^-)$ (see Corollary 9.1.6). Thus the parameter y gives a coordinate of the open set

$$U := \{(x,y,z) \in \check{\Phi}^-_{\nu^-} \mid \beta^- y + (\beta^-)^2 + 1 \neq 0\}$$

of $\check{\Phi}^-_{\nu^-}$. In fact, the equality $(\beta^- y + (\beta^-)^2 + 1)z - iy(y+2\beta^-) = 0$ implies

$$z = \frac{iy(y+2\beta^-)}{\beta^- y + (\beta^-)^2 + 1}.$$

Then it follows that

$$x = iy - \beta^- z = \frac{-iy(y+2\beta^- - 1)}{y + (\beta^- + (\beta^-)^{-1})}.$$

Now, for each $q/p \in \hat{\mathbb{Q}}$, let $V_{q/p} : \Phi \to \mathbb{C}$ be the function defined by $V_{q/p}(\phi) = \phi(q/p)$. Then the above equalities show that the restrictions of the functions $V_{0/1}(\phi) = x$, $V_{1/1}(\phi) = y$ and $V_{1/0}(\phi) = z$ to U are rational functions in y over $\mathbb{C}$ of degree 1. Thus, by Proposition 2.3.4(2) and Sublemma 9.1.17, it follows that the restriction of $V_{q/p}$ to U is a rational function in y of degree p for any pair of coprime integers (p,q) with $0 \le q < p$.

Since all vertices of σ^+ belong to the interval $[0,1]$ by the assumption, we may assume $\langle s_0^+, s_1^+, s_2^+ \rangle = \langle (q_1+q_2)/(p_1+p_2), q_1/p_1, q_2/p_2 \rangle$ where p_1, p_2, q_1 and q_2 are integers such that $0 \le q_j \le p_j$ $(j=1,2)$ and $\begin{vmatrix} q_1 & q_2 \\ p_1 & p_2 \end{vmatrix} = 1$. For each $\phi \in U$, we have

$$\begin{aligned}\zeta^+_{\nu^+,\sigma^+}(\phi) &= \phi(s_0^+) + \alpha^+\phi(s_1^+) + \beta^+\phi(s_2^+)\\ &= V_{(q_1+q_2)/(p_1+p_2)}(\phi) + \alpha^+ V_{q_1/p_1}(\phi) + \beta^+ V_{q_2/p_2}(\phi).\end{aligned}$$

On the other hand, since both p_1 and p_2 are positive, they are smaller than p_1+p_2. Hence, by the observation in the preceding paragraph, $\zeta^+_{\nu^+,\sigma^+} : \check{\Phi}^-_{\nu^-} \to \mathbb{C}$ is a rational function in y of degree $p_1+p_2 > 0$, and hence it is not a constant function.

Case 2. $\sigma^- \neq \sigma^-(\boldsymbol{\nu})$. Then may assume that $\sigma^- = \langle 1/0, 0/1, 1/1\rangle$, $\nu^- \in \langle 0/1, 1/1\rangle$, and all vertices of σ^+ belong to either $[0,1/2]$ or $[1/2,1]$. Then $\check{\Phi}^-_{\nu^-}$ is determined by the equations

$$(\beta^- y + (\beta^-)^2 + 1)z - i(1-(\beta^-)^2)y = 0, \quad x = -(\alpha^- y + \beta^- z),$$

where $\alpha^- = -i\exp(-i\theta_2^-) = i\beta^-$ and $\beta^- = i\exp(i\theta_1^-)$ (see Corollary 9.1.6). Thus, as in the previous case, the parameter y gives a coordinate of some open set of $\check{\Phi}^-_{\nu^-}$, and the parameters x and z are given by the following rational functions in y.

$$z = \frac{i(1-(\beta^-)^2)y}{\beta^- y + (\beta^-)^2 + 1}, \quad x = -(\alpha^- y + \beta^- z) = \frac{-i\beta^- y(\beta^- y + 2)}{\beta^- y + (\beta^-)^2 + 1}.$$

Thus

$$w := yz - x = \frac{-iy(y+2\beta^-)}{\beta^- y + (\beta^-)^2 + 1}.$$

Note that the parameters x, y, z and w are rational functions in y of degree 1, 1, 0 and 1, respectively.

Suppose that all vertices of σ^+ belong to $[0,1/2]$. Then apply a coordinate change so that the triangles σ^- and $\langle 0/1, 1/2, 1/1\rangle$, respectively, become $\langle -1/1, 0/1, 1/0\rangle$ and $\langle 0/1, 1/1, 1/0\rangle$ in the new coordinate. Then all vertices of σ^- belong to $[0,1]$, and the parameters y, z and w, respectively, represent the functions $\check{\Phi}^-_{\nu^-} \ni \phi \mapsto \phi(q/p) \in \mathbb{C}$ where $q/p = 0/1$, $1/0$ and $1/1$. Since they are rational functions in y of degree 1, 0 and 1, respectively, we can prove that $\zeta^+_{\nu^+,\sigma^+} : \check{\Phi}^-_{\nu^-} \to \mathbb{C}$ is a rational function in y of positive degree, by using Sublemma 9.1.17 as in the previous case. The same argument works in the case when all vertices of σ^+ belong to $[1/2,1]$.

This completes the proof of the main assertion of Lemma 9.1.18. The remaining assertion follows from the main assertion and the facts that $\check{\Phi}^-_{\nu^-,\sigma^-}$ is irreducible and $\Phi^+_{\nu^+,\sigma^+} = (\zeta^+_{\nu^+,\sigma^+})^{-1}(0)$.

Proof (Proof of Proposition 9.1.13). Suppose $\nu^- \in \operatorname{int}\sigma^-$. Then $\Phi^-_{\nu^-,\sigma^-}$ is irreducible by Corollary 9.1.6. Since $\Phi^+_{\nu^+,\sigma^+} = (\zeta^+_{\nu^+,\sigma^+})^{-1}(0)$ and since the restriction of $\zeta^+_{\nu^+,\sigma^+}$ to $\Phi^-_{\nu^-,\sigma^-}$ is not a constant function, this implies that $\Phi^-_{\nu^-,\sigma^-}$ and $\Phi^+_{\nu^+,\sigma^+}$ do not share a common component.

Suppose $\nu^- \in \partial\sigma^-$. Then $\Phi^-_{\nu^-,\sigma^-}$ consists of two irreducible components, $\check{\Phi}^-_{\nu^-,\sigma^-}$ and $\Psi^-_{\nu^-,\sigma^-}$, by Corollary 9.1.6. By Lemma 9.1.18, $\check{\Phi}^-_{\nu^-,\sigma^-}$ does not share a common component with $\Phi^+_{\nu^+,\sigma^+}$. Finally we show the other component $\Psi^-_{\nu^-,\sigma^-}$ has the same property. Since the argument is symmetric, we may prove that $\Psi^+_{\nu^+,\sigma^+}$ does not share a component with $\Phi^-_{\nu^-,\sigma^-}$ when $\nu^+ \in \partial\sigma^+$. We may assume $\nu^+ \in \langle s_1^+, s_2^+\rangle$ after a cyclic permutation of vertices. Then $\Psi^+_{\nu^+,\sigma^+}$ coincides with the variety Φ_0 in Lemma 9.1.14 (see Corollary 9.1.6). Thus the restriction of $\zeta^-_{\nu^-,\sigma^-}$ to $\Phi_0 = \Psi^+_{\nu^+,\sigma^+}$ is a non-constant function. Hence we obtain the desired result.

9.2 Unique existence of the geometric root

Throughout this section, σ^ϵ denotes a triangle of $\mathcal{D}$, and ν^ϵ denotes a point in $\sigma^\epsilon \cap \mathbb{H}^2$ for each $\epsilon \in \{-,+\}$. By Proposition 9.1.13, $\Phi^-_{\nu^-,\sigma^-} \cap \Phi^+_{\nu^+,\sigma^+}$ consists of finitely many points, and so is $\check{\Phi}^-_{\nu^-} \cap \check{\Phi}^+_{\nu^+}$. In particular, there are only finitely many geometric roots for $\boldsymbol{\nu} = (\nu^-, \nu^+)$, and they are smooth points of $\check{\Phi}^\epsilon_{\nu^\epsilon}$ for each $\epsilon = \pm$ (Lemma 9.1.12).

Definition 9.2.1. *The* geometric multiplicity, $d_G(\boldsymbol{\nu})$, of $\boldsymbol{\nu}$ *is defined to be the number of geometric roots for $\boldsymbol{\nu}$ counted with multiplicity. Namely, if $\{\phi_1, \cdots, \phi_k\}$ is the set of the geometric roots for $\boldsymbol{\nu}$, then*

$$d_G(\boldsymbol{\nu}) := \sum_{j=1}^{k} \mathrm{Int}(\phi_j, \check{\Phi}^-_{\nu^-} \cap \check{\Phi}^+_{\nu^+}),$$

where $\mathrm{Int}(\phi_j, \check{\Phi}^-_{\nu^-} \cap \check{\Phi}^+_{\nu^+})$ *denotes the intersection multiplicity at ϕ_j of $\check{\Phi}^-_{\nu^-}$ and $\check{\Phi}^+_{\nu^+}$ in Φ. (See for example [31] and [73] for the definition and basic facts concerning the intersection multiplicity.)*

By Lemma 9.1.12, the following hold.

Lemma 9.2.2. $d_G(\boldsymbol{\nu}) = \sum_{j=1}^k \mathrm{Int}(\phi_j, \Phi^-_{\nu^-,\sigma^-} \cap \Phi^+_{\nu^+,\sigma^+})$.

Proposition 6.2.6 (Unique realization) is a direct consequence of the following proposition.

Proposition 9.2.3. *The geometric multiplicity $d_G(\boldsymbol{\nu})$ is equal to 1 for every $\boldsymbol{\nu} \in \mathbb{H}^2 \times \mathbb{H}^2$.*

The remainder of this section is devoted to the proof of this proposition. We assume after a coordinate change that $\sigma^- = \langle 1/0, 0/1, 1/1\rangle$ and $\sigma^+ = \langle q_0/p_0, q_1/p_1, q_2/p_2\rangle$, where p_j and q_j are non-negative integers such that $q_j/p_j \in [0,1] \cup \{1/0\}$ $(j = 0,1,2)$ and $(p_0, q_0) = (p_1 + p_2, q_1 + q_2)$. We fix an affine embedding

$$\Phi \ni \phi \mapsto (x, y, z) := (\phi(1/0), \phi(0/1), \phi(1/1)) \in \mathbb{C}^3$$

and consider its projective completion $\widetilde{\Phi} \subset \boldsymbol{CP}^3$. Let $\widetilde{\Phi}^{\epsilon}_{\nu^{\epsilon},\sigma^{\epsilon}}$ be the subvariety of $\widetilde{\Phi}$ obtained by the projective completion of $\Phi^{\epsilon}_{\nu^{\epsilon},\sigma^{\epsilon}}$. We note that these projective completions depend on the choice of the triangle σ^- defining the affine embedding, because the transformation $(x, y, z) \mapsto (x, y, xy - z)$ does not extend to a map on $\boldsymbol{CP}^3$.

Lemma 9.2.4. *Under the above situation, the intersection number of* $\widetilde{\Phi}^-_{\nu^-,\sigma^-}$ *and* $\widetilde{\Phi}^+_{\nu^+,\sigma^+}$ *in* $\widetilde{\Phi}$ *is equal to* $3p_0$.

Proof. Set $(x, y, z) = (\phi(1/0), \phi(0/1), \phi(1/1))$. Then we can see inductively as in the proof of Lemma 5.3.12 that $\phi(q_j/p_j)$ is an integral polynomial, $W_j(x, y, z)$, in x, y, z of degree p_j and it has a unique term of the maximal degree p_j. Let $\boldsymbol{P}^-$ be the affine variety defined by

$$x + \alpha^- y + \beta^- z = 0,$$

and let $\boldsymbol{P}^+$ be the affine variety defined by

$$W_0(x, y, z) + \alpha^+ W_1(x, y, z) + \beta^+ W_2(x, y, z) = 0.$$

Let $\widetilde{\boldsymbol{P}}^-$ and $\widetilde{\boldsymbol{P}}^+$, respectively, be the subvarieties of $\boldsymbol{CP}^3$ obtained as the completions of $\boldsymbol{P}^-$ and $\boldsymbol{P}^+$. Then we may identify $\widetilde{\boldsymbol{P}}^-$ with the projective completion of the (y, z)-plane via the extension of the linear isomorphism

$$\mathbb{C}^2 \ni (y, z) \mapsto (-\alpha^- y - \beta^- z, y, z) \in \boldsymbol{P}^-.$$

Then $\widetilde{\boldsymbol{P}}^- \cap \widetilde{\Phi}$ is an algebraic curve in the projective plane $\widetilde{\boldsymbol{P}}^-$ defined by the following polynomial of degree 3 in the variables y and z.

$$(\alpha^- y + \beta^- z)^2 + y^2 + z^2 + (\alpha^- y + \beta^- z)yz = 0.$$

On the other hand, $\widetilde{\boldsymbol{P}}^- \cap \widetilde{\boldsymbol{P}}^+$ is the algebraic curve in the projective plane $\widetilde{\boldsymbol{P}}^-$ defined by the polynomial in y, z obtained from the polynomial $W_0(x, y, z) + \alpha^+ W_1(x, y, z) + \beta^+ W_2(x, y, z)$ by substituting x with $-(\alpha^- y + \beta^- z)$. Since the above three-variable polynomial has the unique term of the maximal degree p_0, the resulting two-variable polynomial has degree p_0. Moreover, $\widetilde{\boldsymbol{P}}^- \cap \widetilde{\Phi}$ and $\widetilde{\boldsymbol{P}}^- \cap \widetilde{\boldsymbol{P}}^+$ do not share a component by Proposition 9.1.13, because

$$\left(\widetilde{\boldsymbol{P}}^- \cap \widetilde{\Phi}\right) \cap \left(\widetilde{\boldsymbol{P}}^- \cap \widetilde{\boldsymbol{P}}^+\right) = \left(\widetilde{\boldsymbol{P}}^- \cap \widetilde{\Phi}\right) \cap \left(\widetilde{\boldsymbol{P}}^+ \cap \widetilde{\Phi}\right) = \widetilde{\Phi}^-_{\nu^-,\sigma^-} \cap \widetilde{\Phi}^+_{\nu^+,\sigma^+}.$$

Hence, by Bezout theorem, the intersection number of $\widetilde{\boldsymbol{P}}^- \cap \widetilde{\Phi}$ and $\widetilde{\boldsymbol{P}}^- \cap \widetilde{\boldsymbol{P}}^+$ in $\widetilde{\boldsymbol{P}}^-$ is equal to $3p_0$. Since the above intersection number is equal to that of $\widetilde{\Phi}^-_{\nu^-,\sigma^-} = \widetilde{\boldsymbol{P}}^- \cap \widetilde{\Phi}$ and $\widetilde{\Phi}^+_{\nu^+,\sigma^+} = \widetilde{\boldsymbol{P}}^+ \cap \widetilde{\Phi}$ in $\widetilde{\Phi}$ (see [73, p. 235, Exercise 3]), we obtain the desired result.

The above proof shows that the intersection number of $\widetilde{\Phi}^-_{\nu^-,\sigma^-}$ and $\widetilde{\Phi}^+_{\nu^+,\sigma^+}$ in $\widetilde{\Phi}$ is equal to that of the projective curves $\widetilde{\boldsymbol{P}}^- \cap \widetilde{\Phi}$ and $\widetilde{\boldsymbol{P}}^- \cap \widetilde{\boldsymbol{P}}^+$ in $\widetilde{\boldsymbol{P}}^- \cong \boldsymbol{CP}^2$. Hence, we obtain the following lemma by virtue of Lemma 9.3.3, which is proved in the next section.

Lemma 9.2.5. *Let $\boldsymbol{\nu}_t = (\nu^-, \nu^+_t)$ $(t \in [0,1])$ be a continuous path in $\mathbb{H}^2 \times \mathbb{H}^2$ such that $\nu^+_t \in \sigma^+$ for every $t \in [0,1]$ and $\nu^- \in \sigma^-$. Then there are continuous maps $\varphi_j : [0,1] \to \tilde{\Phi}$ $(1 \le j \le 3p_0)$ such that $\varphi_1(t), \cdots, \varphi_{3p_0}(t)$ form the intersection of $\widetilde{\Phi}^-_{\nu^-,\sigma^-}$ and $\widetilde{\Phi}^+_{\nu^+_t,\sigma^+}$ counted with multiplicity for every $t \in [0,1]$.*

Lemma 9.2.6. *Under the setting of Lemma 9.2.5, for each $j \in \{1, 2, \cdots, 3p_0\}$, if $\varphi_j(t)$ is a geometric root for $\boldsymbol{\nu}_t$ for some $t \in [0,1]$, then $\varphi_j(t)$ is a geometric root for $\boldsymbol{\nu}_t$ for every $t \in [0,1]$.*

Proof. Fix an integer $j \in \{1, 2, \cdots, 3p_0\}$, and let J be the subset of $[0,1]$ consisting of the point t such that $\varphi_j(t)$ is a geometric root for $\boldsymbol{\nu}_t$. Then J is open by Proposition 6.2.1 (Openness). To show the closedness of J, let $t_\infty \in [0,1]$ be a limit point of J, i.e., there is a sequence $\{t_n\}$ in J such that $t_\infty = \lim t_n$. Then, by Proposition 6.2.7 (Convergence), we may assume $\varphi_j(t_\infty) = \lim \varphi_j(t_n)$ is contained in the affine part Φ of $\tilde{\Phi}$. Hence, by Proposition 6.2.4 (Closedness), we see that $(\rho_\infty, \boldsymbol{\nu}_{t_\infty})$ is a good labeled representation, where ρ_∞ is the type-preserving representation induced by $\varphi_j(t_\infty)$. Thus $\varphi_j(t_\infty)$ is a geometric root for $\boldsymbol{\nu}_{t_\infty}$, and therefore J is also closed. Hence we obtain the desired result.

As a corollary to Lemmas 9.2.2, 9.2.5 and 9.2.6, we obtain the following result.

Corollary 9.2.7. *Let $\boldsymbol{\nu}_j = (\nu^-_j, \nu^+_j)$ $(j = 0, 1)$ be elements of $\mathbb{H}^2 \times \mathbb{H}^2$ such that $\nu^-_0 = \nu^-_1$ and that ν^+_0 and ν^+_1 are contained in a common triangle for each $\epsilon \in \{0,1\}$. Then $d_G(\boldsymbol{\nu}_0) = d_G(\boldsymbol{\nu}_1)$.*

We can now complete the proof of Proposition 9.2.3 as follows. Let $\Sigma(\boldsymbol{\nu}) = \{\sigma_1, \cdots, \sigma_m\}$, and pick points $\nu^+_j \in \mathrm{int}(\sigma_j \cap \sigma_{j+1})$ for each $j \in \{1, \cdots, m-1\}$. By repeatedly using Corollary 9.2.7, we have

$$\begin{aligned} d_G(\nu^-, \nu^+) &= d_G(\nu^-, \nu^+_{m-1}) = d_G(\nu^-, \nu^+_{m-2}) = \cdots = d_G(\nu^-, \nu^+_1) \\ &= d_G(\nu^-, \nu^-). \end{aligned}$$

By Proposition 5.1.5 and Lemma 5.2.12, we see $d_G(\nu^-, \nu^-) = 1$. Hence we have $d_G(\boldsymbol{\nu}) = 1$. This completes the proof of Proposition 9.2.3.

9.3 Continuity of roots and continuity of intersections

In this section, we prove Lemma 9.3.3, which is used in Sect. 9.2. Though it is certainly well-known to the experts, we could not find a proof in the literature.

Lemma 9.3.1. *Let $f_t(z) = z^n + a_1(t)z^{n-1} + \cdots + a_n(t)$ $(t \in [0,1])$ be a continuous family of polynomials with complex coefficients of a fixed degree n. Then there are continuous maps $\psi_j : [0,1] \to \mathbb{C}$ $(1 \le j \le n)$ such that $\{\psi_1(t), \cdots, \psi_n(t)\}$ is equal to the set of roots of $f_t(z)$ counted with multiplicity for every $t \in [0,1]$.*

Proof. Let $\mathcal{P}_n$ be the space of monic polynomials of degree n and identify $\mathcal{P}_n$ with $\mathbb{C}^n$ by the correspondence

$$f(z) = z^n + a_1 z^{n-1} + \cdots + a_n \mapsto (a_1, a_2, \cdots, a_n).$$

Recall that if $\omega_1, \cdots, \omega_n$ are roots of $f(z)$, then the coefficient a_d is equal to $(-1)^d s_d(\omega_1, \cdots, \omega_n)$, where s_d is the n-variable elementary symmetric function of degree d, namely

$$s_d(z_1, \cdots, z_n) = \sum_{1 \le j_1 < j_2 < \cdots < j_d \le n} z_{j_1} z_{j_2} \cdots z_{j_d}.$$

Consider the map $\mathbb{C}^n \to \mathbb{C}^n = \mathcal{P}_n$ defined by

$$(z_1, \cdots, z_n) \mapsto (-s_1(z_1, \cdots, z_n), s_2(z_1, \cdots, z_n), \cdots, (-1)^n s_n(z_1, \cdots, z_n)).$$

Then by the fundamental theorem of algebra (and the fact that the polynomial ring $\mathbb{C}[z]$ is a unique factorization domain), it induces a homeomorphism from $\mathbb{C}^n/S_n$, the quotient of $\mathbb{C}^n$ by the canonical action of the symmetric group S_n of degree n, to $\mathcal{P}_n$. By the path lifting theorem [19, Theorem 6.2 in Chap. II], every path in $\mathcal{P}_n \cong \mathbb{C}^n/S_n$ lifts to a path in $\mathbb{C}^n$. The lifted path gives rise to the desired family of continuous roots of the polynomials.

Remark 9.3.2. The authors got the idea of the proof from [62]. However, Theorem 1 in the paper, which claims the existence of a global continuous section to the projection $\mathbb{C}^n \to \mathbb{C}^n/S_n$, is not correct. We thank Michael Heusener for informing us of the paper and pointing out the error. We also thank Norbert A'Campo for proving the path lifting property for this special case, before we found the reference [19].

Lemma 9.3.3. *Let $F_t(x_1, x_2, x_3)$ $(t \in [0,1])$ be a continuous family of homogeneous polynomials with complex coefficients of a fixed degree, and let V_t be the projective curve defined by $F_t(x_1, x_2, x_3)$. Let W be a projective curve which does not share a component with V_t for every $t \in [0,1]$, and assume that the intersection number of V_t and W is a constant number, n, independent of t. Then there are continuous maps $\varphi_j : [0,1] \to \boldsymbol{CP}^2$ $(1 \le j \le n)$ such that $\{\varphi_1(t), \cdots, \varphi_n(t)\}$ is equal to the intersection $V_t \cap W$ counted with multiplicity for every $t \in [0,1]$.*

Proof. We first show that the lemma is valid locally, namely, for every $t_0 \in [0,1]$, there are a connected neighborhood U of t_0 in $[0,1]$ and continuous maps $\varphi_j : U \to \boldsymbol{CP}^2$ $(1 \le j \le n)$ such that $\{\varphi_1(t), \cdots, \varphi_n(t)\}$ is equal to the intersection $V_t \cap W$ counted with multiplicity for every $t \in U$. For notational simplicity we prove this local assertion for $t_0 = 0$. Let p be a point in $V_0 \cap W$, B a branch of W at p, and d be the intersection multiplicity of $V_0 \cap B$ at p. Let w be a local parameter of W at p, namely, w is an injective holomorphic map from $D(\epsilon_1) := \{z \in \mathbb{C} \,|\, |z| < \epsilon\}$ for some $\epsilon_1 > 0$ to $B \subset W \subset \boldsymbol{CP}^2$ such that $w(0) = p$. We may assume that w is given by $w(z) = [x_1(z) : x_2(z) : 1] \in \boldsymbol{CP}^2$. Then $F_t(x_1(z), x_2(z), 1)$ is a continuous function $[0,1] \times D(\epsilon_1) \to \mathbb{C}$ which is holomorphic in the variable z. By (the proof of) Weierstrass preparation theorem, there are positive numbers δ_1 and $\epsilon_2 (< \epsilon_1)$, a continuous map $u : [0, \delta_1] \times D(\epsilon_2) \to \mathbb{C}$ and continuous maps $a_1, \cdots, a_d : [0, \delta_1] \to \mathbb{C}$, such that $F_t(x_1(z), x_2(z), 1) = u(t,z) f_t(z)$ where $f_t(z) = z^d + a_1(t) z^{d-1} + \cdots + a_d(t)$, $u(0,0) \ne 0$, $a_j(0) = 0$ $(1 \le j \le d)$, and $u(t,z)$ is holomorphic in z. By Lemma 9.3.1, there are continuous maps $\psi_j : [0, \delta_1] \to \mathbb{C}$ $(1 \le j \le d)$ such that $\{\psi_1(t), \cdots, \psi_d(t)\}$ is equal to the set of roots of $f_t(z)$ counted with multiplicity for every $t \in [0,1]$. Pick small positive numbers $\delta_2 (< \delta_1)$ and $\epsilon_3 (< \epsilon_2)$ such that (i) $u(t,z) \ne 0$ for every $t \in [0, \delta_2)$ and $z \in D(\epsilon_3)$ and (ii) $\{\psi_1(t), \cdots, \psi_d(t)\} \subset D(\epsilon_3)$ for every $t \in [0, \delta_2)$. Though δ_2 depends on the point $p \in V_0 \cap W$ and the branch B of W at p a priori, we may choose it so that it is common to all such p and B. Then the union of the sets $\{w(\psi_1(t)), \cdots, w(\psi_d(t))\}$ for all p and B form the set of the intersection $V_t \cap W$ counted with multiplicity for every $t \in [0, \delta_2)$. This completes the proof of the local assertion.

To prove the global assertion, let J be the maximal connected subset of $[0,1]$ containing 0 for which the conclusion of the lemma holds, i.e., there are continuous maps $\{\varphi_j : J \to \boldsymbol{CP}^2\}$ $(1 \le j \le n)$ such that $V_t \cap W = \{\varphi_1(t), \cdots, \varphi_n(t)\}$ counted with multiplicity for every $t \in J$. By the local assertion proved in the above, J is open. To show that J is closed, set $t_\infty = \sup J$. Since the intersection number of V_t and W is the constant n, there are n points $p_{j,\infty}$ $(1 \le j \le n)$ of $\boldsymbol{CP}^2$, such that $V_t \cap W = \{p_{1,\infty}, \cdots, p_{n,\infty}\}$ counted with multiplicity. By the local assertion, there are continuous maps $\{\varphi_{j,\infty} : (t_\infty - \delta, t_\infty] \to \boldsymbol{CP}^2\}$ for some $\delta > 0$ such that $\varphi_{j,\infty}(t_\infty) = p_{j,\infty}$ for each j and $V_t \cap W = \{\varphi_{1,\infty}(t), \cdots, \varphi_{n,\infty}(t)\}$ counted with multiplicity for each $t \in (t_\infty - \delta, t_\infty]$. Put $t'_\infty = t_\infty - (\delta/2) \in (t_\infty - \delta, t_\infty]$. Then after changing the indices, we may assume $\varphi_j(t'_\infty) = \varphi_{j,\infty}(t'_\infty)$ for every j. Redefine $\varphi_j : [0, t_\infty] \to \boldsymbol{CP}^2$ so that its restriction to $[0, t'_\infty]$ is equal to the original one and its restriction to $[t'_\infty, t_\infty]$ is equal to the restriction of $\varphi_{j,\infty}$. Then each φ_j is continuous and $V_t \cap W = \{\varphi_1(t), \cdots, \varphi_n(t)\}$ counted with multiplicity for every $t \in [0, t_\infty]$. Thus J is closed. Hence J is equal to the interval $[0,1]$ and we obtain the desired result.

A

Appendix

A.1 Basic facts concerning the Ford domain

In this appendix, we give a proof to some of the basic facts concerning the Ford domain, the proof of which could not be found in the literature. Throughout this appendix, Γ denotes a non-elementary Kleinian group, such that the stabilizer Γ_∞ of ∞ contains parabolic transformations, and H_∞ denotes a fixed horoball centered at ∞ which is precisely (Γ, Γ_∞)-invariant. The following well-known observation plays an important role in this section.

Lemma A.1.1. *(1) The Euclidean radii of the horoballs in $\Gamma H_\infty - \{H_\infty\}$ is bounded from the above.*

(2) For any compact subset K in $\mathbb{H}^3$, only finitely many horoballs in ΓH_∞ intersect K.

Proof. (1) follows from the fact that all horoballs in $\Gamma H_\infty - \{H_\infty\}$ are disjoint from H_∞ and hence their Euclidean radii are less than the half of the "Euclidean height", t, of ∂H_∞, where t is the positive real number such that $\partial H_\infty = \mathbb{C} \times \{t\} \subset \mathbb{H}^3$.

(2) Since K is a compact subset of $\mathbb{H}^3$, its Euclidean height is bounded below, i.e., there is a positive constant c such that $K \subset \mathbb{C} \times [c, \infty) \subset \mathbb{H}^3$. On the other hand, by virtue of (1), there is a compact subset, L, of $\mathbb{C}$ which contains the centers of all horoballs in $\Gamma H_\infty - \{H_\infty\}$ intersecting K. (If r is the upper bound obtained in (1), then we may set L to be the closed r-neighborhood of $\mathrm{proj}(K)$ in $\mathbb{C}$ with respect to the Euclidean metric.) Thus if a horoball in $\Gamma H_\infty - \{H_\infty\}$ intersects K, then its Euclidean radius is $\geq c/2$ and its center is contained in the compact set L. Since ΓH_∞ consists of disjoint horoballs, only finitely many of its members satisfy these conditions. Hence we obtain the conclusion.

We rephrase Proposition 1.1.3. part of which is proved in [72, Lemma 5.20] under the additional assumption that ∞ is a bounded parabolic fixed point.

Proposition A.1.2. *The Ford domain* $Ph(\Gamma)$ *is a "fundamental polyhedron of* Γ *modulo* Γ_∞*", in the following sense.*

1. $\mathbb{H}^3 = \cup\{A(Ph(\Gamma)) \mid A \in \Gamma\}$.
2. $\operatorname{int} Ph(\Gamma)$ *is precisely* (Γ, Γ_∞)*-invariant.*
3. *For any compact set* K *of* $\mathbb{H}^3$*, only finitely many images* $A(Ph(\Gamma))$ $(A \in \Gamma)$ *can intersect* K*, namely the set* $\{A\Gamma_\infty \in \Gamma/\Gamma_\infty \mid A(Ph(\Gamma)) \cap K \neq \emptyset\}$ *is finite.*
4. $Ph(\Gamma)$ *is a closed convex polyhedron (Definition 3.4.1(2)).*

Proof. (1) By applying Lemma A.1.1(1), to a (hyperbolic) closed ball $B(x, r)$ with center $x \in \mathbb{H}^3$ and radius $r > 0$, we see that the minimal distance

$$d(x, \Gamma H_\infty) := \min\{d(x, AH_\infty) \mid A \in \Gamma\}$$

is well-defined, i.e., there is an element $A_x \in \Gamma$ such that $d(x, A_x H_\infty) \leq d(x, AH_\infty)$ for every $A \in \Gamma$. This implies that $A_x^{-1}(x) \in Ph(\Gamma) = \{x \in \mathbb{H}^3 \mid d(x, H_\infty) = d(x, \Gamma H_\infty)\}$. Hence we have $\mathbb{H}^3 = \cup\{A(Ph(\Gamma)) \mid A \in \Gamma\}$.

(2) This is a direct consequence of Lemma 4.1.1(1).

(3) We show that for every $x \in \mathbb{H}^3$ and $\epsilon > 0$, the compact set $B(x, \epsilon)$ satisfies the conclusion. To this end, put $r = d(x, \Gamma H_\infty)$. By Lemma A.1.1(2), $B(x, r + 2\epsilon)$ intersects only finitely many horoballs $A_1(H_\infty), A_2(H_\infty), \cdots, A_n(H_\infty)$ in ΓH_∞. Suppose $A(Ph(\Gamma)) \cap B(x, \epsilon) \neq \emptyset$ for some $A \in \Gamma$. Pick a point y from the intersection. Then

$$\begin{aligned} d(x, A^{-1}(H_\infty)) &\leq \epsilon + d(y, A^{-1}(H_\infty)) = \epsilon + d(y, \Gamma H_\infty) \\ &\leq 2\epsilon + d(x, \Gamma H_\infty) = 2\epsilon + r. \end{aligned}$$

Here the identity in the above follows from the assumption that $y \in A(Ph(\Gamma))$. Hence $B(x, r + 2\epsilon) \cap A(H_\infty) \neq \emptyset$ and therefore $A^{-1}(H_\infty) = A_j(H_\infty)$ for some j. This implies $A^{-1}(Ph(\Gamma)) = A_j(Ph(\Gamma))$. Hence the desired result holds for the compact set $B(x, \epsilon)$.

(4) Let K be a compact subset of $\mathbb{C}$, and consider the isometric hemispheres $Ih(A)$ $(A \in \Gamma - \Gamma_\infty)$ which intersect K. Then as in the proof of Lemma A.1.1(2), we see that their Euclidean radii are bounded below, their centers are contained in a compact subset of $\mathbb{C}$, and that the radii of the corresponding horoballs $A(H_\infty)$ are bounded below. Hence we see that there are only finitely many such isometric hemispheres. (We can also deduce this by using Lemma 2.5.2(2)). Hence $Ph(\Gamma)$ satisfies the condition for a convex polyhedron in Definition 3.4.1(2).

The following lemma is proved by imitating the argument of [55, Proof of Lemma 2.13] for the Dirichlet domain (cf. [72, Lemma 5.37]).

Lemma A.1.3. *For every point* ξ *in the Ford polygon* $P(\Gamma)$*, there is a horoball* H_ξ *centered at* ξ *such that* $H_\xi \cap H_\infty$ *is a singleton and* $(\operatorname{int} H_\xi) \cap \Gamma H_\infty = \emptyset$*. In particular, any point of* $P(\Gamma) \cap \Lambda(\Gamma)$ *is not a horospherical limit point of* Γ *([55, Definition in p.51]).*

Proof. Let ξ be a point in $P(\Gamma)$. Then the vertical geodesic (ξ, ∞) is contained in $Ph(\Gamma)$. Hence, for each $x \in (\xi, \infty) - H_\infty$, $\operatorname{int} B(x, d(x, H_\infty))$ is disjoint from ΓH_∞. Thus the open horoball $\cup\{\operatorname{int} B(x, d(x, H_\infty)) \mid x \in (\xi, \infty) - H_\infty\}$, centered at ξ, is disjoint from ΓH_∞. Moreover its closure, H_ξ, intersect H_∞ precisely at the point $\partial H_\infty \cap (\xi, \infty)$. Thus we obtain the first assertion. To see the second assertion, pick a point x from $\operatorname{int} H_\infty$. Then its orbit Γx is disjoint from H_ξ. Hence ξ is not a horospherical limit point.

Corollary A.1.4. *If Γ is geometrically finite, then any point of $P(\Gamma) \cap \Lambda(\Gamma)$ is the (parabolic) fixed point of a parabolic element of Γ which is not conjugate to an element of Γ_∞. In particular, if Γ is a quasifuchsian punctured torus group, then $P(\Gamma) \subset \Omega(\Gamma)$.*

Proof. Suppose Γ is geometrically finite and let ξ be a point in $P(\Gamma) \cap \Lambda(\Gamma)$. Then ξ is not a horospherical limit point and hence it is a bounded parabolic fixed point (see [55, Theorem 3.7]). On the other hand, we see that the orbit $\Gamma\infty$ is disjoint from $P(\Gamma)$ as follows. Suppose to the contrary that $A(\infty)$ belongs to $P(\Gamma) = \overline{Ph}(\Gamma) \cap \mathbb{C}$. Then $A(\operatorname{int} H_\infty) \cap \operatorname{int} Ph(\Gamma) \neq \emptyset$ and hence $A \in \Gamma_\infty$ by Proposition A.1.2(2). Thus $\infty = A(\infty) \in P(\Gamma)$, a contradiction. Hence ξ does not belong to the orbit $\Gamma\infty$, and we obtain the first assertion. The second assertion follows from the fact that every parabolic transformation of a quasifuchsian punctured torus group is conjugate to an element of Γ_∞.

The following lemma is a refinement of Lemma A.1.3.

Lemma A.1.5. *A point $\xi \in \mathbb{C}$ belongs to $P(\Gamma)$, if and only if there is a horoball H_ξ centered at ξ such that $H_\xi \cap \Gamma H_\infty = \emptyset$ and $d(H_\xi, H_\infty) \le d(H_\xi, A(H_\infty))$ for every $A \in \Gamma$.*

Proof. Suppose that ξ belongs to $P(\Gamma)$. Then by Lemma A.1.3, there is a horoball H_ξ centered at ξ such that $H_\xi \cap H_\infty$ is a singleton and $(\operatorname{int} H_\xi) \cap \Gamma H_\infty = \emptyset$. Reset H_ξ to be a horoball contained in the interior of this horoball. Then we see that this new horoball H_ξ satisfies the desired conditions.

Conversely, suppose that the latter condition is satisfied. We show that $(\xi, \infty) \cap H_\xi \subset Ph(\Gamma)$. To this end, pick a point $x \in (\xi, \infty) \cap H_\xi$ and $A \in \Gamma$. Let γ be the shortest geodesic segment joining x to $A(H_\infty)$. Then

$$\begin{aligned} d(x, A(H_\infty)) &= length(\gamma) \\ &= length(\gamma \cap H_\xi) + length(\gamma \cap (\mathbb{H}^3 - \operatorname{int} H_\xi)) \\ &\ge d(x, \partial H_\xi) + d(H_\xi, A(H_\infty)) \\ &\ge d(x, \partial H_\xi) + d(H_\xi, H_\infty) \\ &= d(x, H_\infty) \end{aligned}$$

(This inequality is easily seen by performing a coordinate change so that H_ξ is centered at ∞.) This implies $x \in Ph(\Gamma)$. Thus we have $(\xi, \infty) \cap H_\xi \subset Ph(\Gamma)$. Hence $\xi \in P(\Gamma)$.

The above lemma immediately implies the following corollary.

Corollary A.1.6. *For each $A_0 \in \Gamma$, a point $\xi \in \mathbb{C}$ belongs to $A_0(P(\Gamma))$, if and only if there is a horoball H_ξ centered at ξ such that $H_\xi \cap \Gamma H_\infty = \emptyset$ and $d(H_\xi, A_0(H_\infty)) \le d(H_\xi, A(H_\infty))$ for every $A \in \Gamma$.*

We now prove the following proposition.

Proposition A.1.7. *The intersection, $P(\Gamma) \cap \Omega(\Gamma)$, of the Ford polygon and the domain of discontinuity is a "fundamental polygon of Γ modulo Γ_∞", for the action of Γ on $\Omega(\Gamma)$ in the following sense.*

1. $\Omega(\Gamma) = \cup\{A(P(\Gamma) \cap \Omega(\Gamma)) \mid A \in \Gamma\}$.
2. $\mathrm{int}\, P(\Gamma) = \mathrm{int}(P(\Gamma) \cap \Omega(\Gamma))$ *is precisely (Γ, Γ_∞)-invariant.*
3. *For any compact set K of $\Omega(\Gamma)$, only finitely many images $A(P(\Gamma))$ ($A \in \Gamma$) can intersect K, namely the set $\{A\Gamma_\infty \in \Gamma/\Gamma_\infty \mid A(P(\Gamma)) \cap K \neq \emptyset\}$ is finite.*

Proof. (1) Let ξ be a point in $\Omega(\Gamma)$. Then some neighborhood of ξ in $\mathbb{C}$ is disjoint from the centers of the horoballs in ΓH_∞, because they are contained in $\Lambda(\Gamma)$. Since the Euclidean radii of the horoballs in $\Gamma H_\infty - \{H_\infty\}$ are bounded above, we can find a closed neighborhood D of ξ in $\overline{\mathbb{H}}^3$ such that $D \cap \Gamma H_\infty = \emptyset$.

Claim A.1.8. Any horoball H_ξ centered at $\xi \in \Omega(\Gamma)$ can intersect only finitely many horoballs in ΓH_∞.

Proof. Let H_ξ be a horoball centered at ξ. Then the closure of $H_\xi - D$ in $\mathbb{H}^3$ is compact and a horoball in ΓH_∞ intersects H_ξ if and only if it intersects the relatively compact set $H_\xi - D$. Hence we have the claim by Lemma A.1.1(2).

Now pick a small horoball H_ξ centered at ξ contained in D. Then by applying the above claim to a closed r-neighborhood of H_ξ in $\mathbb{H}^3$, which is again a horoball centered at ξ, for sufficiently large r, we see that

$$d(H_\xi, \Gamma H_\infty) := \min\{d(H_\xi, AH_\infty) \mid A \in \Gamma\}$$

is a well-defined positive number, i.e., there is an element $A_\xi \in \Gamma$ such that $d(H_\xi, A_\xi(H_\infty)) \le d(H_\xi, A(H_\infty))$ for every $A \in \Gamma$. Thus we see $\xi \in A_\xi(P(\Gamma))$ by Corollary A.1.6, and hence $A_\xi^{-1}(\xi) \in P(\Gamma) \cap \Omega(\Gamma)$. So we have $\Omega(\Gamma) = \cup\{A(P(\Gamma) \cap \Omega(\Gamma)) \mid A \in \Gamma\}$.

(2) This is a direct consequence of Lemma 4.1.1(1).

(3) Let K be a compact subset of $\Omega(\Gamma)$. Then by the argument in the proof of (1), there is a compact neighborhood D of K in $\overline{\mathbb{H}}^3$ such that $D \cap \Gamma H_\infty = \emptyset$. Pick a constant $c > 0$ such that each horoball, H_ξ, with Euclidean radius c centered at a point $\xi \in K$ is contained in $\mathrm{int}(D \cap \mathbb{H}^3)$. Throughout the proof we reserve the symbol H_ξ to denote these horoballs, and set $H_K = \cup\{H_\xi \mid \xi \in K\}$. By using the compactness of K, we can

find a constant $r > 0$ such that $d(H_\xi, \Gamma H_\infty)$, which is well-defined by (the proof of) (1), is at most r for every $\xi \in K$. Consider the closed r-neighborhood $B(H_K, r)$ of H_K in $\mathbb{H}^3$. Then $B(H_K, r) - D$ is relatively compact in $\mathbb{H}^3$. Thus we see, by Lemma A.1.1(2), that $B(H_K, r)$ intersects only finitely many horoballs, $A_1(H_\infty), A_2(H_\infty), \cdots, A_n(H_\infty)$, in ΓH_∞. Now suppose $K \cap A(P(\Gamma)) \neq \emptyset$. Pick a point ξ from the intersection. Then, by Corollary A.1.6, $d(H_\xi, A(H_\infty)) = d(H_\xi, \Gamma H_\infty) \leq r$. Hence $B(H_K, r) \cap A(H_\infty) \neq \emptyset$. Thus $A(H_\infty) = A_j(H_\infty)$ for some j. This implies $A(P(\Gamma)) = A_j(P(\Gamma))$. Hence we obtain the desired result.

Remark A.1.9. (1) Since two "edges" of $\mathrm{fr}\, P(\Gamma)$ may be tangent, $\mathrm{fr}\, P(\Gamma)$ is not necessarily a 1-dimensional manifold.

(2) $\mathrm{fr}\, P(\Gamma) \cap \Omega(\Gamma)$ is locally finite in the sense that any point $x \in \mathrm{fr}\, P(\Gamma) \cap \Omega(\Gamma)$, there is a neighborhood U of x in $\Omega(\Gamma)$ such that $\mathrm{fr}\, P(\Gamma) \cap U$ is a finite union of circular arcs. However, $\mathrm{fr}\, P(\Gamma)$ is not necessarily locally finite around points in $\mathrm{fr}\, P(\Gamma) \cap \Lambda(\Gamma)$.

At the end of this appendix, we prove the following finiteness property for the Ford domain.

Lemma A.1.10. *For a point p in $\overline{Ph}(\Gamma)$, let $[p]$ be the set of points in $\overline{Ph}(\Gamma)$ which are Γ-equivalent to p, namely*

$$[p] = \Gamma p \cap \overline{Ph}(\Gamma) = \{x \in \overline{Ph}(\Gamma) \mid x = A(p) \text{ for some } A \in \Gamma\}.$$

Then the quotient set $[p]/\Gamma_\infty$ is finite provided that $p \in \mathbb{H}^3 \cup \Omega(\Gamma)$ or p is a bounded parabolic fixed point of Γ. In particular, if Γ is geometrically finite, then $[p]/\Gamma_\infty$ is a finite set for every $p \in \overline{Ph}(\Gamma)$.

Proof. We prove the lemma only for the case when p is a point in $\mathbb{C}$ (and hence in $P(\Gamma)$). (The proof for the case $p \in \mathbb{H}^3$ is parallel to this case and is much simpler.) Let ξ be a point in $P(\Gamma)$, and let H_ξ be a horoball centered at ξ satisfying the condition in Lemma A.1.5. Then $r := d(H_\xi, \Gamma H_\infty)$ is well-defined and is equal to $d(H_\xi, H_\infty)$. By Corollary A.1.6, for each $A \in \Gamma$, we have $A(\xi) \in P(\Gamma)$ if and only if $d(H_\xi, A^{-1}(H_\infty)) = r$. The latter condition holds if and only if $A^{-1}(H_\infty)$ intersects the horoball $B(H_\xi, r)$. Hence we see

$$[p] = \{A(p) \mid B(H_\xi, r) \cap A^{-1}(H_\infty) \neq \emptyset\}.$$

Now suppose that $\xi \in \Omega(\Gamma)$. Then by Claim A.1.8, only finitely many horoballs in ΓH_∞ can intersect $B(H_\xi, r)$. Moreover, for two elements A_1 and A_2 of Γ, $A_1^{-1}(H_\infty) = A_2^{-1}(H_\infty)$ if and only if $A_1^{-1}\Gamma_\infty = A_2^{-1}\Gamma_\infty \in \Gamma/\Gamma_\infty$. Hence the set

$$\{A^{-1}\Gamma_\infty \in \Gamma/\Gamma_\infty \mid B(H_\xi, r) \cap A^{-1}(H_\infty) \neq \emptyset\}$$

is finite. By the observation in the previous paragraph, the correspondence $A^{-1}\Gamma_\infty \mapsto \Gamma_\infty A(\xi)$ determines a surjective map from the above finite set to the quotient set $[p]/\Gamma_\infty$. (To be precise, the symbol $[p]/\Gamma_\infty$ should be denoted by $\Gamma_\infty\backslash[p]$, because the action of Γ_∞ on $[p]$ is a left action.) Hence $[p]/\Gamma_\infty$ is finite.

Next suppose that ξ is a bounded parabolic fixed point. Then we can easily see that, modulo the action of the parabolic stabilizer Γ_ξ of ξ, only finitely many horoballs in ΓH_∞ can intersect $B(H_\xi, r)$. Thus the set

$$\{\Gamma_p A^{-1}\Gamma_\infty \in \Gamma_p\backslash\Gamma/\Gamma_\infty \mid B(H_\xi, r) \cap A^{-1}(H_\infty) \neq \emptyset\}$$

is finite. Hence we obtain the finiteness of $[p]/\Gamma_\infty$ as in the previous case.

References

1. Adams, C. C.: Hyperbolic 3-manifolds with two generators. *Comm. Anal. Geom.* **4** (1996), *no. 1-2*, 181–206
2. Akiyoshi, H.: On the Ford domains of once-punctured torus groups. Hyperbolic spaces and related topics (Kyoto, 1998). *Sūrikaisekikenkyūsho Kōkyūroku No. 1104* (1999), 109–121
3. Akiyoshi, H.: End invariants and Jorgensen's angle invariants of punctured torus groups. Perspectives of Hyperbolic Spaces II (Kyoto, 2003). *Sūrikaisekikenkyūsho Kōkyūroku No. 1387* (2004), 59–69
4. Akiyoshi, H.: Punctured torus groups surviving surgery. Talk at the international workshop First KOOK Seminar International for Knot Theory and Related Topics, July 2004, Awaji-shima, Japan
5. Akiyoshi, H.; Miyachi, H.; Sakuma, M.: A refinement of McShane's identity for quasifuchsian punctured torus groups. *In the tradition of Ahlfors and Bers, III*, 21–40, *Contemp. Math.*, 355, *Amer. Math. Soc., Providence, RI*, 2004
6. Akiyoshi, H.; Miyachi, H.; Sakuma, M.: Variations of McShane's identity for punctured surface groups. *Spaces of Kleinian groups*, 151–185, *London Math. Soc. Lecture Note Ser.*, 329, *Cambridge Univ. Press, Cambridge*, 2006
7. Akiyoshi, H.; Sakuma, M.: Comparing two convex hull constructions for cusped hyperbolic manifolds. *Kleinian groups and hyperbolic 3-manifolds (Warwick, 2001)*, 209–246, *London Math. Soc. Lecture Note Ser.*, 299, *Cambridge Univ. Press*, Cambridge, 2003
8. Akiyoshi, H.; Sakuma, M.; Wada, M.; Yamashita, Y.: Punctured torus groups and two-parabolic groups. Analysis and geometry of hyperbolic spaces (Kyoto, 1997). *Sūrikaisekikenkyūsho Kōkyūroku No. 1065* (1998), 61–73
9. Akiyoshi, H.; Sakuma, M.; Wada, M.; Yamashita, Y.: Ford domains of punctured torus groups and two-bridge knot groups. Knot Theory. Proceedings of the workshop dedicated to 70th birthday of Prof. K. Murasugi, (Toronto, 1999)
10. Akiyoshi, H.; Sakuma, M.; Wada, M.; Yamashita, Y.: Jørgensen's picture of punctured torus groups and its refinement. *Kleinian groups and hyperbolic 3-manifolds (Warwick, 2001)*, 247–273, *London Math. Soc. Lecture Note Ser.*, 299, *Cambridge Univ. Press, Cambridge*, 2003
11. Akiyoshi, H.; Sakuma, M.; Wada, M.; Yamashita, Y.: Punctured torus groups and 2-bridge knot groups (II). In preparation

12. Alestalo, P.; Helling, H.: On torus fibrations over the circle. Preprint, Univ. Bielefeld
13. Alperin, R. C.; Dicks, W.; Porti, J.: The boundary of the Gieseking tree in hyperbolic three-space. *Topology Appl.* **93** (1999), *no. 3*, 219–259
14. Beardon, A. F.: The geometry of discrete groups. Graduate Texts in Mathematics, 91. *Springer-Verlag, New York*, 1983
15. Benedetti, R.; Petronio, C.: Lectures on hyperbolic geometry. Universitext. *Springer-Verlag, Berlin*, 1992
16. Berger, M.: Geometry. II. Translated from the French by M. Cole and S. Levy. Universitext. *Springer-Verlag, Berlin*, 1987
17. Bowditch, B. H.: Markoff triples and quasi-Fuchsian groups. *Proc. London Math. Soc. (3)* **77** (1998), *no. 3*, 697–736
18. Bowditch, B. H.: The Cannon-Thurston map for punctured-surface groups. *Math. Z.* **255** (2007), *no. 1*, 35–76
19. Bredon, G. E.: Introduction to compact transformation groups. Pure and Applied Mathematics, Vol. 46. *Academic Press, New York-London*, 1972
20. Brenner, J. L.: Quelques groupes libres de matrices. *C. R. Acad. Sci. Paris* **241** (1955), 1689–1691
21. Brock, J. F.: The Weil-Petersson metric and volumes of 3-dimensional hyperbolic convex cores. *J. Amer. Math. Soc.* **16** (2003), *no. 3*, 495–535
22. Burde, G.; Zieschang, H.: Knots. de Gruyter Studies in Mathematics, 5. *Walter de Gruyter & Co., Berlin*, 1985
23. Cannon, J. W.; Dicks, W.: On hyperbolic once-punctured-torus bundles. Proceedings of the Conference on Geometric and Combinatorial Group Theory, Part I (Haifa, 2000). *Geom. Dedicata* **94** (2002), 141–183
24. Cannon J. W.; Thurston W. P.: Group invariant Peano curves. Preprint, 1989
25. Drumm, T. A.; Poritz, J. A.: Ford and Dirichlet domains for cyclic subgroups of $\mathrm{PSL}_2(\mathbf{C})$ acting on $\mathbf{H}^3_{\mathbf{R}}$ and $\partial\mathbf{H}^3_{\mathbf{R}}$. *Conform. Geom. Dyn.* **3** (1999), 116–150
26. Epstein, D. B. A.; Marden, A.: Convex hulls in hyperbolic space, a theorem of Sullivan, and measured pleated surfaces. *Analytical and geometric aspects of hyperbolic space (Coventry/Durham, 1984)*, 113–253, London Math. Soc. Lecture Note Ser., 111, *Cambridge Univ. Press, Cambridge*, 1987
27. Epstein, D. B. A.; Penner, R. C.: Euclidean decompositions of noncompact hyperbolic manifolds. *J. Differential Geom.* **27** (1988), *no. 1*, 67–80
28. Epstein, D. B. A.; Petronio, C.: An exposition of Poincare's polyhedron theorem. *Enseign. Math. (2)* **40** (1994), *no. 1-2*, 113–170
29. Floyd, W.; Hatcher, A.: Incompressible surfaces in punctured-torus bundles. *Topology Appl.* **13** (1982), *no. 3*, 263–282
30. Ford L. R.: Automorphic functions. *Chelsea Pub., New York*, 1951
31. Fulton, W.: Introduction to intersection theory in algebraic geometry. CBMS Regional Conference Series in Mathematics, 54. *Published for the Conference Board of the Mathematical Sciences, Washington, DC; by the American Mathematical Society, Providence, RI*, 1984
32. Gordon, C. McA.; Luecke, J.: Knots are determined by their complements. *J. Amer. Math. Soc.* **2** (1989), *no. 2*, 371–415
33. Guéritaud, F. with an appendix by Futer, D.: On canonical triangulations of once-punctured torus bundles and two-bridge link complements. *Geom. Topol.* **10** (2006), 1239–1284
34. Guéritaud, F.: Géométrie hyperbolique effective et triangulations idéals canoniques en dimension 3. These, *Univ. Paris-sud, Orsay*, 2006

35. Haas, A.: Diophantine approximation on hyperbolic orbifolds. *Duke Math. J.* **56** (1988), *no. 3*, 531–547
36. Helling, H.; Sakuma, M.: Unfinished draft
37. Hoste, J.; Shanahan, P. D.: Trace fields of twist knots. *J. Knot Theory Ramifications* **10** (2001), *no. 4*, 625–639
38. Imayoshi, Y.; Taniguchi, M.: An introduction to Teichmüller spaces. Translated and revised from the Japanese by the authors. *Springer-Verlag, Tokyo*, 1992
39. Jørgensen, T.: On cyclic groups of Möbius transformations. *Math. Scand.* **33** (1973), 250–260 (1974)
40. Jørgensen, T.: On pairs of once-punctured tori. Unfinished manuscript available in *Kleinian groups and hyperbolic 3-manifolds (Warwick, 2001)*, 183–207, London Math. Soc. Lecture Note Ser., 299, *Cambridge Univ. Press, Cambridge*, 2003
41. Jørgensen, T.: Compact 3-manifolds of constant negative curvature fibering over the circle. *Ann. Math. (2)* **106** (1977), *no. 1*, 61–72
42. Jørgensen, T.: Lecture at Osaka University, 1997
43. Jørgensen, T.; Marden, A.: Two doubly degenerate groups. *Quart. J. Math. Oxford Ser. (2)* **30** (1979), *no. 118*, 143–156
44. Keen, L.; Series, C.: Pleating coordinates for the Maskit embedding of the Teichmüller space of punctured tori. *Topology* **32** (1993), *no. 4*, 719–749
45. Keen, L.; Series, C.: The Riley slice of Schottky space. *Proc. London Math. Soc. (3)* **69** (1994), *no. 1*, 72–90
46. Keen, L.; Series, C.: How to bend pairs of punctured tori. *Lipa's legacy (New York, 1995)*, 359–387, Contemp. Math., 211, *Amer. Math. Soc., Providence, RI*, 1997
47. Keen, L.; Series, C.: Pleating invariants for punctured torus groups. *Topology* **43** (2004), *no. 2*, 447–491
48. Knapp, A. W.: Doubly generated Fuchsian groups. *Michigan Math. J.* **15** (1969), 289–304
49. Komori, Y.; Series, C.: The Riley slice revisited. *The Epstein birthday schrift*, 303–316, Geom. Topol. Monogr., 1, *Geom. Topol. Publ., Coventry*, 1998
50. Komori, Y.; Sugawa, T.; Wada, M.; Yamashita, Y.: Drawing Bers embeddings of the Teichmüller space of once-punctured tori. *Experiment. Math.* **15** (2006), *no. 1*, 51–60
51. Komori, Y.; Yamashita, Y.: Linear slices of the quasifuchsian space of punctured tori. Preprint
52. Lackenby, M.: The canonical decomposition of once-punctured torus bundles. *Comment. Math. Helv.* **78** (2003), *no. 2*, 363–384
53. Maskit, B.: On Poincaré's theorem for fundamental polygons. *Advances in Math.* **7**, 219–230. (1971)
54. Maskit, B.: Kleinian groups. Grundlehren der Mathematischen Wissenschaften, 287. *Springer-Verlag, Berlin*, 1988
55. Matsuzaki, K.; Taniguchi, M.: Hyperbolic manifolds and Kleinian groups. Oxford Mathematical Monographs. Oxford Science Publications. *The Clarendon Press, Oxford University Press, New York*, 1998
56. McMullen, C. T.: Local connectivity, Kleinian groups and geodesics on the blowup of the torus. *Invent. Math.* **146** (2001), *no. 1*, 35–91
57. Mednykh, A. D.; Parker, J. R.; Vesnin, A. Yu.: On hyperbolic polyhedra arising as convex cores of quasi-Fuchsian punctured torus groups. *Bol. Soc. Mat. Mexicana (3)* **10** (2004), Special Issue, 357–381

58. Minsky, Y. N.: The classification of punctured-torus groups. *Ann. of Math. (2)* **149** (1999), *no. 2*, 559–626
59. Morimoto, K.; Sakuma, M.: On unknotting tunnels for knots. *Math. Ann.* **289** (1991), *no. 1*, 143–167
60. Mumford, D.; McMullen, C. T.; Wright, D.: Limit sets of free two-generator kleinian groups. Unpublished
61. Mumford, D.; Series, C.; Wright, D.: Indra's pearls. The vision of Felix Klein. *Cambridge University Press, New York*, 2002
62. Naulin, R.; Pabst, C.: The roots of a polynomial depend continuously on its coefficients. *Rev. Colombiana Mat.* **28** (1994), *no. 1*, 35–37
63. Otal, J.-P.: Le théorème d'hyperbolisation pour les variétés fibrées de dimension 3. *Astérisque No. 235* (1996)
64. Parker, J. R.: Tetrahedral decomposition of punctured torus bundles. *Kleinian groups and hyperbolic 3-manifolds (Warwick, 2001)*, 275–291, London Math. Soc. Lecture Note Ser., 299, *Cambridge Univ. Press, Cambridge*, 2003
65. Parker, J. R.; Stratmann, B. O.: Kleinian groups with singly cusped parabolic fixed points. *Kodai Math. J.* **24** (2001), *no. 2*, 169–206
66. Parkkonen, J.: The outside of the Teichmüller space of punctured tori in Maskit's embedding. *Ann. Acad. Sci. Fenn. Math.* **24** (1999), *no. 2*, 305–342
67. Riley, R.: A quadratic parabolic group. *Math. Proc. Cambridge Philos. Soc.* **77** (1975), 281–288
68. Riley, R.: Algebra for Heckoid groups. *Trans. Amer. Math. Soc.* **334** (1992), *no. 1*, 389–409
69. Riley, R.: Groups generated by two parabolics A and $B(w)$ for w in the first quadrant. Output of an computer experiment
70. Sakuma, M.: Unknotting tunnels and canonical decompositions of punctured torus bundles over a circle. Analysis of discrete groups (Kyoto, 1995). *Sūrikaisekikenkyūsho Kōkyūroku No. 967* (1996), 58–70
71. Sakuma, M.; Weeks, J.: Examples of canonical decompositions of hyperbolic link complements. *Japan. J. Math. (N.S.)* **21** (1995), *no. 2*, 393–439
72. Schedler T. J.: Troels Jorgensen's once-punctured theory. Thesis, 2002, Harvard University
73. Shafarevich, I. R.: Basic algebraic geometry. 1. Varieties in projective space. Second edition. Translated from the 1988 Russian edition and with notes by Miles Reid. *Springer-Verlag, Berlin*, 1994
74. Sheingorn, M.: Characterization of simple closed geodesics on Fricke surfaces. *Duke Math. J.* **52** (1985), *no. 2*, 535–545
75. Sullivan, D.: Travaux de Thurston sur les groupes quasi-fuchsiens et les variétés hyperboliques de dimension 3 fibrées sur S^1. *Bourbaki Seminar, Vol. 1979/80*, pp. 196–214, Lecture Notes in Math., 842, *Springer, Berlin-New York*, 1981
76. Thurston, W. P.: The Geometry and Topology of Three-Manifolds. Electronic version 1.0 - October 1997, http://msri.org/publications/books/gt3m/
77. Thurston, W. P.: Hyperbolic structures on 3-manifolds II. Preprint, 1987
78. Wada, M.: OPTi. http://vivaldi.ics.nara-wu.ac.jp/~wada/OPTi/index.html
79. Wada, M.: OPTi's algorithm for discreteness determination. *Experiment. Math.* **15** (2006), *no. 1*, 61–66
80. Wada, M.: From once-punctured torus to twice-punctured torus. Preprint
81. Weeks, J. R.: Convex hulls and isometries of cusped hyperbolic 3-manifolds. *Topology Appl.* **52** (1993), *no. 2*, 127–149

82. Yamashita, Y.: Computer experiments on the discreteness locus in projective structures. *Spaces of Kleinian groups*, 375–390, London Math. Soc. Lecture Note Ser., 329, *Cambridge Univ. Press, Cambridge*, 2006

Notation

- General topology
  - $\operatorname{int} X$: the interior of X
  - $\operatorname{fr} X$: the frontier of X
  - $\overline{X}$: the closure of X
  - $\pi_1(X)$: the fundamental group of X
- Cellular complex
  - $C^{(k)}$: the k-skeleton of a complex C
  - $\operatorname{lk}(\xi, \mathcal{L})$: the link of ξ in $\mathcal{L}$
  - $\operatorname{st}_0(\xi, \mathcal{L})$: the subcomplex of $\mathcal{L}$ spanned by ξ and $\operatorname{lk}(\xi, \mathcal{L})$
- Abstract group
  - $[X, Y] := XYX^{-1}Y^{-1}$
  - $X^Y := YXY^{-1}$
  - $\langle\langle X \rangle\rangle$: the normal closure of X
- Numbers
  - $\mathbb{N}$: the set of natural numbers
  - $\mathbb{Z}$: the set of integral numbers
  - $\mathbb{Q}$: the set of rational numbers
  - $\mathbb{R}$: the set of real numbers
  - $\mathbb{C}$: the set of complex numbers
  - $\hat{\mathbb{Q}}$: the union $\mathbb{Q} \cup \{\infty\}$
  - $\widehat{\mathbb{C}}$: the Riemann sphere $\mathbb{C} \cup \{\infty\}$
  - $[j]$: the integer in $\{0, 1, 2\}$ such that $[j] \equiv j \pmod 3$
  - ϵ: a sign $-$ or $+$
- Surfaces
  - T: the once-punctured torus
  - S: the four-times punctured sphere
  - $\mathcal{O}$: the $(2, 2, 2, \infty)$-orbifold over S^2
  - $\operatorname{Teich}(X)$: the Teichmüller space of a surface X
- Model spaces
  - $\mathbb{H}^n$: the hyperbolic n-space

  - $\overline{\mathbb{H}^n}$: the closure, $\mathbb{H}^n \cup \partial\mathbb{H}^n$, of $\mathbb{H}^n$
  - $\overline{\mathbb{H}}^3$: the closure, $\mathbb{H}^3 \cup \mathbb{C}$, of the upper half space model of $\mathbb{H}^3$
  - $\mathbb{E}^{1,n}$: the $(n+1)$-dimensional Minkowski space
  - $\mathcal{P}(\mathbb{H}^3)$: the space of convex polyhedra in $\mathbb{H}^3$
  - $\mathcal{P}(\overline{\mathbb{H}}^3)$: the space of convex polyhedra in $\overline{\mathbb{H}}^3$
  - $\mathbb{E}$: the symbol to mean something which is Euclidean
  - $\mathcal{C}(X)$: the closed convex hull of X
- Matrix groups: for $R = \mathbb{Z}, \mathbb{R}$ or $\mathbb{C}$
  - $SL(2,R)$: the set of all 2×2-matrices with determinant 1
  - $PSL(2,R) := SL(2,R)/\{\pm 1\}$
- Farey triangulation
  - $\mathcal{D}$: the Farey triangulation (the modular diagram)
  - $\langle v_0, \ldots, v_k \rangle$: the k-simplex in $\mathcal{D}$ spanned by vertices $v_0, \ldots, v_k \in \mathcal{D}^{(0)}$
  - $s(X) \in \mathcal{D}^{(0)}$: the slope of X, where X is a generator of $\pi_1(T)$ or an elliptic generator of $\pi_1(\mathcal{O})$
  - $\mathcal{T}$: the binary tree dual to $\mathcal{D}$
  - $\vec{E}(\mathcal{T})$: the set of directed edges of $\mathcal{T}$
  - $\vec{e}$: a directed edge of $\mathcal{T}$
  - $\mathcal{EG}$: the set of elliptic generators
  - $\Sigma = (\sigma_1, \ldots, \sigma_m)$: a chain of triangles
  - σ^ϵ: the ϵ-terminal triangle of a chain
  - $\Sigma(\boldsymbol{\nu})$: the chain of triangles determined by a label $\boldsymbol{\nu}$
  - $\sigma^\epsilon(\boldsymbol{\nu})$: the ϵ-terminal triangle of $\Sigma(\boldsymbol{\nu})$
- Ideal triangulation of $T \times [-1,1]$
  - $\mathrm{trg}(\sigma)$: the topological ideal triangulation of T determined by σ
  - $\widetilde{\mathrm{trg}}(\sigma)$: the topological ideal triangulation of $\mathbb{R}^2 - \mathbb{Z}^2$ determined by σ
  - $\mathrm{Trg}(\boldsymbol{\nu})$: the layer of topological ideal triangulations of T determined by $\boldsymbol{\nu}$
  - $\widetilde{\mathrm{Trg}}(\boldsymbol{\nu})$: the layer of topological ideal triangulations of $\mathbb{R}^2 - \mathbb{Z}^2$ determined by $\boldsymbol{\nu}$
  - $\mathrm{spine}(\sigma)$: the spine of T determined by σ
  - $\mathrm{Spine}(\delta^-, \delta^+)$: the "trace of spines" of T in $T \times [-1,1]$
- For an element X of $SL(2,\mathbb{C})$ or $PSL(2,\mathbb{C})$
  - $\mathrm{tr}\, X$: the trace of X
  - $\mathrm{Axis}\, X$: the axis of $X \subset \mathbb{H}^3$
  - $\overline{\mathrm{Axis} X}$: the closure of $\mathrm{Axis}\, X$ in $\overline{\mathbb{H}}^3$
  - $\mathrm{Fix}\, X$: the fixed point of X in $\mathbb{C}$
  - $I(X)$: the isometric circle of X
  - $E(X)$: the exterior of $I(X)$
  - $Ih(X)$: the isometric hemisphere of X
  - $Eh(X)$: the exterior of $Ih(X)$
  - $\overline{Ih}(X)$: the closure of $Ih(X)$ in $\overline{\mathbb{H}}^3$
  - $\overline{Eh}(X)$: the closure of $Eh(X)$ in $\overline{\mathbb{H}}^3$
  - $\overline{Dh}(X)$: the closure of $Dh(X)$ in $\overline{\mathbb{H}}^3$

  - $c(X)$: the center of $I(X)$
  - $r(X)$: the radius of $I(X)$
- For a Kleinian group Γ
  - $\mathrm{Stab}_\Gamma(x)$: the stabilizer of x with respect to the action of Γ
  - $\Omega(\Gamma)$: the domain of discontinuity of Γ
  - $\Omega^\epsilon(\Gamma)$: the ϵ-component of $\Omega(\Gamma)$
  - $\Lambda(\Gamma)$: the limit set of Γ
  - $M(\Gamma)$: the hyperbolic manifold (or orbifold) $\mathbb{H}^3/\Gamma$
  - $\bar{M}(\Gamma)$: the quotient manifold (or orbifold) $(\mathbb{H}^3 \cup \Omega(\Gamma))/\Gamma$
  - $P(\Gamma)$: the Ford polygon of Γ in $\widehat{\mathbb{C}}$
  - $Ph(\Gamma)$: the Ford domain of Γ in $\mathbb{H}^3$
  - $\overline{Ph}(\Gamma) := Ph(\Gamma) \cup P(\Gamma)$
  - $\mathrm{Ford}(\Gamma)$: the Ford complex of Γ in $\bar{M}(\Gamma)$
  - $\Delta_{\mathbb{E}}(\Gamma)$: the Euclidean decomposition of $M(\Gamma)$
- Spaces of representations
  - $\mathrm{Hom}_{\mathrm{tp}}(\pi_1(X), PSL(2,\mathbb{C}))$: the space of all type-preserving $PSL(2,\mathbb{C})$-representations of $\pi_1(X)$ for $X = T$, $\mathcal{O}$ and S
  - $\mathcal{X}$: $= \mathrm{Hom}_{\mathrm{tp}}(\pi_1(X), PSL(2,\mathbb{C}))/PSL(2,\mathbb{C})$
  - ρ: a type-preserving representation
  - $\mathcal{QF}$: the space of quasifuchsian representations $\subset \mathcal{X}$
  - $\overline{\mathcal{QF}}$: the closure of $\mathcal{QF}$ in $\mathcal{X}$
  - Φ: the space of Markoff maps
  - ϕ: a Markoff map
  - Ψ: the space of complex probabilities
  - $\zeta^\epsilon_{\nu,\sigma}$: the polynomial function $\Phi \to \mathbb{C}$ defined by $\zeta^\epsilon_{\nu,\sigma}(\phi) = \phi(s_0) + \alpha^\epsilon\phi(s_1) + \beta^\epsilon\phi(s_2)$
  - $\Phi^\epsilon_{\nu,\sigma}$: the subvariety $(\zeta^\epsilon_{\nu,\sigma})^{-1}(0)$ of Φ
  - Φ^ϵ_ν: the "geometric" irreducible component of $\Phi^\epsilon_{\nu,\sigma}$
  - $d_G(\boldsymbol{\nu})$: the geometric degree of $\boldsymbol{\nu}$
- Side parameter
  - $\boldsymbol{\nu}$: the side parameter
  - $\nu^\epsilon(\rho)$: the ϵ-component of $\boldsymbol{\nu}$
  - ν: used to denote a label $\nu \in \mathcal{D}$
  - $\theta^\epsilon(\rho,\sigma)$: the ϵ-angle invariant of ρ at σ
  - $\theta^\epsilon(\rho,\sigma; s_{[j]})$: the $s_{[j]}$-component of $\theta^\epsilon(\rho,\sigma)$
  - $\mathcal{J}[\mathcal{QF}] \subset \mathcal{X} \times (\mathbb{H}^2 \times \mathbb{H}^2)$: the space of good labeled representations
  - $\mu_1 : \mathcal{J}[\mathcal{QF}] \to \mathcal{X}$: the natural projection
  - $\mu_2 : \mathcal{J}[\mathcal{QF}] \to \mathbb{H}^2 \times \mathbb{H}^2$: the natural projection
  - $\boldsymbol{\rho} = (\rho, \boldsymbol{\nu})$: a labeled representation
- For a pair (ρ,σ)
  - $I(j) = I(\rho(P_j))$: the isometric circle of $\rho(P_j)$
  - $D(j) = D(\rho(P_j))$: the disk bounded by $I(\rho(P_j))$
  - $E(j) = E(\rho(P_j))$: the exterior of $I(\rho(P_j))$
  - $Ih(j) = Ih(\rho(P_j))$: the isometric hemisphere of $\rho(P_j)$

- $Dh(j) = Dh(\rho(P_j))$: the half space of $\mathbb{H}^3$ bounded by $Ih(\rho(P_j))$ whose closure contains $D(\rho(P_j))$
- $Eh(j) = Eh(\rho(P_j))$: the half space $\mathbb{H}^3 - \mathrm{int}\, Dh(\rho(P_j))$
- $c(j) = c(\rho(P_j))$: the center of $I(\rho(P_j))$
- $\overrightarrow{c}(j, j+1) = \overrightarrow{c}(\rho; P_j, P_{j+1})$: the vector or oriented line $c(\rho(P_{j+1})) - c(\rho(P_j))$
- $\mathrm{Fix}^{\epsilon}(j) = \mathrm{Fix}^{\epsilon}_{\sigma}(\rho(P_j))$: the fixed point of $\rho(P_j)$ which lies in the ϵ-side of $\mathcal{L}(\rho, \sigma)$
- $\mathrm{Axis}(j) = \mathrm{Axis}(\rho(P_j))$: the axis of $\rho(P_j)$
- $\overrightarrow{f}(j)$: the oriented line $\overrightarrow{\mathrm{Fix}^{-}_{\sigma}(\rho(P_j))\,\mathrm{Fix}^{+}_{\sigma}(\rho(P_j))}$.
- $v^{\epsilon}(j, j+1) = v^{\epsilon}(\rho; P_j, P_{j+1})$: the point of $I(\rho(P_j)) \cap I(\rho(P_{j+1}))$ which lies in the ϵ-side of $\mathcal{L}(\rho, \sigma)$
- $e^{\epsilon}(j) = e^{\epsilon}(\rho, \sigma; P_j)$: the ϵ-ideal edge in $I(\rho(P_j))$ (Notation 4.3.7)
- $\Delta^{\epsilon}_j = \Delta^{\epsilon}_j(\rho, \sigma)$: "the j-th triangle" in the ϵ-side of $\mathcal{L}(\rho, \sigma)$ (Definition 4.2.10)
- $\Delta(\rho, \sigma)$: the model triangle for $\Delta^{\epsilon}_j(\rho, \sigma)$'s
- $f^{\epsilon}_{\boldsymbol{\rho}}(\xi)$: the ϵ-ideal face determined by $\boldsymbol{\rho}$ and ξ
- $\alpha(\rho, \sigma; s_j)$: the (inner) angle of the triangle at the vertex w_j

- Elliptic generator complex
  - $\mathcal{L}(\rho, \sigma)$: a bi-infinite broken line in $\mathbb{C}$ determined by ρ and σ
  - $\mathcal{L}(\rho, \Sigma)$: the union of bi-infinite broken lines $\mathcal{L}(\rho, \sigma)$ in $\mathbb{C}$ for triangles σ in a chain Σ
  - $\mathcal{L}(\boldsymbol{\rho})$: the set $\mathcal{L}(\rho, \Sigma(\boldsymbol{\nu}))$ for $\boldsymbol{\rho} = (\rho, \boldsymbol{\nu})$
  - $\mathcal{L}(\sigma)$: an abstract bi-infinite broken line determined by σ
  - $\mathcal{L}(\Sigma)$: the elliptic generator complex associated with Σ
  - $\mathcal{L}(\boldsymbol{\nu})$: the elliptic generator complex associated with $\Sigma(\boldsymbol{\nu})$
  - $\mathcal{L}^*(\boldsymbol{\nu})$: the augmentation of $\mathcal{L}(\boldsymbol{\nu})$
  - $\partial^{\epsilon}_{aug}\mathcal{L}^*(\boldsymbol{\nu})$: the "$\epsilon$-boundary" of $\mathcal{L}^*(\boldsymbol{\nu})$
- Dual map from $\mathcal{L}$ to ∂Eh
  - $F_{\boldsymbol{\rho}} : \mathcal{L}(\boldsymbol{\nu})^{(\le 2)} \to \mathcal{P}(\mathbb{H}^3)$: the dual map to $\mathbb{H}^3$
  - $F_{\boldsymbol{\rho}}(\xi)$: the image of ξ by $F_{\boldsymbol{\rho}}$
  - $\overline{F}_{\boldsymbol{\rho}} : \mathcal{L}(\boldsymbol{\nu})^{(\le 2)} \to \mathcal{P}(\overline{\mathbb{H}}^3)$: the dual map to $\overline{\mathbb{H}}^3$
  - $\overline{F}_{\boldsymbol{\rho}}(\xi)$: the image of ξ by $\overline{F}_{\boldsymbol{\rho}}$
- Quotient by the action of $\langle K \rangle$
  - $\mathrm{Cusp}(K)$: $= \mathbb{H}^3/\langle \rho(K) \rangle$
  - $\overline{\mathrm{Cusp}}(K)$: $= \overline{\mathbb{H}}^3/\langle \rho(K) \rangle$
  - $\partial \mathrm{Cusp}(K)$: $= \mathbb{C}/\langle \rho(K) \rangle$
  - $q_K : \overline{\mathbb{H}}^3 \to \overline{\mathrm{Cusp}}(K)$: the projection
  - $Eh_K(\boldsymbol{\rho}) := q_K(Eh(\boldsymbol{\rho})) \subset \mathrm{Cusp}(K)$
  - $E_K(\boldsymbol{\rho}) := q_K(E(\boldsymbol{\rho})) \subset \partial\overline{\mathrm{Cusp}}(K)$
  - $F_{K,\boldsymbol{\rho}}(\xi) := q_K(F_{\boldsymbol{\rho}}(\xi))$

Index

Lecture Notes in Mathematics

For information about earlier volumes
please contact your bookseller or Springer
LNM Online archive: springerlink.com

Vol. 1764: A. Cannas da Silva, Lectures on Symplectic Geometry (2001)
Vol. 1765: T. Kerler, V. V. Lyubashenko, Non-Semisimple Topological Quantum Field Theories for 3-Manifolds with Corners (2001)
Vol. 1766: H. Hennion, L. Hervé, Limit Theorems for Markov Chains and Stochastic Properties of Dynamical Systems by Quasi-Compactness (2001)
Vol. 1767: J. Xiao, Holomorphic Q Classes (2001)
Vol. 1768: M. J. Pflaum, Analytic and Geometric Study of Stratified Spaces (2001)
Vol. 1769: M. Alberich-Carramiñana, Geometry of the Plane Cremona Maps (2002)
Vol. 1770: H. Gluesing-Luerssen, Linear Delay-Differential Systems with Commensurate Delays: An Algebraic Approach (2002)
Vol. 1771: M. Émery, M. Yor (Eds.), Séminaire de Probabilités 1967-1980. A Selection in Martingale Theory (2002)
Vol. 1772: F. Burstall, D. Ferus, K. Leschke, F. Pedit, U. Pinkall, Conformal Geometry of Surfaces in S^4 (2002)
Vol. 1773: Z. Arad, M. Muzychuk, Standard Integral Table Algebras Generated by a Non-real Element of Small Degree (2002)
Vol. 1774: V. Runde, Lectures on Amenability (2002)
Vol. 1775: W. H. Meeks, A. Ros, H. Rosenberg, The Global Theory of Minimal Surfaces in Flat Spaces. Martina Franca 1999. Editor: G. P. Pirola (2002)
Vol. 1776: K. Behrend, C. Gomez, V. Tarasov, G. Tian, Quantum Comohology. Cetraro 1997. Editors: P. de Bartolomeis, B. Dubrovin, C. Reina (2002)
Vol. 1777: E. García-Río, D. N. Kupeli, R. Vázquez-Lorenzo, Osserman Manifolds in Semi-Riemannian Geometry (2002)
Vol. 1778: H. Kiechle, Theory of K-Loops (2002)
Vol. 1779: I. Chueshov, Monotone Random Systems (2002)
Vol. 1780: J. H. Bruinier, Borcherds Products on O(2,1) and Chern Classes of Heegner Divisors (2002)
Vol. 1781: E. Bolthausen, E. Perkins, A. van der Vaart, Lectures on Probability Theory and Statistics. Ecole d' Eté de Probabilités de Saint-Flour XXIX-1999. Editor: P. Bernard (2002)
Vol. 1782: C.-H. Chu, A. T.-M. Lau, Harmonic Functions on Groups and Fourier Algebras (2002)
Vol. 1783: L. Grüne, Asymptotic Behavior of Dynamical and Control Systems under Perturbation and Discretization (2002)
Vol. 1784: L. H. Eliasson, S. B. Kuksin, S. Marmi, J.-C. Yoccoz, Dynamical Systems and Small Divisors. Cetraro, Italy 1998. Editors: S. Marmi, J.-C. Yoccoz (2002)
Vol. 1785: J. Arias de Reyna, Pointwise Convergence of Fourier Series (2002)
Vol. 1786: S. D. Cutkosky, Monomialization of Morphisms from 3-Folds to Surfaces (2002)
Vol. 1787: S. Caenepeel, G. Militaru, S. Zhu, Frobenius and Separable Functors for Generalized Module Categories and Nonlinear Equations (2002)
Vol. 1788: A. Vasil'ev, Moduli of Families of Curves for Conformal and Quasiconformal Mappings (2002)
Vol. 1789: Y. Sommerhäuser, Yetter-Drinfel'd Hopf algebras over groups of prime order (2002)
Vol. 1790: X. Zhan, Matrix Inequalities (2002)
Vol. 1791: M. Knebusch, D. Zhang, Manis Valuations and Prüfer Extensions I: A new Chapter in Commutative Algebra (2002)
Vol. 1792: D. D. Ang, R. Gorenflo, V. K. Le, D. D. Trong, Moment Theory and Some Inverse Problems in Potential Theory and Heat Conduction (2002)
Vol. 1793: J. Cortés Monforte, Geometric, Control and Numerical Aspects of Nonholonomic Systems (2002)
Vol. 1794: N. Pytheas Fogg, Substitution in Dynamics, Arithmetics and Combinatorics. Editors: V. Berthé, S. Ferenczi, C. Mauduit, A. Siegel (2002)
Vol. 1795: H. Li, Filtered-Graded Transfer in Using Non-commutative Gröbner Bases (2002)
Vol. 1796: J.M. Melenk, hp-Finite Element Methods for Singular Perturbations (2002)
Vol. 1797: B. Schmidt, Characters and Cyclotomic Fields in Finite Geometry (2002)
Vol. 1798: W.M. Oliva, Geometric Mechanics (2002)
Vol. 1799: H. Pajot, Analytic Capacity, Rectifiability, Menger Curvature and the Cauchy Integral (2002)
Vol. 1800: O. Gabber, L. Ramero, Almost Ring Theory (2003)
Vol. 1801: J. Azéma, M. Émery, M. Ledoux, M. Yor (Eds.), Séminaire de Probabilités XXXVI (2003)
Vol. 1802: V. Capasso, E. Merzbach, B. G. Ivanoff, M. Dozzi, R. Dalang, T. Mountford, Topics in Spatial Stochastic Processes. Martina Franca, Italy 2001. Editor: E. Merzbach (2003)
Vol. 1803: G. Dolzmann, Variational Methods for Crystalline Microstructure – Analysis and Computation (2003)
Vol. 1804: I. Cherednik, Ya. Markov, R. Howe, G. Lusztig, Iwahori-Hecke Algebras and their Representation Theory. Martina Franca, Italy 1999. Editors: V. Baldoni, D. Barbasch (2003)
Vol. 1805: F. Cao, Geometric Curve Evolution and Image Processing (2003)
Vol. 1806: H. Broer, I. Hoveijn. G. Lunther, G. Vegter, Bifurcations in Hamiltonian Systems. Computing Singularities by Gröbner Bases (2003)
Vol. 1807: V. D. Milman, G. Schechtman (Eds.), Geometric Aspects of Functional Analysis. Israel Seminar 2000-2002 (2003)
Vol. 1808: W. Schindler, Measures with Symmetry Properties (2003)
Vol. 1809: O. Steinbach, Stability Estimates for Hybrid Coupled Domain Decomposition Methods (2003)
Vol. 1810: J. Wengenroth, Derived Functors in Functional Analysis (2003)
Vol. 1811: J. Stevens, Deformations of Singularities (2003)
Vol. 1812: L. Ambrosio, K. Deckelnick, G. Dziuk, M. Mimura, V. A. Solonnikov, H. M. Soner, Mathematical Aspects of Evolving Interfaces. Madeira, Funchal, Portugal 2000. Editors: P. Colli, J. F. Rodrigues (2003)
Vol. 1813: L. Ambrosio, L. A. Caffarelli, Y. Brenier, G. Buttazzo, C. Villani, Optimal Transportation and its Applications. Martina Franca, Italy 2001. Editors: L. A. Caffarelli, S. Salsa (2003)
Vol. 1814: P. Bank, F. Baudoin, H. Föllmer, L.C.G. Rogers, M. Soner, N. Touzi, Paris-Princeton Lectures on Mathematical Finance 2002 (2003)
Vol. 1815: A. M. Vershik (Ed.), Asymptotic Combinatorics with Applications to Mathematical Physics. St. Petersburg, Russia 2001 (2003)
Vol. 1816: S. Albeverio, W. Schachermayer, M. Talagrand, Lectures on Probability Theory and Statistics. Ecole d'Eté de Probabilités de Saint-Flour XXX-2000. Editor: P. Bernard (2003)
Vol. 1817: E. Koelink, W. Van Assche (Eds.), Orthogonal Polynomials and Special Functions. Leuven 2002 (2003)

Vol. 1818: M. Bildhauer, Convex Variational Problems with Linear, nearly Linear and/or Anisotropic Growth Conditions (2003)
Vol. 1819: D. Masser, Yu. V. Nesterenko, H. P. Schlickewei, W. M. Schmidt, M. Waldschmidt, Diophantine Approximation. Cetraro, Italy 2000. Editors: F. Amoroso, U. Zannier (2003)
Vol. 1820: F. Hiai, H. Kosaki, Means of Hilbert Space Operators (2003)
Vol. 1821: S. Teufel, Adiabatic Perturbation Theory in Quantum Dynamics (2003)
Vol. 1822: S.-N. Chow, R. Conti, R. Johnson, J. Mallet-Paret, R. Nussbaum, Dynamical Systems. Cetraro, Italy 2000. Editors: J. W. Macki, P. Zecca (2003)
Vol. 1823: A. M. Anile, W. Allegretto, C. Ringhofer, Mathematical Problems in Semiconductor Physics. Cetraro, Italy 1998. Editor: A. M. Anile (2003)
Vol. 1824: J. A. Navarro González, J. B. Sancho de Salas, $\mathscr{C}^\infty$ – Differentiable Spaces (2003)
Vol. 1825: J. H. Bramble, A. Cohen, W. Dahmen, Multiscale Problems and Methods in Numerical Simulations, Martina Franca, Italy 2001. Editor: C. Canuto (2003)
Vol. 1826: K. Dohmen, Improved Bonferroni Inequalities via Abstract Tubes. Inequalities and Identities of Inclusion-Exclusion Type. VIII, 113 p, 2003.
Vol. 1827: K. M. Pilgrim, Combinations of Complex Dynamical Systems. IX, 118 p, 2003.
Vol. 1828: D. J. Green, Gröbner Bases and the Computation of Group Cohomology. XII, 138 p, 2003.
Vol. 1829: E. Altman, B. Gaujal, A. Hordijk, Discrete-Event Control of Stochastic Networks: Multimodularity and Regularity. XIV, 313 p, 2003.
Vol. 1830: M. I. Gil', Operator Functions and Localization of Spectra. XIV, 256 p, 2003.
Vol. 1831: A. Connes, J. Cuntz, E. Guentner, N. Higson, J. E. Kaminker, Noncommutative Geometry, Martina Franca, Italy 2002. Editors: S. Doplicher, L. Longo (2004)
Vol. 1832: J. Azéma, M. Émery, M. Ledoux, M. Yor (Eds.), Séminaire de Probabilités XXXVII (2003)
Vol. 1833: D.-Q. Jiang, M. Qian, M.-P. Qian, Mathematical Theory of Nonequilibrium Steady States. On the Frontier of Probability and Dynamical Systems. IX, 280 p, 2004.
Vol. 1834: Yo. Yomdin, G. Comte, Tame Geometry with Application in Smooth Analysis. VIII, 186 p, 2004.
Vol. 1835: O.T. Izhboldin, B. Kahn, N.A. Karpenko, A. Vishik, Geometric Methods in the Algebraic Theory of Quadratic Forms. Summer School, Lens, 2000. Editor: J.-P. Tignol (2004)
Vol. 1836: C. Năstăsescu, F. Van Oystaeyen, Methods of Graded Rings. XIII, 304 p, 2004.
Vol. 1837: S. Tavaré, O. Zeitouni, Lectures on Probability Theory and Statistics. Ecole d'Eté de Probabilités de Saint-Flour XXXI-2001. Editor: J. Picard (2004)
Vol. 1838: A.J. Ganesh, N.W. O'Connell, D.J. Wischik, Big Queues. XII, 254 p, 2004.
Vol. 1839: R. Gohm, Noncommutative Stationary Processes. VIII, 170 p, 2004.
Vol. 1840: B. Tsirelson, W. Werner, Lectures on Probability Theory and Statistics. Ecole d'Eté de Probabilités de Saint-Flour XXXII-2002. Editor: J. Picard (2004)
Vol. 1841: W. Reichel, Uniqueness Theorems for Variational Problems by the Method of Transformation Groups (2004)
Vol. 1842: T. Johnsen, A. L. Knutsen, K_3 Projective Models in Scrolls (2004)
Vol. 1843: B. Jefferies, Spectral Properties of Noncommuting Operators (2004)
Vol. 1844: K.F. Siburg, The Principle of Least Action in Geometry and Dynamics (2004)
Vol. 1845: Min Ho Lee, Mixed Automorphic Forms, Torus Bundles, and Jacobi Forms (2004)
Vol. 1846: H. Ammari, H. Kang, Reconstruction of Small Inhomogeneities from Boundary Measurements (2004)
Vol. 1847: T.R. Bielecki, T. Björk, M. Jeanblanc, M. Rutkowski, J.A. Scheinkman, W. Xiong, Paris-Princeton Lectures on Mathematical Finance 2003 (2004)
Vol. 1848: M. Abate, J. E. Fornaess, X. Huang, J. P. Rosay, A. Tumanov, Real Methods in Complex and CR Geometry, Martina Franca, Italy 2002. Editors: D. Zaitsev, G. Zampieri (2004)
Vol. 1849: Martin L. Brown, Heegner Modules and Elliptic Curves (2004)
Vol. 1850: V. D. Milman, G. Schechtman (Eds.), Geometric Aspects of Functional Analysis. Israel Seminar 2002-2003 (2004)
Vol. 1851: O. Catoni, Statistical Learning Theory and Stochastic Optimization (2004)
Vol. 1852: A.S. Kechris, B.D. Miller, Topics in Orbit Equivalence (2004)
Vol. 1853: Ch. Favre, M. Jonsson, The Valuative Tree (2004)
Vol. 1854: O. Saeki, Topology of Singular Fibers of Differential Maps (2004)
Vol. 1855: G. Da Prato, P.C. Kunstmann, I. Lasiecka, A. Lunardi, R. Schnaubelt, L. Weis, Functional Analytic Methods for Evolution Equations. Editors: M. Iannelli, R. Nagel, S. Piazzera (2004)
Vol. 1856: K. Back, T.R. Bielecki, C. Hipp, S. Peng, W. Schachermayer, Stochastic Methods in Finance, Bressanone/Brixen, Italy, 2003. Editors: M. Fritelli, W. Runggaldier (2004)
Vol. 1857: M. Émery, M. Ledoux, M. Yor (Eds.), Séminaire de Probabilités XXXVIII (2005)
Vol. 1858: A.S. Cherny, H.-J. Engelbert, Singular Stochastic Differential Equations (2005)
Vol. 1859: E. Letellier, Fourier Transforms of Invariant Functions on Finite Reductive Lie Algebras (2005)
Vol. 1860: A. Borisyuk, G.B. Ermentrout, A. Friedman, D. Terman, Tutorials in Mathematical Biosciences I. Mathematical Neurosciences (2005)
Vol. 1861: G. Benettin, J. Henrard, S. Kuksin, Hamiltonian Dynamics – Theory and Applications, Cetraro, Italy, 1999. Editor: A. Giorgilli (2005)
Vol. 1862: B. Helffer, F. Nier, Hypoelliptic Estimates and Spectral Theory for Fokker-Planck Operators and Witten Laplacians (2005)
Vol. 1863: H. Führ, Abstract Harmonic Analysis of Continuous Wavelet Transforms (2005)
Vol. 1864: K. Efstathiou, Metamorphoses of Hamiltonian Systems with Symmetries (2005)
Vol. 1865: D. Applebaum, B.V. R. Bhat, J. Kustermans, J. M. Lindsay, Quantum Independent Increment Processes I. From Classical Probability to Quantum Stochastic Calculus. Editors: M. Schürmann, U. Franz (2005)
Vol. 1866: O.E. Barndorff-Nielsen, U. Franz, R. Gohm, B. Kümmerer, S. Thorbjønsen, Quantum Independent Increment Processes II. Structure of Quantum Lévy Processes, Classical Probability, and Physics. Editors: M. Schürmann, U. Franz, (2005)
Vol. 1867: J. Sneyd (Ed.), Tutorials in Mathematical Biosciences II. Mathematical Modeling of Calcium Dynamics and Signal Transduction. (2005)

Vol. 1868: J. Jorgenson, S. Lang, $Pos_n(R)$ and Eisenstein Series. (2005)
Vol. 1869: A. Dembo, T. Funaki, Lectures on Probability Theory and Statistics. Ecole d'Eté de Probabilités de Saint-Flour XXXIII-2003. Editor: J. Picard (2005)
Vol. 1870: V.I. Gurariy, W. Lusky, Geometry of Müntz Spaces and Related Questions. (2005)
Vol. 1871: P. Constantin, G. Gallavotti, A.V. Kazhikhov, Y. Meyer, S. Ukai, Mathematical Foundation of Turbulent Viscous Flows, Martina Franca, Italy, 2003. Editors: M. Cannone, T. Miyakawa (2006)
Vol. 1872: A. Friedman (Ed.), Tutorials in Mathematical Biosciences III. Cell Cycle, Proliferation, and Cancer (2006)
Vol. 1873: R. Mansuy, M. Yor, Random Times and Enlargements of Filtrations in a Brownian Setting (2006)
Vol. 1874: M. Yor, M. Émery (Eds.), In Memoriam Paul-André Meyer - Séminaire de probabilités XXXIX (2006)
Vol. 1875: J. Pitman, Combinatorial Stochastic Processes. Ecole d'Eté de Probabilités de Saint-Flour XXXII-2002. Editor: J. Picard (2006)
Vol. 1876: H. Herrlich, Axiom of Choice (2006)
Vol. 1877: J. Steuding, Value Distributions of L-Functions (2007)
Vol. 1878: R. Cerf, The Wulff Crystal in Ising and Percolation Models, Ecole d'Eté de Probabilités de Saint-Flour XXXIV-2004. Editor: Jean Picard (2006)
Vol. 1879: G. Slade, The Lace Expansion and its Applications, Ecole d'Eté de Probabilités de Saint-Flour XXXIV-2004. Editor: Jean Picard (2006)
Vol. 1880: S. Attal, A. Joye, C.-A. Pillet, Open Quantum Systems I, The Hamiltonian Approach (2006)
Vol. 1881: S. Attal, A. Joye, C.-A. Pillet, Open Quantum Systems II, The Markovian Approach (2006)
Vol. 1882: S. Attal, A. Joye, C.-A. Pillet, Open Quantum Systems III, Recent Developments (2006)
Vol. 1883: W. Van Assche, F. Marcellàn (Eds.), Orthogonal Polynomials and Special Functions, Computation and Application (2006)
Vol. 1884: N. Hayashi, E.I. Kaikina, P.I. Naumkin, I.A. Shishmarev, Asymptotics for Dissipative Nonlinear Equations (2006)
Vol. 1885: A. Telcs, The Art of Random Walks (2006)
Vol. 1886: S. Takamura, Splitting Deformations of Degenerations of Complex Curves (2006)
Vol. 1887: K. Habermann, L. Habermann, Introduction to Symplectic Dirac Operators (2006)
Vol. 1888: J. van der Hoeven, Transseries and Real Differential Algebra (2006)
Vol. 1889: G. Osipenko, Dynamical Systems, Graphs, and Algorithms (2006)
Vol. 1890: M. Bunge, J. Funk, Singular Coverings of Toposes (2006)
Vol. 1891: J.B. Friedlander, D.R. Heath-Brown, H. Iwaniec, J. Kaczorowski, Analytic Number Theory, Cetraro, Italy, 2002. Editors: A. Perelli, C. Viola (2006)
Vol. 1892: A. Baddeley, I. Bárány, R. Schneider, W. Weil, Stochastic Geometry, Martina Franca, Italy, 2004. Editor: W. Weil (2007)
Vol. 1893: H. Hanßmann, Local and Semi-Local Bifurcations in Hamiltonian Dynamical Systems, Results and Examples (2007)
Vol. 1894: C.W. Groetsch, Stable Approximate Evaluation of Unbounded Operators (2007)
Vol. 1895: L. Molnár, Selected Preserver Problems on Algebraic Structures of Linear Operators and on Function Spaces (2007)
Vol. 1896: P. Massart, Concentration Inequalities and Model Selection, Ecole d'Eté de Probabilités de Saint-Flour XXXIII-2003. Editor: J. Picard (2007)
Vol. 1897: R. Doney, Fluctuation Theory for Lévy Processes, Ecole d'Eté de Probabilités de Saint-Flour XXXV-2005. Editor: J. Picard (2007)
Vol. 1898: H.R. Beyer, Beyond Partial Differential Equations, On linear and Quasi-Linear Abstract Hyperbolic Evolution Equations (2007)
Vol. 1899: Séminaire de Probabilités XL. Editors: C. Donati-Martin, M. Émery, A. Rouault, C. Stricker (2007)
Vol. 1900: E. Bolthausen, A. Bovier (Eds.), Spin Glasses (2007)
Vol. 1901: O. Wittenberg, Intersections de deux quadriques et pinceaux de courbes de genre 1, Intersections of Two Quadrics and Pencils of Curves of Genus 1 (2007)
Vol. 1902: A. Isaev, Lectures on the Automorphism Groups of Kobayashi-Hyperbolic Manifolds (2007)
Vol. 1903: G. Kresin, V. Maz'ya, Sharp Real-Part Theorems (2007)
Vol. 1904: P. Giesl, Construction of Global Lyapunov Functions Using Radial Basis Functions (2007)
Vol. 1905: C. Prévôt, M. Röckner, A Concise Course on Stochastic Partial Differential Equations (2007)
Vol. 1906: T. Schuster, The Method of Approximate Inverse: Theory and Applications (2007)
Vol. 1907: M. Rasmussen, Attractivity and Bifurcation for Nonautonomous Dynamical Systems (2007)
Vol. 1908: T.J. Lyons, M. Caruana, T. Lévy, Differential Equations Driven by Rough Paths, Ecole d'Eté de Probabilités de Saint-Flour XXXIV-2004. (2007)
Vol. 1909: H. Akiyoshi, M. Sakuma, M. Wada, Y. Yamashita, Punctured Torus Groups and 2-Bridge Knot Groups (I) (2007)
Vol. 1910: V.D. Milman, G. Schechtman (Eds.), Geometric Aspects of Functional Analysis. Israel Seminar 2004-2005 (2007)

Recent Reprints and New Editions

Vol. 1618: G. Pisier, Similarity Problems and Completely Bounded Maps. 1995 – 2nd exp. edition (2001)
Vol. 1629: J.D. Moore, Lectures on Seiberg-Witten Invariants. 1997 – 2nd edition (2001)
Vol. 1638: P. Vanhaecke, Integrable Systems in the realm of Algebraic Geometry. 1996 – 2nd edition (2001)
Vol. 1702: J. Ma, J. Yong, Forward-Backward Stochastic Differential Equations and their Applications. 1999 – Corr. 3rd printing (2007)
Vol. 830: J.A. Green, Polynomial Representations of GL_n, with an Appendix on Schensted Correspondence and Littelmann Paths by K. Erdmann, J.A. Green and M. Schocker 1980 – 2nd corr. and augmented edition (2007)